Benchmark Papers in Geology Series

Editor: Rhodes W. Fairbridge, Columbia University

A selection from the published volumes in this series
SPITS AND BARS / Maurice L. Schwartz
COASTAL SEDIMENTATION / Donald J. P. Swift and Harold D. Palmer
OVERWASH PROCESSES / Stephen P. Leatherman
EROSION AND SEDIMENT YIELD / J. B. Laronne and M. P. Mosley
GEOSYNCLINES: Concept and Place within Plate Tectonics / F. L. Schwab
DOLOMITIZATION / Donald H. Zenger and S. J. Mazzullo
MODERN CARBONATE ENVIRONMENTS / Ajit Bhattacharyya and
    G. M. Friedman
RIVER NETWORKS / Richard S. Jarvis and Michael J. Woldenberg
MODERN AND ANCIENT ALLUVIAL FAN DEPOSITS / Tor H. Nilsen

Related titles of interest
THE ENCYCLOPEDIA OF BEACHES AND COASTAL ENVIRONMENTS /
    Maurice L. Schwartz
THE ENCYCLOPEDIA OF SEDIMENTOLOFY / Rhodes W. Fairbridge and
    Joanne Bourgeois

MODERN AND ANCIENT ALLUVIAL FAN DEPOSITS

Edited by
TOR H. NILSEN
Menlo Park, California

A Hutchinson Ross Benchmark® Book

VAN NOSTRAND REINHOLD COMPANY
New York

Copyright © 1985 by **Van Nostrand Reinhold Company Inc.**
Benchmark Papers in Geology, Volume 87
Library of Congress Catalog Card Number: 84-15167
ISBN: 0-442-26753-3

All rights reserved. No part of this work covered by the copyrights
hereon may be reproduced or used in any form or by any means—
graphic, electronic, or mechanical, including photocopying,
recording, taping, or information storage and retrieval systems—
without permission of the publisher.

Manufactured in the United States of America.

Published by Van Nostrand Reinhold Company Inc.
135 West 50th Street
New York, New York 10020

Van Nostrand Reinhold Company Limited
Molly Millars Lane
Wokingham, Berkshire RG11 2PY, England

Van Nostrand Reinhold
480 Latrobe Street
Melbourne, Victoria 3000, Australia

Macmillan of Canada
Division of Gage Publishing Limited
164 Commander Boulevard
Agincourt, Ontario M1S 3C7, Canada

15 14 13 12 11 10 9 8 7 6 5 4 3 2 1

Library of Congress Cataloging in Publication Data
Main entry under title:
Modern and ancient alluvial fan deposits.
  (Benchmark papers in geology ; 87)
  "A Hutchinson Ross benchmark book."
  Includes indexes.
  1. Alluvial fans—Addresses, essays, lectures.
2. Alluvium—Addresses, essays, lectures. I. Nilsen,
Tor Helge. II. Series.
GB591.M59   1985   551.4   84-15167
ISBN 0-442-26753-3

CONTENTS

Series Editor's Foreword … vii
Preface … ix
Contents by Author … xi

Introduction … 1

PART I: EARLY STUDIES OF ALLUVIAL FAN DEPOSITS

Editor's Comments on Papers 1 Through 5 … 22

1. **SHARP, R. P.:** Early Tertiary Fanglomerate, Big Horn Mountains, Wyoming … 30
   Jour. Geology **56:**1-15 (1948)

2. **BLISSENBACH, E.:** Geology of Alluvial Fans in Semiarid Regions … 49
   Geol. Soc. America Bull. **65:**175-190 (1954)

3. **BEATY, C. B.:** Origin of Alluvial Fans, White Mountains, California and Nevada … 66
   Assoc. Am. Geographers Annals **53:**516-535 (1963)

4. **BULL, W. B.:** Alluvial-Fan Deposits in Western Fresno County, California … 86
   Jour. Geology **71:**243-251 (1963)

5. **BLUCK, B. J.:** Sedimentation of an Alluvial Fan in Southern Nevada … 97
   Jour. Sed. Petrology **34:**395-400 (1964)

PART II: MODERN ALLUVIAL FAN DEPOSITS

Editor's Comments on Papers 6 Through 12 … 104

6. **WINDER, C. G.:** Alluvial Cone Construction by Alpine Mudflow in a Humid Temperate Region … 108
   Canadian Jour. Earth Sci. **2:**270-277 (1965)

7. **LEGGETT, R. F., R. J. E. BROWN and G. H. JOHNSTON:** Alluvial Fan Formation Near Aklavik, Northwest Territories, Canada … 118
   Geol. Soc. America Bull. **77:**15-30 (1966)

8. **DENNY, C. S.:** Fans and Pediments … 137
   Am. Jour. Sci. **265:**81-105 (1967)

9. **HOOKE, R. LeB.:** Processes on Arid-Region Alluvial Fans … 162
   Jour. Geology **75:**438-460 (1967)

v

Contents

10	WASSON, R. J.: Intersection Point Deposition on Alluvial Fans: An Australian Example *Geog. Annaler* **56**:83-92 (1974)	187
11	WASSON, R. J.: Catchment Processes and the Evolution of Alluvial Fans in the Lower Derwent Valley, Tasmania *Zeitschr. Geomorphologie* **21**:147-168 (1977)	197
12	HOOKE, R. LeB., and W. L. ROHRER: Geometry of Alluvial Fans: Effect of Discharge and Sediment Size *Earth Surf. Process.* **4**:147-166 (1979)	219

PART III: ANCIENT ALLUVIAL FAN DEPOSITS

Editor's Comments on Papers 13 Through 16 — 240

13	NILSEN, T. H.: Old Red Sedimentation in the Buelandet-Vaerlandet Devonian District, Western Norway *Sed. Geology* **3**:35-57 (1969)	245
14	STEEL, R. J., S. MAEHLE, H. NILSEN, S. L. RØE, and A. SPINNANGR: Coarsening-Upward Cycles in the Alluvium of Hornelen Basin (Devonian) Norway: Sedimentary Response to Tectonic Events *Geol. Soc. America Bull.* **88**:1124-1134 (1977)	268
15	HEWARD, A. P.: Alluvial Fan Sequence and Megasequence Models: with Examples from Westphalian D—Stephanian B Coalfields, Northern Spain *Fluvial Sedimentology*, A. D. Miall, ed., Canadian Society of Petroleum Geologists Memoir No. 5, 1978, pp. 669-702	279
16	GLOPPEN, T. G., and R. J. STEEL: The Deposits, Internal Structure and Geometry in Six Alluvial Fan-Fan Delta Bodies (Devonian-Norway)—A Study of the Significance of Bedding Sequences in Conglomerates *Recent and Ancient Nonmarine Depositional Environments: Models for Exploration*, F. G. Ethridge and R. M. Flores, eds., Soc. Econ. Paleontologists and Mineralogists Spec. Pub. 31, 1981, pp. 49-69.	313

PART IV: COMPARATIVE STUDIES OF MODERN AND ANCIENT ALLUVIAL FAN DEPOSITS

Editor's Comments on Paper 17 — 336

17	BULL, W. B.: Recognition of Alluvial-Fan Deposits in the Stratigraphic Record *Recognition of Ancient Sedimentary Environments*, J. K. Rigby and W. K. Hamblin, eds., Soc. Econ. Paleontologists and Mineralogists Spec. Pub. 16, 1972, pp. 63-83	340

Author Citation Index	361
Subject Index	366
About the Editor	372

SERIES EDITOR'S FOREWORD

The philosophy behind the Benchmark Papers in Geology is one of collection, sifting, and rediffusion. Scientific literature today is so vast, so dispersed, and, in the case of old papers, so inaccessible for readers not in the immediate neighborhood of major libraries that much valuable information has been ignored by default. It has become just so difficult, or so time consuming, to search out the key papers in any basic area of research that one can hardly blame a busy person for skimping on some of his or her "homework."

This series of volumes has been devised, therefore, as a practical solution to this critical problem. The geologist, perhaps even more than any other scientist, often suffers from twin difficulties—isolation from central library resources and immensely diffused sources of material. New colleges and industrial libraries simply cannot afford to purchase complete runs of all the world's earth science literature. Specialists simply cannot locate reprints or copies of all their principal reference materials. So it is that we are now making a concerted effort to gather into single volumes the critical materials needed to reconstruct the background of any and every major topic of our discipline.

We are interpreting "geology" in its broadest sense: the fundamental science of the planet Earth, its materials, its history, and its dynamics. Because of training in "earthy" materials, we also take in astrogeology, the corresponding aspect of the planetary sciences. Besides the classical core disciplines such as mineralogy, petrology, structure, geomorphology, paleontology, and stratigraphy, we embrace the newer fields of geophysics and geochemistry, applied also to oceanography, geochronology, and paleoecology. We recognize the work of the mining geologists, the petroleum geologists, the hydrologists, and the engineering and environmental geologists. Each specialist needs a working library. We are endeavoring to make the task of compiling such a library a little easier.

Each volume in the series contains an introduction prepared by a specialist (the volume editor)—a "state of the art" opening or a summary of the object and content of the volume. The articles, usually some twenty to fifty reproduced either in their entirety or in significant extracts, are selected in an attempt to cover the field, from the key papers of the last century to fairly recent work. Where the original works are in foreign languages, we

Series Editor's Foreword

have endeavored to locate or commission translations. Geologists, because of their global subject, are often acutely aware of the oneness of our world. The selections cannot therefore be restricted to any one country, and whenever possible an attempt is made to scan the world literature.

To each article, or group of kindred articles, some sort of "highlight commentary" is usually supplied by the volume editor. This commentary should serve to bring that article into historical perspective and to emphasize its particular role in the growth of the field. References, or citations, wherever possible, will be reproduced in their entirety—for by this means the observant reader can assess the background material available to that particular author, or, if desired, he or she too can double check the earlier sources.

A "benchmark," in surveyor's terminology, is an established point on the ground that is recorded on our maps. It is usually anything that is a vantage point, from a modest hill to a mountain peak. From the historical viewpoint, these benchmarks are the bricks of our scientific edifice.

RHODES W. FAIRBRIDGE

PREFACE

Alluvial fan deposits have been of great interest for many years to sedimentologists, geomorphologists, stratigraphers, groundwater hydrologists, engineering geologists, environmental geologists, soil scientists, structural geologists, and other earth scientists. Modern alluvial fans have been the subject of intensive study in order to understand depositional processes, morphological development, tectonic history of adjacent faults and uplifted source areas, environmental aspects of land use, sources of unconsolidated materials, and groundwater resources. Ancient alluvial fan deposits have also been the subject of much study, chiefly because they record so clearly in their stratigraphic sequences the uplift of land areas and subsidence of basins; they are particularly common in some orogenic belts, and their recognition is of great importance to paleogeographic and paleotectonic studies. Experimental studies in which small fanlike bodies have been constructed in flumes and small rock quarries have also been of considerable interest, both as models for the larger scale processes of alluvial fans and as vehicles for the study of the dynamics of sediment transport and deposition. The purpose of this volume is to bring together in one place the significant papers that have contributed to the development of our ideas on modern and ancient alluvial fan deposits.

The literature on alluvial fan deposits is voluminous, and I have had to be very selective in determining the articles for this volume. I have probably omitted a number of important papers, but space limitations prohibited the incorporation of more than a sampling of the papers that I consider to be most important and critical in the development of concepts related to alluvial fan deposits. The entire text of each paper is included, and I have divided them into four groups: (1) early studies of both modern and ancient alluvial fan deposits that span the interval 1948–1964; in these papers, many of our fundamental concepts of fan deposits were developed, and themes common to most subsequent studies were first introduced; (2) studies of modern alluvial fans, published during the interval 1965–1979; in these papers, alluvial fans from different geographic areas and climatic regimes were analyzed to determine variations in depositional processes, fan morphologies, and geometry of the deposits; (3) studies of ancient alluvial fan deposits, published during the interval 1969–1981; in these papers, examples of ancient fans of Devonian and Carboniferous age are used to

Preface

show how models for alluvial fan sedimentation are constructed from sedimentologic, stratigraphic, and tectonic data; and (4) a comparative study of modern and ancient alluvial fan deposits, published in 1972, that incorporates grain size, depositional processes, sedimentary structures, stratigraphy, and geometry; although several comparative papers that are of much greater length and detail have been published since 1972, the paper by Bull selected for this volume is still the most succinct and possibly most useful comparative study.

The papers selected for this volume are those that emphasize depositional processes and morphology, in the case of modern alluvial fan deposits, and paleogeography and depositional models, in the case of ancient alluvial fan deposits. I have not attempted to incorporate papers that emphasize land-use patterns, environmental geology, soil stratigraphy, groundwater hydrology, or engineering geology of alluvial fans. Readers concerned chiefly with these aspects of alluvial fan deposits can refer to some of the bibliographies and lengthy summary papers cited in the introduction to this volume.

I thank the authors of the papers reprinted herein for their courtesy, promptness, and cooperation in the preparation of this volume. They have supplied me with original materials, in many cases original photographs, and this has greatly enhanced the quality of the reproduction in this volume. I also thank many authors for their helpful suggestions regarding organization of the volume and selection of papers for the volume. I apologize to those workers on modern and ancient alluvial fan deposits whose papers, because of space limitations, do not appear herein. I thank James C. Yount for helpful review and Jan L. Zigler for helpful editing of the text.

TOR H. NILSEN

CONTENTS BY AUTHOR

Beaty, C. B., 66
Blissenbach, E., 49
Bluck, B. J., 97
Brown, R. J. E., 118
Bull, W. B., 86, 340
Denny, C. S., 137
Gloppen, T. G., 313
Heward, A. P., 279
Hooke, R. LeB., 162, 219
Johnston, G. H., 118
Leggett, R. F., 118

Maehle, S., 268
Nilsen, H., 268
Nilsen, T. H., 245
Røe, S. L., 268
Rohrer, W. L., 219
Sharp, R. P., 30
Spinnangr, A., 268
Steel, R. J., 268, 313
Wasson, R. J., 187, 197
Winder, C. G., 108

INTRODUCTION

DEFINITIONS

Alluvial fans are fan- or cone-shaped sedimentary bodies that accumulate at the base of a mountain front or other upland area downslope from the point where streams emerge from the uplands (Paper 2; Bull, Paper 17, 1977). The slopes of alluvial fans depend upon many variables and range from less than a degree to as much as 25°; however, they rarely exceed 10° and probably average less than 5° (Denny, 1965; Bull, 1977). Smaller, more steeply sloping sedimentary bodies deposited by small streams have been called alluvial cones rather than fans (Fairbridge, 1968). The cones commonly grade into talus or scree cones deposited by falling debris rather than by stream transport (Bull, 1968b; Rapp and Fairbridge, 1968).

Fan deltas are fan-shaped deltas built by rivers into standing bodies of water without the presence of either an adjacent highland or steep slopes (McGowen, 1971; Sheh, 1979; Wescott and Ethridge, 1980). The distal facies of fan deltas are commonly retransported by lake or ocean currents and thus do not resemble the distal facies of alluvial fans. Glacial outwash plains, locally referred to as outwash fans or sandur plains, are generally flat and have radiating distributaries that emerge from a melting glacier rather than a mountain front (Krigstrom, 1962; Gustavson, 1974; Bluck, 1974; Boothroyd and Ashley, 1975; Boothroyd, 1976; Boothroyd and Nummedal, 1978).

Pediments are sloping surfaces cut into bedrock by periodic flooding; they consist of either bedrock or a thin veneer of loose sediment over bedrock (Tator, 1952, 1953; Paper 8; Hadley, 1967; Twidale, 1968). Pediments may morphologically resemble fans, and in some arid regions they flank alluvial fans.

Where the mountain front is relatively straight, a series of adjacent alluvial fans will coalesce to form a bajada or piedmont slope (Tolman, 1909; Fairbridge, 1968). The term "clastic wedge" has been applied to these deposits, especially where the mountain front is uplifted along faults.

Introduction

CHARACTERISTICS OF ALLUVIAL FAN DEPOSITS

Alluvial fan deposits may be subdivided into two basic types, those resulting from streamflow and those resulting from debris-flow and related processes. In streamflow processes, sediment is transported in suspension, saltation, and traction by channelized or nonchannelized flowing water. In debris-flow, mudflow, and related processes, a muddy or clayey matrix helps to support and transport clasts and fragments (Blackwelder, 1928; Davis, 1938; Prior, Stephens, and Douglas, 1970; see Johnson, 1970 and Pierson, 1980 for discussions of the mechanics of this process). Streamflow processes involve movement of essentially Newtonian fluids, whereas debris flows and related processes involve movement of plastico-viscous fluids. Streamflow deposits are generally well stratified, contain a variety of sedimentary structures indicative of different hydraulic flow regimes, have small amounts of clay-size matrix, are clast-supported, and contain clasts that are imbricated and oriented relative to flow direction. Debris-flow and related deposits, in contrast, are generally poorly stratified, contain few sedimentary structures, have large amounts of clay-size matrix, are matrix-supported, and contain clasts with poorly developed to no imbrication or orientation relative to flow direction.

The abundance of debris-flow and related deposits on alluvial fans is one of the more important features that permits alluvial fan deposits to be recognized in the ancient stratigraphic record (Rust, 1977, 1979). Although both streamflow and debris-flow deposits can be found in any part of a fan and are typically intermixed, streamflow deposits are generally more characteristic of distal fan facies and humid-climate fans, and debris-flow deposits more characteristic of proximal fan facies and arid-climate fans (Hooke, 1968a; Bull, 1977; Nilsen, 1982). Proximal alluvial fan deposits contain the coarsest sediments and accumulate on the upper or inner parts of the fan, close to where the stream emerges from the upland area. Distal alluvial fan deposits contain finer grained sediments and accumulate on the lower or outer parts of the fan. Proximal and distal facies can be recognized in ancient alluvial fan deposits through the use of paleocurrent data, maps showing the distribution of the coarsest clasts and thickest beds, and, in some cases, downfan changes in clast roundness and other textural parameters. In some cases, the proximal deposits in ancient alluvial fan deposits are characterized by highly channeled and lenticular beds and the distal facies by less channeled, sheetlike beds (e.g., Kerr, Pappajohn, and Peterson, 1979). This results where flows emerge from the channels and spread out to deposit laterally extensive sheets in the distal parts of alluvial fans.

The composition of alluvial fans correlates closely with that of the source area because of the relatively short transport distances and general lack of sorting within the fans. However, the composition may become more complex distally and laterally as a result of coalescence of adjacent lithologically dissimilar fans and mixing with marginal alluvial-plain deposits.

The sediment deposited on alluvial fans ranges in size from clay to boulder. Breccia and conglomerate are very common, and fan deposits are characterized by abrupt vertical and lateral changes in sorting as well as in maximum and mean grain size. Debris-flow and mudflow deposits are generally poorly sorted and streamflow deposits better sorted. Clast size and bedding thickness typically decrease downfan, although alternations of debris-flow and streamflow deposits may yield irregular trends. Clasts in streamflow deposits tend to be better rounded than those in debris-flow or mudflow deposits because they have been subjected to more clast-to-clast collisions. Although clast roundness increases downfan, clast sphericity shows little downfan change and seems unrelated to the depositional process (e.g., Nilsen, 1968, Paper 13).

The size of alluvial fans ranges from less than 100 m to more than 150 km in radius, although most are probably less than 10 km in radius. Very small fans, less than a meter in radius, can be constructed where suitable slopes, sources, and drainages are present. The size and shape of fans are controlled by several factors, including the size of drainage area, the amount of sediment carried by the feeding stream, and the relief and rate of tectonic uplift at the basin margin. For the same type of sediment, fans with larger drainage areas generally have gentler slopes than those with smaller drainage areas (Hooke, 1968a, 1968b). As the size of debris and concentration of sediment in flows increase, fans become steeper (Hooke, 1968a, 1968b; Paper 12). Because debris-flow deposits form steeper slopes than streamflow deposits, the slope of alluvial fans is generally lower in areas of greater precipitation. Fans that drain areas underlain by fine-grained sedimentary rocks such as shale or argillite are larger and steeper, whereas those that drain areas underlain by coarser grained sedimentary rocks such as sandstone or crystalline rocks are smaller and more gently sloping (Paper 12).

The thickness of alluvial fan deposits ranges from a few meters to perhaps as much as 25,000 m, the stratigraphic thickness of the Devonian Old Red Sandstone in the Hornelen basin of western Norway (Steel, 1976). Because fans are deposited close to the source area, they are particularly sensitive to tectonic activity, and their thickness is strongly influenced by tectonism (Steel and Wilson, 1975; Steel et al., 1979, Paper 14; Paper 15; Steel and Gloppen, 1980). Alluvial fans may undergo several major cycles of growth that normally begin as a result of tectonic uplift of the upland source terrane.

The geometry and facies distribution of alluvial fan sequences are controlled primarily by the rate of differential vertical movement between the mountains and adjacent basin, by the rate of erosion in the source area, and by the rate of sedimentation in the basin. Fan growth continues as long as uplift adjacent to the mountain front exceeds or equals the sum of channel downcutting in the mountains and deposition on the fan apex (Bull, 1977, p. 250). New sedimentation can take place in either proximal or distal areas. Rapid uplift and generation of debris flows may yield new deposition in upper- and middle-fan areas, whereas slower uplift and downcutting may yield channel entrenchment and new streamflow deposition in lower-fan areas. Syndepositional deformation may result in intraformational angular unconformities, broad structural warps, local thinning, and stacked sedimentary megacycles within the basin fill (Miall, 1978, 1981).

Bull (Paper 17) recognized three main types of fans based on radial or longitudinal cross-section: (1) wedge-shaped bodies that are thick close to the mountain front and that thin or wedge out away from the mountain front, reflecting major uplift of the mountains before initiation of fan sedimentation; (2) lens-shaped bodies that thin both toward and away from the mountain front, reflecting continued uplift of the mountains during fan sedimentation; and (3) wedge-shaped bodies that are thin adjacent to the mountain front and thicker away from it, reflecting a lengthy interval of erosion and redistribution of proximal fan deposits farther downfan, commonly associated with pediment formation.

The internal geometry of alluvial fans can be very complex. As alluvial fans build outward toward the basin, lateral shifting of channels gradually constructs a depositional body that is fan-shaped in plan view, convex upward and lens-shaped in transverse cross-section, and concave upward in radial profile. Coalescing of adjacent fans can result in a series of merging lens-shaped bodies, although larger and more rapidly prograding fans will encroach upon and overlap the distal parts of smaller fans. New fans typically begin in topographically low interfan areas or zones of coalescence and interfingering of adjacent fans. By the process of entrenchment and abandonment, a complex overlapping sequence of lens-shaped bodies is gradually constructed. Weathering of inactive parts of alluvial fan surfaces usually produces soil horizons, which can be used to differentiate different ages of alluvial fan deposits in many basins (Davis and Hall, 1959; Gile and Hawley, 1966; Marchand and Allwardt, 1981). Because of poor fossil control, great thicknesses of coarse-grained deposits, and lack of laterally extensive marker beds, the geometry of both individual fans as well as ancient, coalesced fan deposits is commonly difficult to delineate, especially where adjacent fans are similar compositionally.

Alluvial fans generally grade sourceward into channeled fluvial deposits of the river or stream that is feeding the fan. However, the stream deposits commonly are very thin and not preserved in ancient systems because of continued uplift of or erosion into the source area. The fans may rest on older pediment surfaces or on parts of older fans, inheriting previously established fan morphology. At the base of and on the lower parts of the range-front slope, the fans may grade laterally into talus deposits and associated slope wash, colluvial deposits, glacial deposits of various types, terrace accumulations, and various types of landslide deposits, including flows, falls, slumps and slides (e.g., Paper 2; Drewes, 1963; Paper 13). Talus accumulations can generally be recognized locally by their underlying contact with bedrock or soil developed on bedrock, increase in clast size downslope, and skewness toward coarser clasts (Paper 17).

Alluvial fans can grade laterally into other alluvial fan deposits derived from adjacent feeding streams. Because of the many possible ages, stages of activity, and sizes of adjacent fans, facies relations between coalescing fans can be complex. There is a complex interfingering of adjacent fans along most range fronts to form a laterally extensive piedmont. Bull (Paper 17, p. 78) examined electric logs of many fans in the western Great Valley of California and concluded that the boundaries of adjacent coalesced fans should change little in position through time, because the rates of sediment yield from adjacent drainages would be equally affected by tectonic and climatic changes. Longer term tectonic effects and changes in size of drainage systems, however, can yield major changes in the size of adjacent fans.

Alluvial fans most commonly grade distally into other types of alluvial deposits, typically those of the bounding alluvial plain. The marginal river system may erode the distal parts of the fans, which yields complex depositional and erosional interfingering of the alluvial fan and alluvial-plain facies (i.e., Larsen and Steel, 1978). The alluvial-plain deposits may be distinguishable by their better sorting and rounding, more extensive floodplain and levee deposits, larger channel systems, and general lack of mudflow, debris flow, and landslide units.

Alluvial fans can have various types of cement that develop at different times in the depositional and postdepositional history of the fan. Carbonate is the cementing agent in the majority of fans studied.

Introduction

Iron-oxide, petroliferous, and siliceous cements have also been noted, the latter particularly in silica-rich fan sediments. Caliche zones are common in alluvial fan deposits (Lattman, 1973). However, neither cementation nor diagenesis have been extensively studied in ancient alluvial fan deposits.

Because alluvial fans are typically highly porous and permeable, they form excellent aquifers. The deposits characteristically consist of alternating zones of porous and permeable streamflow deposits and relatively nonporous and less permeable debris-flow and mudflow deposits. Sieve deposits, which consist of angular but well-sorted gravel, are highly porous and permeable on modern alluvial fans. Flow of ground water is commonly aided by paleochannels, which act as conduits for flow. However, aquifer characteristics vary greatly with the type of deposit and their location on a fan (Cehrs, 1979).

Various tectonic settings on continents yield abundant alluvial fan deposition. Fans are generated along intracontinental rifts such as the Basin and Range province of the United States and the east African rift system. Basement uplifts in continental interiors can result in alluvial fan deposition, such as during the late Paleozoic in Colorado (Howard, 1966) and the Cretaceous and early Tertiary in the Rocky Mountains. Fan deposits are preserved in fault-bounded, postorogenic basins of the Devonian Old Red Sandstone, the Carboniferous, and the Permian and Triassic New Red Sandstone of western Europe, Greenland, and eastern North America. Alluvial fan deposits are associated with collisional orogenies, and the modern fans in India are examples that record suturing of India and Asia. Alluvial fan deposits are also characteristic of sedimentation in strike-slip regions, having been reported from the Devonian Hornelen basin of western Norway, the Carboniferous basins of northern Spain, and in the various basins of California adjacent to the San Andreas fault.

MODERN ALLUVIAL FAN DEPOSITS

For many years, the only modern alluvial fans that were studied in detail were from arid and semiarid regions of the world. As prominent and conspicuous as these fans are, they may only be representative of deposition in a limited climatic range, whereas a large number of ancient alluvial fans may have been deposited in temperate and humid climatic conditions. Starting in the 1960s, studies of alluvial fans deposited in cold climates, including polar regions, contributed greatly to our understanding of this group of fans. Because of vegetation cover, soil formation, and human development, less work has been done on alluvial fans in humid regions. In addition, many modern fans may have developed primarily during wetter and colder intervals of the Pleistocene.

Deposition on modern arid-region fans of the western United States has been studied chiefly by Blissenbach (Paper 2), Beaty (Paper 3, 1970), Bluck (Paper 5), Ruhe (1964), Lustig (1965), Denny (Paper 8), Hooke (Paper 9, 1968a), Reeves (1977), and Wells (1977). Other important studies of modern arid-region alluvial fans include those of Beaumont (1972) in Iran and Schick (1977) in Israel. Deposition on modern Arctic-, temperate- and humid-region fans of the western United States has been studied chiefly by Bull (1961, 1962a, 1962b, Paper 4, 1964a, 1964b, 1964c), Tricart (1963), Tricart and Cailleux (1965), Anderson and Hussey (1962), Hoppe and Ekman (1964), Ray (1964), Winder (Paper 6), Anstey (1965), Carryer (1966), Leggett, Brown, and Johnston (Paper 7), Ryder (1971a, 1971b), McPherson and Hirst (1972), Wasson (Papers 10 and 11, 1977, 1979) Sanlaville (1977), Ori (1982) and Budel (1982). Experimental studies and modeling of alluvial fans have been conducted by Hooke (Paper 9), Price (1974), Schumm (1977, p. 255-264) and colleagues, Hooke and Rohrer (Paper 12), and Rachocki (1981).

There have been few published geophysical studies of alluvial fans (Bull, 1977; Nilsen, 1982). The three-dimensional geometry, continuity of reflecting surfaces, subsurface extent of channels, and thickness variations of modern fans are poorly known.

Late Cenozoic climatic and sea-level changes have drastically altered the depositional setting of many modern alluvial fans. Because many modern fans developed chiefly during the Pleistocene under different climatic conditions and when base level was lower as a result of lowered sea level, modern processes may not be wholly indicative of events that led to construction of most of the fan. Conversely, growth of fans has been accelerated in some areas by Pleistocene to Holocene climate changes. Areas largely unaffected by either climatic, sea-level, or lake-level changes may have experienced little change from Pleistocene to Holocene, but these are probably few in number.

ANCIENT ALLUVIAL FAN DEPOSITS

The great majority of significant studies of ancient alluvial fan deposits has been restricted to a few regions and time intervals, chiefly the Old Red Sandstone (Devonian) and New Red Sandstone (Permian and Triassic) in the Caledonian-Appalachian chain, and scattered Upper Cretaceous and Cenozoic sequences in the western

Introduction

United States. These alluvial fan deposits, although generally well exposed and studied in some detail, may not be typical or characteristic of most ancient alluvial fan deposits. They are commonly very thick and generally form redbeds, where deposited in both arid and humid paleoclimates.

Ancient alluvial fan deposits have been most thoroughly described from the Devonian Old Red Sandstone (Bluck, 1967, 1969, 1978, 1980; Allen and Friend, 1968; Nilsen, 1968, Paper 9, 1973; Miall, 1970; Stephenson, 1972; Schluger, 1973; Horne, 1975; Steel, 1976; Paper 14; Steel and Aasheim, 1978) and Carboniferous to Jurassic redbeds that include the New Red Sandstone (Krynine, 1950; Klein, 1962; Bluck, 1965; Meckel, 1967; Laming, 1966; Belt, 1968; Wessel, 1969; Deegan, 1973; Teisseyre, 1973; Steel, 1974; Steel and Wilson, 1975; Hubert, Reed and Carey, 1976; Hubert et al., 1978; Brookfield, 1980) in the Caledonian, Appalachian, and Hercynian orogens of North America and Europe. Other well-described ancient fan accumulations include Cretaceous and Tertiary deposits of the western North American Cordillera (Paper 1; Carter and Gualtieri, 1965; Soister, 1968; Wilson, 1970; Steidtmann, 1971; Love, 1973; Ryder, Fouch, and Elison, 1976), and late Cenozoic deposits associated with strike-slip pull-apart basins along the San Andreas fault (Van de Kamp, 1973; Crowell, 1973, 1974, 1982; Link and Osborne, 1978; Kerr, Pappajohn, and Peterson, 1979). Alluvial-fan deposits of Precambrian age have been studied in some detail because many are associated with ore deposits (Williams, 1969; McGowen and Groat, 1971; Pretorius, 1974a, 1974b, 1975, 1976; Reid, 1974; Minter, 1976, 1978, 1981; Robertson, 1976, 1981).

The recognition of ancient alluvial fan deposits has been greatly influenced by clast-size considerations. In general, most workers have interpreted conglomerate as ancient alluvial fan deposits and sandstone, siltstone, and mudstone as alluvial plain deposits. In most paleogeographic studies, this subdivision has been convenient and has generally been accepted. However, it is clear that alluvial fans can be constructed of material of any grain size, depending simply upon what type of material is brought to the depositional basin by feeding streams. As a result, the literature on ancient alluvial fan deposits contains scarcely a single reference to nonconglomeratic fan sequences, and descriptions of many fan deposits may be hidden in the literature under the assumption that they are alluvial plain deposits.

Because alluvial fans develop at the base of slopes adjacent to mountainous terrain and uplifted blocks, alluvial fan deposits have been most commonly reported from mobile belts. They are most typically synorogenic or postorogenic. Within major mountain chains, they may be intramontane, intermontane, or extramontane. Intramontane fans are perhaps most common, but their chances of preservation in the stratigraphic record is low because of continued uplift of the mountains and eventual erosion of the fan deposits. Intermontane and extramontane fans have a greater chance of preservation but in many cases may have been interpreted as alluvial plain rather than alluvial fan deposits. Small fans, of course, may develop at the base of slopes adjacent to small streams. Many thin fan deposits may result from minor climatic changes and represent brief depositional intervals during long periods of erosion; some of these fans may include and may not be easily distinguishable in the rock record from talus, scree, slope wash, and landslide deposits.

ECONOMIC AND ENVIRONMENTAL ASPECTS OF ALLUVIAL FAN DEPOSITS

Alluvial fan deposits are good sources of sand and gravel and also may contain heavy-mineral placers (Gross, 1968; Minter, 1976, 1978, 1981; McGowen, 1979; Robertson, 1976, 1981; Roscoe, 1981; Armstrong, 1981), may be host rocks for roll-front uranium deposits (Ethridge and Thompson, 1978), and may contain minable coal (Heward, 1978). They are generally not good sources of ore minerals, except perhaps where metamorphosed or subjected to faulting and emplacement of ore-bearing fluids that emanate from fault zones. Placer deposits are common in alluvial fan and associated bajada, pediment, arroyo, and fluvial deposits in many parts of the world (i.e., Johnson, 1972a, 1972b, 1973; Pretorius, 1974b). Ancient alluvial fan placers have not always been clearly distinguished from alluvial plain placers. Alluvial fan placers are generally developed in channels on bedrock near the heads of fans or in sandy depositional lows in the outer parts of fans. Placers have been produced experimentally by Schumm (1977) and colleagues, who concluded that sediment reworking and sorting is required for their formation.

Alluvial fan deposits are generally not reservoir rocks for petroleum because they are not laterally connected to source rocks, have not been deeply enough buried, are not extensive enough laterally, do not have proper seals, may have low permeability and porosity following diagenesis, and generally do not contain facies that are good source rocks. Fan deltas, however, can form suitable stratigraphic traps for oil and gas, especially the outer-fan facies if reworked into bars by marine currents (Fisher and Brown, 1972).

Alluvial fans are critically important to man as sources of groundwater, especially in arid regions. Alluvial fan deposits form part of the groundwater reservoir in alluviated basins throughout the world

Introduction

(Bindeman and Farengol'ts, 1975), and in many arid and semiarid regions they are the only sources of water. In the western United States, groundwater pumped from late Cenozoic alluvial fan deposits is critically important in places such as the Santa Clara Valley in northern California (Poland, 1971; Rantz, 1972), the San Joaquin Valley in central California (Paper 4; Magleby and Klein, 1965; Croft, 1972), the Los Angeles basin in southern California (Eckis, 1928; Poland et al., 1956; Poland, 1959; California Department of Water Resources, 1966), and the basin surrounding Tucson, Arizona (Pashley, 1966). Groundwater recharge takes place mainly through alluvial fans on the margins of basins; extensive covering of fan surfaces by urban development, which increases runoff and decreases infiltration, has caused problems in groundwater recharge in several areas.

Alluvial fans are suitable sites for farming and grazing. The lower parts of large fans, which commonly consist of weathered sandy deposits, may form particularly good soil that typically overlies good aquifers (Bull, 1977). Alluvial fans present serious environmental hazards, especially as sites of flash floods, landslides, deep and near-surface land subsidence, and ground displacement and shaking from seismicity produced by range-front faults.

Flooding is perhaps the major environmental hazard and is often more serious than flooding of river systems because the rapidity and unpredictability of flooding presents a greater hazard to life (Chawner, 1935; Dawdy, 1979). Active channels can shift abruptly, even during the same flood event, and sheetfloods, debris flows, and mudflows can quickly cover broad areas with coarse debris. Renwick (1977) showed the effect of intense rainfall on a small fan in northern New York State in which sediment was deposited rapidly in channels, levees, and interchannel areas. Flooding on fans yields both erosional and depositional hazards—roads on fans can be either eroded away if constructed above the natural fan surface or covered with debris if constructed below it (Schick, 1974). Aqueducts, such as those in the Great Valley of California, must be constructed so that flows pass either under or over covered portions of them. Flood hazards associated with fans in various areas have been summarized by Bull (1977).

Landslides develop on the bedrock and colluvium-mantled slopes of the adjacent range front and typically descend out onto alluvial fan surfaces or become entrained in fan channels. These locally sudden and often unpredictable events result in deposition of debris flows, mudflows, rockfalls, slumps, block glides, and other types of landslide debris on the fans (Beaty, 1974; Eisbacher, 1979). Because of the rapid movement, some of these features can be hazardous to life as well as property, as demonstrated by debris flows in southern California

(Sharp and Nobles, 1953; Scott, 1971; Morton and Campbell, 1978).

Subsidence on alluvial fans can affect manmade structures. Near-surface subsidence, resulting in settling and cracking of valuable farmland, can result from wetting of clayey fan deposits, as has been described from the western San Joaquin Valley by Bull (1972). Subsidence caused by compaction of deeper levels of fans because of groundwater withdrawal, especially from clay-rich or fine sandy reservoirs, has caused surface fissuring that has damaged aqueducts, railroads, airport runways, and roads (Schumann and Poland, 1969). More than 2 m of subsidence has been documented for Arizona (Holzer, 1977; Jachens and Holzer, 1982) and as much as 8 m for central California (Miller, Green, and Davis, 1971; Bull, 1975). Because of range-front faulting, seismic hazards are commonly associated with alluvial-fan deposits. The hazards are especially great near rapidly accumulating, coarse-grained, thick and steep alluvial fans.

COMPARATIVE STUDIES OF MODERN AND ANCIENT ALLUVIAL FAN DEPOSITS

A number of general summaries of modern and ancient alluvial fan deposits have been published. Allen (1965) discussed the origin and characteristics of alluvial sediment and included a brief summary of alluvial fans. Bull (1968a) briefly summarized the general characteristics of alluvial fan deposits. Yazawa, Toya, and Kaizuka (1971) discussed most of the processes and morphologic aspects of alluvial fans. Fisher and Brown (1972) presented a general model of alluvial fan sedimentation. Bull (Paper 17, 1977) prepared summaries of modern alluvial fans and compared their characteristics with some ancient alluvial fan deposits. Spearing (1974) provided a useful, well-illustrated summary sheet of the chief sedimentary characteristics of alluvial fan deposits. Schumm (1977), Miall (1977), Collinson (1978), and Rust (1978, 1979) discussed varied aspects of fluvial sedimentation, including that of alluvial fans and fan deltas. Ethridge and Thompson (1978) and McGowen (1979), in summarizing the characteristics of fluvial sediment, discussed mineral deposits associated with fans, especially placer gold and uranium. A selected list of references to alluvial fan deposits was prepared by Nilsen and Moore (1980) and a bibliography by Nilsen and Moore (1984). Rachocki (1981), in a book about small alluvial fans produced by erosion and sedimentation in quarries in northern Poland, included a general summary of alluvial fan deposits. Nilsen (1982) summarized stratigraphic, sedimentologic, and morphologic characteristics of alluvial fan deposits.

REFERENCES

Allen, J. R. L., 1965, A Review of the Origin and Characteristics of Recent Alluvial Sediments, *Sedimentology* **5**:89-191.

Allen, J. R. L., and P. F. Friend, 1968, Deposition of the Catskill Facies, Appalachian Region—with notes on some other Old Red Sandstone Basins, in *Late Paleozoic and Mesozoic Continental Sedimentation, northeastern North America,* G. deV. Klein, ed., Geol. Soc. America Special Paper 106, p. 21-74.

Anderson, G. S., and K. M. Hussey, 1962, Alluvial Fan Development at Franklin Bluffs, Alaska, *Iowa Acad. Sci. Proc.* **69**:310-322.

Anstey, R. L., 1965, *Physical Characteristics of Alluvial Fans,* U.S. Army Natick Laboratories, Technical Report ES-20, 109p.

Armstrong, F. C., ed., 1981, Genesis of Uranium- and Gold-Bearing Precambrian Quartz-Pebble Conglomerates, *U.S. Geol. Survey Prof. Paper 1161,* variably paginated.

Beaty, C. B., 1970, Age and Estimated Rate of Accumulation of an Alluvial Fan, White Mountains, California, *Am. Jour. Sci.* **268**:50-77.

Beaty, C. B., 1974, Debris Flows, Alluvial Fans and a Revitalised Catastrophism, *Zeitschr. Geomorphologie,* Supplementband, **21**:39-51.

Beaumont, P., 1972, Alluvial Fans along the Foothills of the Elburz Mountains, Iran, *Palaeogeography, Palaeoclimatology, Palaeoecology,* **12**: 251-273.

Belt, E. S., 1968, Carboniferous Continental Sedimentation, Atlantic Provinces, Canada, in *Late Paleozoic and Mesozoic Continental Sedimentation, Northwestern North America,* G. deV. Klein, ed., Geol. Soc. America Special Paper 106, pp. 127-176.

Bindeman, N. N., and Z. D. Farengol'ts, 1975, Ground water in colluvial slopes and alluvial fans, in *Regional'naya otsenka resursov podzemnykh vod,* Izd. Nauka, Moscow, pp. 86-94.

Blackwelder, E., 1928, Mudflow as a Geological Agent in Semi-Arid Mountains, *Geol. Soc. America Bull.* **39**:465-484.

Bluck, B. J., 1965, The Sedimentary History of some Triassic Conglomerates in the Vale of Glamorgan, South Wales, *Sedimentology* **4**:225-245.

Bluck, B. J., 1967, Deposition of Some Upper Old Red Sandstone Conglomerates in the Clyde Area: A Study of the Significance of Bedding, *Scottish Jour. Geology* **3**:139-167.

Bluck, B. J., 1969, Old Red Sandstone and Other Paleozoic Conglomerates of Scotland, in *North Atlantic—Geology and Continental Drift,* M. Kay, ed., Am. Assoc. Petroleum Geologists Mem. 12, pp. 711-723.

Bluck, B. J., 1974, Structure and Directional Properties of Some Valley Sandur Deposits in Southern Iceland, *Sedimentology* **21**:533-554.

Bluck, B. J., 1978, Sedimentation in a Late Orogenic Basin: The Old Red Sandstone of the Midland Valley of Scotland, in *Crustal Evolution in Northwest Britain and Adjacent Regions,* D. R. Bowes and B. E. Leake, eds., Geol. Jour. Special Issue No. 10, pp. 249-278.

Bluck, B. J., 1980, Evolution of a Strike-Slip Fault-Controlled Basin, Upper Old Red Sandstone, Scotland, in *Sedimentation in Oblique-Slip Mobile Zones,* P. F. Ballance and H. G. Reading, eds., International Association of Sedimentologists Special Publication No. 4, p. 63-78.

Boothroyd, J. C., 1976, Sandur Plains, Northeast Gulf of Alaska, A Model for Alluvial Fan-Fan Delta Sedimentation in Cold-Temperature Environments, in *Recent and ancient sedimentary environments in Alaska,* T. P. Miller, ed., Alaska Geological Society, Anchorage, Alaska, p. N1–N13.

Boothroyd, J. C., and Ashley, G. M., 1975, Process, Bar Morphology and Sedimentary Structures on Braided Outwash Fans, Northeastern Gulf of Alaska, in *Glaciofluvial and Glacio-Lacustrine Sedimentation,* A. V. Jopling, and B. C. McDonald, eds., Soc. Econ. Paleontologists and Mineralogists Spec. Pub. 23, pp. 193–222.

Boothroyd, J. C., and D. Nummedal, 1978, Proglacial Braided Outwash—a Model for Humid Alluvial-Fan Deposits, in *Fluvial Sedimentology,* A. D. Miall, ed., Canadian Society of Petroleum Geologists Memoir No. 5, pp. 641–668.

Brookfield, M. E., 1980, Permian Intermontane Basin Sedimentation in Southern Scotland, *Sed. Geol.* **27:**167–194.

Budel, J., 1982, *Climatic Geomorphology,* Princeton University Press, Princeton, New Jersey, 443p.

Bull, W. B., 1961, Tectonic significance of radial profiles of alluvial fans in western Fresno County, California, *U.S. Geol. Survey Prof. Paper 424B,* pp. 182–184.

Bull, W. B., 1962a, Relations of Alluvial-Fan Size and Slope to Drainage-Basin Size and Lithology in Western Fresno County, California, *U.S. Geol. Survey Prof. Paper 450-B,* pp. B51–B53.

Bull, W. B., 1962b, Relation of Textural (CM) Patterns of Depositional Environment of Alluvial-Fan Deposits, *Jour. Sed. Petrology,* **32:**211–216.

Bull, W. B., 1964a, Alluvial Fans and Near-Surface Subsidence in Western Fresno County, California, *U.S. Geol. Survey Prof. Paper 437-A,* 70p.

Bull, W. B., 1964b, Geomorphology of Segmented Alluvial Fans in Western Fresno County, California, *U.S. Geol. Survey Prof. Paper 352-E,* pp. 89–129.

Bull, W. B., 1964c, History and Causes of Channel Trenching in Western Fresno County, California, *Am. Jour. Sci.* **262:**249–258.

Bull, W. B., 1968a, Alluvial Fans, *Jour. Geol. Education* **16:**101–106.

Bull, W. B., 1968b, Alluvial Fan, Cone, in *Encyclopedia of Geomorphology,* R. W. Fairbridge, ed., Reinhold Book Corporation, New York, p. 7–10.

Bull, W. B., 1972, Prehistoric Near-Surface Subsidence Cracks in Western Fresno County, California, *U.S. Geol. Survey Prof. Paper 437-C,* 86p.

Bull, W. B., 1975, Land Subsidence in the Los Banos–Kettleman City Area, California, Part 2, Subsidence and Compaction of Deposits, *U.S. Geol. Survey Prof. Paper 437-F,* 90p.

Bull, W. B., 1977, The Alluvial Fan Environment, *Prog Phys. Geog.* **1:**222–270.

California Department of Water Resources, 1966, Planned Utilization of Ground Water Basins—San Gabriel Valley, Appendix A: Geohydrology, *California Dept. Water Resources Bull. 104-2,* 230p.

Carryer, S. J., 1966, A Note on the Formation of Alluvial Fans, *New Zealand Jour. Geol. Geophys.* **9:**91–94.

Carter, W. D., and J. L. Gualtieri, 1965, Geyser Creek Fanglomerate (Tertiary), La Sal Mountains, Eastern Utah, *U.S. Geol. Survey Bull. 1244-E,* 11p.

Cehrs, D., 1979, Depositional Control of Aquifer Characteristics in Alluvial Fans, Fresno County, California: Summary, *Geol. Soc. America Bull.* **90** pt. 1, p. 709–711; pt. 2, p. 1282–1309.

Chawner, W. D., 1935, Alluvial Fan Flooding, the Montrose, California Flood of 1934, *Geog. Rev.* **25:**77–88.

Collinson, J. D., 1978, Alluvial Sediments, in *Sedimentary Environments and*

Facies, H. G. Reading, ed., Blackwell Scientific Publications, London, pp. 15-60.

Croft, M. G., 1972, Subsurface Geology of the Late Tertiary and Quaternary Water-Bearing Deposits of the Southern San Joaquin Valley, California, *U.S. Geol. Survey Water-Supply Paper 1999-H,* pp. H1-H29.

Crowell, J. C., 1973, Ridge Basin, Southern California, in *Sedimentary Facies Change in Tertiary rocks, California Transverse and Southern Coast Ranges,* Soc. Econ. Paleontologists and Mineralogists Guidebook, Field Trip no. 2, p. 17.

Crowell, J. C., 1974, Sedimentation along the San Andreas Fault, California, in *Modern and Ancient Geosynclinal Sedimentation,* R. H. Dott, Jr., and R. H. Shaver, eds., Soc. Econ. Paleontologists and Mineralogists Spec. Pub. 19, pp. 292-303.

Crowell, J. C., 1982, The Violin Breccia, Ridge Basin, Southern California, in *Geologic History of Ridge Basin, Southern California: Pacific Section,* J. C. Crowell, and M. H. Link, eds., Soc. Econ. Paleontologists and Mineralogists, pp. 89-98.

Davis, S. N., and F. R. Hall, 1959, Water Quality of Eastern Stanislaus and Northern Merced Counties, California, *Stanford Univ. Pub. Geol. Sci.* **6**(1):112.

Davis, W. M., 1938, Sheet Floods and Streamfloods, *Geol. Soc. America Bull.* **49:**1337-1416.

Dawdy, D. R., 1979, Flood-Frequency Estimates on Alluvial Fans, *Am. Soc. Civil Engineers, Jour. Hydraulics Div.* **HY11:**1407-1413.

Deegan, C. E., 1973, Tectonic Control of Sedimentation at the Margin of a Carboniferous Depositional Basin in Kirkudbrightshire, *Scottish Jour. Geology* **9:**1-28.

Denny, C. S., 1965, Alluvial Fans in the Death Valley Region, California and Nevada, *U.S. Geol. Survey Prof. Paper 466,* 62p.

Drewes, H., 1963, Geology of the Funeral Peak Quadrangle, California, on the East Flank of Death Valley. *U.S. Geol. Survey Prof. Paper 413,* 78p.

Eckis, R., 1928, Alluvial Fans in the Cucamonga District, Southern California, *Jour. Geology* **36:**111-141.

Eisbacher, G. H., 1979, Cliff Collapse and Rock Avalanches (sturzstroms) in the Mackenzie Mountains, Northwestern Canada, *Canadian Geotechnical Jour.* **16**(2):309-334.

Ethridge, F. G., and T. B. Thompson, 1978, *Lecture Notes for the Short Course on the Fluvial System,* Colorado State University, Fort Collins, Part A, 221p.; Part B, 101p.

Fairbridge, R. W., ed., 1968, *The Encyclopedia of Geomorphology,* Reinhold Book Corporation, New York, 1295p.

Fisher, W. L., and L. F. Brown, Jr., 1972, *Clastic Depositional Systems—a Genetic Approach to Facies Analysis,* Texas Bureau of Economic Geology, Austin, Tx., 211p.

Gile, L. H., and J. W. Hawley, 1966, Periodic Sedimentation and Soil Formation on an Alluvial-Fan Piedmont in Southern New Mexico, *Soil Sci. Soc. America Proc.* **30:**261-268.

Gross, W. H., 1968, Evidence for a Modified Placer Origin for Auriferous Conglomerates, Canavieiras Mine, Jacobina, Brazil, *Econ. Geology* **63**(3):271-276.

Gustavson, T. C., 1974, Sedimentation on Gravel Outwash Fans, Malaspina Glacier Foreland, Alaska, *Jour. Sed. Petrology* **44:**374-389.

Hadley, R. F., 1967, Pediments and Pediment Forming Processes, *Jour. Geol. Education* **15:**83-89.

Heward, A. P., 1978, Alluvial Fan and Lacustrine Sediments from the Stephanian A and B (La Magdalena, Cinera-Matallana and Sabero) Coalfields, Northern Spain, *Sedimentology* **25:**451-488.

Holzer, T. L., 1977, *Ground Failure in Areas of Subsidence Due to Ground-Water Decline in the United States,* Proceedings of 2nd International Association of Hydrological Sciences and Land Subsidence Symposium, International Association of Hydrological Sciences, Anaheim, California (Pub. no. 121), pp. 423-433.

Hooke, R. LeB., 1968a, Steady-State Relationships on Arid-Region Alluvial Fans in Closed Basins, *Am. Jour. Sci.* **266:**609-629.

Hooke, R. LeB., 1968b, Slopes of Alluvial Fans (abs.), *Geol. Soc. America Spec. Paper 101,* pp. 97-98.

Hoppe, G., and S. R. Ekman, 1964, A Note on the Alluvial Fans of Ladtjovagge, Swedish Lapland, *Geog. Annaler* **46:**338-342.

Horne, R. R., 1975, The Association of Alluvial Fan, Eolian, and Fluviatile Facies in the Caherbla Group (Devonian), Dingle Peninsula, Ireland, *Jour. Sed. Petrology* **45:**535-540.

Howard, J. D., 1966, Patterns of Sediment Dispersal in the Fountain Formation of Colorado, *Mountain Geologist* **3:**147-153.

Hubert, J. F., A. A. Reed, and P. J. Carey, 1976, Paleogeography of the East Berlin Formation, Newark Group, Connecticut Valley, *Am. Jour. Sci.* **276:**1183-1207.

Hubert, J. F., A. A. Reed, W. L. Dowdall, and J. M. Gilchrist, 1978, Guide to the Mesozoic Redbeds of Central Connecticut, *Connecticut Geol. and Nat. History Survey Guide book No. 4,* 129p.

Jachens, R. C., and T. L. Holzer, 1982, Differential Compaction Mechanism for Earth Fissures near Casa Grande, Arizona, *Geol. Soc. America Bull.* **93:**998-1012.

Johnson, A. M., 1970, *Physical Processes in Geology,* Freeman, Cooper and Company, San Francisco, 577p.

Johnson, M. G., 1972a, Placer Gold Deposits of New Mexico, *U.S. Geol. Survey Bull. 1348,* 46p.

Johnson, M. G., 1972b, Placer Gold Deposits of Arizona, *U.S. Geol. Survey Bull. 1355,* 103p.

Johnson, M. G., 1973, Placer Gold Deposits of Nevada, *U.S. Geol. Survey Bull. 1356,* 118p.

Kerr, D. R., S. Pappajohn, and G. L. Peterson, 1979, Neogene Stratigraphic Section at Split Mountain, Eastern San Diego County, California, in *Tectonics of the Juncture between the San Andreas Fault System and Salton Trough, southeastern California,* J. C. Crowell, and A. G. Sylvester, eds., Department of Geological Sciences, University of California at Santa Barbara, pp. 111-124. [Field trip guidebook for 1980 annual meeting, Geological Society of America, San Diego, California.]

Klein, G. DeV., 1962, Triassic Sedimentation, Maritime Provinces, Canada, *Geol. Soc. America Bull.* **73:**1127-1146.

Krigstrom, A., 1962, Geomorphological Studies of Sandur Plains and Their Braided Rivers in Iceland, *Geog. Annaler* **44:**328-346.

Krynine, P. D., 1950, Petrology, Stratigraphy, and Origin of Triassic Sedimentary Rocks of Connecticut, *Connecticut Geol. and Nat. History Survey Bulletin No. 73,* 247p.

Introduction

Laming, D. J. C., 1966, Imbrication, Palaeocurrents and Other Sedimentary Features in the Lower New Red Sandstone, Devonshire, England, *Jour. Sed. Petrology* **36:**940-959.

Larsen, V. and R. J. Steel, 1978, The Sedimentary History of a Debris Flow-Dominated, Devonian Alluvial Fan—a Study of Textural Inversion, *Sedimentology* **25:**37-59.

Lattman, L. H., 1973, Calcium carbonate cementation of alluvial fans in southern Nevada, *Geol. Soc. America Bull.* **84:**3013-3028.

Link, M. H., and R. H. Osborne, 1978, Lacustrine Facies in the Pliocene Ridge Basin Group, Ridge Basin, California, in *Modern and Ancient Lake Sediments,* A. Matter and M. E. Tucker, eds., International Association of Sedimentologists Spec. Pub. 2, pp. 169-187.

Love, J. D., 1973, Harebell Formation (Upper Cretaceous) and Pinyon Conglomerate (Uppermost Cretaceous and Paleocene), Northwest Wyoming, *U.S. Geol. Survey Prof. Paper 743-A,* 54p.

Lustig, L. K., 1965, Clastic Sedimentation in Deep Springs Valley, California, *U.S. Geol. Survey Prof. Paper 352-F,* pp. 131-192.

Magleby, D. C., and I. E. Klein, 1965, *Ground-Water Conditions and Potential Pumping Resources above the Corcoran Clay—an Addendum to the Ground-Water Geology and Resources Definite Plan Appendix, 1963,* U.S. Bureau of Reclamation Open-File Report.

Marchand, D. E., and A. Allwardt, 1981, Late Cenozoic Stratigraphic Units, Northeastern San Joaquin Valley, California, *U.S. Geol. Survey Bull. 1470,* 70p.

McGowen, J. H., 1971, Gum Hollow Fan Delta, Nueces Bay, Texas, *Texas Univ. Bur. of Econ. Geology Rept. Inv. No. 69,* 91p.

McGowen, J. H., 1979, Alluvial Fan Systems, in *Depositional and Ground-Water Flow Systems in the Exploration for Uranium,* W. E. Galloway, C. W. Kreitler, and J. H. McGowen, Texas Univ. Bureau of Economic Geology Research Colloquium, pp. 43-79.

McGowen, J. H., and C. G. Groat, 1971, Van Horn Sandstone, West Texas, an Alluvial Fan Model for Mineral exploration: *Texas Univ. Bur. Econ. Geology Rept. Inv. No. 72,* 57p.

McPherson, H. J., and F. Hirst, 1972, Sediment Changes on Two Alluvial Fans in the Canadian Cordillera, in *Mountain Geomorphology: Geomorphological Processes in the Canadian Cordillera,* H. O. Slaymaker and H. J. McPherson, eds., Tantalus Research Limited, Vancouver, no. 14, pp. 161-175.

Meckel, L. D., 1967, Origin of Pottsville Conglomerates (Pennsylvanian) in the Central Appalachians, *Geol. Soc. America Bull.* **78:**223-258.

Miall, A. D., 1970, Devonian Alluvial Fans, Prince of Wales Island, Arctic Canada, *Jour. Sed. Petrology* **40:**556-571.

Miall, A. D., 1977, *Fluvial Sedimentology,* Canadian Society of Petroleum Geologists, Calgary, Canada, Lecture notes for short course, variably paginated.

Miall, A. D., 1978, Tectonic Setting and Syndepositional Deformation of Molasse and other Nonmarine-Paralic Sedimentary Basins, *Canadian Jour. Earth Sci.* **15**(10):1613-1632.

Miall, A. D., 1981, Alluvial Sedimentary Basins: Tectonic Setting and Basin Architecture, in *Sedimentation and Tectonics in Alluvial Basins,* A. D. Miall, ed., Geol. Assoc. Canada Spec. Paper 23, pp. 1-33.

Miller, R. E., J. H. Green, and G. H. Davis, 1971, Geology of the Compacting

Deposits in the Los Banos-Kettleman City Subsidence Area, California, *U.S. Geol. Survey Prof. Paper 497-E,* 46p.

Minter, W. E. L., 1976, Detrital Gold, Uranium, and Pyrite Concentrations Related to Sedimentology in the Precambrian Vaal Reef Placer, Witwatersrand, South Africa, *Econ. Geol.* **71**(no. 1):157-176.

Minter, W. E. L., 1978, A Sedimentological Synthesis of Placer Gold, Uranium and Pyrite Concentrations in Proterozoic Witwatersrand Sediments, in *Fluvial Sedimentology,* A. D. Miall, ed., Canadian Soc. Petroleum Geologists Memoir 5, pp. 801-829.

Minter, W. E. L., 1981, Examples that Illustrate Sedimentological Aspects of the Proterozoic Placer Model on the Kaap-Vaal Craton, Witwatersrand, South Africa, *U.S. Geol. Survey Prof. Paper 1161-E,* pp. E1-E6.

Morton, D. M., and R. H. Campbell, 1978, Cyclic Landsliding at Wrightwood Southern California—A preliminary report, *U.S. Geol. Survey Open-File Report 78-1079,* 23p.

Nilsen, T. H., 1968, The Relationship of Sedimentation to Tectonics in the Solund District of Southwestern Norway, *Norges Geol. Undersokelse No. 359,* 108p.

Nilsen, T. H., 1973, Devonian (Old Red Sandstone) Sedimentation and Tectonics of Norway, in *Arctic Geology,* M. D. Pitcher, ed., Am. Assoc. Petroleum Geologists Mem. 19, pp. 471-481.

Nilsen, T. H., 1982, Alluvial Fan Deposits, in *Sandstone Depositional Environments,* P. A. Scholle and D. Spearing, eds., Am. Assoc. Petroleum Geologists Mem. 31, pp. 49-86.

Nilsen, T. H., and T. E. Moore, 1980, Selected List of References to Modern and Ancient Alluvial Fan Deposits, *U.S. Geol. Survey Open-File Report 80-658,* 58p.

Nilsen, T. H., and T. E. Moore, 1984, *Bibliography of Alluvial-Fan Deposits,* GeoBooks, Norwich, England, 96p.

Ori, G. G., 1982, Braided to Meandering Channel Patterns in Humid-Region Alluvial Fan Deposits, River Reno, Po Plain (Northern Italy), *Sed. Geology* **31**:231-248.

Pashley, E. F., Jr., 1966, Structure and Stratigraphy of the Central Northern, and Eastern parts of the Tucson Basin, Arizona, Ph.D. thesis, Arizona University, Tucson, 273p.

Pierson, T. C., 1980, Erosion and Deposition by Debris Flows at Mt. Thomas, North Canterbury, New Zealand, *Earth Surf. Process.* **5**:227-247.

Poland, J. F., 1959, Hydrology of the Long Beach-Santa Ana Area, California, *U.S. Geol. Survey Water-Supply Paper 1471,* 257p.

Poland, J. F., 1971, Land Subsidence in the Santa Clara Valley, Alameda, San Mateo, and Santa Clara Counties, California, *U.S. Geol. Survey Open-File Report,* one sheet.

Poland, J. F., A. M. Piper, and others, 1956, Ground-Water Geology of the Coastal Zone, Long Beach-Santa Ana area, California, *U.S. Geol. Survey Water-Supply Paper 1109,* 162p.

Pretorius, D. A., 1974a, The Nature of the Witwatersrand Gold-Uranium Deposits, *Witwatersrand Univ. Econ. Geology Research Unit Inf. Circ. 86,* 50p.

Pretorius, D. A., 1974b, Gold in the Proterozoic Sediments of South Africa—Systems, Paradigms, and Models, *Witwatersrand Univ. Econ. Geology Research Unit Inf. Circ. 87,* 22p.

Pretorius, D. A., 1975, The Depositional Environments of the Witwatersrand

Goldfields—a Chronological Review of Speculations and Observations, *Witwatersrand Univ. Econ. Geology Research Unit Inf. Circ. 95,* 47p.

Pretorius, D. A., 1976, The Stratigraphic, Geochronologic, Ore-Type, and Geologic-Environment Sources of Mineral Wealth in the Republic of South Africa, *Econ. Geology* **71**:5-15.

Price, W. E., Jr., 1974, Simulation of Alluvial Fan Deposition by a Random Walk Model, *Water Resources Research* **10**:263-274.

Prior, D. B., N. Stephens, and G. R. Douglas, 1970, Some Examples of Modern Debris Flows in Northeast Ireland, *Zeitschr. Geomorphologie* **14**:275-288.

Rachocki, A. H., 1981, *Alluvial Fans,* John Wiley and Sons, New York, 161p.

Rantz, S. E., 1972, A Summary View of Water Supply and Demand in the San Francisco Bay Region, California, *U.S. Geol. Survey Open-File Report,* 41p.

Rapp, A., and R. W. Fairbridge, 1968, Talus Fan or Cone; Scree and Cliff Debris, *Encyclopedia of Geomorphology,* R. W. Fairbridge, ed., Reinhold Book Corporation, New York, pp. 1106-1109.

Ray, L. L., 1964, The Charleston, Missouri Alluvial Fan, *U.S. Geol. Survey Prof. Paper 501-B,* pp. B130-B134.

Reeves, C. C., Jr., 1977, Intermontane Basins of the Arid Western United States, in *Geomorphology in Arid Regions,* D. O. Doehring, ed., Eighth Annual Geomorphology Symposium Proceedings, State University of New York at Binghamton, p. 7-25.

Reid, J. C., 1974, Hazel Formation, Culberson and Hudspeth Counties, Texas, M.S. thesis, Texas University, Austin, 88p.

Renwick, W. H., 1977, Erosion Caused by Intense Rainfall in a Small Catchment in New York State, *Geology* **5**:361-364.

Robertson, J. A., 1976, The Blind River Uranium Deposits—the Ores and Their Setting, *Ontario Dept. Mines Misc. Paper 65,* 45p.

Robertson, J. A., 1981, The Blind River Uranium Deposits: The Ores and Their Setting, *U.S. Geol. Survey Prof. Paper 1161-U,* pp. U1-U23.

Roscoe, S. M., 1981, Temporal and Other Factors Affecting Deposition of Uraniferous Conglomerates, *U.S. Geol. Survey Prof. Paper 1161-W,* pp. W-1-W17.

Ruhe, R. V., 1964, Landscape Morphology and Alluvial Deposits in Southern New Mexico, *Am. Assoc. Geographers Annals* **54**:147-159.

Rust, B. R., 1977, The Interpretation of Ancient Alluvial Successions in the Light of Modern Investigations, in *Research in Fluvial Geomorphology,* R. Davidson-Arnott, and W. Nickling, eds., Fifth Guelph Symposium on Geomorphology Proceedings, Geo Abstracts Ltd., Norwich, England, pp. 67-105.

Rust, B. R., 1978, Depositional Models for Braided Alluvium, in *Fluvial Sedimentology,* A. D. Miall, ed., Canadian Soc. Petroleum Geologists Mem. 5, pp. 605-625.

Rust, B. R., 1979, Facies models, Chapter 2—Coarse Alluvial Deposits, in *Facies Models,* R. G. Walker, ed., Geoscience Canada Reprint series 1, pp. 9-21.

Ryder, J. M., 1971a, The Stratigraphy and Morphology of Paraglacial Alluvial Fans in South-Central British Columbia, *Canadian Jour. Earth Sci.* **8**:279-298.

Ryder, J. M., 1971b, Some Aspects of the Morphometry of Paraglacial Alluvial Fans in South-Central British Columbia, *Canadian Jour. Earth Sci.* **8**:1252-1264.

Ryder, R. T., T. D. Fouch, and J. H. Elison, 1976, Early Tertiary Sedimentation in the Western Uinta Basin, Utah, *Geol. Soc. America Bull.* **87:**496–512.

Sanlaville, P., 1977, *Etude geomorphologique de la region littorale du Liban,* 2 vols., Publications de L'Universite Libanaise, Beyrouth, 859p.

Schick, A. P., 1974, Alluvial Fans and Desert Roads—a Problem in Applied Geomorphology, *Akad. Wiss. Gottingen Abh. Math. Phys. Kl. III F.,* **29:**418–425.

Schick, A. P., 1977, A Tentative Sediment Budget for an Extremely Arid Watershed in the Southern Negev in *Geomorphology in Arid Regions,* D. O. Doehring, ed., Eighth Annual Geomorphology Symposium Proceedings, State University of New York at Binghamton, pp. 139–163.

Schluger, P. R., 1973, Stratigraphy and Sedimentary Environments of the Devonian Perry Formation, New Brunswick, Canada, and Maine, U.S.A., *Geol. Soc. America Bull.* **84:**2533–2548.

Schumann, H., and J. F. Poland, 1969, Land Subsidence, Earth Fissures, and Ground-Water Withdrawal in South-Central Arizona, in *Land Subsidence,* L. J. Tison, ed., AIHS/UNESCO Pub. No. 88, pp. 295–302.

Schumm, S. A., 1977, *The Fluvial System,* John Wiley and Sons, New York, 338p.

Scott, K. M., 1971, Origin and Sedimentology of 1969 Debris Flows near Glendora, California, *U.S. Geol. Survey Prof. Paper 750-C,* pp. C242–C247.

Sharp, R. P., and L. H. Nobles, 1953, Mudflow of 1941 at Wrightwood, Southern California, *Geol. Soc. America Bull.* **64:**547–560.

Sheh, A., 1979, Late Pleistocene Fan-Deltas along the Dead Sea Rift, *Jour. Sed. Petrology* **49:**541–552.

Soister, P. E., 1968, Stratigraphy of the Wind River Formation in South-Central Wind River Basin, Wyoming, *U.S. Geol. Survey Prof. Paper 594-A,* pp. A1–A50.

Spearing, D. R., 1974, *Alluvial Fan Deposits: Summary Sheets of Sedimentary Deposits,* Sheet I, Geological Society of America, Boulder, Colorado.

Steel, R. J., 1974, New Red Sandstone Floodplain and Piedmont Sedimentation in the Hebridean Province, Scotland, *Jour. Sed. Petrology* **44:**336–357.

Steel, R. J., 1976, Devonian Basins of Western Norway—Sedimentary Response to Tectonism and to Varying Tectonic Context, *Tectonophysics* **36:**207–224.

Steel, R. J., and S. M. Aasheim, 1978, Alluvial Sand Deposition in a Rapidly Subsiding Basin (Devonian, Norway), in *Fluvial Sedimentology,* A. D. Miall, ed., Canadian Soc. Petroleum Geologists Mem. 5, pp. 385–412.

Steel, R. J., and T. G. Gloppen, 1980, Late Caledonian (Devonian) Basin Formation, Western Norway: Signs of Strike-Slip Tectonics during Infilling, in *Sedimentation in Oblique-Slip Mobile Zones,* P. F. Ballance and H. G. Reading, eds., Internat. Assoc. Sedimentologists Spec. Publ. 4, pp. 79–103.

Steel, R. J., S. Maehle, H. Nilsen, S. L. Roe, and A. Spinnangr, 1979, Reply to discussion by H. F. Garner, *Geol. Soc. America Bull.* part 1, **90:**122–124.

Steel, R. J., and A. C. Wilson, 1975, Sedimentation and Tectonism (Permo-Triassic) on the Margin of the North Minch Basin, Lewis, *Geol. Soc. London Jour.* **131:**183–202.

Steidtmann, J. R., 1971, Origin of the Pass Peak Formation and Equivalent Early Eocene Strata, Central Western Wyoming, *Geol. Soc. America Bull.* **82:**156–176.

Stephenson, C., 1972, Middle Old Red Sandstone Alluvial Fan and Talus Deposits at Foyers, Omvernessshire, *Scottish Jour. Geology* **8,** pt. 2, pp. 121-127.

Tator, B. A., 1952, Pediment Characteristics and Terminology, *Assoc. Am. Geographers Annals* **42:**295-317.

Tator, B. A., 1953, Pediment Characteristics and Terminology, *Assoc. Am. Geographers Annals* **43:**37-53.

Teisseyre, A. K., 1973, Carboniferous Fans and Fanglomerates in the Central Sudetes 1: marginal faults, downfaulting and sedimentation, *Acad. Polonaise Sci. Bull.* Serie des Sci. de Terre, **21:**147-155.

Tolman, C. F., 1909, Erosion and Deposition in Southern Arizona Bolson Region, *Jour. Geology* **17:**136-163.

Tricart, J., 1963, *Geomorphologie des regions froides,* Orbis Series, Presses Universitaires de France, Paris, 289p.

Tricart, J., and A. Cailleux, 1965, *Introduction a la Geomorphologie Climatique,* SEDES, Paris, 306p.

Twidale, C. R., 1968, Pediments, in *Encyclopedia of Geomorphology,* R. W. Fairbridge, ed., Reinhold Publishing Corporation, New York, pp. 817-818.

Van de Kamp, P. C., 1973, Holocene Continental Sedimentation in the Salton Basin, California, a Reconnaissance: *Geol. Soc. America Bull.* **84:** 827-848.

Wasson, R. J., 1977, Late-Glacial Alluvial Fan Sedimentation in the Lower Derwent Valley, Tasmania, *Sedimentology* **24:**781-799.

Wasson, R. J., 1979, Sedimentation History of the Mundi Mundi Alluvial Fans, Western New South Wales, *Sed. Geology* **22:**21-51.

Wells, S. G., 1977, Geomorphic Controls of Alluvial Fan Deposition in the Sonoran Desert, Southwestern Arizona, in *Geomorphology in arid regions,* D. O. Doehring, ed., Eighth Annual Geomorphology Symposium Proceedings, State University of New York at Binghamton, pp. 27-50.

Wescott, W. A., and F. G. Ethridge, 1980, Fan-Delta Sedimentology and Tectonic Setting—Yallahs Fan Delta, Southeast Jamaica, *Am. Assoc. Petroleum Geologists Bull.* **64:**374-399.

Wessel, J. M., 1969, Sedimentary History of Upper Triassic Alluvial Fan Complexes in North-Central Massachusetts, *Massachusetts Univ. Dept. Geology Contributions,* No. 2, 157p.

Williams, G. E., 1969, Characteristics and Origin of a Precambrian Pediment, *Jour. Geology* **77:**183-207.

Wilson, M. D., 1970, Upper Cretaceous-Paleocene Synorogenic Conglomerates of Southwestern Montana, *Am. Assoc. Petroleum Geologists Bull.* **54:** 1843-1867.

Yazawa, D., H. Toya, and S. Kaizuka, 1971, *Alluvial Fans,* Kokon Shoin, Toyko, 318p. (In Japanese).

Part I

EARLY STUDIES OF ALLUVIAL FAN DEPOSITS

Editor's Comments
on Papers 1 Through 5

1   **SHARP**
    Early Tertiary Fanglomerate, Big Horn Mountains, Wyoming

2   **BLISSENBACH**
    Geology of Alluvial Fans in Semiarid Regions

3   **BEATY**
    Origin of Alluvial Fans, White Mountains, California and Nevada

4   **BULL**
    Alluvial-Fan Deposits in Western Fresno County, California

5   **BLUCK**
    Sedimentation of an Alluvial Fan in Southern Nevada

Alluvial fan deposits began to attract attention and to be described in some detail in the late 1800s. Six major published studies from that time stand out: (1) descriptions by Drew (1873) of alluvial fans of the upper Indus Basin; (2) Dutton's (1880) discussion of fans in the Colorado Plateau; (3) a general investigation of the manner in which alluvial fans are constructed in the western United States by Gilbert (1882); (4) descriptions of modern alluvial fans and flooding in the Mustagh Mountains of Kashmir by Conway (1893); (5) descriptions by McGee (1897) of sheetflood erosion associated with major flooding events on alluvial fans in western Sonora and southwestern Arizona; and (6) a general summary of the physical geography of alluvial fans, based on much study in arid regions, by Davis (1898). These papers drew attention to the morphology and shape of fan deposits, to their presence at the base of many mountain ranges, and to the coarse-grained nature of these deposits. In addition, the nature of alternating cycles of deposition and erosion on alluvial fans attracted much attention.

In the early 1900s, North American geomorphologists began to study alluvial fan deposits of the arid and semiarid western United States in more detail, adding progressively to our knowledge of both the geomorphologic and sedimentologic aspects of alluvial fans. Lawson (1906) studied alluvial fans on the flanks of the Tehachapi Mountains in southern California; Tolman (1909), numerous alluvial fans in southern Arizona; Trowbridge (1911), large alluvial fans of the Owens Valley in California; Paige (1912), fans and pediments in New Mexico; Vaughan (1922), alluvial fans in the San Bernardino Mountains of southern California; Bryan (1922, 1925), alluvial fans adjacent to Cucamonga in southern California; Longwell (1930), alluvial fans of the Sheep Range in southern Nevada; Blackwelder (1931), alluvial fans and pediments of the Great Basin; Rich (1935), alluvial fans of the arid southwestern United States; Eardley (1938), alluvial fans surrounding the Great Salt Lake; and Denny (1941), alluvial fans in New Mexico.

Lawson (1913) provided a lithologic terminology for the deposits of alluvial fans, suggesting the name "fanglomerate" for consolidated alluvial fan gravels. Much study of the types of deposition that occurred on alluvial fans took place concurrently with study of particular groups of fans. The deposits of huge volumes of water that spread out rapidly over large parts of fans or entire fans were defined as sheetflood deposits (McGee, 1897), whereas flood deposits confined to channels were termed streamflood deposits (Davis, 1938). Flooding on modern alluvial fans was studied by Pack (1923) along the western flank of the Wasatch Mountains in northern Utah, where blocks as large as 95 tons were transported by debris flows; Chawner (1935) in southern California; Ives (1936) in the Sonoyta Valley, Sonora, Mexico; Troxell and Peterson (1937) and Troxell et al. (1942) in southern California; Krumbein (1942) in Arroyo Seco, southern California; and Woolley (1946) in Utah.

Mudflow and debris-flow deposition was observed and studied by Rickmers (1913) in Turkestan; Blackwelder (1928) in the Western United States; Taylor (1934) in Los Angeles County, southern California; Fryxell and Horberg (1941) in Grand Teton National Park; Gleason and Amidon (1941) in Wrightwood, southern California; Sharp (1942) in the Western United States; and Sharp and Nobles (1953) in Wrightwood, southern California. General studies of alluvial fans, piedmont gravels, and pediments of arid and semiarid regions were made by Lawson (1915), Gilbert (1928), Johnson (1932), Teillard de Chardin (1933), Davis (1938), and Tator (1952, 1953). Important early studies of alluvial fans in other parts of the world were made by Cotton (1922) in New Zealand, Anderson (1936) in Algeria, Bohlin (1940) in western China, and Bond (1953) in northern Nigeria.

The study of ancient deposits of clastic sedimentary rocks that were interpreted to be of alluvial fan origin developed more slowly.

Editor's Comments on Papers 1 Through 5

Rich (1910) studied the Bishop Conglomerate of southwestern Wyoming and interpreted it to be an ancient alluvial fan deposit. Tieje (1923) interpreted some of the red beds of the Front Range of the Rocky Mountains in Colorado to be alluvial fan deposits; subsequent study of sediment dispersal patterns in Pennsylvanian strata by Howard (1966) confirmed some of Tieje's conclusions.

The first study of ancient alluvial fan deposits that deals directly with the interpretation of stratigraphic, sedimentologic, and petrographic aspects of a suite of rocks thought to be of alluvial-fan origin is Paper 1 by Sharp, a study of a sequence of conglomerates on the east flank of the Big Horn Mountains in Wyoming. Largely on the basis of his ongoing studies of depositional processes on modern alluvial fans, particularly of mudflows and debris flows (Sharp, 1942; Sharp and Nobles, 1953), Sharp is able to show that these ancient strata, named by him the Moncrief gravel, had been deposited in a similar manner. Sharp's paper is a masterpiece of careful description and logical, step-by-step deductive reasoning. It demonstrates beautifully that "the present is also the key to the past," and that stratigraphers and sedimentologists working in the ancient record need to examine modern depositional processes and environments to be able to interpret successfully the past history of sedimentary deposits. Previous workers had interpreted the Moncrief gravel to have been deposited as glacial drift, as terminal moraines, as glacial outwash, or as alluvial bench gravels. Because, like most alluvial fan deposits, it lacked fossils and its age was difficult to determine, most previous workers suggested a late Tertiary to Holocene age. Sharp, as a result of careful mapping and description of the deposits, is able to demonstrate an early Tertiary age and deposition as alluvial fans. The fans were deposited on the eastern flank of the mountain range, marginal to lower Tertiary coalfields located further east. Sharp was Professor of Geology at the California Institute of Technology in Pasadena; he retired a few years ago, but remains very active in a number of geomorphology programs.

Paper 2 is the first systematic attempt to define, describe, and characterize alluvial fan deposits of modern semiarid regions. Blissenbach briefly examines examples of ancient alluvial fan deposits and provides criteria for their recognition. He incorporates the results of almost all previous studies and discusses in review fashion most aspects of alluvial fan deposits, including composition, sorting, roundness and sphericity of clasts, sedimentary structures, cementation, color, porosity, and permeability. Blissenbach provides a splendid summary of the chief characteristics of semiarid-region fans; he concludes that mudflow deposits are more common in arid regions than humid regions. Blissenbach has resided in Germany for many years, working for Preussag AG in Hannover.

One of the first wholly integrated studies of modern alluvial fans is Paper 3, a classic study of alluvial fans of the White Mountains of California and Nevada. Beaty discusses in detail the history of growth, geometry, and geomorphology of alluvial fans deposited along the north and west margins of the White Mountains, along the east side of the upper Owens Valley, north of the area previously considered by Trowbridge (1911). The White Mountains form an impressive range at the west edge of the Basin and Range, reaching an altitude of 4,343 m; the upper Owens Valley ranges in elevation from about 1,220 m to 1,980 m. Beaty's detailed maps and observations of depositional processes reveal much about how debris flows, sheetfloods, and perennial streams transport and deposit material on different parts of fans, particularly between channelized and nonchannelized parts of the same fan. He examines downfan and across-fan variations in bed thickness and grain size, the sorting of deposits, the manner in which distributary channels shift and become entrenched on fans, and the processes responsible for transport and deposition of very large boulders. Beaty (1961a, 1961b, 1968, 1970, 1974) continued his work on alluvial fans in the White Mountains and several other areas in western North America; at the present time, he is at the Department of Geography, University of Lethbridge in Alberta.

Paper 4 describes fine-grained alluvial fan deposits along the western margin of the San Joaquin Valley. This paper is one of a long series of papers on various aspects of these fans (Bull, 1961a, 1961b, 1962a, 1962b, 1963, 1964a, 1964b, 1964c, 1965a, 1965b, 1972, 1975). I have selected the present paper for this volume because it is short and contains an excellent description of a coalesced group of fans deposited by mudflows, streamflows (waterlaid deposits), and intermediate types of flows. Because Bull worked chiefly with subsurface samples, he did not use the generally applied sheetflood-streamflood-stream classification of fan deposits. Bull provides an excellent description of the differences between mudflow and streamflow deposits in terms of their sorting, thickness, imbrication, fabric, sedimentary structures, grading, void space, mudcracks, bubble cavities, and porosity and permeability. The results of this study were incorporated into Bull's comparison of modern and ancient alluvial fan deposits, reproduced herein as Paper 17. Bull moved from the U.S. Geological Survey to the Department of Geosciences at the University of Arizona, where he currently teaches.

Bluck records the depositional characteristics of a single arid-region alluvial fan in the Arrow Canyon Range of southern Nevada in Paper 5. Bluck's study emphasizes downfan changes in maximum clast size in streamflow and mudflow deposits; he found that the clast size in mudflow deposits decreased more rapidly downfan than in streamflow deposits. He found no systematic downfan changes in sphericity of clasts, but some systematic variations in the distribution of differently shaped clasts. He concluded that the change in depositional regime on the fan from mudflow to streamflow may have been responsible for fanhead entrenchment. Bluck continued his work for many years on modern and ancient coarse-grained fluvial deposits, studying the sandur plains of southern Iceland and the Old Red Sandstone of Great Britain in great detail (Bluck, 1965, 1967, 1969, 1974, 1978, 1980). He has been at Glasgow University for about 20 years.

REFERENCES

Anderson, R. V., 1936, Geology in the Coastal Atlas of Western Algeria, *Geological Society of America Memoir 4*, 450p.

Beaty, C. B., 1961a, Topographic Effects of Faulting—Death Valley, California, *Assoc. Am. Geographers Annals* **51**(2):234-240.

Beaty, C. B., 1961b, Boulder Deposit in the Flint Creek Valley, Western Montana, *Geol. Soc. America Bull.* **72**:1015.

Beaty, C. B., 1968, *Sequential Study of Desert Flooding in the White Mountains of California and Nevada*, U.S. Army Natick Laboratories, Earth Science Laboratory, Technical Report 68-31-ES, 96p.

Beaty, C. B., 1970, Age and Estimated Rate of Accumulation of an Alluvial Fan, White Mountains, California, *Am. Jour. Sci.* **268**:50-77.

Beaty, C. B., 1974, Debris Flows, Alluvial Fans and a Revitalized Catastrophism, *Zeitschr. Geomorphologie,* Supplementband, **21**:39-51.

Blackwelder, E., 1928, Mudflow as a Geological Agent in Semi-Arid Mountains, *Geol. Soc. America Bull.* **39**:465-484.

Blackwelder, E., 1931, Desert Plains, *Jour. Geology* **39**:133-140.

Bluck, B. J., 1965, The Sedimentary History of Some Triassic Conglomerates in the Vale of Glasmorgan, South Wales, *Sedimentology* **4**:225-245.

Bluck, B. J., 1967, Deposition of Some Upper Old Red Sandstone Conglomerates in the Clyde Area: A Study of the Significance of Bedding, *Scottish Jour. Geology* **3**:139-167.

Bluck, B. J., 1969, Old Red Sandstone and Other Paleozoic Conglomerates of Scotland, in *North Atlantic—Geology and Continental Drift*, M. Kay, ed., American Association of Petroleum Geologists Memoir 12, pp. 711-723.

Bluck, B. J., 1974, Structure and Directional Properties of Some Valley Sandur Deposits in Southern Iceland, *Sedimentology* **21**:533-554.

Bluck, B. J., 1978, Sedimentation in a Late Orogenic Basin: The Old Red Sandstone of the Midland Valley of Scotland, in *Crustal Evolution in Northwest Britain and Adjacent Regions,* D. R. Bowes, B. E. Leake, Geological Journal Special Issue No. 10, pp. 249-278.

Bluck, B. J., 1980, Evolution of a Strike-Slip Fault-Controlled Basin, Upper Old Red Sandstone, Scotland, in *Sedimentation in Oblique-Slip Mobile Zones,* P. F. Ballance and H. G. Reading, eds., International Association of Sedimentologists Special Publication No. 4, p. 63-78.

Bohlin, B., 1940, *Notes on the Hydrography of Western Kansu: Reports from the Scientific Expedition to the North-western Provinces of China Under the Leadership of Dr. Sven Hedin,* Publication No. 10, III, Geology, 3, 54p.

Bond, G., 1953, *A Preliminary Account of Pleistocene Geology of the Plateau Tin Fields Region of Northern Nigeria,* Proceedings of the Third West African Conference, Abidjan, v. 3, p. 187-201.

Bryan, K., 1922, Erosion and Sedimentation in the Papago Country, *U.S. Geol. Survey Bull.* **730-B:**19-90.

Bryan, K., 1925, The Papago Country, Arizona; a Geographic, Geologic, and Hydrologic Reconnaissance with a Guide to Desert Watering Places, *U.S. Geol. Survey Water-Supply Paper 499,* 436p.

Bull, W. B., 1961a, Tectonic Significance of Radial Profiles of Alluvial Fans in Western Fresno County, California, *U.S. Geol. Survey Prof. Paper 424-B,* pp. 182-184.

Bull, W. B., 1961b, Causes and Mechanics of Near-Surface Subsidence in Western Fresno County (abs.), *Geol. Soc. America Spec. Paper 68,* p. 11-12.

Bull, W. B., 1962a, Relation of Textural (CM) Patterns of Depositional Environment of Alluvial-Fan Deposits, *Jour. Sed. Petrology* **32:**211-216.

Bull, W. B., 1962b, Alluvial Fans and Near-Surface Subsidence, Western Fresno County, California (abs.), *Dissert. Abs.* **22**(10):3602-3603.

Bull, W. B., 1963, Tectonic History as Related to Terraces and Alluvial-Fan Segments in Western Fresno County, California (abs.), *Geol. Soc. America Spec. Paper 73,* p. 29.

Bull, W. B., 1964a, Alluvial Fans and Near-Surface Subsidence in Western Fresno County, California, *U.S. Geol. Survey Prof. Paper 437-A,* 70p.

Bull, W. B., 1964b, Geomorphology of Segmented Alluvial Fans in Western Fresno County, California, *U.S. Geol. Survey Prof. Paper 352-E,* pp. 89-129.

Bull, W. B., 1964c, History and Causes of Channel Trenching in Western Fresno County, California, *Am. Jour. Sci.* **262:**249-258.

Bull, W. B., 1965a, *The Alluvial Fans of Western Fresno County, California,* Geol. Soc. America Cordilleran Section, Guidebook Field Trip No. 2, 21p.

Bull, W. B., 1965b, The Tumey Gulch Fanhead Trench, Western Fresno County, California, *Northern Great Basin and California,* 7th Congress of International Association for Quaternary Research, Guidebook for field conference I, pp. 137-139.

Bull, W. B., 1972, Prehistoric Near-Surface Subsidence Cracks in Western Fresno County, California, *U.S. Geol. Survey Prof. Paper 437-C,* 86p.

Bull, W. B., 1975, Land Subsidence in the Los Banos-Kettleman City Area, California, Part 2, Subsidence and Compaction of Deposits, *U.S. Geol. Survey Prof. Paper 437-F,* 90p.

Chawner, W. D., 1935, Alluvial Fan Flooding, the Monrose, California Flood of 1934, *Geog. Rev.* **25:**77-88.

Conway, W. M., 1893, Exploration in the Mustagh Mountains, *Geog. Jour.* **2**(4):289-303.

Cotton, C. A., 1922, *Geomorphology of New Zealand, Part 1,* Systematic, Wellington, New Zealand.

Davis, W. M., 1898, *Physical Geography* Ginn & Co., Boston and London, 428p.

Davis, W. M., 1938, Sheet Floods and Streamfloods, *Geol. Soc. America Bull.* **49:**1337-1416.

Denny, C. S., 1941, Quaternary geology of the San Acacia area, New Mexico, *Jour. Geology* **49:**225-260.

Drew, F., 1873, Alluvial and Lacustrine Deposits and Glacial Records of the Upper Indus Basin, *Geol. Soc. London Quart. Jour.* **29:**441-471.

Dutton, C. E., 1880, *Report on the Geology of the High Plateaus of Utah,* U.S. Government Printing Office, Washington, D. C.

Eardley, A. J., 1938, Sediments of Great Salt Lake, Utah, *Am. Assoc. Petroleum Geologists Bull.* **22:**1305-1411.

Fryxell, F. M., and Horberg, L., 1941, Alpine Mudflows in Grand Teton National Park, Wyoming, *Geol. Soc. America Bull.* **54:**457.

Gilbert, G. K., 1882, Contributions to the History of Lake Bonneville, *U.S. Geol. Survey Second Annual Report,* pp. 167-200.

Gilbert, G. K., 1928, Studies of the Basin-Range Structure, *U.S. Geol. Survey Prof. Paper 153,* 92p.

Gleason, C. H., and Amidon, R. E., 1941, *Landslide and mudflow, Wrightwood, California,* California Forest and Range Experiment Station, unpublished report, pp. 1-7.

Howard, J. D., 1966, Patterns of Sediment Dispersal in the Fountain Formation of Colorado, *Mountain Geologist* **3:**147-169.

Ives, R. L., 1936, Desert Floods in the Sonoyta Valley, *Am. Jour. Sci.* **32:** 349-360.

Johnson, D. W., 1932, Rock Fans of Arid Regions, *Am. Jour. Sci.* **23:**389-416.

Krumbein, W. C., 1942, Flood Deposits of Arroyo Seco, Los Angeles County, California, *Geol. Soc. America Bull.* **53:**1355-1402.

Lawson, A. C., 1906, The Geomorphogeny of the Tehachapi Valley system, *California Univ. Pubs. Geol. Sci.* **4:**431-462.

Lawson, A. C., 1913, The Petrographic Designation of Alluvial Fan Formations, *California Univ. Pubs. Geol. Sci.* **7:**325-334.

Lawson, A. C., 1915, The Epigene Profiles of the Desert, *California Univ. Pubs. Geol. Sci.* **9:**23-48.

Longwell, C. R., 1930, Faulted Fans West of the Sheep Range, Southern Nevada, *Am. Jour. Sci.* **220:**1-13.

McGee, W. J., 1897, Sheetflood Erosion, *Geol. Soc. America Bull.* **8:**87-112.

Pack, F. J., 1923, Torrential Potential of Desert Waters, *Pan-American Geologist* **40:**349-356.

Paige, S., 1912, Rock-Cut Surfaces in the Desert Ranges, *Jour. Geol.* **20:**442-450.

Rich, J. L., 1910, The Physiography of the Bishop Conglomerate, Southwestern Wyoming, *Jour. Geol.* **18:**601-632.

Rich, J. L., 1935, Origin and Evolution of Rock Fans and Pediments, *Geol. Soc. America Bull.* **46:**999-1024.

Rickmers, W. R., 1913, *The Duab of Turkestan,* Cambridge University Press, New York, 563p.

Sharp, R. P., 1942, Mudflow Levees, *Jour. Geomorphology* **5:**222-227.

Sharp, R. P., and Nobles, L. H., 1953, Mudflow of 1941 at Wrightwood, Southern California, *Geol. Soc. America Bull.* **64:**547-560.

Tator, B. A., 1952, Pediment Characteristics and Terminology, *Assoc. Am. Geographers Annals* **42:**295-317.

Tator, B. A., 1953, Pediment Characteristics and Terminology, *Assoc. Am. Geographers Annals* **43:**37-53.

Taylor, C. A., 1934, Debris Flow From Canyons in Los Angeles County Flood, *Eng. News-Record* **112:**439-440.

Teillard de Chardin, P., 1933, *The Significance of Piedmont Gravels in Continental Geology,* International Geological Congress, Section XVI, U.S.A., v. 2, Washington, D.C., pp. 1031-1039.

Tieje, A. J., 1923, The Red Beds of the Front Range of Colorado: A Study in Sedimentation, *Jour. Geology* **31:**192-207.

Tolman, C. F., 1909, Erosion and Deposition in Southern Arizona Bolson Region, *Jour. Geology* **17:**136-163.

Trowbridge, A. C., 1911, The Terrestrial Deposits of Owens Valley, California, *Jour. Geology* **19:**736-740.

Troxell, H. C., et al., 1942, Floods of March 1938 in Southern California, *U.S. Geol. Survey Water-Supply Paper 844,* 399p.

Troxell, H. C., and Peterson, J. Q., 1937, Flood in La Canada Valley, California, *U.S. Geol. Survey Water-Supply Paper 796-C,* pp. 53-98.

Vaughan, F. E., 1922, Geology of the San Bernardino Mountains North of the San Gorgonio Pass, California, *California Univ. Pubs. Geol. Sci.* **13**(9):319-411.

Woolley, R. R., 1946, Cloudburst Floods in Utah, 1850-1938, *U.S. Geol. Survey Water-Supply Paper 994,* 128p.

Copyright © 1948 by The University of Chicago
Reprinted from *Jour. Geology* **56**:1-15 (1948), by permission of The University of Chicago Press

EARLY TERTIARY FANGLOMERATE, BIG HORN MOUNTAINS, WYOMING

ROBERT P. SHARP

California Institute of Technology, Pasadena

ABSTRACT

Great accumulations of coarse bouldery gravel along the east base of the central Big Horn Mountains are composed almost wholly of debris derived from the pre-Cambrian core of the range. The name "Moncrief gravel" is proposed for this deposit. It has previously been described as Pleistocene glacial material, late Tertiary to earliest Quaternary bench gravels, or the coarse phase of an early Tertiary basin fill. The gravel was found to be gradational into fine-grained early Tertiary beds and to be separated from the pre-Tertiary rocks of the mountains by thrust faults. For these reasons the Moncrief gravel is identified as an early Tertiary, probably Eocene, fan deposit, formed as the Big Horn Mountains were progressively uplifted and thrust eastward during the Laramide Revolution. Early Tertiary glaciation may have played a part, but this is largely speculation.

INTRODUCTION

GENERAL STATEMENT

Thick deposits of coarse bouldery gravel with interbedded silt, sand, and arkose layers compose large piedmont ridges and spurs along the east front of the Big Horn Mountains. This gravel consists predominantly of granitic rock fragments derived from the pre-Cambrian core of the range and has its best development in Bald, North, and Moncrief ridges (fig. 3), which extend 2-3 miles east from the mountains and rise 1,500 feet above the general piedmont level.

Max Demorest (1938, p. 18) referred to these deposits as the "Clear Creek gravels," but Clear Creek is preoccupied several times over (Wilmarth, 1938, pp. 457-459), so the name "Moncrief gravel," from excellent exposures on Moncrief Ridge (fig. 3) in Sheridan County, is substituted. The best exposures are in high north-facing cliffs at the top of the ridge and on the slopes below; hence the north face of Moncrief Ridge (secs. 34 and 35, T. 54 N., R. 84 W., Sheridan quadrangle) is designated as the type locality. The present topographic surface determines the top of the unit, and its basal and lateral margins are gradational into finer-grained "Wasatch" beds except where the gravel lies with angular unconformity on the Eocene Kingsbury conglomerate and older formations or is in fault contact with pre-Tertiary rocks of the mountains. More detailed work may subsequently provide a better basis of delineation, but for the present the boundaries of the Moncrief may be determined by the lowermost and outermost gravel bed, consisting primarily of pre-Cambrian rock debris. The age is early Tertiary, with Eocene a strong probability; and the maximum thickness observed on Moncrief Ridge is 1,400 feet.

Brief investigation of these deposits was made in 1940 and 1946, incidental to study of erosion surfaces along the east flank of the Big Horn Mountains. Field time totals 7 weeks.

of a major anticlinal uplift trending northwest, with an exposed pre-Cambrian core flanked by upturned Paleozoic and Mesozoic strata. Laramide thrusting is directed eastward in the central section of the range, westward in the north and south sections (Demorest, 1941, pp. 165–166; Chamberlin, 1940, p. 680), and southward at the southern end (Love, 1940, p. 1934; Tourtelot, 1946). In the adjoining basins Mesozoic and Paleozoic rocks are largely buried beneath early Tertiary continental deposits. Topo-

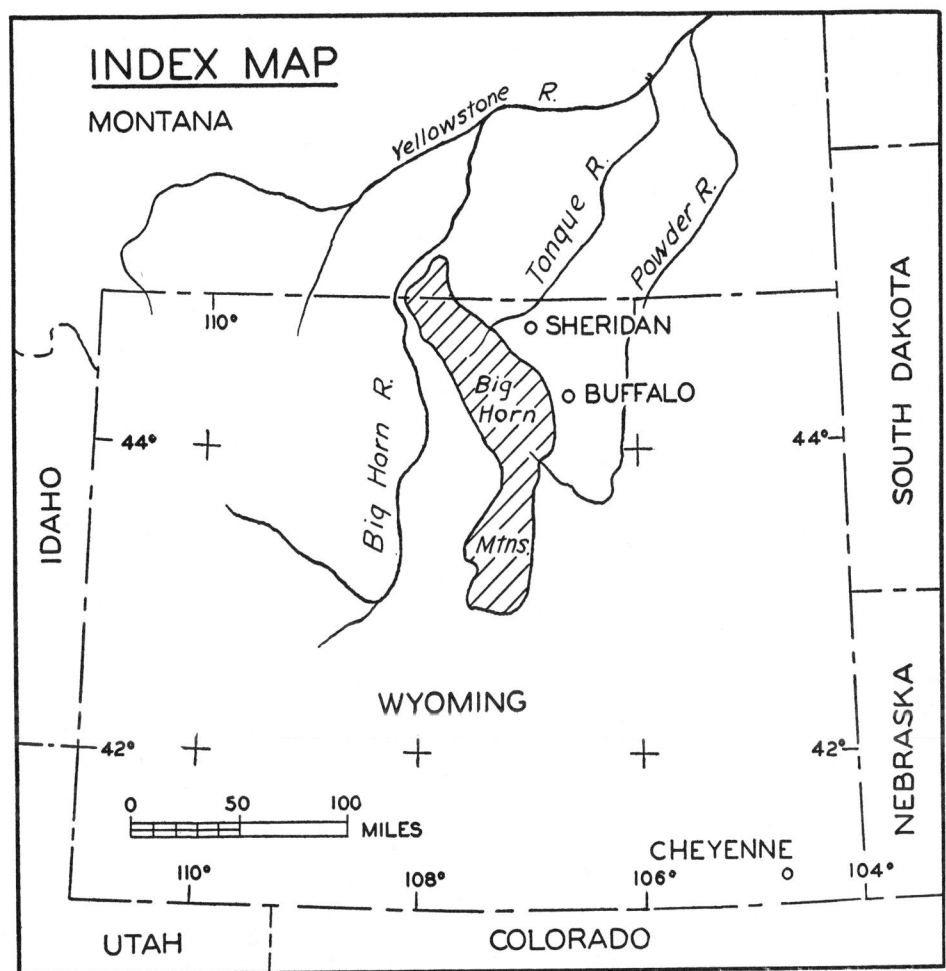

FIG. 1.—Area described lies at east base of Big Horn Mountains between Sheridan and Buffalo

GEOLOGICAL SETTING

The Big Horn Mountains are the front range of the Middle Rockies in north-central Wyoming (fig. 1). They consist

graphically the central part of the range consists of a backbone of high peaks (Cloud Peak, 13,165 feet) flanked by a broad rolling subsummit upland at 7,500–9,000 feet (fig. 2). Imposing parapets developed by erosion of sharply upturned Paleozoic beds form the east front of the range and separate the subsummit upland from the piedmont slope, which lies at 4,500–6,000 feet and truncates tilted Mesozoic and more nearly horizontal Tertiary strata. The backbone of the range has been heavily scoured by late Pleistocene glaciers, which advanced onto the subsummit upland but failed to reach the east base of the range (Darton, 1906a, pl. 26).

REVIEW OF PREVIOUS WORK

R. D. Salisbury and Eliot Blackwelder (1903, pp. 221–223) first described the bouldery accumulations at Bald and North ridges and tentatively suggested that they might be glacial deposits of earlier date than the late Pleistocene glaciation recognized higher on the mountains. Blackwelder (1915, pp. 313–315) later compared the deposits of Bald Ridge with coarse fan or outwash gravel on the Black Rock erosion surface (early Pleistocene) in western Wyoming. N. H. Darton (1906a, pp. 68–70, pl. 47; 1906b, pp. 8–9) mapped all occurrences of the Moncrief gravel and concluded that they were remnants of a late Tertiary or earliest Quaternary bench gravel which at one time extended all along the east-central front of the range. W. C. Alden (1932, pp. 41–44) gives the most thorough treatment of the deposits composing Bald, North, and Moncrief ridges and suggests that they are remnants of great lobate terminal moraines built by early Pleistocene glaciers which flowed from mountains onto the surface of the Flaxville Plain (No. 1 bench). The late Max Demorest (1938, pp. 22–23) suggested two possible interpretations for the Moncrief gravel: (1) It may be a coarse, mountain facies of the widespread Tertiary deposits which filled the intermontane basins and built up the Great Plains, or (2) it is the remnant of local fans deposited in response to local and temporary conditions after the plains surface was largely developed, perhaps in the early Pleistocene. As a result of later field work, he[1] dismissed the second possibility and adopted the view that the Moncrief gravel was gradational into the underlying Eocene Kingsbury conglomerate and also continuous in its uppermost part with Tertiary(?) gravel on the subsummit upland (fig. 2) of the Big Horn Mountains. His published report

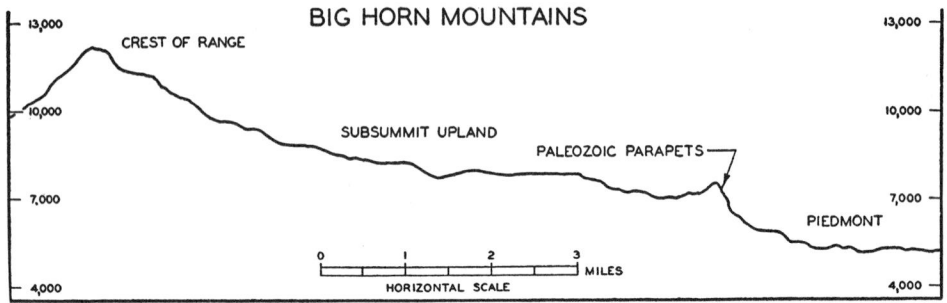

FIG. 2.—Topographic profile, east side of Big Horn Mountains

---

[1] Personal letters (1940).

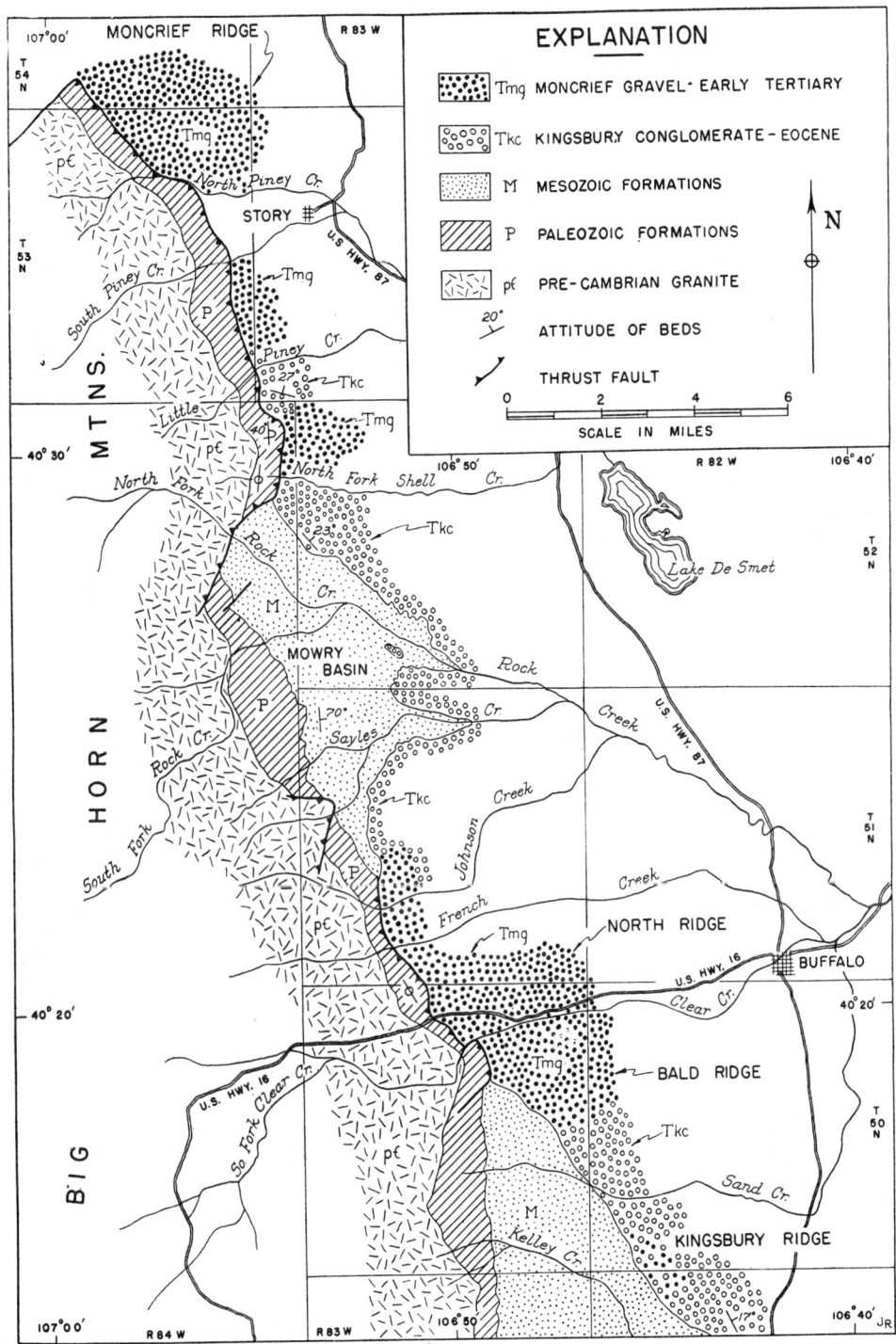

FIG. 3.—Geological map along east base of central Big Horn Mountains, modified from Darton and Demorest.

PLATE I

A

B

A, Boulder beds at top of Moncrief gravel in type section on north side of Moncrief Ridge. Arkose, sandstone, and siltstone in middle of exposure. About 200 feet of beds shown.

B, Large boulders of pre-Cambrian granite and gneiss in Moncrief gravel north of Little Piney Creek. View northeastward.

PLATE II

A

B

A, Large boulders of pre-Cambrian gneissic granite from Moncrief gravel on top of Bald Ridge.

B, Looking southeast near mouth of Clear Creek with Bald Ridge on left skyline. Moncrief gravel in left half, separated from Paleozoic beds of mountain front by a west-dipping thrust fault.

PLATE III

Perched boulder formed by weathering in fanglomeratic phase of Kingsbury conglomerate, north fork of Rock Creek.

PLATE IV

View northward at mouth of French Creek showing Moncrief gravel (*Tmg*) abutting against Paleozoic strata (*P*)

(1941, pp. 167–168), although not explicit, seems to represent the same view. In 1909 J. A. Taff (p. 131) described the gravel composing Moncrief Ridge as gradational into the fine-grained early Tertiary coal-bearing beds of the Sheridan coal field; but his contribution seems to have escaped the notice of subsequent workers. Taff's views and, with some modification, those of Demorest approach most closely the interpretation favored in this study.

composed of pre-Cambrian granite and gneissic granite. They lie as much as 1 mile east of the mountains and an additional mile from the nearest exposure of pre-Cambrian bedrock. Other lithologies represented, but in smaller amounts, are pre-Cambrian pegmatite, diabase, hornfels, gneiss, and schist, all of which are exposed in the core of the Big Horn Range. Stones derived from Paleozoic formations are extremely rare in the higher beds, but boulders of Paleozoic limestone compose up to 15 per cent of the lower strata, 900–1,400 feet below the top of Moncrief Ridge. The limestone boulders are subrounded to rounded and 6–18 inches in diameter at the most.

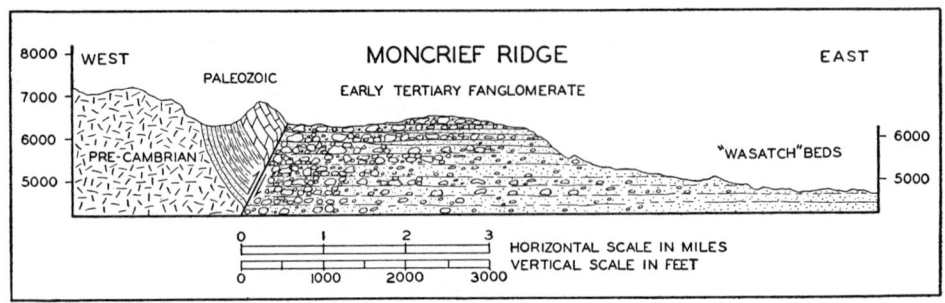

FIG. 4.—Sketch showing gradation of early Tertiary fanglomerate (Moncrief gravel) downward and eastward into "Wasatch" beds.

DESCRIPTION OF THE MONCRIEF GRAVEL

LITHOLOGY AND CONSTITUTION

This deposit features beds composed of subangular to rounded boulders of pre-Cambrian granitic rocks 1–5 feet in diameter; and larger stones are usually present. For example, on Moncrief Ridge are boulders up to 15 feet long; north of Little Piney Creek (pl. 1, B); and between Johnson and French creeks are boulders 25 feet long; south of Little Piney Creek one boulder was measured at 20 × 12 × 10 feet; and on the west end of Bald Ridge is one measuring 27 × 22 × 10 feet. Numerous boulders 10–15 feet in diameter are not exceptional at most exposures (pl. 2, A). These larger stones are subrounded, and most are

Matrix in the gravel is predominantly arkosic and usually sparse, particularly in the upper part where the boulder beds are coarsest, thickest, and most numerous. However, even here fines are not entirely lacking, for beds of arkose and fine micaceous siltstone outcrop in the midst of great boulder layers within 155 feet of the top of Moncrief Ridge (pl. 1, A). The deposits become finer downward and eastward, with an increase in number and thickness of arkose and micaceous arkosic sandstone layers. Boulder beds become fewer and thinner and their constituents smaller in the same directions (fig. 4); and near the base cobble and pebble beds are common.

The finer materials contain crudely spherical arkosic concretions, 1–2 feet in diameter, formed by local cementation. These are identical with concretions in the so-called "Wasatch" of this area. The micaceous sandstone contains smaller limonitic concretions. Cementation of the gravel is poor except locally, where some arkose layers are well cemented and weather out in large fragments or outcrop as ledges. The gravel is white to light-gray and tan. Beds containing stones rich in ferromagnesian minerals are brownish, and micaceous sandy layers are colored grayish-green by chloritized biotite. Bedding is nearly indistinguishable in the coarser phases of the deposit (pl. 1, A), and at best it is crude and irregular with many scour channels and some cross bedding in the fine layers. Bits of lignitized woody material are included in fine beds low in the section,[2] and one decomposed lignitized log $1\frac{1}{2}$ feet in diameter was found in the river bank behind the powerhouse on Clear Creek (NE. $\frac{1}{4}$ of SE. $\frac{1}{4}$ sec. 1, T. 50 N., R. 83 W.). Disintegration is so far advanced in many exposures that 50–90 per cent of the smaller stones can be chopped away with a geological pick, and some boulders up to 5 feet in diameter are friable to the core. Disintegration is most extensive in the upper 75 feet of the deposit.

Many features of the Moncrief gravel almost exactly duplicate S. H. Knight's (1937, p. 84), description of giant conglomerates at Green and Crook's mountains, Wyoming, and closely resemble conglomerates in the Wind River formation at the south end of the Big Horn Mountains (Tourtelot, 1946).

[2] Demorest (1938, p. 21) also reports coalified wood in these deposits.

THICKNESS

Thickness of the Moncrief gravel has been variously given as: probably exceeding 100 feet (Salisbury and Blackwelder, 1903, p. 22); at least 300 feet and possibly several hundred feet (Darton, 1906a, p. 69); 700–800 feet (Alden, 1932, p. 42); and more than 1,000 feet (Taff, 1909, p. 131). In this study, 485 feet of coarse bouldery gravel were measured by hand-leveling in continuous exposures on the north face of Moncrief Ridge (pl. 1, A), and similar boulder beds outcrop at least 900 feet below the top of the ridge on its north side. Along the south side of Moncrief Ridge, gravel beds with pre-Cambrian boulders $2\frac{1}{2}$ feet in diameter outcrop along the bank of North Piney Creek, 1,400 feet vertically below the highest exposure of similar material. New cuts on U.S. Highway 16 along Clear Creek west of Buffalo show that the coarse boulder beds composing North Ridge are at least 1,200 feet thick.[3] It is safe to say that, close to the mountains, gravel composed predominantly of pre-Cambrian boulders is at least 1,200 feet thick on Clear Creek and 1,400 feet thick at Moncrief Ridge.

DISTRIBUTION

The largest accumulations of Moncrief gravel are in Bald and North ridges, 5 miles west of Buffalo, and in Moncrief Ridge, $2\frac{1}{2}$ miles northwest of Story. However, as shown in figure 3, other sizable deposits of the gravel lie along the base of the range near French, Johnson, Shell, and the several branches of Piney creeks.

Boulders, up to 5 feet in diameter, of pre-Cambrian gneissic granite on Kingsbury Ridge, 4 miles south-southeast of Bald Ridge and 3 miles from the moun-

[3] Demorest (1938, p. 21) recognized that the gravel composing North Ridge was about this thick but gave no figures.

tains, have been mapped by Darton (1906a, pl. 47; 1906b, areal geology map) and Demorest (1938, fig. 5, p. 18a) as part of the Moncrief gravel and are so shown in figure 3. However, this occurrence is regarded with some skepticism because actual exposures of the gravel could not be found and large boulders of Bighorn dolomite (Ordovician) are also present. This material might be coarse debris washed out from Bald Ridge and the mountains during late Cenozoic degradation of the piedmont area.

In all other places Moncrief gravel lies directly against the upturned Paleozoic beds of the range front, which rise several hundred feet higher (pl. 2, B, and pl. 4). The gravel attains its greatest elevation, 6,900 feet, in Bald Ridge; and the lowest exposure is along North Piney Creek at 5,100 feet. It extends at least $2\frac{1}{2}$ miles eastward from the mountain front, discounting the questionable occurrence on Kingsbury Ridge.

The subsummit upland of the Big Horn Mountains (fig. 2) is mantled with "Tertiary" gravels, which Demorest (1938, p. 21; ftn. 1) thought were to be correlated with the uppermost beds of the Moncrief. It is not unlikely that the Moncrief gravel formerly overlapped the lower flanks of the mountains, but the gravels on the subsummit upland are thought to be much younger, on the basis of structural and age relations to be detailed shortly. No deposits of the Moncrief were found on the subsummit upland.

Especially worthy of note are the map relations (fig. 3), which show that the Moncrief gravel occurs only where early Tertiary formations lie close to the base of the range. Where such formations are lacking, there is no gravel, and this holds even locally, as in Mowry Basin (fig. 3).

STRATIGRAPHIC AND STRUCTURAL RELATIONS OF THE MONCRIEF GRAVEL

WITH THE "WASATCH"

Essentially horizontal beds of arkose and gray-green micaceous arkosic sandstone along the east front of the central Big Horn Mountains are shown on the latest maps (U.S. Geol. Survey, 1925, 1932) as "Wasatch,"[4] presumably on the basis of revisions (Taff, 1909, pp. 127–131; Wegemann, 1917, p. 60; Thom and Dobbin, 1924, pp. 494–498) of Darton's 1906a, p. 66; 1906b, p. 8) original treatment, in which they were dated as Tertiary or Cretaceous. Layers of identical material are interbedded with the Moncrief gravel even in its highest and coarsest part, and similar fine materials become more abundant downward and eastward in the section. This relation is particularly clear on Moncrief Ridge, where there can be little doubt that the gravel is but a coarse phase of the so-called "Wasatch"[5] (fig. 4). The boulder beds can be seen to lens out eastward along North Piney Creek, so that exposures of the "Wasatch" at the same level several miles east of the mountains consist principally of coarse arkose and micaceous arkosic sandstone. Fine beds in the Moncrief gravel contain coalified plant debris, a further similarity with the "Wasatch," which is coal-bearing in the Sheridan and Buffalo coal fields.

Previous workers, except Taff (1909, p. 131) and Demorest (1941, p. 168; ftn. 1),

[4] R. L. Nace (1936, p. 121) suggests that the name "Wasatch" be avoided as unnecessary and meaningless. Proper identification and naming of Paleocene and Eocene units east of the Big Horn Mountains remains for future detailed paleontological and stratigraphic studies, and the word "Wasatch" as used here is simply a quotation from earlier maps and publications.

[5] Taff (1909, p. 131) recognized and properly interpreted this relation in 1909.

have inferred or postulated an unconformity between the gravel and the "Wasatch," but extended search failed to reveal any discernible break between these units.

WITH THE KINGSBURY CONGLOMERATE

The Kingsbury is a limestone conglomerate, locally fanglomeratic, defined and named by Darton (1906a, p. 60) from Kingsbury Ridge (fig. 3). As he mapped the formation, it is a lens-shaped deposit, 2–5 miles wide, extending 40 miles along the east base of the central Big Horns. Wegemann (1917, p. 59) speaks of it as a "fan deposit," and Darton (1906b, p. 8) states that it was produced by uplift of the range. Demorest (1938, p. 20) also emphasizes that the Kingsbury is a fanglomerate, formed by uparching of the central segment of the Big Horn Mountains.

One of the best exposures of the Kingsbury is in prominent bluffs northeast of the North Fork of Rock Creek (sec. 19, T. 52 N., R. 83 W., Fort McKinney quadrangle), where it is chiefly a heterogeneous, poorly bedded, massive fanglomerate of subangular to subrounded fragments of Paleozoic limestone up to 5 feet in diameter (pl. 3). No boulders of pre-Cambrian origin were found here, although a few have been seen in other exposures. Cementation, at least locally, is good. Not less than 400 feet of beds are exposed north of Rock Creek; and Darton (1906b, p. 8) gives the maximum thickness of the formation as 2,500 feet (fig. 5).

In other localities farther from the mountains, the Kingsbury is predominantly a conglomerate containing well-rounded to subrounded pebbles and cobbles of Paleozoic limestones and occasional stones of pre-Cambrian origin. The Kingsbury is unconformable on tilted Mesozoic beds (Knowlton, 1909, p. 209; Gale and Wegemann, 1910, p. 144; Wegemann, 1917, p. 60; Demorest, 1938, p. 19), although Darton (1906b, p. 8) thought this contact was conformable and dated the formation as later Cretaceous. Wegemann (1917, p. 60) placed the Kingsbury in the "Wasatch," but Knowlton (1909, p. 210) later summarized fossil evidence indicating that it is probably Fort Union, and it is so shown on the latest maps (U.S. Geol. Survey, 1925, 1932). Nace (1936, p. 100) includes the Kingsbury in the "Wasatch Group" for purposes of discussion and summarizes the conflicting evidence as to its age, which he tentatively puts as Paleocene(?); but Jepsen (1940, p. 242) cites fossil evidence indicating that the Kingsbury is Early Eocene. Roland W. Brown, of the United States Geological Survey, has collected vertebrate remains from the Kingsbury conglomerate which definitely fix its age as Eocene.[6]

Bedding in the Moncrief gravel, where discernible, is essentially horizontal, with eastward dips not exceeding 2°–3°. All dips measured in the Kingsbury conglomerate are 17°–23° northeastward. The actual contact between the Moncrief and the Kingsbury has not been seen; but, because of this discordance in dips, it is thought to be an angular unconformity, and the marked difference in composition of the two formations also suggests a break between them (fig. 5). This was Darton's (1906b, p. 8) original interpretation, as well as Alden's (1932, p. 42); but Demorest (ftn. 1; 1941, pp. 167–168) described this contact on the north side of Moncrief Ridge as gradational, although in earlier work (1938, p. 21) he treated the Moncrief-Kingsbury contact in the area of Clear Creek

[6] Personal letter (1947).

as an angular unconformity. It now appears that the gradation described by Demorest is the Moncrief-"Wasatch" relation treated in the preceding pages.

In places, outcrops of the Kingsbury conglomerate have greater elevations than the lowest near-by exposures of Moncrief gravel. This relation is well shown on Little Piney Creek and is thought to indicate that the Moncrief was deposited on a land surface of considerable local relief (fig. 5). Ridges and hills of this early landscape, held up by firmly cemented Kingsbury conglomerate, have been exhumed in the modern cycle of erosion.

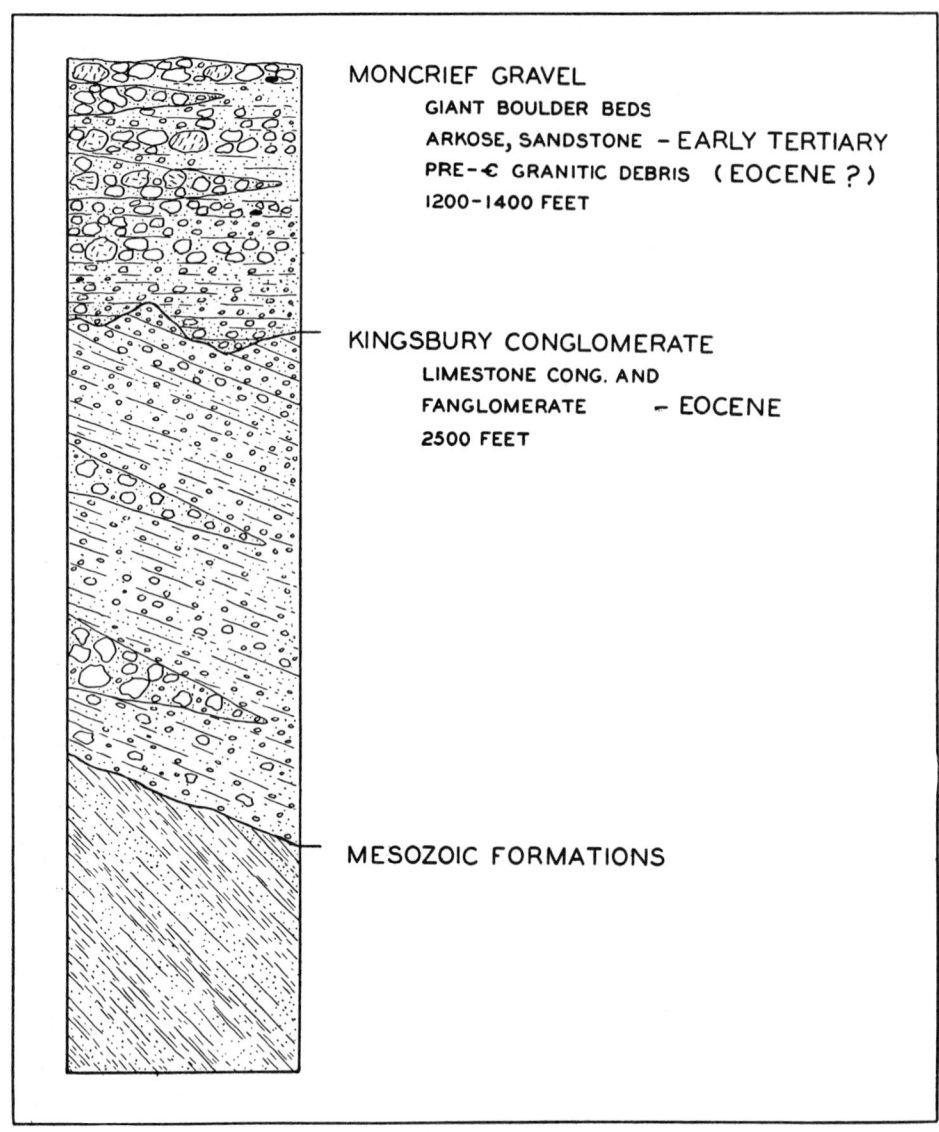

Fig. 5.—Columnar section of Moncrief gravel and Kingsbury conglomerate

WITH PRE-TERTIARY ROCKS

The Moncrief gravel rests on tilted Mesozoic strata with a marked angular unconformity, as recognized by Darton (1906a, pl. 47; 1906b, areal geology map) and as clearly demonstrated by the field and map relations (fig. 3). The contact between gravel and upturned Paleozoic beds forming the mountain front has also been interpreted as depositional (Salisbury and Blackwelder, 1903; Darton, 1906a, pl. 47; Alden, 1932; Demorest, 1941, pls. 1 and 4); but such an interpretation meets insurmountable obstacles. One has but to see the great coarse-boulder layers, composed wholly of pre-Cambrian rock fragments, abutting directly against the upturned Paleozoic limestone beds (pl. 2, B, and pl. 4) to realize that the Paleozoic rocks could not possibly have had their present position when the gravel was being deposited, unless the gravel came from the east. This possibility is most remote, for the nearest known source of pre-Cambrian debris to the east is the Black Hills, 145

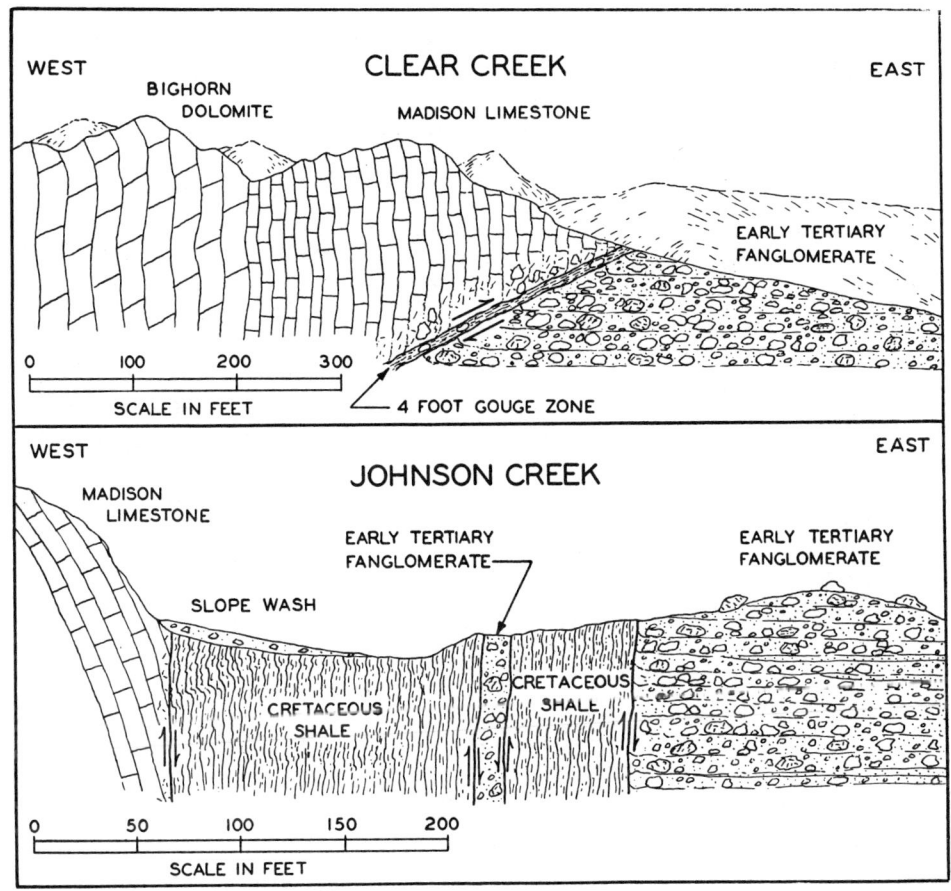

FIG. 6.—Field sketches of fault relations between early Tertiary fanglomerate (Moncrief gravel) and pre-Tertiary rocks of the Big Horn Mountains. Exposure at Clear Creek is in a road cut on U.S. Highway 16. The Johnson Creek exposure is in a gully on the first spur to the north.

miles away. The relations are those expected of a fault; and actual exposures of a fault contact between Moncrief gravel and Paleozoic rocks have been observed at Clear Creek, north of Johnson Creek, and on North Piney Creek. Details of the fault exposures on Clear and Johnson creeks are sketched in figure 6. In the summer of 1946 a new road cut on U.S. Highway 16 along Clear Creek gave a superb exposure of the fault, showing Paleozoic limestone thrust eastward over Moncrief gravel along a plane dipping 25° west. On Johnson Creek the fault is a complex zone, consisting of several nearly vertical planes of displacement involving Paleozoic limestone, distorted Cretaceous shale, and the Moncrief gravel. Before faulting, the gravel probably overlay the Cretaceous shale with an angular unconformity; and in the subsequent displacement a thin sliver or wedge of the gravel was dropped down into the shale, which, in turn, forms a larger slice between the Paleozoics and the gravel. The fault planes appear nearly vertical in the exposure, but they may dip westward at depth. On North Piney Creek the fault is a thrust dipping westward.

On the basis of these exposures and from the fact that Paleozoic strata are steeply tilted or even overturned wherever the Moncrief gravel abuts against them, the contact is interpreted as a fault in all places. Nowhere was a depositional contact observed between these two units. Stratigraphic throw on the fault must have been at least 2,000 feet to give the present relations. Thrust faults have long been recognized (Darton, 1906b; Demorest, 1941; Bucher, Thom, and Chamberlin, 1934, p. 169) along the east base of the central Big Horns, but it has not been shown previously that the Moncrief gravel is involved in some of the thrusting.[7] Faulting occurred at more than one period, for Demorest (1941, pp. 165–166) and Darton (1906a, p. 93) recognize at least two episodes of faulting. Darton (1906b, pp. 60–61, 93) also speaks of the Kingsbury conglomerate–Paleozoic contact as being a fault in some places and an overlap elsewhere. Some skepticism is entertained with respect to a few of Darton's fault relations, as he appears to have mapped great landslide masses as fault blocks.

AGE OF THE GRAVEL

The Moncrief gravel has previously been dated as early Pleistocene (Salisbury and Blackwelder, 1903, pp. 222–223; Alden, 1932, p. 42), late Tertiary or earliest Quaternary (Darton, 1906a; 1906b), and has been compared with the Bishop conglomerate (Atwood and Atwood, Jr., 1938, p. 966; Rich, 1910, pp. 601–632) of the Uinta Mountain region, which is thought to be Miocene or Oligocene by W. H. Bradley (1936, p. 163, 185) or about Pliocene by W. W. Atwood (1940, p. 313). Demorest (1941, pp. 167–168) includes the Moncrief beds in his Tertiary(?) gravels, which may range from Eocene well up into the Tertiary. Taff (1909, p. 131) thought the gravel at Moncrief Ridge was equivalent to nearly all his upper member in the Sheridan coal field, which includes the Tongue River member of the Fort Union (Paleocene) and the Intermediate and Ulm coal groups, which may be Eocene (Thom and Dobbin, 1924, pp. 494–498;

[7] The 1947 Bighorn Basin field conference guidebook prepared by the University of Wyoming, the Wyoming Geological Association, and the Yellowstone-Bighorn Research Association, which appeared since this paper was prepared for publication, makes note (p. 95) of the fault between Moncrief gravel and Paleozoic rocks exposed on U.S. Highway 16 along Clear Creek.

Baker, 1929, p. 24; Nace, 1936, pp. 91, 95, 104).

An early Tertiary age is favored in this paper because the gravel appears to be part of the "Wasatch" on two counts. First, it contains typical "Wasatch" material, including coalified plant remains, and passes by gradation downward and eastward into "Wasatch" beds. Second, the gravel occurs only where the "Wasatch" lies along the base of the range. Structural relations also suggest an early Tertiary age, for the gravel is older than some of the eastward thrusting of the Big Horn Mountains, which is presumably Laramide. If the gravel were early Pleistocene, we should have evidence of at least 2,000 feet of Pleistocene thrust-fault displacement—a new and unlikely chapter in the evolution of the Middle Rocky Mountains.

Exact dating of the Moncrief gravel awaits detailed paleontologic and stratigraphic study of early Tertiary beds in the Sheridan coal field. As it now stands, Taff's (1909, p. 131) original description, as modified by more recent work (Thom and Dobbin, 1924; Baker, 1929), makes the lower part of the Moncrief gravel Paleocene and the upper part Eocene; but the evidence of angular unconformity between Moncrief gravel and Kingsbury conglomerate, already described, renders a Paleocene age for any part of the Moncrief unlikely. The resemblance to conglomerates described by Tourtelot (1946) in the late lower Eocene Wind River formation at the south end of the Big Horn Range strengthens the probability that the Moncrief gravel is Eocene.

ORIGIN

Proponents (Salisbury and Blackwelder, 1903; Alden, 1932) of a Pleistocene glacial origin for the Moncrief gravel have based their arguments chiefly on relations at Bald, North, and Moncrief ridges. However, the equally coarse gravel near Johnson, Shell, and Little Piney creeks is part of the same deposit, and it is noteworthy that none of these latter streams shows the slightest evidence of glaciation in their lower parts. Furthermore, neither Johnson Creek nor Little Piney Creek extends far enough into the range to reach areas of known glaciation (Darton, 1906a, pl. 26; 1906b, areal geology map). A Pleistocene age is further ruled out by stratigraphic and structural relations already detailed. These relations likewise render untenable Darton's (1906a, pp. 69–70; 1906b, p. 9) interpretation that the boulder deposits are late Tertiary or earliest Quaternary bench gravels. In this study the present outcrops of Moncrief gravel are interpreted as remnants of early Tertiary alluvial fans built up along the east base of the Big Horn Mountains as they were being uplifted during the Laramide Revolution.[8]

The composition of the Moncrief gravel clearly indicates derivation from the pre-Cambrian core of the Big Horn Mountains; and the crude bedding, angularity, and size of boulders, sparseness of matrix, and occasional fine beds are all consistent with fan deposition (Trowbridge, 1911, pp. 706–747). The huge

[8] This concept was outlined by Professor W. T. Thom, Jr., in a conversation in 1940; and Demorest had essentially the same idea. It was viewed with skepticism during the early stages of this work, until the accumulating field evidence rendered other interpretations untenable. Love (1940) and Tourtelot (1946) report boulder beds in lower Eocene deposits along the south flank of the Big Horn Mountains; and coarse fanlike deposits in early Tertiary beds bordering other Middle Rocky Mountain ranges have been described by the following: Eldridge (1894, pp. 25–26); Westgate and Branson (1913, p. 144); Blackwelder (1915, pp. 106–108); Bauer (1934, pp. 678–679); Love (1939, pp. 60, 106–107); Branson and Branson (1941, pp. 142–143); Van Houten (1944, pp. 179–181).

boulders furnish no objection, for equally large boulders are found on alluvial fans at the east base of the Sierra Nevada (Trowbridge, 1911, pp. 716–717, 740–743). Conditions causing fan deposition were presumably brought about by uplift of the mountains, and this uplift was probably by stages.[9] Initially Mesozoic and Paleozoic strata were exposed in the up-arched area; and deposits composed primarily of Paleozoic fragments, such as the Kingsbury conglomerate, were formed along the flanks of the range. Darton (1906b, p. 8), Wegemann (1917, pp. 59–60), and Demorest (1938, p. 20; 1941, p. 168) all recognize the Kingsbury as a product of mountain uplift. A subsequent uplift which extended into the piedmont area deformed the Kingsbury mildly and initiated erosion. This was followed by deposition of coarse fan materials, the Moncrief gravel, derived from the mountains which were being raised abruptly above the piedmont, presumably by faulting. At first, the fans consisted of mixed Paleozoic and pre-Cambrian rock fragments; but, as stripping of the Paleozoic rocks became more complete, a larger percentage of the debris was derived from the pre-Cambrian core, until, finally, practically all the detritus was of pre-Cambrian origin.

Increase in coarseness of the gravel upward in the section may have been due to the more massive nature of the pre-Cambrian rocks at depth, to an increasing rate of uplift, or possibly to early Tertiary glaciation in the high parts of the range, like that recognized elsewhere in the Rocky Mountains (W. W. Atwood, 1915, pp. 13–26; W. W. and W. R. Atwood, 1926, pp. 612–622; W. W. and W. W. Atwood, Jr., 1938, p. 961; Drysdale, 1915, pp. 65–66; Scott, 1938, pp. 628–636). C. J. Hares (1926a, pp. 174–175; 1926b, p. 175; 1926c, pp. 175–176) has suggested that somewhat similar deposits along the flanks of various Rocky Mountain ranges are the product of a widespread mid-Tertiary glaciation, but this has apparently not been confirmed by subsequent investigators. The Moncrief gravel is certainly not a till; but it may consist of coarse outwash debris, although there is no direct evidence indicating a glacial source. Alden (1932, p. 43) found one boulder with something like chatter marks; and one small stone with striae-like scratches was collected during the present investigation. An early Tertiary glacial source for the gravel is not established by such meager evidence, and it is not essential to the major thesis proposed. In fact, the map relations (fig. 3) showing that the Moncrief gravel is best developed along those parts of the range uplifted by thrust faults suggest that the gravel was derived from rising thrust blocks as proposed by Knight (1937) and Tourtelot (1946) for similar gravels elsewhere in Wyoming. The fault contact between Moncrief gravel and pre-Tertiary rocks indicates that movement on the faults also occurred after deposition of the gravel.

CONCLUSIONS

A series of coarse bouldery fans composed primarily of pre-Cambrian debris was built up along the east base of the central Big Horn Mountains in early Tertiary time, probably the Eocene. This was the section of greatest uplift, and the fan debris was presumably coarser and thicker here than elsewhere. Subsequent faulting thrust the Paleozoic beds of the mountain front eastward against the gravel, and erosion during the remainder of the Cenozoic has gradually etched out

---

[9] Love (1939, pp. 116–117) recognizes eight pulsations of the Laramide Revolution in western Wyoming.

the finer materials, leaving the thickest and coarsest parts of the fan deposits as prominent ridges in the present landscape. At least three episodes of Laramide deformation are indicated: (1) An uplift which produced the Kingsbury conglomerate. Demorest (1941, p. 168) cites evidence indicating that faulting probably occurred during this uplift. (2) A second uplift, also probably attended by faulting, which deformed the Kingsbury and produced the coarse Moncrief gravel. This episode may have been accompanied or closely followed by alpine glaciation. (3) A third, post-Moncrief, period of thrust faulting toward the east in the central segment of the range.

ACKNOWLEDGMENTS.—Numerous residents of the Big Horn Mountain area, particularly U. J. Post, facilitated the field study. Professor W. T. Thom, Jr., has kindly made suggestions in the course of the work and has read the manuscript. Much profit has been derived from the suggestions of S. H. Knight, D. L. Blackstone, J. D. Love, and H. A. Tourtelot. Preparation of manuscript and illustrations was aided by John Townley and Kenneth Thompson working under a research grant from the Graduate School of the University of Minnesota, and this report was written while the author was a staff member of that institution.

REFERENCES CITED

ALDEN, W. C. (1932) Physiography and glacial geology of eastern Montana and adjacent areas: U.S. Geol. Survey Prof. Paper 174.

ATWOOD, W. W. (1915) Eocene glacial deposits in southwestern Colorado: U.S. Geol. Survey Prof. Paper 95-B.

—— (1940) The physiographic provinces of North America, New York, Ginn & Co.

—— and ATWOOD, W. R. (1926) Gunnison tillite of Eocene age: Jour. Geology, vol. 34.

—— and ATWOOD, W. W., JR. (1938) Working hypothesis for the physiographic history of the Rocky Mountain region: Geol. Soc. America Bull. 49.

BAKER, A. A. (1929) The northward extension of the Sheridan coal field, Big Horn and Rosebud counties, Montana: U.S. Geol. Survey Bull. 806, pt. 2.

BAUER, C. M. (1934) Wind River Basin: Geol. Soc. America Bull. 45.

BLACKWELDER, ELIOT (1915) Post-Cretaceous history of the mountains of central western Wyoming: Jour. Geology, vol. 23.

BRADLEY, W. H. (1936) Geomorphology of the north flank of the Uinta Mountains: U.S. Geol. Survey Prof. Paper 185-I.

BRANSON, E. B., and BRANSON, C. C. (1941) Geology of Wind River Mountains, Wyoming: Am. Assoc. Petroleum Geologists Bull. 25.

BUCHER, W. H.; THOM, W. T., JR.; and CHAMBERLIN, R. T. (1934) Geologic problems of the Beartooth-Bighorn region: Geol. Soc. America Bull. 45.

CHAMBERLIN, R. T. (1940) Diastrophic behavior around the Bighorn Basin: Jour. Geology, vol. 48.

DARTON, N. H. (1906a) Geology of the Bighorn Mountains: U.S. Geol. Survey Prof. Paper 51.

—— (1906b) Cloud Peak–Fort McKinney folio: U.S. Geol. Survey Atlas, Folio 142.

DEMOREST, MAX (1938) Structural geology of a part of the east front of the Bighorn Mountains near Buffalo, Wyoming, unpublished Ph.D. thesis, Princeton University Library.

—— (1941) Critical structural features of the Bighorn Mountains, Wyoming: Geol. Soc. America Bull. 52.

DRYSDALE, C. W. (1915) Geology of Franklin Mining Camp, British Columbia: Canadian Geol. Survey Mem. 56.

ELDRIDGE, G. H. (1894) A geological reconnaissance in northwest Wyoming: U.S. Geol. Survey Bull. 119.

GALE, H. S., and WEGEMANN, C. H. (1910) The Buffalo coal field, Wyoming: U.S. Geol. Survey Bull. 381, pt. 2.

HARES, C. J. (1926a) Glacial origin of the Bishop conglomerate of Wyoming, Colorado, and Utah (abst.): Geol. Soc. America Bull. 37.

—— (1926b) What is the Denver formation? (abst.): ibid.

—— (1926c) Post-Eocene–pre-Miocene glaciation in the Rocky Mountains (abst.): ibid.

JEPSEN, G. L. (1940) Paleocene faunas of the Polecat Bench formation, Park County, Wyoming: Am. Philos. Soc. Proc., vol. 83, pt. 1.

KNIGHT, S. H. (1937) Origin of the giant conglomerates of Green Mountain and Crook's Mountain, central Wyoming: Proc. Geol. Soc. America for 1936.

KNOWLTON, F. H. (1909) The stratigraphic relations and paleontology of the "Hell Creek Beds," "Ceratops Beds," and equivalents, and their reference to the Fort Union formation: Proc. Washington Acad. Sci., vol. 11.

Love, J. D. (1939) Geology along the southern margin of the Absaroka Range, Wyoming: Geol. Soc. America Special Paper 20.

——— (1940) Thrust faulting at the southern end of the Bighorn Mountains, Wyoming: Geol. Soc. America Bull. 51, pt. 2.

Nace, R. L. (1936) Summary of the late Cretaceous and early Tertiary stratigraphy of Wyoming: Wyoming Geol. Survey Bull. 26.

Rich, J. L. (1910) The physiography of the Bishop conglomerate, southwestern Wyoming: Jour. Geology, vol. 18.

Salisbury, R. D., and Blackwelder, Eliot (1903) Glaciation in the Bighorn Mountains: Jour. Geology, vol. 11.

Scott, H. W. (1938) Eocene glaciation in southwestern Montana: Jour. Geology, vol. 46.

Taff, J. A. (1909) The Sheridan coal field, Wyoming: U.S. Geol. Survey Bull. 341, pt. 2.

Thom, W. T., Jr., and Dobbin, C. E. (1924) Stratigraphy of Cretaceous-Eocene transition beds in eastern Montana and the Dakotas: Geol. Soc. America Bull. 35.

Tourtelot, H. A. (1946) Tertiary stratigraphy and its bearing on oil and gas possibilities in the northeastern part of the Wind River Basin, Wyoming: U.S. Geol. Survey Oil and Gas Inv. Preliminary Chart 22.

Trowbridge, A. C. (1911) The terrestrial deposits of Owens Valley, California: Jour. Geology, vol. 19.

U.S. Geol. Survey (1925) Geologic map of Wyoming.

——— (1932) Geologic map of the United States.

van Houten, F. B. (1944) Stratigraphy of the Willwood and Tatman formations in northwestern Wyoming: Geol. Soc. America Bull. 55.

Wegemann, C. H. (1917) Wasatch fossils in so-called Fort Union beds of the Powder River Basin, Wyoming, and their bearing on the stratigraphy of the region: U.S. Geol. Survey Prof. Paper 108-D.

Westgate, L. G., and Branson, E. B. (1913) The later Cenozoic history of the Wind River Mountains, Wyoming: Jour. Geology, vol. 21.

Wilmarth, M. G. (1938) Lexicon of geologic names of the United States: U.S. Geol. Survey Bull. 896, pt. 1.

Reprinted from *Geol. Soc. America Bull.* **65**:175-190 (1954), courtesy of the Geological Society of America

GEOLOGY OF ALLUVIAL FANS IN SEMIARID REGIONS

By Erich Blissenbach

Abstract

An alluvial fan is a body of detrital sediments built up at a mountain base by a mountain stream. Bold relief is essential, moderately arid to semiarid climate favorable for the development of fans. The depositing agents are sheet floods, stream floods, and streams. Compound alluvial fans result from lateral coalescence of single fans.

Development of alluvial fans is affected by changes in the course of a cycle, varying base level, climatic changes, tectonic movements, and slumping of fan deposits. Telescoped or superimposed structure may be developed.

Fan deposits are arkosic or graywacke. Sorting and roundness of particles range widely. The matrix is primary or secondary. In general alluvial-fan deposits are stratified. Channel cut-and-fill is pronounced. Individual strata in fans are up to 20 feet thick. Particles in stream deposits are imbricated.

Talus-slope deposits at the apex of a fan and floodplain deposits at its base can be separated from those of an alluvial fan by particle sizes, angularity and orientation of fragments, sorting, and original dip of strata. Mudflow deposits in an alluvial fan indicate certain climatic conditions during its formation.

Many ancient fan deposits may have escaped recognition because of the common misconception that fan deposits are necessarily unstratified, composed of angular fragments, poorly sorted, and without distinctive sedimentary structures.

CONTENTS

TEXT

	Page
Introduction	176
Purpose of investigation	176
Areas of field study	176
Acknowledgments	176
Geology of alluvial fans	176
Definitions	176
General statement	176
Process of forming alluvial fans	177
Late-stage development	179
Changes in normal course of a cycle	179
Effects of varying base level	179
Effects of climatic changes	180
Effects of tectonic movements	180
Effects of slumping of unconsolidated fan material	181
Secondary alluvial fans	181
Geology of alluvial-fan deposits	181
Facies	181
Particle sizes and particle-size distribution	182
Composition	182
Sorting	183
Roundness	183
Sphericity	184
Matrix	184
Cement	185
Color	185
Porosity	185
Permeability	185
Sedimentary structures	185
Organic contents	187

	Page
Ancient alluvial-fan deposits	187
Occurrence	187
Relative abundance	187
Criteria for recognition of ancient alluvial-fan deposits	188
References cited	189

ILLUSTRATIONS

Figure	Page
1.—Compound alluvial fan formed by lateral coalescence of single alluvial fans	177
2.—Alluvial fan with telescope structure	180
3.—Superimposed alluvial fan	181
4.—Pseudo-telescope structure (modified after Gilbert)	181
5.—Distribution of maximum particle sizes and surface angles along a radial profile on an alluvial fan of the Santa Catalina Mountains, Arizona	182
6.—Distribution of roundness and sphericity of alluvial fan particles along the same radial profile as Figure 5. Base of Santa Catalina Mountains, Arizona	184
7.—Characteristic sedimentary structures in an alluvial fan of the Santa Catalina Mountains, Arizona. Main direction of transport from right to left	186

Plate	Facing page
1.—Alluvial fans	186
2.—Alluvial-fan deposits	187

INTRODUCTION

Purpose of Investigation

This paper gives the results of a comprehensive study of alluvial fans. Though detailed field work was carried out only in the southwestern United States, most of the data may be applied to alluvial fans in other areas where similar conditions prevail. Thus, through detailed study of recent alluvial-fan deposits, geologists may be aided in determining whether ancient rocks were formed under similar conditions.

Areas of Field Study

Alluvial fans studied in southern Arizona include those at the southern base of the Santa Catalina Mountains, north and northeast of Tucson, Arizona, and fans of the Tucson Mountains, west and northwest of Tucson, Arizona.

Investigations in the Mammoth area, Arizona, were carried out on alluvial fans on both sides of an intermontane valley of the Black Hills, Arizona, west of Mammoth.

In northern Arizona alluvial fans were studied along the Aubrey Cliffs, northwest of Seligman, Arizona, and along the Bright Angel Trail, north of the Grand Canyon Village, Arizona.

Studies were also carried out on alluvial fans in southern California and southern New Mexico.

Certain alluvial fans were studied in every detail, whereas only certain properties were examined in other fans.

Acknowledgments

The aid and counsel of Professor Edwin D. McKee of the University of Arizona, under whose guidance this work was undertaken, is sincerely appreciated.

GEOLOGY OF ALLUVIAL FANS

Definitions

An *alluvial fan* is a body of detrital sediments built up by a mountain stream at the base of a mountain front. It develops because all streams tend to form a graded course. A talus slope differs from an alluvial fan in that it is built up predominantly by gravitational sliding.

The *apex* of an alluvial fan develops at the point where the stream emerges from the mountain. It is the point of highest elevation on the alluvial fan. *Fanhead* applies to the area on the alluvial fan close to the apex; *midfan* designates the area between the fanhead and the outer, lower margins of the fan. *Base* of an alluvial fan is the term applied to the outermost or lowest zone of the fan.

If the fanhead area reaches far into a mountain canyon the term *fan-bay* (Davis, 1938, p. 1374) is applied to it. *Fan-mesa* (Eckis, 1928, p. 243–244) is the term applied to an alluvial fan remnant left standing in the process of degradation of a fan.

General Statement

The occurrence of alluvial fans has been reported from all continents. However, due to conspicuous development and easy accessibility, alluvial fans in California have been studied in more detail than any others (Lawson, 1913, p. 332; Eckis, 1928, p. 232–246; Krumbein, 1937, p. 586–594; Buwalda, 1951, p. 1491).

Investigators generally agree that an alluvial fan resembles geometrically the segment of a cone. From the apex of the fan the surface dips toward the base in which direction the angles of dip gradually become flatter. Thus, a radial profile through the fan is concave upward; a profile at right angles convex. The steepest angle of dip on the alluvial fan is encountered close to its apex.

The angle of dip of the fan surface rarely exceeds 10° (Dana, 1894, p. 194–195; Eckis, 1928, p. 223; Scott, 1932, p. 269; Eardley, 1938, p. 1408). Some authors report no angle greater than 5° or 6° (Lawson, 1915, p. 25; Vaughan, 1922, p. 341). The maximum angles of slope of alluvial fans studied by the writer range from 5° observed on fans of the Aubrey Cliffs, Arizona, to 9° on fans of the Black Hills of Arizona.

Surface angles greater than 5° are characteristic of the upper half or more of small alluvial fans with a radial extent of a few hundred feet like those of the Black Hills of Arizona.

Large alluvial fans such as those of the Santa Catalina Mountains, Arizona, with a radial extent of about 4 miles, exhibit surface angles greater than 5° only within the upper one twentieth or less of their extent.

In geologic literature small alluvial fans are commonly called alluvial cones. As no definition specifies the difference between an alluvial fan and a cone, this classification is not followed. Instead the terms steep, gentle, and flat angles of dip are proposed to denote the degree of slope of alluvial fans or parts of them.

Because angles of dip greater than 5° are rare these are called steep. Angles of dip between 2° and 5° are termed gentle, and sloping angles below 2° flat. On the basis of these definitions the slope of the surface of a particular alluvial fan can be described as follows:

Alluvial fans of the Santa Catalina Mountains with a radial extent of about 4 miles have steep slopes in the uppermost 0.2 mile of the radial extent of the fans. In the adjoining 0.7 mile surface angles are gentle, while the remaining part, about 3.1 miles, has flat angles of dip.

The radius of an alluvial fan may be as great as 40 miles under exceptional conditions (Grabau, 1913, p. 584). The radii of alluvial fans studied by the writer range from 4 miles at the Catalina Mountains, Arizona, to about 500 feet in the Black Hills, Arizona.

By lateral coalescence of single alluvial fans, a *compound alluvial fan* may result (Miller, 1926, p. 164–166) (Fig. 1; Pl. 1, fig. 1). The compound alluvial fan is equivalent to the alluvial piedmont slope (Lahee, 1941, p. 101) and the "bajada" (Blackwelder, 1931, p. 136).

Alluvial fans may be 1000 or more feet in thickness (Eckis, 1928, p. 224). In general, the thickness of alluvial-fan deposits is greatest at the apex of a fan (Scott, 1932, p. 269). Under exceptional conditions, however, a fan may be thicker at the base than at its apex. This is illustrated in an alluvial fan of the Santa Catalina Mountains, Arizona; the thickness of the fan deposits increases from between 0 and 100 feet at the apex to 200 or 300 feet at the base of this fan.

Alluvial fans occur in areas of bold relief, and their development is most conspicuous under moderately arid to semiarid conditions. The influence of bold relief seems obvious as only under such conditions will there be profound erosion and transportation together with a strong tendency for deposition as the mountain streams reach areas of low gradient.

Practically all reductions of rock slope and contributions to alluvial embankment are

FIGURE 1.—COMPOUND ALLUVIAL FAN FORMED BY LATERAL COALESCENCE OF SINGLE ALLUVIAL FANS

affected by brief and infrequent periods of downpour (Lawson, 1915, p. 28). This type of precipitation is characteristic of areas with arid and semiarid climate. Under humid conditions also alluvial fans are developed as seen along the Alps and the Himalayas. Fans formed in humid environment commonly are flatter than those of arid environment owing to the abundance of running water which favors the development of gentler gradients. Under conditions of extreme aridity, mountains are buried under their own debris (Lawson, 1906, p. 449), and the deposits formed are in the nature of talus slopes.

Thus, the climatic conditions most favorable for the development of alluvial fans appear to range from moderately arid to semiarid. The annual precipitation in the region of an alluvial fan studied by Eckis (1928, p. 225) averages about 17 inches. Alluvial fans studied by the writer in Arizona receive a mean annual precipitation ranging from 10 to 19 inches.

Process of Forming Alluvial Fans

In the stage of deposition the surface of an alluvial fan shows a system of radiating channels focused in the main stream at the apex of the fan. This physiographic pattern is referred to as a braiding stream system. Stream channels are distributaries. They commonly are shallow, entrenched but little beneath the gently sloping surface of an alluvial fan (Davis, 1898, p. 291).

Alluvial fans in process of degradation normally are cut by streams that tend to form deep and narrow channels. These streams are tributaries, and the resulting physiographic pattern constitutes a reversal of the braiding stream system focused in the apex of the fan.

Streams emerging from steep mountain canyons commonly are loaded with detrital material, especially if the times of flow are infrequent and separated by long, dry intervals. At the alluvial fan a pronounced tendency for deposition is immediately developed, because of a less steep gradient on the alluvial fan and the pervious nature of the alluvial-fan deposits which effects a continuous decrease in the volume of water.

Thus, fan channels frequently are silted up causing overflows and forming new distributaries. When one sector of a fan has been built up, the streams shift to another, lower section of the fan and build that up (Longwell, Knopf, and Flint, 1948, p. 88). The process is repeated again and again until the mountain stream and the alluvial fan have reached graded conditions.

In arid and semiarid regions short, violent flows normally constitute the transporting and depositing agents responsible for the development of alluvial fans. The depositing agents on alluvial fans may be classified into three types:

(1) Sheetfloods. These occur when an exceedingly large amount of water and detritus emerges from a mountain canyon. The flow, acting like a viscous medium, tends to spread out in the form of a sheet covering the alluvial fan or parts of it. Deep channels on alluvial fans tend to prevent sheetfloods, in the opinion of Davis (1938, p. 1348). Furthermore, he believed that the graded state of compound alluvial fans is effected largely by sheetfloods. A striking peculiarity of sheetfloods is the shortness in distance as well as in time of their flows. Great floods of this type may come only at intervals of decades or centuries (McGee, 1897, p. 108).

(2) Streamfloods. These are confined to definite channels on alluvial fans. They are formed where a large amount of water and detrital material emerges from a mountain canyon. Streamfloods may also form because channels on alluvial fans are too deep to allow a sheetflood to develop. The spasmodic and impetuous character of these floods is such that the term streamflood rather than streams is applied (Davis, 1938, p. 1347). The deposits of violent streamfloods tend to be identical with those of sheetfloods except that, instead of being blanket-shaped, they are linear in plan view.

(3) Streams. These are formed if both the amount of water and the quantity of detritus are less than the requirements for sheetfloods or streamfloods. A steady, rather than abundant supply and recharge of water from the mountains must be maintained. The action of streams is of minor significance in arid or semiarid climates for these climatic conditions do not favor the development of streams. In relatively humid regions, however, as in the Alps or the Himalayas, alluvial-fan deposits laid down by streams are of considerable magnitude.

Every possible gradation between these types of depositing agents must be expected on alluvial fans. Streamflood deposits exhibit gradations ranging from those formed during violent flows to others formed by floods of moderate intensity.

The deposits of sheetfloods and violent streamfloods tend to have a high percentage of mud-sized particles. The term mudflow deposits is commonly applied to these sediments, but this term should not be employed if the deposits can be ascribed to sheetflood or streamflood.

Any one alluvial fan may not show the effect of all three types of depositing agents. Furthermore, certain types of depositing agents may predominate at a particular stage in the development of an alluvial fan. Thus, the action of sheetfloods seems to be most pronounced when a graded stage has been reached—i.e., with completion of the maturity of a fan (Davis, 1938, p. 1340).

The accumulation of flood waters in a mountain area may first give rise to a sheetflood and later, with decreasing recharge of water, change to streamfloods and finally to streams. The duration of sheetfloods is measured in terms of seconds or minutes (McGee, 1897, p. 101), that of streamfloods in minutes or hours, and the flow of streams in hours or even days.

Pack (1923, p. 349–356) described a flood along the west front of the Wasatch Mountains, Utah, in which he recognized three stages of flooding: (1) torrential streamflood, (2) mudflow, (3) dwindling streamflood. These stages compare well with the classification of depositing agents adopted by the writer.

Alluvial fans studied by the writer in southwestern United States suggest that all three depositing agents—(1) sheetfloods, (2) streamfloods, and (3) streams—participated to a varying degree in their formation. In the Black Hills of Arizona, with a mean annual precipitation of 19 inches, mudflow deposits—*i.e.*, those formed by sheetfloods and violent streamfloods—constitute about 5 to 10 per cent of the fan deposits; moderate streamfloods or streams are responsible for the other 90 to 95 per cent. In the Tucson area with a mean annual precipitation of about 11 inches, mudflow deposits form from about 20 to 40 per cent of the alluvial fans, while streams and moderate streamfloods are responsible for the deposition of the other sediments in these fans.

LATE-STAGE DEVELOPMENT

Changes in Normal Course of a Cycle

Until graded conditions are reached, erosion predominates along a mountain canyon, deposition on an alluvial fan. Upon development of a graded course in a mountain stream and an alluvial fan, erosion will reach over the apex of the fan and affect the fanhead area. Thus, the fanhead area undergoes dissection while a depositional tendency prevails farther downfan. The importance of such changes in the cycle of fan aggradation and degradation was first realized by Eckis (1928, p. 237).

Recession of a mountain front results in formation of a rock pediment normally covered by the fanhead part of an alluvial fan. Dissection of the fanhead area due to changes in the normal course of a cycle or to other causes may strip the rock pediment of its alluvial cover and expose it. A rock pediment may be an important indicator of the extent of an alluvial fan at its maturity.

The results attained by a single alluvial fan may also be developed on a compound type of fan. The normal change in a cycle, however, need not take place on adjacent units within a compound fan at the same time. Accomplishment of the maturity of a fan and the beginning of late-stage development is mainly a function of the size of the mountain stream and its gradient, factors which may vary considerably among different units of the same compound fan.

The compound fan at the southern base of the Santa Catalina Mountains, Arizona, is more heavily dissected along its eastern extent than in the west. In the absence of evidence for tectonic activities, the writer believes that the heavier erosion in the eastern part of the compound fan results from larger streams in the eastern part of the range whereby the maturity of fans and the beginning of late-stage development is accelerated. Slow development of alluvial fans and similarly slow changes in the cycle must be expected along the western extent of the compound fan where mountain streams are comparatively small.

Effects of Varying Base Level

Changes in the base level of a fan are commonly effected by a stream flowing along the fan base. Degradation of a fan base may also be caused by wind deflation as seen on some alluvial fans of the Aubrey Cliffs, Arizona (Blissenbach, 1952, p. 123).

Aggrading base level may be reflected in accelerated deposition on an alluvial fan. Under such conditions, the apex of a fan may migrate into the mountain canyon, and fan-bays may be developed.

Degrading base level results in erosion on an alluvial fan. Available evidence indicates that several of the fans studied by the writer have been affected by strong dissection due in part to the cutting in of a stream at the base of these fans.

The effects of the lateral swing of a river at the base of a fan is a combination of the effects of aggrading and degrading base level. The Colorado River has swung westward against the base of an alluvial fan, about 20 miles north of Needles. As a consequence of this undercutting, this alluvial fan is strongly dissected (Davis, 1938, p. 1349). In contrast, a lateral swing of a river away from the fan base favors depositional tendencies on a fan, and the graded conditions may be restored.

A small alluvial fan of recent age, at the base of the compound fan of the Santa Catalina Mountains, Arizona, is built up of alternating layers of stream or streamflood and sheetflood deposits. Development of this fan is controlled in part by the lateral swinging of the Rillito Creek at the fan base. Each stream or streamflood deposit and the overlying sheetflood deposit constitute a depositional cycle; deep channels cut in the sheetflood deposits represent an erosional cycle. At times when the Rillito Creek swung toward the base of this fan, erosion was active on the fan, and deep channels were formed. When the river migrated away from the base, these channels were filled by stream and streamflood deposits. Upon accomplishment of graded conditions, sheetfloods may have once more contributed to the building up of this fan.

As a result of the lateral swinging of a river at the base of a fan a peculiar overall structure may be developed in the fan. It is characterized by younger fans with flatter gradients spreading out from between fan mesas of older fans with steeper gradients; the name *telescope structure* is proposed for it (Fig. 2). Alluvial fans with telescope structure in the Indus basin are illustrated by Drew (1873, p. 454). Conspicuous development of telescope structure is shown in fans of the Santa Catalina Mountains, Arizona, the Black Hills, near Mammoth, Arizona (Pl. 1, fig. 2), and near Indian Garden Springs, Grand Canyon, Arizona.

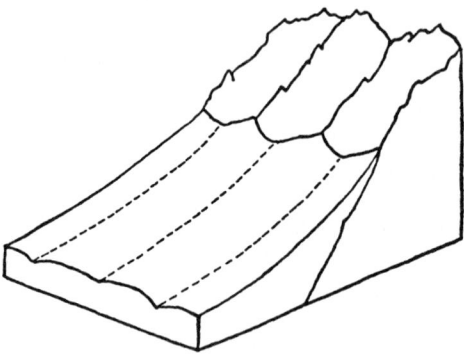

FIGURE 2.—ALLUVIAL FAN WITH TELESCOPE STRUCTURE

Effects of Climatic Changes

Of all possible changes in climatic conditions, a variation in the amount of precipitation has the most profound effect on the development of alluvial fans. Increased precipitation results in dissection of fans and the development of a gentler gradient (Eckis, 1928, p. 237; Davis, 1938, p. 1349). In contrast, a decrease in precipitation may be responsible for a period of aggradation and the development of steeper gradients.

Telescope structure may be developed as a result of climatic changes. Individual levels of the telescope structure correspond to arid periods with their tendency for aggradation; the erosion cycles separating aggradation periods mark the humid stages with their tendency for erosion.

The development of telescope structure in alluvial fans at the southern base of the Santa Catalina Mountains, Arizona, is believed to have resulted from climatic changes. These fans formed in Pleistocene and Recent time during which climatic changes were numerous.

Climatic changes may exert an indirect influence on the development of alluvial fans by allowing a through-flowing stream to form at the base of fans. This stream may, in turn, control development of an alluvial fan by changing its base level. Thus, the San Pedro River, north of the Santa Catalina Mountains, Arizona, was a through-flowing stream during a late Pleistocene stage. Alluvial fans extending north from the Santa Catalina Mountains into the San Pedro Valley were dissected by the erosive action of the main river (Davis, 1938).

Effects of Tectonic Movements

Alluvial fans occur commonly at the base of up-faulted mountain blocks. Rejuvenation of tectonic activities along the same fault lines is a frequent phenomenon. Tectonic disturbance is believed the most important reason for changes on alluvial fans, in the opinion of Davis (1938, p. 1348).

If the relief between mountains and fans is increased by tectonic movements, a new stage of deposition may set in; alluvial fans with steeper gradients begin to form. For these newly deposited fans with steeper gradients the name *superimposed fans* is proposed. Superimposed and original fans do not exhibit distinct boundary lines, and morphologically they cannot be separated. In sections, however, a separation of the two different generations of alluvial fans may be possible (Fig. 3).

Slumping of unconsolidated fan deposits is responsible for the development of *pseudo-telescope structure*. The difference from true telescope structure may be obvious only in favorable exposures (Fig. 4).

FIGURE 3.—SUPERIMPOSED ALLUVIAL FAN

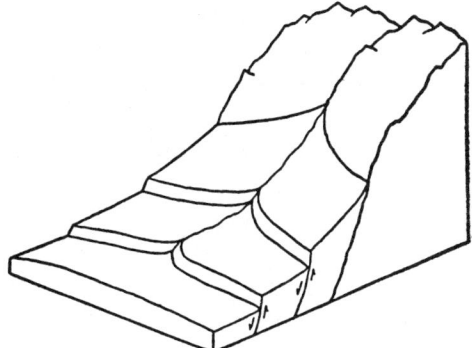

FIGURE 4.—PSEUDO-TELESCOPE STRUCTURE
(Modified After Gilbert)

If the relief between mountains and alluvial fans is decreased by tectonic activities, erosion will be active on the fans, and flatter gradients will be developed. Telescope structure may result. Recognition of rock pediments along the mountain front and their correlation with fan levels may indicate whether telescope structure in fans is due to tectonic movements or to other influences.

In cases where individual parts of an alluvial fan are subjected to tectonic movements, no generalization can be made, and the effects on fans depend on the location, the extent, and the nature of tectonic activities.

Effects of Slumping of Unconsolidated Fan Material

Slumping of unconsolidated fan deposits is a common cause for changes on alluvial fans (Gilbert, 1928, p. 12). An alluvial fan in Nevada on which the effects of slumping were observed is described by Longwell (1930, p. 8).

Slumping of alluvial-fan deposits is favored by the comparatively steep, original dip of strata, the presence of ground water acting as lubricant, the presence of layers with mud-sized constituents acting as sliding planes, and ample voids in coarse-grained layers allowing for considerable settling of strata.

These are the main causes for changes taking place on alluvial fans. Several of these influences may be at work at the same time; their effects may enforce each other, or their tendencies may be opposite and tend to compensate each other. Therefore, the investigator may not always be able to ascribe a particular change in the development of an alluvial fan to any definite cause.

Secondary Alluvial Fans

These form at the base of primary fans. The deposits of secondary alluvial fans are mainly reworked primary fan deposits.

At the base of the large, primary fans at the southern flank of the Santa Catalina Mountains, Arizona, small, secondary fans are in the process of forming (Pl. 1, fig. 3).

GEOLOGY OF ALLUVIAL-FAN DEPOSITS

Facies

The facies of unconsolidated fan deposits are fan gravel, fan sand, and fan mud. The lithified equivalents are fanglomerate (Lawson, 1913, p. 17), fan sandstone, and fan mudstone.

Only one or two of these sedimentary facies normally are developed in an alluvial fan. Many alluvial fans in Arizona, southern California, and southwestern New Mexico show a unifacial, fanglomeratic development. In contrast, in small alluvial fans of the Black Hills of Arizona, only the sand facies is developed.

If alluvial fans are built up in dry regions, eolian deposits may be incorporated in the fluviatile deposits. In the Aubrey Valley, Arizona, sand dunes rest on the lower parts of some alluvial fans. Further building up of these fans may eventually incorporate the sand dunes within the alluvial-fan deposits (Blissenbach, 1952, p. 123).

Particle Sizes and Particle-Size Distribution

The particles in alluvial-fan deposits range from boulder to clay size. In general, coarse gravels predominate near the apex of an alluvial fan; material of intermediate particle size may occupy a central zone, and silts and clays in the area close to the base of a fan (Lawson, 1913, p. 330–331; Vaughan, 1922, p. 340; Troxell and others, 1942, p. 322).

Particle-size distribution curves along radial profiles of alluvial fans have been established by the writer for the fans examined. The distribution curve of particle sizes of an alluvial fan of the Santa Catalina Mountains, Arizona, was compiled from a great number of measurements of the maximum particle size (Fig. 5). This distribution curve is typical for alluvial fans; it shows the rapid change in particle sizes close to the apex of a fan. Available evidence indicates that the mean particle size behaves similarly.

Composition

The composition of alluvial-fan deposits is determined by:

(1) the composition of the parent rock from which the fan deposits are derived;
(2) the type and degree of weathering to which the parent rock is subjected;
(3) syngenetic alterations during transport from the source to the site of deposition;
(4) epigenetic alterations effected after deposition.

In arid and semiarid regions, the action of mechanical weathering predominates over the action of chemical weathering. Therefore the composition of the detritus prepared for transport differs only slightly from the composition of the parent rock. In humid regions, however, considerable allowance must be made for alterations effected by the action of chemical weathering.

Syngenetic alterations have little influence on changes in the composition of fan deposits because both the distance of transport and the time during which the detritus is in contact with water are short. Epigenetic alterations are more effective, however. They include the

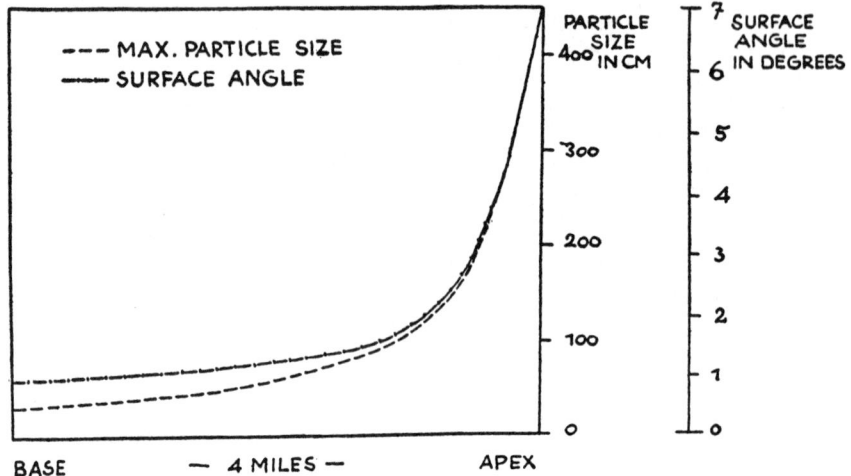

FIGURE 5.—DISTRIBUTION OF MAXIMUM PARTICLE SIZES AND SURFACE ANGLES ALONG A RADIAL PROFILE ON AN ALLUVIAL FAN OF THE SANTA CATALINA MOUNTAINS, ARIZONA

addition of mineral constituents treated later under Cementation.

Based on the classification of compositional groups proposed by Short, McKee, and Jahns (1953), most alluvial-fan deposits belong to one of two compositional groups, the arkose and the graywacke. Mountains exposing plutonic or coarsely crystalline metamorphic rocks commonly exhibit alluvial fans of arkosic composition. In some places the fan deposits of the gneissic Santa Catalina Mountains, Arizona, show a percentage of feldspars twice as high as the minimum requirement of 20 per cent. Mountains composed of fine-grained metamorphic (mostly volcanic and sedimentary) rocks have alluvial fans with graywacke composition. Fans of the volcanic Tucson Mountains, Arizona, and those of the Aubrey Cliffs, Arizona, exposing sedimentary rocks are of this compositional type.

The composition of fan gravels does not appear to change considerably along a radial profile of a fan. Thus, at the apex of an alluvial fan of the Aubrey Cliffs, Arizona, the composition of fan gravels is 45 per cent sandstones and 55 per cent limestones; at the base of this fan the fan gravels are composed of 48 per cent sandstones and 52 per cent limestones. The slight difference in composition may be accidental, and no definite tendency in compositional change can be recognized.

In some areas fans of contrasting compositional types occur side by side because their apices head in different source rocks. This is illustrated in alluvial fans of the Aubrey Cliffs, Arizona, where large fans are fed from the Kaibab and Toroweap formations and adjacent small fans from the Coconino sandstone.

Sorting

Sorting of alluvial-fan deposits varies widely. In general sorting of alluvial-fan deposits may be regarded as a function of:

(1) the range in particle sizes of the detritus prepared for transport in a mountain area;

(2) the type of transporting and depositing agent;

(3) the distance of transport.

Detritus prepared for transport in a mountain area normally has a wide range in particle sizes. An exception to this trend is found in the Black Hills of Arizona, where the source rock of some alluvial fans is a plutonic rock of granitic texture which weathers to equigranular detritus. The breakdown of this parent rock to uniform particle sizes is responsible for the comparatively high degree of sorting in the resulting fan deposits.

There is considerable variation in the degree of sorting of alluvial-fan deposits depending on whether sheetfloods, streamfloods, or streams are responsible for forming a particular deposit.

Sheetfloods tend to accomplish the least amount of sorting (McGee, 1897, p. 106); Davis, 1938, p. 1344). The coefficient of sorting (Trask, 1932, p. 70) of a sheetflood deposit in a small alluvial fan of the Santa Catalina Mountains, Arizona, is about 3.0. Stream deposits, in contrast, show fair to good sorting. The coefficient of sorting of a stream deposit in a small alluvial fan of the Santa Catalina Mountains, Arizona, is 1.7. Streamflood deposits occupy a position in degree of sorting intermediate between that of sheetflood and of stream deposits.

Most alluvial-fan deposits have been transported only short distances and therefore are not well sorted. However, fairly good sorting of fan deposits may be expected near the base of large fans. Comparatively good sorting is also exhibited in secondary fans due to reworking and selective transport of detritus from primary fans. Thus, deposits of a small, secondary fan at the base of the large, primary Santa Catalina Mountain fans (Pl. 1, fig. 3) show a fair to good degree of sorting.

Roundness

A study of the roundness of fragments in alluvial-fan deposits was carried out on a fan at the southern base of the Santa Catalina Mountains, Arizona. Samples of 25 fragments of about equal size were obtained from 13 points along the radial profile of this fan; determination of the roundness was based on Krumbein's charts (1941, p. 68).

The minimum roundness of fragments in the Santa Catalina Mountain fan studied is 0.2, at the point closest to the apex of this fan; the maximum roundness is 0.7, at the point closest to the fan base (Fig. 6). The roundness distribution curve (Fig. 6) suggests that the function between the roundness of fan particles and the distance from the apex is approximately linear for this fan.

ing observed in these gravels was attained by long transport in a floodplain environment (Koons, 1948, p. 59). Later faulting brought these gravels into a position from which they could be redeposited as fan gravels. Thus, the

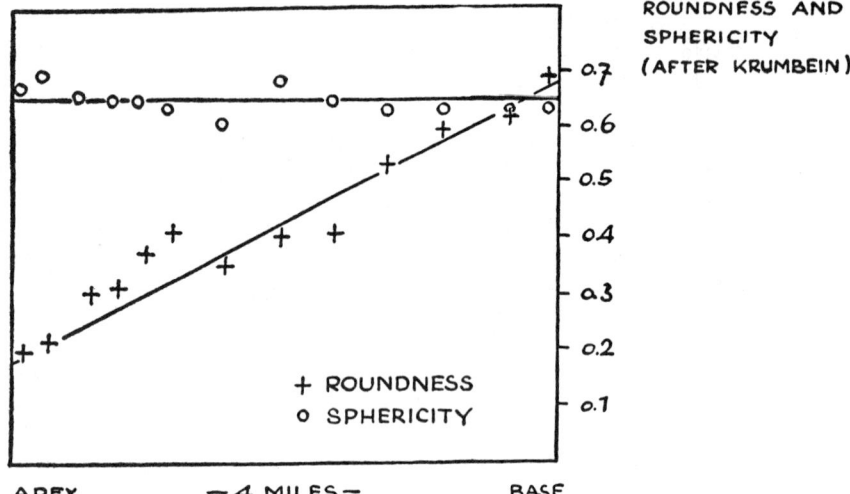

FIGURE 6.—DISTRIBUTION OF ROUNDNESS AND SPHERICITY OF ALLUVIAL FAN PARTICLES ALONG THE SAME RADIAL PROFILE AS FIGURE 5
Base of Santa Catalina Mountains, Arizona

As gneiss particles predominate in alluvial fans of the Santa Catalina Mountains, and as gneiss occupies an intermediate position between the extremely resistant and the highly soluble and friable rocks, a roundness distribution of fan particles in the Santa Catalina Mountain fans may serve as illustration for the normal change in roundness with distance from the apex. Based on the classification of roundness proposed by Short, McKee, and Jahns (1953), fragments of this fan may be described as angular, subangular, and subrounded. As the radial extent of alluvial fans may greatly exceed the 4-mile extent of the Santa Catalina Mountain fans, rounded fragments must be expected on larger fans. The common conception of alluvial-fan deposits being necessarily characterized by angularity of their fragments (Chawner, 1935, p. 259) must therefore be abandoned.

Pliocene gravel deposits at Blue Mountain, northwest of Seligman, Arizona, are described in a recent study by the writer (1952, p. 120). The Blue Mountain gravels were deposited in part on alluvial fans. The high degree of rounding in these fan gravels was not accomplished on an alluvial fan but in a floodplain environment.

Sphericity

Sphericity meansurements in alluvial-fan deposits of the Santa Catalina Mountains, Arizona, were made of the samples obtained for the determination of roundness. For the determination of the sphericity of fan particles, Krumbein's charts (1941, p. 66) were employed. The mean sphericity of particles in this fan is 0.65. The maximum and the minimum values of the sphericity of fan particles along the 4-mile extent of this fan deviate but 7 per cent from the value for the mean sphericity (Fig. 6).

No major change in the sphericity of fan particles is expected on any particular fan as long as there is no change in composition or facies of the fan deposits.

Matrix

Particles forming the matrix of alluvial-fan gravels range from sand to mud size. Finer particles predominate in the matrix of sheetflood deposits, coarser particles in that of stream deposits. A typical sample from the matrix of a sheetflood deposit of a small alluvial fan of the Santa Catalina Mountains, Arizona, shows a mean particle size of 0.15 mm. The mean particle size of the matrix of a stream deposit in the same alluvial fan is 0.7 mm.

The composition of the matrix in a fan deposit determines its compositional classification. Commonly the matrix is arkosic or graywacke.

A secondary matrix is a common feature in fan gravels. This results from removal of an original interstitial fill between fan gravels and later filling of interstices. This process may be seen in action on alluvial fans of the Santa Catalina Mountains, Arizona.

Cement

Calcium carbonate is the commonest cement in alluvial-fan deposits in most areas. It is precipitated from ascending or descending ground water and coats fan particles as solid layers or concretions, or is disseminated as minute calcite crystals in the matrix. If all available pore space is filled with calcium carbonate the cement may constitute up to 30 per cent of the volume of the composite fan deposit. Limonite cement is encountered in a few places in alluvial-fan deposits.

Color

The colors of alluvial-fan deposits are due largely to the types of rock of which they are composed, but they vary somewhat with the climate.

Mudflow deposits of alluvial fans of the Santa Catalina Mountains, Arizona, are light yellow; the interbedded stream deposits are a darker yellow. Where calcium carbonate cement is abundant, alluvial-fan deposits range from gray to white.

Porosity

The porosity in pebble lenses with loose packing, observed in a small, recent fan of the Santa Catalina Mountains, Arizona, was estimated to be 30 per cent. Owing to cementation the porosity in the deposits of older fans along these mountains does not exceed 15 per cent. The porosity of alluvial-fan deposits may attain any value between the above extremes and zero, as controlled by sorting, degree of packing, roundness of particles, compaction, and cementation.

In general poorly sorted mudflow deposits have low porosity; well-sorted stream deposits have high porosity.

Permeability

The permeability of alluvial-fan deposits depends on their porosity and, in addition, is a function of the size of interstitial space. In addition to their low porosity, mudflow deposits have small interstitial openings. Molecular forces therefore tend to retain water, and ground water cannot circulate freely. Thus, mudflow deposits are ideal aquicludes. In contrast, stream deposits in alluvial fans have comparatively large pore spaces which account for the ease with which ground water can move through them. Stream deposits therefore are good aquifers.

Alluvial fans built up by streams and mudflows are, in terms of ground-water geology, "alternating aquifers and aquicludes". This property together with the original dip of the fan strata tends to make an alluvial fan an ideal site for the recovery of artesian water (Tolman, 1937, p. 364–365; Eardley, 1938, p. 1408).

Sedimentary Structures

Alluvial-fan deposits are laid down in beds approximately parallel to the surface of a fan. Thus, the surface angles of alluvial fans are repeated within the fan strata. The classification adopted for the angles of dip of a fan surface may also be applied to the dip of the strata in an alluvial fan; the dip may be described as steep, gentle, or flat.

The overall stratification of most alluvial fans is fair. In fans built up through their entire extent by mudflows, stratification may be poor. In general mudflow deposits form one solid layer. Mudflow deposits show distinct contacts with stream deposits (Fig. 7). Deposits laid down by streams or moderate streamfloods show good stratification; they are subdivided into thinner layers (Pl. 2, fig. 2). Where the mean particle size of these deposits allows such development, lamination is common.

Cross-stratification is well developed in deposits formed by streams on alluvial fans.

feet. Most deposits laid down by streams on alluvial fans are in layers from 1 inch to 1 foot in thickness. Successive stream deposits may attain a vertical extent of several tens of feet.

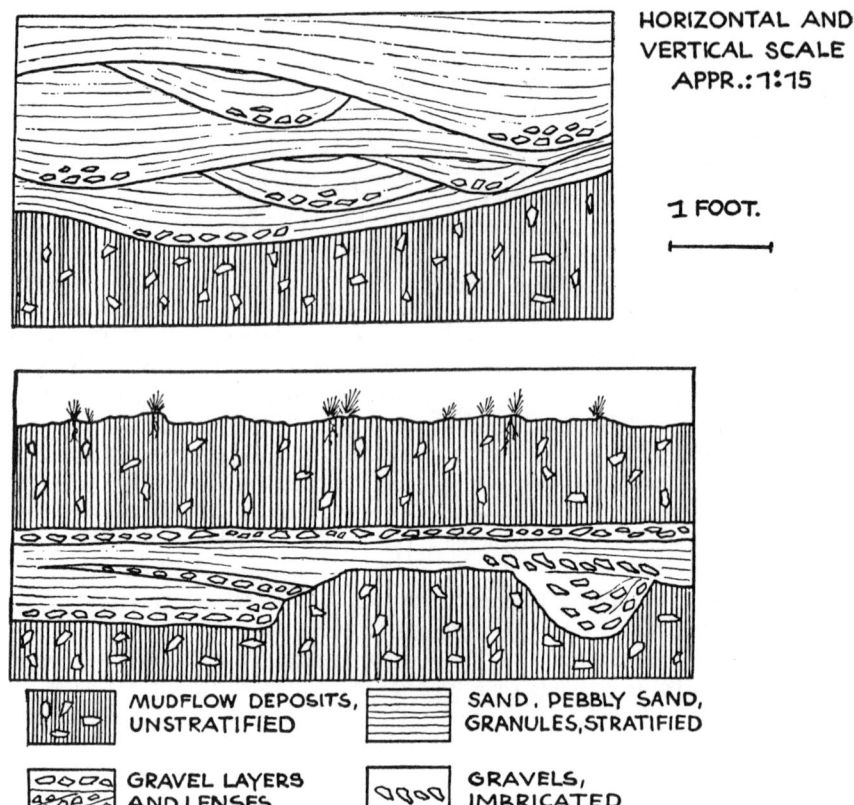

FIGURE 7.—CHARACTERISTIC SEDIMENTARY STRUCTURES IN AN ALLUVIAL FAN OF THE SANTA CATALINA MOUNTAINS, ARIZONA
Main direction of transport from right to left.

Current ripple marks may be present (Twenhofel, 1950, p. 71). The most common and conspicuous type of cross-stratification is due to the filling of channels that have been cut into the underlying deposits (Fig. 7; Pl. 2, fig. 2).

The thickness of individual strata in fans ranges from a fraction of an inch to 15 or 20

Individual mudflow deposits range in thickness from 1 foot on small fans to 15 or 20 feet on large fans. Alternating mudflow deposits and series of stream deposits commonly are of comparable thickness.

Particles in mudflow deposits frequently stand on edge, and in all other normal positions

---

PLATE 1.—ALLUVIAL FANS
FIGURE 1.—COMPOUND ALLUVIAL FAN ON THE SOUTH FLANK OF THE SANTA CATALINA MOUNTAINS, ARIZONA
FIGURE 2.—SMALL ALLUVIAL FAN WITH TELESCOPE STRUCTURE, BLACK HILLS, ARIZONA
FIGURE 3.—SECONDARY ALLUVIAL FANS (S) AT THE BASE OF PRIMARY FANS (P), SANTA CATALINA MOUNTAINS, ARIZONA

FIGURE 1

FIGURE 2

FIGURE 3

ALLUVIAL FANS

FIGURE 1

FIGURE 2

FIGURE 3

ALLUVIAL-FAN DEPOSITS

of repose (Fig. 7; Pl. 2, fig. 1) they do not exhibit any preferred orientation. Imbrication is pronounced in deposits laid down by streams or moderate streamfloods wherever the shape of the particles allows such development. The gneissic bedrock of the Santa Catalina Mountains, Arizona, tends to break down to platy or disc-shaped fragments; therefore imbrication is distinct in certain fan deposits at the base of these mountains (Fig. 7). The disc-shaped fragments are inclined toward the site of derivation as might be expected within a fluviatile environment. The angle of inclination reaches a maximum of 30° among particles of pebble and cobble size; boulders, 100 to 200 cm in diameter, have no inclination greater than 25° (Pl. 2, fig. 3).

Stream and streamflood deposits are rudely lenticular. They are deposited in channels and follow in their extent the winding courses of the former channels. Sheetflood deposits, in contrast, are laid down in definite sheets. They terminate, however, with sharp, abrupt edges (Chawner, 1935, p. 256).

Organic Contents

Fossils are rare in the deposits of an alluvial fan, but there may be local occurrences of abundant fossils due to sudden floods which caused wholesale killing and rapid burial (Twenhofel, 1950, p. 71). Fossil animals were not found within the alluvial fans studied by the writer. Fossil plants or plant fragments, however, appear to be abundant in certain fan deposits.

ANCIENT ALLUVIAL-FAN DEPOSITS

Occurrence

Numerous occurrences of ancient alluvial-fan deposits are listed by Twenhofel (1950, p. 72). According to him, the earliest recorded alluvial-fan deposits are represented by some of the coarse sediments of the Huronian and some of the Keweenawan conglomerates and sandstones of the Lake Superior region. Alluvial-fan deposits have been recognized in many younger sedimentary formations in all parts of the world.

Relative Abundance

The failure to recognize alluvial-fan deposits in the geologic column to the extent they are known among Recent sediments may be explained by the following factors:

(1) The bold relief necessary for the development of alluvial fans was not present in many parts of the geologic past (Barrell, 1925, p. 292).

(2) A combination of bold relief and aridity favorable for the building of alluvial fans was not common in the geologic past (Lawson, 1913, p. 334).

(3) The dominance of land, as in the Quaternary, is exceptional in geologic history (Lawson, 1913, p. 334).

(4) The process of peneplanation eroding the mountains to their roots also affects the alluvial fans adjacent to the mountains and tends to remove them. Only where some unusual feature such as extensive downfaulting, like that involving the Newark series, has protected the alluvial-fan deposits from further erosion may they be preserved.

Alluvial-fan deposits have not always been described as such but have been classified as angular conglomerates or sedimentary breccias (Lawson, 1913, p. 333). In the opinion of the present writer, the misconception common in geologic literature that fan deposits are necessarily unstratified, composed of highly angular particles, poorly sorted, and without distinctive

---

PLATE 2.—ALLUVIAL-FAN DEPOSITS

FIGURE 1.—FANGLOMERATE DEPOSITED BY MUDFLOW
Poor sorting, random orientation of particles. Largest particles in picture are 10 centimeters in diameter. Santa Catalina Mountain fans, Arizona

FIGURE 2.—FANGLOMERATE DEPOSITED BY STREAM OR MODERATE STREAM-FLOOD
The effect of channel cut-and-fill is pronounced. Ruler is 15 centimeters. Main direction of deposition from left to right. Santa Catalina Mountain fans, Arizona

FIGURE 3.—IMBRICATION IN LARGE PARTICLES
Up to 15° centimeters. Direction of transport from right to left. Close to the apex of an alluvial fan of the Santa Catalina Mountains, Arizona

sedimentary structures may in part be responsible for such classification.

Criteria for Recognition of Ancient Alluvial-Fan Deposits

The following criteria can be applied where erosion or tectonic movements have obscured the original surface of a deposit. The recognition of an ancient alluvial-fan deposit must rest on a number of criteria.

The facies of alluvial-fan deposits is coarse detrital as a rule but may be fine detrital under exceptional conditions. Individual particles in these deposits may range from boulder to clay size. A rapid change in the maximum or the mean particle size suggests an origin on an alluvial fan. The presence of sheetflood deposits makes such an assumption more probable. Arkoses and graywackes are the compositional groups in which alluvial-fan deposits commonly range.

The coefficient of sorting (Trask, 1932, p. 70) of fan deposits may range from about 1.5 to 3.0. Fragments in alluvial-fan deposits are angular, subangular, subrounded, and rounded. The change in roundness is rapid close to the area of derivation.

The matrix of fan gravels is sand or mud in size; arkose or graywacke in composition; of primary or secondary origin. The original dip of fan strata may be steep, gentle, or flat. Alluvial-fan deposits have good to poor stratification depending on the relative abundance of unstratified mudflow deposits and stratified stream deposits. Channel cut-and-fill is common in the deposits formed by streams and moderate streamfloods. In these sediments imbrication is pronounced; particles are inclined toward the site of derivation; the inclination commonly does not exceed 30°.

The thickness of individual fan strata ranges from a fraction of an inch to 15 or 20 feet. Stream and streamflood deposits follow the winding courses of former channels; sheetflood deposits form definite sheets.

In some places the fanhead area may be in contact with a talus slope, and in these places fan deposits may be expected to interfinger with talus-slope deposits. The following characteristics of talus-slope deposits may help to separate these sediments from the strata of an alluvial fan. Talus-slope deposits are almost devoid of sedimentary structures as deposition takes place by gravitational sliding. Particles are larger than those found in the adjacent fan deposits. Roundness of talus-slope particles is expected to be of a low degree. Platy fragments may be arranged parallel to the former surface of a talus slope whose original dip may range from 10° to 30°.

At the base of a fan the deposits of an alluvial fan may interfinger with the sediments of a floodplain or a playa lake. The deposits of a playa lake are easily distinguished from the fan deposits as they comprise sediments of clay size and chemical deposits both of which are uncommon or absent in alluvial-fan strata. Floodplain deposits have better stratification than those of alluvial fans; mudflow deposits are uncommon in a floodplain environment. Sorting of the deposits and rounding of sedimentary particles are likely to attain better values in a floodplain than on an alluvial fan. Floodplain deposits are laid down at angles commonly lower than 1°.

This discussion offers some suggestions for the recognition of the climatic conditions under which an ancient fan deposit was formed. With increased precipitation, mudflow deposits become progressively less numerous. From findings on recent alluvial fans it appears that no mudflow deposits are expected in alluvial-fan strata if the mean annual precipitation exceeds 20 or 25 inches. If mudflow deposits make up 50 per cent of an alluvial fan, 5 to 15 inches of annual rainfall appears to be a conservative estimate for the climate under which these fans were formed. The types of rainfall and their distribution through the year as well as the relative abundance of vegetation are factors which control the development of certain types of deposits, in addition to the total amount of precipitation.

Abundance of mudflow deposits in an alluvial fan does not necessarily imply the forming of this fan in a tropical or hot, arid region; it may have been formed in zones with moderate temperatures or even in periglacial environment. Conspicuous development of mudflow deposits may be expected especially along the margins of glaciated areas where they is no movement of water during the colder season and a sudden release of abundant water with the beginning of a warmer season. Fragments in these deposits probably bear ample evidence in their surface texture as to their source within the glaciated area.

Humid conditions during the formation of an ancient alluvial fan may be assumed if the composition of the deposits of the fan is considerably different from that of the parent rock indicating a predominance of chemical over mechanical weathering.

References Cited

Barrell, J., 1925, Marine and terrestrial conglomerates: Geol. Soc. Am. Bull., v. 36, p. 279–342.
Blackwelder, E., 1931, Desert plains: Jour. Geology, v. 39, p. 133–140.
Blissenbach, Erich, 1952, Geology of the Aubrey Valley, south of the Hualpai Indian Reservation, Northwest Arizona: Plateau, v. 24, no. 4, p. 119–127.
Buwalda, J. F., 1951, Transportation of coarse material on alluvial fans (Abstract): Geol. Soc. America Bull., v. 62, p. 1491.
Chawner, W. D., 1935, Alluvial fan flooding: The Montrose, California, flood of 1934: Geog. Rev., v. 25, p. 255–263.
Dana, J. D., 1894, Manual of geology: New York, 4th ed., 1087 pages.
Davis, W. M., 1898, Physical geography: Boston, 428 pages.
—— 1938, Sheetfloods and streamfloods: Geol. Soc. America Bull., v. 49, p. 1337–1416.
Drew, F., 1873, Alluvial and lacustrine deposits and glacial records of the upper Indus basin: Geol. Soc. London Quart. Jour., v. 29, p. 441–471.
Eardley, A. J., 1938, Sediments of Great Salt Lake, Utah: Am. Assoc. Petroleum Geologists Bull., v. 22, p. 1305–1411.
Eckis, R., 1928, Alluvial fans in the Cucamonga District, Southern California: Jour. Geology, v. 36, p. 224–247.
Gilbert, G. K., 1928, Studies of the Basin-Range structure: U. S. Geol. Survey Prof. Paper 153, 92 pages.
Grabau, A. W., 1913, Principles of stratigraphy: New York, A. G. Seiler Co., 1185 pages.

Koons, D., 1948, Geology of the eastern Hualpai Reservation: Plateau, v. 20, no. 4, p. 53–60.
Krumbein, W. C., 1937, Sediments and exponential curves: Jour. Geology, v. 45, no. 6, p. 577–601.
—— 1941, Measurement and geologic significance of shape and roundness of sedimentary particles: Jour. Sed. Petrology, v. 11, no. 2, p. 64–72.
Lahee, F. H., 1941, Field geology: 4th ed., McGraw-Hill Book Co., New York, 853 pages.
Lawson, A. C., 1906, The geomorphogeny of the Techachapi Valley System: Univ. Calif. Pub., Dept. Geology, Bull., v. 4, p. 431–462.
—— 1913, The petrographic designation of alluvial fan formations: Univ. Calif. Pub., Dept. Geology, Bull., v. 7, p. 325–334.
—— 1915, The epigene profiles of the desert: Univ. Calif. Pub., Dept. Geology, Bull., v. 9, p. 23–48.
Longwell, C. R., 1930, Faulted fans west of the Sheep Range, South Nevada: Am. Jour. Sci., 5th ser., v. 20, p. 1–13.
Longwell, C. R., Knopf, A., and Flint, R. F., 1948, Physical geology: 3 rd ed, New York, John Wiley and Sons, 602 pages.
McGee, W. J., 1897, Sheetflood erosion: Geol. Soc. America Bull., v. 8, p. 87–112.
Miller, W. J., 1926, Geology: New York, D. Van Nostrand Co., 555 pages.
Pack, F. J., 1923, Torrential potential of desert waters: Pan-Am. Geologist, v. 40, p. 349–356.
Scott, W. B., 1932, An introduction to geology, vol. 1: New York, Macmillan Co., 604 pages.
Short, M. N., McKee, E. D., and Jahns, Richard, 1953, Hand specimen petrology: New York, John Wiley and Sons, (In press).
Tolman, C. T., 1937, Groundwater: 1st ed. New York, McGraw-Hill Book Co., 593 pages.
Trask, P. D., 1932, Origin and environment of source sediments of petroleum: Gulf Pub. Co., Houston, Texas, 323 pages.
Troxell, H. C., et al., 1942, Floods of March 1938 in southern California: U. S. Geol. Survey Water-Supply Paper 844, 399 pages.
Twenhofel, W. H., 1950, Principles of sedimentation: 2d ed., New York, McGraw-Hill Book Co., 673 pages.
Vaughan, F. E., 1922, Geology of the San Bernardino Mountains north of San Gorgonio Pass, California: Calif. Univ., Dept. Geol. Sci., Bull., v. 13, no. 9, p. 319–411.

Landshut/Bayern, Lebühlstrasse 2, Germany
Manuscript Received by the Secretary of the Society, October 29, 1952

[*Editor's Note:* The figure captions for Figures 1 and 2 have been reversed.]

ERRATUM

Page 188, the second line from the bottom in the righthand column should read: "margins of glaciated areas where there . . . ."

3

Copyright © 1963 by the Association of American Geographers
Reprinted from *Assoc. Am. Geographers Annals* **53**:516–535 (1963)

ORIGIN OF ALLUVIAL FANS, WHITE MOUNTAINS, CALIFORNIA AND NEVADA[1]

CHESTER B. BEATY
Montana State University

AT the base of the flanks of a desert mountain, conditions are propitious for deposition by streams coming from the highlands. There is usually a decrease in gradient at canyon mouths, and all streams undergo a reduction of velocity and transporting power at these sites. Velocity loss is due both to lessened gradient and to decrease in volume by seepage into loose, permeable materials. Deposition is the natural result, and alluvial fans are the landforms produced. Alluvial fans built by concentration and movement of debris in trunk canyons accumulate at the foot of the mountain mass. Their upper portions are separated by interfan depressions, but their lower surfaces coalesce to form piedmont alluvial plains, or bajadas.

This study is a contribution toward an understanding of depositional processes and the landforms produced by them in the southwestern United States. The evidence comes from the White Mountains of California and Nevada, a high-standing desert range at the western margin of the Great Basin. Particular attention was given to the western flank of the mountains, but the larger fans and associated drainage basins on the east side were also investigated. Conclusions reached about alluvial fan formation are thus based upon study of depositional features of diverse morphology and size.

GENERAL SETTING

The White Mountains are the westernmost of the Great Basin ranges in southeastern California and western Nevada (Fig. 1). The northern end of the range is at Montgomery Pass, elevation 7,200 feet. Westgard Pass, also 7,200 feet, marks the southern limit. The length of the highland between the passes is approximately 45 miles.

The northern two-thirds of the range is bounded on the west by the northern extension of Owens Valley, California, designated "Upper Owens Valley" in this study. Fish Lake Valley, Nevada, is immediately east of the range. Between the two lowlands the mountains have a width which increases from about 10 miles in the north to 20 miles in the south.

White Mountain Peak, culminating summit of the range, has an elevation of 14,246 feet, only 250 feet lower than Mount Whitney. Average elevations of the crest vary from 10,000 to approximately 13,000 feet. Upper Owens Valley has elevations ranging from 4,100 feet near Bishop, California, to 6,500 feet at the western foot of Montgomery Pass. Fish Lake Valley has an average elevation of 5,000 feet. Local relief is of the order of 6,000 to 8,000 feet in horizontal distances of 4 to 8 miles, and slopes on the mountain flanks and within the range are correspondingly steep.

The White Mountains consist essentially of a granitic core partially covered by masses of sedimentary, metamorphic, and volcanic rocks. The granitic core underlies the crest in the northern third; in the southern part of the range the summit and western flank are underlain by metamorphic and sedimentary rocks.

Major valleys in the northern end of the mountains are cut primarily in granitic rocks. Slopes are steep and considerable bedrock is exposed. In the southern two-thirds of the range, drainage basins along the west flank are cut dominantly or exclusively in metamorphic and sedimentary rock. Slopes tend to be less rugged, with fewer bedrock exposures and more gentle inclination.

Marked differences exist in the size of sur-

---

[1] This paper is an outgrowth of a contract study of desert flooding made by the writer and J. E. Kesseli for the Research and Engineering Command, U.S. Army Quartermaster Corps; see J. E. Kesseli and C. B. Beaty, *Desert Flood Conditions in the White Mountains of California and Nevada* (Natick, Mass.: QM Research and Engineering Command, U.S. Army, Technical Report EP-108, 1959). During the prosecution of the contract the writer spent 15 months in the White Mountains area in 1956 and 1957. This period of field work was financed by the Quartermaster Corps, and a debt of gratitude to that organization is recognized, particularly to Dr. Peveril Megs, whose interest in desert terrain made possible the original study.

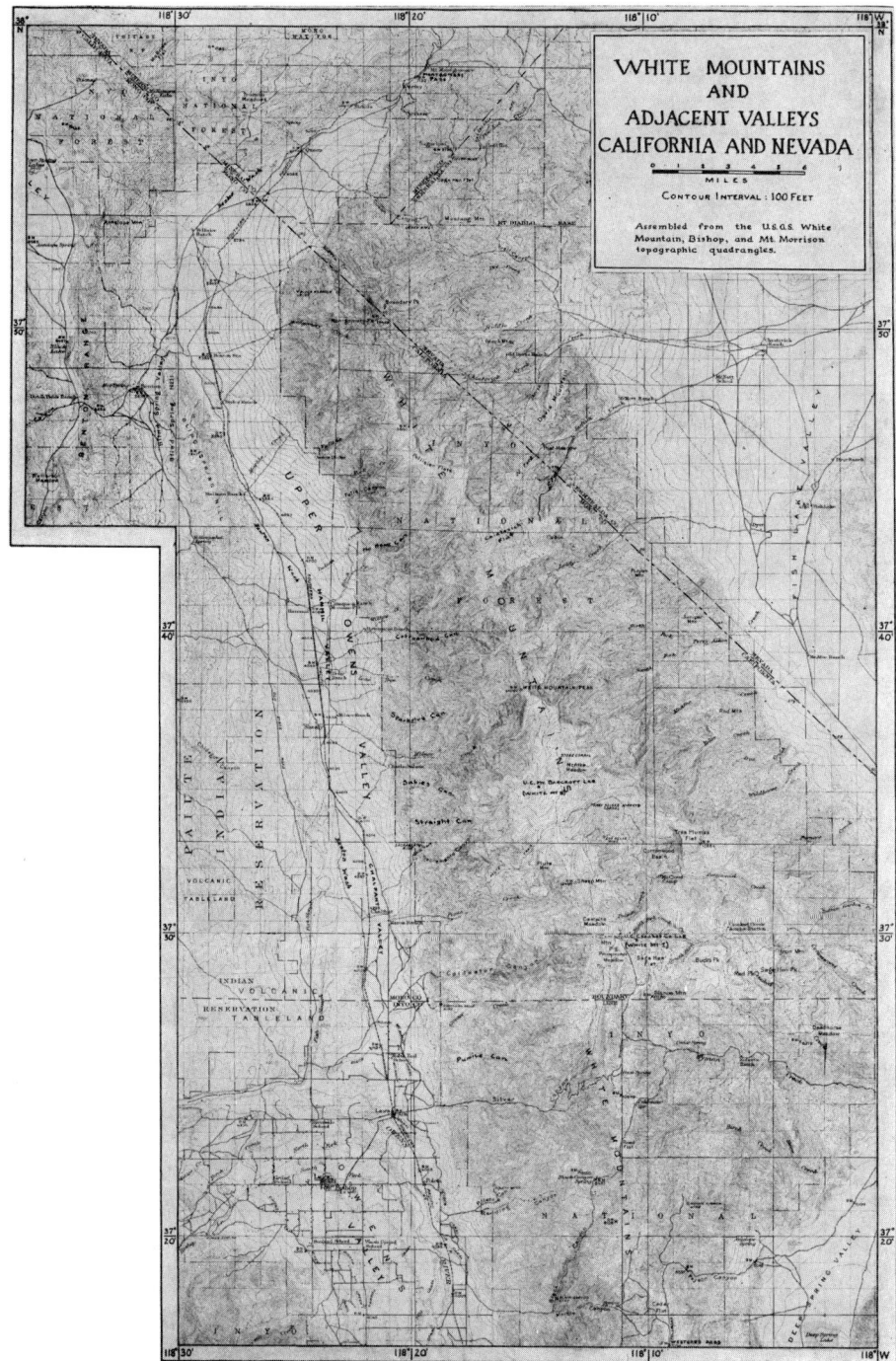

Fig. 1. The White Mountains and adjacent valleys, California and Nevada.

face debris on alluvial fans built by streams coming from areas of different bedrock. Debris in fans below granitic sections of the range consists in the main of sand, cobbles, and numerous larger boulders. The fan surfaces tend to be rough, with occasional boulders of extreme size far from canyon mouths. Fans below metamorphic and sedimentary sectors of the mountains contain a large proportion of debris in the sand, pebble, and cobble ranges, and are characterized by comparatively smooth surfaces with few large boulders.

Winters are wet and summers generally dry in the White Mountains area. Wintertime precipitation is associated with the west-to-east passage of cyclonic storms, which dump a large proportion of their moisture over the Sierra Nevada to the west. As a result, depth of winter precipitation is only moderate, even at higher elevations.

During the summer months occasional thunderstorms occur, especially over the higher parts of the region. Thunderstorm precipitation is sporadic, both in time and place, and reliable records of intensities and absolute totals are scarce. Nevertheless, it is apparent that precipitation amounts from individual summertime cloudbursts in the mountains are much greater than those from wintertime frontal disturbances. The heaviest precipitation recorded anywhere within the general area occurred on July 19, 1955, when more than 8 inches of rain fell in slightly more than 2 hours on Chiatovich Flat, on the east flank of the mountains.[2] Nowhere on valley floors in the region have totals even approaching this figure been observed.

Temperatures are relatively moderate throughout the year in the lowlands. Winter minima are infrequently below zero, and summertime maxima occasionally exceed 100° F. during July and August, but equable sensible temperatures are the rule in all seasons.

Most of the crest of the range has a Steppe Climate–Cold Steppe, BSk, according to the Koeppen classification—while the adjacent valleys are climatically true desert–Cold Desert, BWk.

Three well-defined vegetation associations are found in and adjacent to the White Mountains, while above timberline is a zone of comparatively bare regolith. An association of valley shrubs blankets valley floors and alluvial fans and continues up the mountain flanks to about 6,000 feet. From 6,000 to 9,000 feet a pinyon pine–sagebrush association covers the slopes. Extending from around 9,000 feet to the upper timberline at 10,500 to 12,000 feet is the zone of high-altitude pines. Bristlecone pine (*Pinus aristata*) is the dominant species, but groves or individual specimens of limber pine (*Pinus flexilis*) are found intermixed with the bristlecone pine. Finally, reaching to the very crest of the range is a zone of shattered rock and bedrock outcrops on which a few dwarfed but hardy plants manage to survive.

Although a vegetation cover of one sort or another is found in virtually all parts of the study area, in few places does it appear to exert a significant control over runoff from rains of brief duration and high intensity.

FAN ORIGIN

The alluvial fan, as a distinctive element of the landscape, has long been recognized as a major form in arid and semiarid regions. The literature on fan formation is not voluminous, but a number of significant contributions may be mentioned. The relationship between erosion in highlands and deposition in adjacent lowlands was acknowledged during the 19th century, especially by students of landform evolution and engineers in France and Italy, to whom the work of alpine torrents was particularly impressive. In America, more than 80 years ago, G. K. Gilbert indicated in general terms the processes by which fans are constructed.[3] Noting that drainage systems in the mountains of the southwest become integrated within the highlands, Gilbert wrote: ". . . when water leaves the margin of the rocky mass it is always united into a comparatively small number of streams, and it is by these that the entire volume of detritus [from the range] is deposited. About the mouth of each gorge a symmetric heap of alluvium is produced—a conical mass of low slope, descending equally in all directions

---

[2] Statement in personal interview with Douglas Powell, University of California, Berkeley.

[3] G. K. Gilbert, *Contributions to the History of Lake Bonneville* (U.S. Geological Survey Second Annual Report, 1882), pp. 167–200.

from the point of issue; and the base of each mountain exhibits a series of such alluvial cones, each with its apex at the mouth of a gorge and with its broad base resting upon the adjacent plain or valley."[4]

Another major contribution to an understanding of fan formation was made by A. C. Trowbridge in 1911, with the publication of a valuable study of alluvial deposits in Owens Valley, California.[5] Trowbridge described fans of both the Sierra Nevada and the Inyo Range in the southern part of the valley. He thought their origin—which he attributed to deposition by running water—was clearly revealed by both surface features and internal structure.[6]

A. C. Lawson, in 1915, drew further attention to the areal importance of alluvial deposits in the American Great Basin; he proposed that alluvial fan formations be termed "fanglomerate," and described ancient and recent deposits in central Nevada and elsewhere.[7] Lawson called attention to the general transition in alluvial deposits from coarse blocks near fan apexes to finer material on the lower surfaces, a transition clearly related to deposition by running water. The presence of large boulders far down fan surfaces was ascribed to "exceptional rushes of water from the mountain cañons."[8]

Physiographic effects of major floods in the Wasatch Range of Utah were graphically described by F. J. Pack in 1923,[9] and the addition of debris to alluvial fan surfaces as a result of serious flooding and debris flowage was stressed.

Still another contribution to the literature on fan formation appeared in 1928, with the publication of Eliot Blackwelder's investigation of mudflows in the semiarid environment.[10] Blackwelder noted that mudflow deposits constitute the greater bulk of some alluvial fans,[11] and he concluded that the geologic importance of this particular agent of gradation had been badly slighted in the literature.

In the same year, 1928, R. Eckis described the morphology of alluvial fans of the Cucamonga district of southern California.[12] Eckis was primarily concerned with surface features and possible causes of changes in surface characteristics; his work provides a useful account of the later history of older fans and the several processes by which dissection and deposition may be brought about on fan surfaces.

Finally, in 1954, E. Blissenbach's study of the geology of alluvial fans in the American southwest appeared;[13] the necessary climatic and topographic conditions for fan formation were discussed in detail, and criteria for recognizing and differentiating ancient fan deposits were enumerated. Blissenbach suggested that there is a correlation between aridity of climate and type of deposition on fans; mudflow deposits, he noted, seem to occur predominantly under more arid conditions, while streamflow deposits appear to be more common in a wetter climatic situation.[14]

Against the background of this brief account of earlier reports, the present investigation is set.

The geographic and geologic characteristics of the White Mountains area make it especially appropriate for the study of alluvial fan formation. Steep slopes, sparse vegetation cover, and a climate in which summer thunderstorms are experienced all appear to favor alluvial fan development. Evidence of origin is primarily of two kinds, that represented by surface morphologic features and that revealed in stream cuts on the fans. Few facts concerning composition and structure at depth are available. The field evidence strongly suggests debris-flow deposition as the major element in fan formation.

---

[4] *Ibid.*, p. 184.
[5] Arthur C. Trowbridge, "The Terrestrial Deposits of Owens Valley, California," *Journal of Geology*, Vol. 19 (1911), pp. 706–44.
[6] *Ibid.*, pp. 738–44.
[7] A. C. Lawson, "The Petrographic Designation of Alluvial-Fan Formations," University of California Publications, *Bulletin*, Department of Geology, Vol. 7 (1913), pp. 325–34.
[8] *Ibid.*, p. 327.
[9] F. J. Pack, "Torrential Potential of Desert Waters," *Pan-American Geologist*, Vol. 40 (1923), pp. 349–56.
[10] Eliot Blackwelder, "Mudflow as a Geologic Agent in Semiarid Mountains," *Bulletin*, Geological Society of America, Vol. 39 (1928), pp. 465–84.
[11] *Ibid.*, p. 474.
[12] Rollin Eckis, "Alluvial Fans of the Cucamonga District, Southern California," *Journal of Geology*, Vol. 36 (1928), pp. 224–47.
[13] Erich Blissenbach, "Geology of Alluvial Fans in Semiarid Regions," *Bulletin*, Geological Society of America, Vol. 65 (1954), pp. 175–89.
[14] *Ibid.*, p. 188.

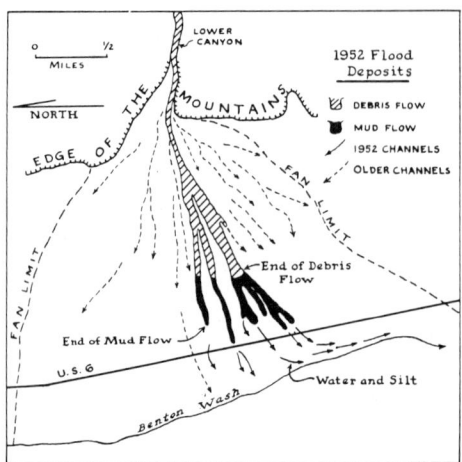

Fig. 2. Alluvial fan of Milner Creek, showing deposits of the 1952 debris flow. Traced from an aerial photograph.

EVIDENCE FROM SURFACE MORPHOLOGY

Recent debris flows.—Major floods and debris flows occurred in 1952 on the fans of Milner Creek, Cottonwood Canyon, and Lone Tree Creek, all on the west flank of the range (Figs. 2, 3, and 4). Two eyewitnesses of the floods and debris flows on the fans of Lone Tree Creek and Cottonwood Canyon were contacted. Their testimony provides invaluable information about the behavior of large rubble flows on alluvial fans. The witnesses are stockmen whose ranches are on fan margins from which almost the whole extent of the fans is visible. The lower end of the flood on the Cottonwood Canyon fan passed over part of the property of one of the witnesses; on the Lone Tree Creek fan the flood passed within a half mile of the ranch.

The following reconstruction of events is possible:[15]

1. Two hours after a heavy thunderstorm in the mountains on the afternoon of July 26, 1952, loud rumbling and roaring noises were heard emanating from the lower canyons of the affected drainages.

2. About 30 minutes later, masses of debris

[15] This account is based upon statements in personal interviews with William Symons, Jr., and Robert Cashbaugh, ranchers in Upper Owens Valley.

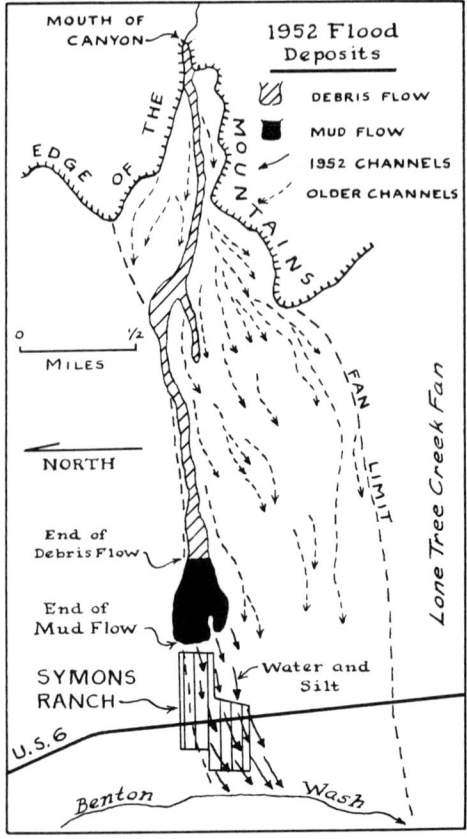

Fig. 3. Alluvial fan of Cottonwood Canyon, showing deposits of the 1952 debris flow. Traced from an aerial photograph.

were noticed advancing downslope on the upper parts of the fans. From a distance of a mile the leading edge of the debris appeared to be a low wall of boulders and thick mud, without visible water.

3. The moving material was accompanied by a rolling cloud of dust, presumably thrown up from the dry fan surface.

4. The debris appeared to advance downslope in a series of waves, or surges, each wave overtaking and submerging the preceding one.

5. The flows were accompanied by noises likened to "the sound of a thousand freight cars bumping together simultaneously."

6. The material moved over the lower part of the Cottonwood Canyon fan "about as

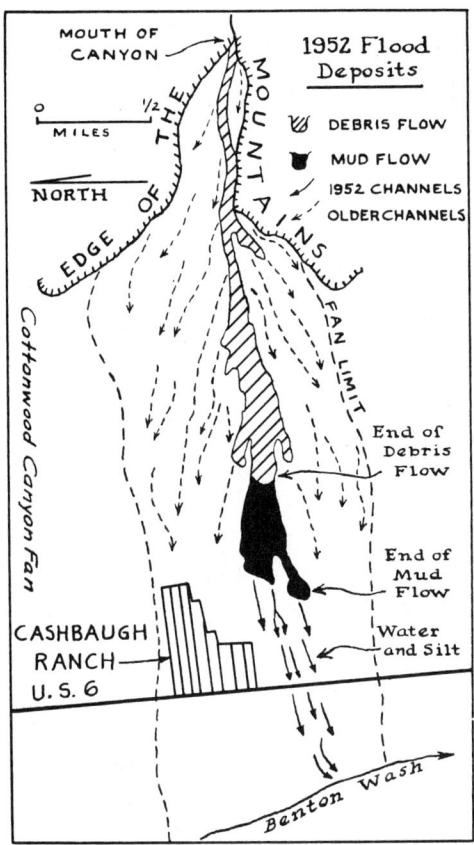

Fig. 4. Alluvial fan of Lone Tree Creek, showing deposits of the 1952 debris flow. Traced from an aerial photograph.

fast as a man can dog-trot," perhaps 400 or 500 feet per minute; it was 1 to 2 feet thick and seemed to be more fluid than when first observed on the upper part of the fan.

7. Flowage of rubble on the fans lasted about 45 minutes to an hour. The lower ends of the debris flows consisted primarily of silts and fine sands with intermixed cobbles and pebbles.

8. After the debris flows had come to rest, high water flow continued in both drainages for 24 to 48 hours.

9. Deposition of 8 to 10 inches of silt on parts of the Symons ranch resulted from this excessive discharge from Cottonwood Canyon.

An addition of surface debris and a deepening of active channels[16] on the fans were significant physiographic consequences of the 1952 floods. Evidence of the former is much the more conspicuous, and the contemporary surface morphology is dominated by the readily recognizable debris-flow deposits (Figs. 5 and 6).

Debris deposition was in the form of long, relatively narrow strips extending radially from apexes toward margins (Figs. 2, 3, and 4). Immediately below canyon mouths the rubble flows were confined to preexistent active channels. The average width of the flows near the mountains was 100 to 150 feet. In a few places debris spilled over channel walls, and short tongues spread diagonally away from the main rubble masses. The largest diverging debris mass left the active channel near the apex of the Lone Tree Creek fan, and, maintaining a width of about 200 feet, advanced nearly 700 feet in a southwesterly direction (Fig. 4). A narrower and longer tongue branched off from the Cottonwood Canyon debris flow on the upper portion of the fan (Fig. 3).

In the middle parts of the fans, where active channels shallow, the debris masses overtopped channel walls and either spread laterally or divided into two or more individual strands. On the Cottonwood Canyon fan (Fig. 3), the increase in width was gradual, except for a sudden widening at a major bend in the active channel. At this point the flow widened to 400 feet, but immediately below, it narrowed to 100 feet and in the next mile gradually spread to 400 feet. On the Lone Tree Creek fan (Fig. 4), the flow increased in width from 200 to 500 feet within less than a quarter mile, maintaining this width for a half mile. Six short and two longer lobes branched off from the main flow in this sector. On the Milner Creek fan (Fig. 2), the flow divided first into two, then three arms; these several tongues collectively cover an area nearly 1 mile in length and attain a maximum width of almost a half mile.

On all three fans the blocky debris flows

[16] The *active channel* on a fan is the channel leading directly from the canyon mouth onto the fan surface. Most White Mountain fans display a radial pattern of channels, the individual strands of which diverge from apexes. The majority of the channels are inactive; the active channel is the feature of most recent origin.

Fig. 5. General view of the Lone Tree Creek fan. Deposits of the 1952 debris flow are conspicuous by virtue of their lighter color. Compare with Figure 4.

Fig. 6. General view of the Cottonwood Canyon alluvial fan. The lighter-colored deposits of the 1952 debris flow make it a readily recognizable element of the contemporary surface morphology. Symons ranch is at the left by the clump of trees at the lower end of the flow. Compare with Figure 3.

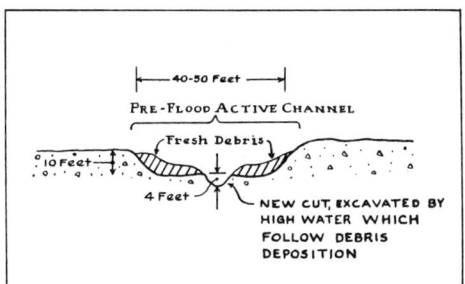

Fig. 7. Cross-sectional sketch of the active channel on the upper Milner Creek fan showing deposits of the 1952 debris flow and channel cutting which followed debris deposition.

ended about two-thirds of the way down the surface. Beyond the poorly defined lower margins of the debris flows proper, larger boulders are scarce. The condition of the deposits suggests that the fluid material here had the consistency of a thick mixture of mud and smaller rubble. In this form the flows advanced another quarter or half mile across the fan margins in numerous narrow tongues (Milner Creek, Fig. 2), or in two broad lobes (Cottonwood Canyon and Lone Tree Creek, Figs. 3 and 4). The change in consistency from debris flow to mudflow was gradual on the fans and much less sudden than is suggested on the maps.

Depth of debris deposited by the flows decreases downslope from fan apexes. On upper surfaces the deposits are 6 to 8 feet thick. The height of channel walls where occasional spilling out took place indicates that the flows attained depths of 15 to 20 feet during the passage of some of the surges described by the eyewitnesses. Within the mountains, in narrow lower trunk canyons, depths of fluid debris and water of 50 to 60 feet were attained in a few sites.

Thickness of the blocky deposits in the middle parts of the fans, where lateral spreading took place, is 1 to 3 feet, and the material left in the mudflow-like deposits on fan margins is thinner still, a foot or less on the average.

Depths of active channels were sufficient to insure confinement of much of the fluid debris for one-third to two-thirds of the radial length of the fans. A deeply incised active channel on the upper part of a fan inhibits flooding and debris deposition there and extends downslope the confined flow of water and rubble. In this way, large boulders may be transported far from canyon mouths on gentle fans. On the three White Mountain fans affected by the floods of 1952, 5-foot quartzite boulders were moved in channels for 1 to 2 miles.

One can only speculate concerning water content of the debris flows. According to eyewitness accounts, the masses appeared to be highly fluid. Yet where smaller tongues of rubble spilled out of main channels they did not flow for any great distance. There is little evidence that surface water drained from the subsidiary lobes after they had come to rest. Apparently they simply stopped moving and "set," much as concrete does when poured. Probably a sizable proportion of the contained water was lost by percolation into the permeable fan surfaces.

Water content of the 1941 Wrightwood, California, mudflow was measured during its occurrence and found to vary from 15 to 30 per cent by weight.[17] It is reasonable to suppose that comparable amounts were present in the White Mountain flows of 1952, since in both cases the described behavior of the material during the floods and its later appearance were similar.

After deposition of the main rubble masses, surges of muddy water came from the canyons. This is evidently a fairly common occurrence, since it has been reported that water floods have often followed debris deposition in other parts of arid western America.[18] The temporarily enlarged White Mountain streams dissected the fresh debris and cut into the older material underlying the new deposits. It is difficult to determine with accuracy how much channel deepening occurred. It is possible that high-water flooding *preceded* the rubble flows, and some channel deepening may have occurred then. If that did happen, however, the eyewitnesses were not in a position to see it.

---

[17] R. P. Sharp and L. H. Nobles, "Mudflow of 1941 at Wrightwood, Southern California," *Bulletin,* Geological Society of America, Vol. 64 (1953), p. 552.

[18] F. J. Pack, "Torrential Potential of Desert Waters," *Pan-American Geologist,* Vol. 40 (1923), p. 353; E. Blackwelder, "Mudflow as a Geologic Agent in Semiarid Mountains," *Bulletin,* Geological Society of America, Vol. 39 (1928), pp. 469–70.

FIG. 8. Lateral ridges along the margins of the upper part of the 1952 debris flow on the Cottonwood Canyon fan. The ridge in the foreground is about 3–4 feet high and 6–7 feet wide at the base. Symons ranch and lower end of flow in background.

The high discharge which *followed* debris deposition dissected the fresh material and cut into the underlying channel floors. Estimates by local residents of the depth of this erosion give an average of 4 to 5 feet in apex regions of the three fans. Field determination of the position of the preflood channel floor revealed approximately this depth of entrenchment on the upper part of the Milner Creek fan (Fig. 7). The excavated material was transported beyond the debris flow and deposited as tongues of silt and sand in shallow channels on the lower fan or in thin sheets of mud on the adjacent valley floor.

The appearance of the debris-flow deposits is distinctive more than a decade after the event. They have been invaded by vegetation along lower margins, but their lighter color makes them still quite conspicuous. In cross section they are concave upward, with well-defined lateral ridges along parts of their margins (Fig. 8). Evidently the axial portions of the flows had greater fluidity and suffered less frictional retardation than the margins, and coarser rubble was left stranded along the edges. Subsequent runoff has followed the axes of the flows and removed a certain amount of debris, accentuating the ridge-like characteristics of parts of the margins.

Lateral ridges on Alaskan mudflows have been called "mudflow levees,"[19] and the term "moraine" has been used to describe similar features on high-altitude flows in Russian Turkestan.[20] Ridges along margins of the White Mountain flows resemble both natural levees and glacial moraines. They are levee-like in that they are represented in plan by paired, linear ridgelets, and they display characteristics of moraines in the steepness of their sides and heterogeneous composition.

---

[19] R. P. Sharp, "Mudflow Levees," *Journal of Geomorphology*, Vol. 5 (1942), pp. 222–27.

[20] W. R. Rickmers, *The Duab of Turkestan: A Physiographic Sketch and Account of Some Travels* (London: Cambridge University Press, 1913), p. 197.

Fig. 9. Looking out of lower Cottonwood Canyon. The approximate position of the old canyon floor is indicated by the base of the trunks of the dead cottonwood trees; the trees have died since removal of debris by the 1952 flood. Cultivated fields on Symons ranch are visible in the background.

The three creeks which flooded severely to produce the debris flows of 1952 are presently flowing on or very close to bedrock throughout most of their length. One of the headwater branches of each drainage system is also flowing on or near bedrock; these appear to have been the upper sources of the debris flows. The affected headwater branch of each stream rises in close proximity to White Mountain Peak, suggesting an extreme localization of the cell of heavy rain responsible for the floods.

In the headwater areas of all three drainages there is no evidence of landsliding as a source of debris during the floods. There is no morphologic evidence at junctions of tributary with trunk canyons to indicate that tributaries contributed notable amounts of rubble to the main canyons. The evidence suggests that virtually all of the debris in these massive flows came from the floors of the trunk canyons and one of the upper branches of each drainage system.

There is excellent evidence of debris removal from the floor of lower Cottonwood Canyon. A stand of dead cottonwood trees covers alluvial and colluvial terraces 15 to 20 feet above the creek bottom in the lower half mile of canyon (Fig. 9). Prior to 1952, these trees were alive and growing on a flat extending across the floor of the lower canyon; the creek then occupied a shallow cut in the center of the flat.[21] During the debris flow of 1952, a body of material roughly 20 to 30 feet wide, 20 feet thick, and about 1,500 to 2,000 feet long was removed from the lower canyon and spread over the alluvial

---

[21] Statement in personal interview with R. Springer, resident of area.

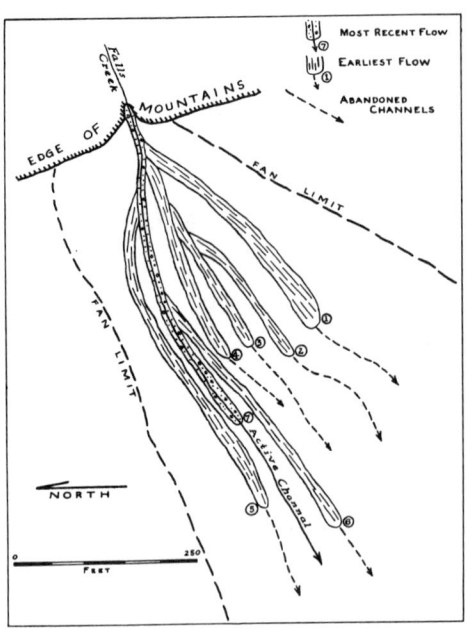

Fig. 10. Sketch map of apex region of Falls Creek fan showing the seven recognizable older debris flows. Traced and enlarged from an aerial photograph.

fan. Its estimated volume is 35,000 to 45,000 cubic yards. Lowering of the stream bed has caused the death of the trees.

Narrow terraces, 5 to 15 feet above the creek bottom, are present for more than a mile above the mouths of Milner and Lone Tree creeks. The freshness of the walls of the cuts indicates a recent removal of unconsolidated fill. It is reasonable to suppose that all or the greater part of the presumed removal in Milner and Lone Tree creek canyons was accomplished by the floods of 1952. The excavated debris makes up the bulk of the deposits on the fans.

Unconsolidated alluvial and colluvial accumulations on trunk canyon floors are believed to be the primary sources of debris shifted from the mountains and deposited on fans by major rubble flows. Canyons which lack alluvial and colluvial deposits are not apt to produce morphologically significant debris flows for many years. A buildup of fresh material on their floors must take place before they can again yield large amounts of rubble. Debris flowage is undoubtedly a highly irregular process, dependent upon the

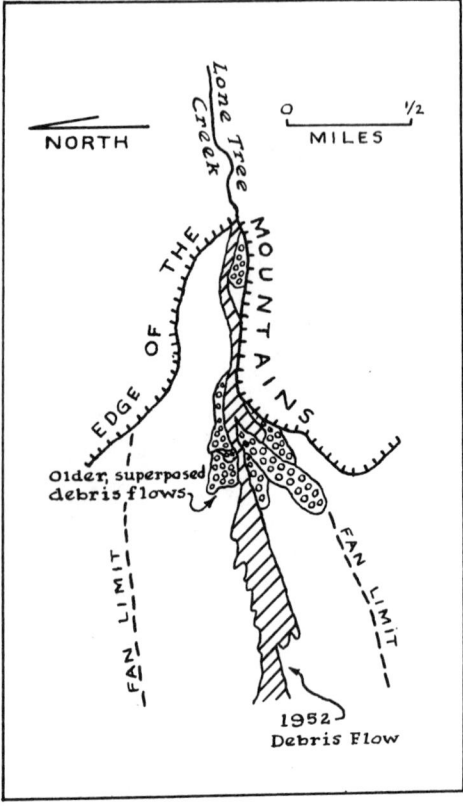

Fig. 11. Upper half of Lone Tree Creek alluvial fan, showing older debris-flow deposits. Traced from an aerial photograph.

chance concentration of a cell of heavy rain in a drainage basin the trunk canyon of which is floored with unconsolidated materials.

Older debris flows.—Although the recent deposits on the alluvial fans of Milner Creek, Cottonwood Canyon, and Lone Tree Creek are the freshest and best preserved in the area, other fans display equally convincing evidence of the major role assumed by debris-flow deposition in their construction. One such fan is that of Falls Creek, on the western flank of the northern part of the range. Seven debris flows can be distinguished on the upper portion of this fan (Fig. 10). All of the deposits are morphologically similar, each consisting of an elongated pair of ridges between which is a water-cut channel. The individual masses of rubble extend

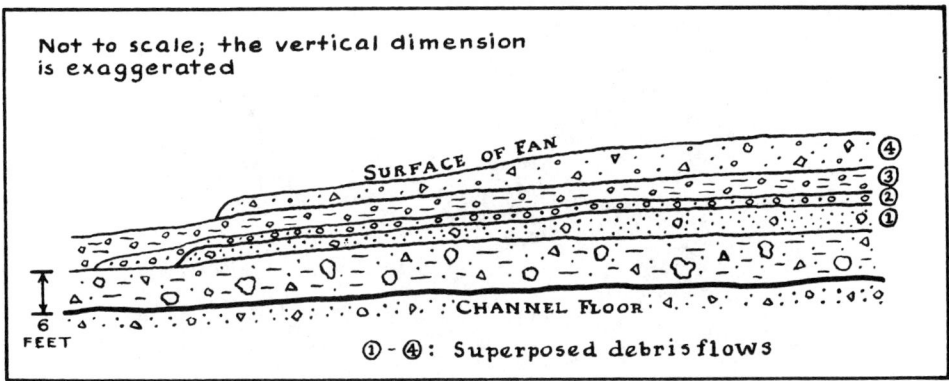

FIG. 12. Radial cross section of part of the upper Lone Tree Creek alluvial fan showing superimposed debris-flow deposits.

downslope in comparatively narrow heaps that end in blunt-nosed snouts. The flows are 300 to 700 feet in length, with widths of 30–40 feet. Thickness ranges from 2 to 6 feet. The central channels have been cut through the deposits and into the older surface to a depth of 1 or 2 feet. The present active channel is in the middle of the most recent flow. What is apparent here is that successive debris flows have added appreciably to the volume of the Falls Creek fan. They have created a corrugated surface on its upper half, and the central channels have enhanced the ruggedness of that surface.

Recognizable older debris flows are also found on the upper part of the Lone Tree Creek fan (Fig. 11). They are superposed and are five to ten times as wide as those on the Falls Creek fan. The 1952 debris flow is the most conspicuous surface element, but the older flows are easy enough to identify, especially on aerial photographs. A radial cross section reveals clearly the structure of the uppermost layers of the fan (Fig. 12). Individual strata of rubble are 3 to 6 feet thick and show a progressive thinning downslope. That these are debris-flow deposits and not stream deposits is indicated by a lack of sorting in the heterogeneous mixture of rubble in each distinct layer. In section they closely resemble the deposits of 1952.

Again on the west side of the range, debris flows occurred on the alluvial fans of Sabies and Straight canyons in the summer of 1918.[22]

[22] Statement in personal interview with Evelyn Buckley, resident of area.

From a distance the deposits are still recognizable by their somewhat lighter color, but at close range they are much harder to distinguish. Surface washing has erased the originally sharp marginal contacts with the underlying material, and vegetation on the deposits resembles that on adjacent parts of the fans. However, the rubble masses can be identified as debris-flow deposits by their distinctive morphologic characteristics, which include well-preserved lateral ridges in places. The lateral ridges of the Straight Canyon flow just inside the mouth of the canyon are particularly striking (Fig. 13).

Channel changes.—On many of the White Mountain fans, channel changes have been brought about by the plugging of active channels by large boulders. Evidence of this process of channel change is particularly good on the Rock Creek fan in Upper Owens Valley. A number of channel-plugging boulders were found, behind which rubble had piled up until channel walls were overtopped and the floods and debris flows swerved to one side or the other and continued down the fan along new courses (Fig. 14). Below the sites at which the large boulders were stranded the abandoned channels are still identifiable.

On some fans on which comparable channel changes have taken place, large boulders are not located at the upper end of the abandoned stretch of channel. Rather often an assortment of boulders will be found near the point of change, some imbedded in channel walls, others lying on the surface. One channel change of this sort on the Marble Creek fan

Fig. 13. Channel and debris-flow ridges in lower Straight Canyon. The channel was cut and the ridges deposited by the debris flow of 1918. Despite their man-made appearance, the ridges are natural features. The channel is 8–10 feet deep; the ridges are 3–5 feet high.

will serve to illustrate the type (Fig. 15). For some reason deposition of the large boulders was favored at this site. A logical cause could have been the formation of a temporary debris dam in the channel down which a flood or rubble flow was coming. A dam of this sort might exist long enough to permit a certain amount of deposition and to divert at least part of a flood or debris flow to a new course, after which it might be destroyed by fresh surges of water or debris. This kind of behavior by a debris flow has been observed on an alluvial fan in southeastern Utah.[23] Evidence of such a course of events would be represented by the existence of a secondary channel diverging from an active channel, with an accumulation of larger boulders nearby. One example has been given, and others are to be seen in the area of study. It seems reasonable to ascribe at least some of the observed channel changes on White Mountain fans to this process.

The point being stressed here is that aggradation throughout the entire length of a channel on a fan is *not* necessary for that channel to be abandoned. Channel changes and abandonments produced by debris dams or stranded boulders are common features on many White Mountain fans. Indeed, it is the frequent changes of channel that account for the numerous unused flood courses on these fans. The rugged microrelief and radial pattern of abandoned channels, the two most outstanding morphologic characteristics of the White Mountain fans, are the direct consequence of frequent channel changes during major debris flows and floods.

Large boulders on fan surfaces.—A notable

[23] Statement in personal interview with Charles B. Hunt, geologist.

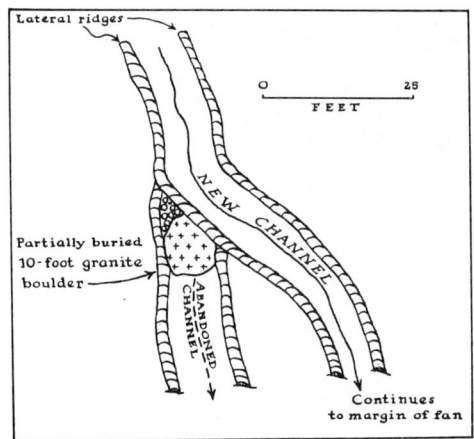

Fig. 14. Sketch of channel-plugging boulder and channel change on the lower Rock Creek fan.

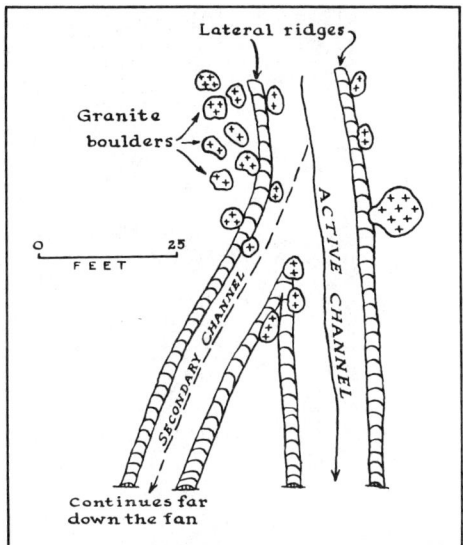

Fig. 15. Sketch of accumulation of granite boulders near point of channel change on the upper Marble Creek fan.

morphologic characteristic of many of the White Mountain fans is the presence on their surfaces of large boulders far from canyon mouths. These are particularly conspicuous in the northern part of Upper Owens Valley. The largest boulder discovered is on the Marble Creek fan (Fig. 16). This massive granite block measures about $30 \times 25 \times 25$ feet and has an estimated weight of 1,500 tons. The location of such huge blocks has stimulated much speculation about how they could have arrived at their present sites. Most are found close to or actually imbedded in the walls of either active or abandoned channels. The fan of Rock Creek in Upper Owens Valley provides an excellent case in point (Fig. 17).

The larger boulders on this fan are almost invariably located in or near the active channel or one of the abandoned channels. The implications are clear: either the boulders were transported to their present positions in preexistent channels in which floods or debris flows were confined, or they were moved by, and as part of, large rubble flows, the central channels of which were cut after the debris had come to rest. Both processes would be effective, and there seems to be no other conceivable way by which transportation of large boulders could be accomplished on the relatively gentle slopes of many fans.

The effect of entrenchment of the active channel on a fan on the carrying capacity of its stream when in flood has been noted elsewhere.[24] There is little doubt that many of the larger blocks have reached locations on lower White Mountain fans as a direct result of confinement of floods or debris flows to deep, narrow channels.

Movement of large boulders in the viscous mass of a major debris flow is also apparently an efficient process of transportation. Evidence is available in the study area and other parts of western America that boulders up to 15 or 20 feet long are readily carried by such agents of gradation.[25] Pieces of rock as large as 10 feet long were moved by the most recent major debris flows on White Mountain fans, and similar flows in the past have obviously contained and transported even larger boulders.

---

[24] A. C. Trowbridge, "The Terrestrial Deposits of Owens Valley, California," *Journal of Geology*, Vol. 19 (1911), pp. 736–40; H. C. Troxell and J. Q. Peterson, *Flood in La Cañada Valley, California* (U.S. Geological Survey Water Supply Paper 796-C, 1937), p. 37; J. P. Buwalda, "Transportation of Coarse Material on Alluvial Fans" (abstract), *Bulletin,* Geological Society of America, Vol. 62 (1951), p. 1497.

[25] Blackwelder, *op. cit.*, p. 473; R. R. Woolley, *Cloudburst Floods in Utah 1850–1938; With a Chapter on Physiographic Features* (U.S. Geological Survey Water Supply Paper 994, 1946), pp. 75–83; Sharp and Nobles, *op. cit.*, pp. 551–53.

Fig. 16. Massive granite boulder on Marble Creek alluvial fan approximately 1½ miles from fan apex. Man standing at base of boulder gives scale. This boulder is the largest of numerous blocks imbedded in the matrix of a huge, ancient debris flow on this particular fan.

Deposition by perennial streams.—Deposition by perennial streams on White Mountain alluvial fans appears to be of minor significance. All the fans have discontinuous or isolated accumulations of water-sorted silt, sand, and fine gravel in sections of their active channels, but the total volume of such materials is limited. Streams in the area appear to accomplish little erosional or depositional work during periods of normal discharge, either within the mountains or on the fans. Their theoretical capacity for load may be high, but the caliber of most of the available debris is such that they are incompetent to move it. During a major flood or debris flow, however, a certain amount of finer material is shifted downstream and out onto the fans along with the coarser rubble. As high water recedes, the streams quickly become clear, and it is during such periods of falling water that deposition of the water-sorted fines occurs.

These deposits are often seen on the beds of normally dry stream cuts which have recently undergone minor flooding. Patches of silt and fine sand are found flanking the channel, usually in the form of isolated, narrow terraces. Deposits of this sort, for example, are found on the alluvial fans of Pumice and Sparkplug canyons along the west flank of the range; both experienced minor flooding in 1952.

Although it is apparent that a limited amount of water-sorted debris is deposited during stages of falling water following major and minor floods, the field evidence supports the conclusion that deposition by perennial streams is of secondary importance in fan construction.

Sheetfloods.—There is also little evidence

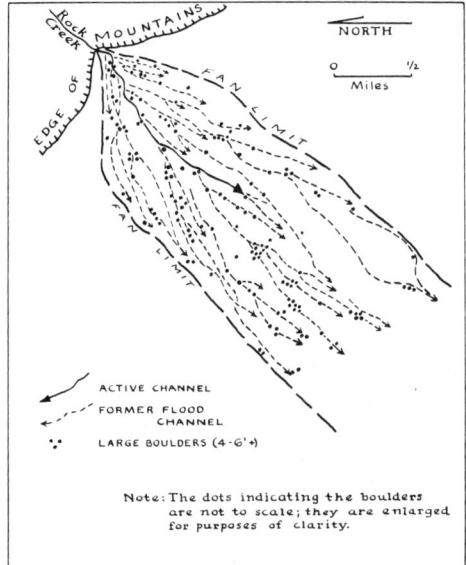

FIG. 17. Map of the Rock Creek fan showing location of large boulders and former flood channels. Traced from an aerial photograph.

of deposition by sheetfloods in the area of study. Ruggedness of surface on the upper parts of most fans precludes even the existence of a *bona fide* sheetflood, unless the sheet of water and debris were to be of an impossible depth. Local relief on the rougher fans is from 5 to 30 feet. Morphologic evidence indicates that large volumes of water or fluid debris emerging from canyons tend either to remain in active channels, or, where a spilling out occurs, to follow older channels. Only on the peripheral parts of the fans are there sufficiently smooth surfaces on which true sheetflooding could take place. Some spreading of more fluid debris and water was noted during the most recent major rubble flows, but there was a pronounced tendency for the material to follow shallow, preexistent channels on the lower fan surfaces. It is therefore concluded that the true sheetflood, as described before the turn of the century by McGee[26] and later by others, notably W. M. Davis,[27] is not an important agent of deposition on alluvial fans flanking the White Mountains.

Thunderstorm rain.—Debris-flow deposition in this part of the Great Basin appears to be intimately associated with summertime thunderstorms. The historical record reveals that nearly three-fourths of the recorded floods in the White Mountains and adjacent desert regions have occurred during the summer months (Fig. 18). All of the large debris flows mentioned in the press or remembered by long-time local residents have followed summertime cloudbursts in the mountains. It seems evident that only as a result of torrential thunderstorm rain can sufficient water accu-

[26] W J McGee, "Sheetflood Erosion," *Bulletin,* Geological Society of America, Vol. 8 (1897), pp. 87–112.

[27] W. M. Davis, "Sheetfloods and Streamfloods," *Bulletin,* Geological Society of America, Vol. 49 (1938), pp. 1337–1416.

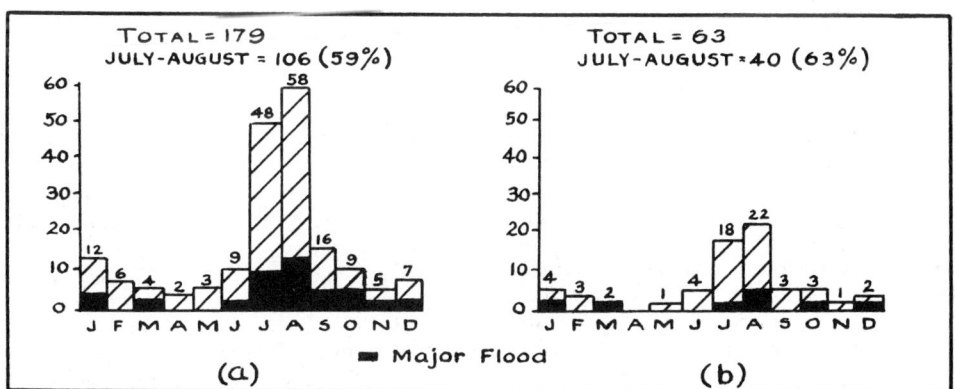

FIG. 18. Occurrence of floods in each month of the year during the period 1872–1957 in (a) part of southwestern Nevada and southeastern California; and (b) the White Mountains alone. Based upon newspaper records and personal accounts of major and minor floods.

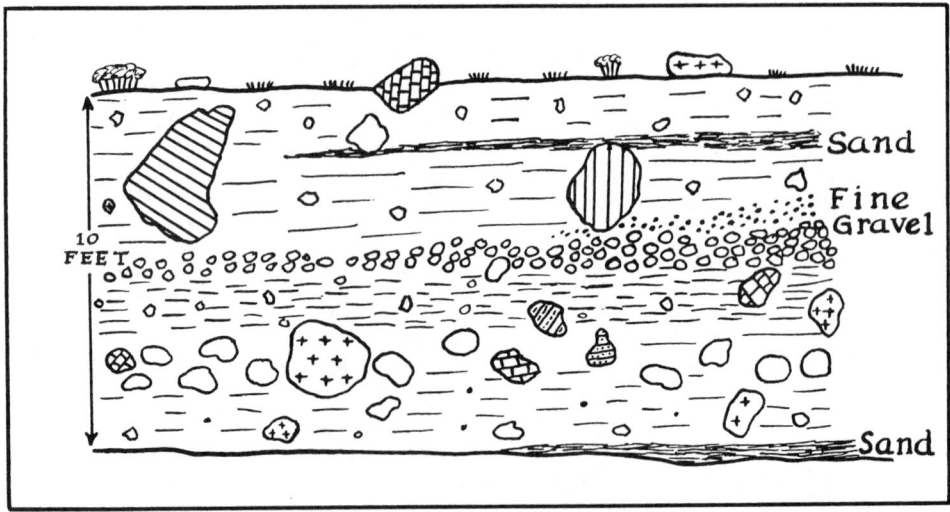

Fig. 19. Wall of active channel near the apex of the Milner Creek fan, showing heterogeneous layering of successive debris-flow deposits. Sketched from a photograph.

mulate rapidly enough in trunk canyons to bring about the true debris or rubble flow. Higher-than-normal runoff from snowmelt or prolonged frontal rain is rarely, if ever, of the necessary intensity.

Evidence From Channel Walls

Good vertical exposures from which some idea of internal construction can be gained are rare on White Mountain alluvial fans. Prolonged weathering and surface washing have operated to round off formerly sharp stream cuts, the walls of most of which are smooth, gentle in slope, and partially or completely covered with vegetation.

Nevertheless, a few stream cuts were examined on both sides of the range, and these furnish information concerning the origin of the upper 15 to 30 feet of the fans on which they were found. One good exposure is seen near the apex of the Milner Creek fan (Fig. 19). A crude layering is visible in the section, and individual strata containing boulders and cobbles are discernible. The material in the cut is typical, unconsolidated fanglomerate.[28] The bedding which can be identified in this and other stream cuts is discontinuous when traced up- or downslope from any given site. Large boulders are irregularly distributed in channel walls, as they are on the surface.

Since the older layers exposed in stream cuts resemble the deposits of the 1952 debris flow in internal structure, it is concluded that they represent the same sort of deposition. Lack of well-defined strata, discontinuous bedding, and an assortment of debris of all sizes are taken to indicate debris- or rubble-flow deposition as the source of most of the material in the upper layers of White Mountain fans.

Evidence From Wells

Few accounts of materials encountered in the process of well drilling are available for the White Mountains area. Those that were found are from wells drilled on peripheral parts of the fans or on adjacent valley flats. No well logs were obtained; the information gained was that remembered by ranchers on whose property the wells had been sunk.

In the vicinity of Benton Station, in the northern end of Upper Owens Valley, a number of wells have been dug into the toe of the Montgomery Creek fan and the adjacent valley floor. None is deeper than 60 feet, since the water table stands at about 50 feet

---

[28] A. C. Lawson, "The Petrographic Designation of Alluvial-Fan Formations," *Bulletin*, University of California Department of Geological Science, Vol. 7 (1913), pp. 325–34.

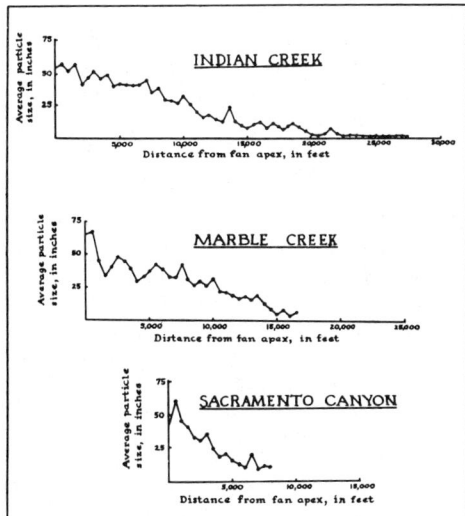

FIG. 20. Plots of average size of ten largest pieces of rock in five-foot squares against distance from apex on three White Mountain alluvial fans.

below the surface in this part of the valley. According to one informant, alternating layers of sand and fine gravel 1 to 2 feet thick have been encountered in all of the wells from the surface to the greatest depths reached.[29] One 55-foot well was investigated by the writer. There appeared to be little change in texture and thickness of the sand and gravel layers from the surface to the bottom. The debris is dominantly of granitic origin, with an average size of pebbles in the gravel layers of slightly more than 1 inch. Three 2-foot boulders were found, one 16 feet below the surface, one at 32 feet, and one near the bottom. Most of the material at depth resembles that at the surface, which is primarily coarse granitic sand with a scattering of small granite pebbles.

Farther south, several wells have been drilled on the floor of Upper Owens Valley between the alluvial fans of Milner and Marble creeks. Depths range from 450 to 1,000 feet. All of the wells penetrated alternating layers of fine gravel, sand, and silt from the surface to the bottom.[30] Depth of

---

[29] Statement in personal interview with Ray Steffen, resident of area.

[30] Statement in personal interviews with William Symons, Jr., rancher in Upper Owens Valley, and George

the water table in this part of Upper Owens Valley is evidently quite variable, and producing wells are pumping from 440, 550, and 800 feet. Whatever the configuration of the bedrock floor of this part of the valley may be, it is buried beneath a considerable accumulation of unconsolidated debris. As is the case in the vicinity of Benton Station, surface debris and detritus at depth appear to be similar.

Information from wells on lower fan surfaces and adjacent valley floors can give but a limited indication of the internal structure of the fans. However, although the suggestion can be only tentative, the similarity between surface debris and material at depth indicates comparability of processes of deposition.

Size Distribution of Surface Materials

The size of the debris out of which the fans have been constructed shows an irregular decrease downslope from apexes. Size distribution of surface debris on three fans was recorded by noting the average diameter of the ten largest pieces of rock within 5-foot squares at 500-foot intervals along straight-line traverses from apexes to perimeters (Fig. 20). The fans of Marble Creek and Sacramento Canyon are on the west side of the range, while that of Indian Creek is in northern Fish Lake Valley. A general decrease in size away from canyon mouths is notable, but the decrease is far from regular. The irregular distribution of larger boulders and cobbles on fan surfaces suggests deposition primarily by debris flows. A progressive sorting and deposition by running water appears not to have taken place.

An especially interesting feature on some of the White Mountain fans is the occasional occurrence, far from the edge of the mountains, of fields of coarse blocks with little finer interstitial material. Such accumulations may be 50 to 100 feet wide and several hundred feet long. The blocks range from fist-sized cobbles to boulders 3 or 4 feet in diameter. Such deposits of coarse rubble are almost invariably found near a channel, either the active one or an abandoned flood course.

Clarkson, former ranch foreman in Upper Owens Valley.

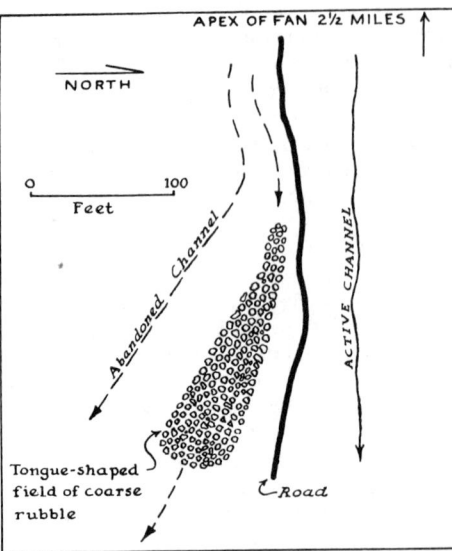

Fig. 21. Block field on the compound alluvial fan of Chiatovich and Middle creeks in northern Fish Lake Valley.

They usually have a shape in plan of a lobe or tongue jutting diagonally away from the channel. One block field on the compound fan of Chiatovich and Middle creeks in northern Fish Lake Valley is located 2½ miles below the apex (Fig. 21). Its general form and the proximity of a nearby abandoned channel suggest that it represents a lobe of debris that branched off from a large rubble flow.

Lobes or tongues split away from the main masses of the most recent major rubble flows on the west flank of the range. Material of all sizes, including 2- to 3-foot boulders, is visible on the surfaces of these most recent flows. It is postulated that the fortuitous occurrence of a heavy rain shortly after a major debris flow had taken place could bring about the washing away of most of the fines in a subsidiary lobe. Such an origin is suggested for the occasional isolated boulder field found on the White Mountain fans.

SUMMARY AND CONCLUSIONS

Morphologic characteristics of the surface and shallower subsurface features of the White Mountain alluvial fans provide a reasonably consistent picture of the manner of their construction. Fan surfaces clearly are covered with numerous ancient and recent debris flows. Frequent channel changes during floods and debris flows are readily detectable and account for the corrugated nature of the upper surfaces of many of the fans. Stream cuts reveal superposed debris-flow deposits. Available well records, which are admittedly few, at least do not seriously negate the idea that debris-flow deposition has been of importance. And finally, the size distribution of surface materials on the fans suggests strongly that deposition by running water has been of only minor significance. The inference is therefore drawn that debris flows have contributed by far the greater bulk of the material in the fans.

Such debris flows have occurred primarily as a result of cloudbursts in the mountains, the surface runoff from which is concentrated in larger trunk canyons. The turbulent, surging mass of water and rubble that bursts from the mouth of a steep canyon may follow any of many possible courses down the surface of the fan. Blocking by large boulders and the formation of temporary debris dams in the active channel cause course changes of successive debris flows, sometimes near the apex of the fan, sometimes farther down the slope. In time, all parts of the surface will have been traversed by and built up by a series of superimposed debris flows. A cut into the fan parallel to the mountain front ought to reveal a series of distinctive individual deposits, long and comparatively narrow in the radial direction, thick and stubby in transverse section.

An internal structure of this sort was envisioned many years ago by C. E. Dutton, one of the pioneer students of desert landforms. Describing alluvial fans in the semiarid Colorado Plateau country, he wrote: "In the vicissitudes to which a stream . . . is subject it occasionally happens that indirect causes have set it at work cutting into its cone; dissecting it, so to speak, by a deep cut and laying bare its anatomy. Our surprise is often great at finding the cone wonderfully well stratified, but in a peculiar way. The most perfect stratification is present when the dissecting cut is made radially. But when a cut transverse to the radius is made by excavation of another stream, the stratification, though still conspicuous, is much less uniform and harmonious. The cone appears to be built of long radial or sectorial slabs superposed like a series of shingles or thatches."[31] The structure of the White Mountain fans, so far as can be ascertained and as described above, appears closely to accord with Dutton's description.

Alluvial fans of the White Mountains, and, by analogy, those of other Great Basin ranges, appear not to have been built by the ordinary processes of stream deposition. Instead, spectacular episodes of debris-flow deposition have been irregularly interspersed with periods of quiescence, during which unconsolidated material has accumulated on canyon floors. From time to time cloudbursts have occurred, and enormous volumes of debris have been shifted from the mountains to the subjacent fans. During intervals between major rubble flows, while mass movements and other slope processes have been transferring debris from canyon walls to trunk canyon floors within the mountains, morphologic changes on the fans have been minor. In the Great Basin environment the spectacular seems to have been the normal, for it is by spectacular processes that a large share of the grand work of gradation has been carried out.

[31] C. E. Dutton, *Report on the Geology of the High Plateaus of Utah* (Washington: U.S. Government Printing Office, 1880), pp. 221–22.

ALLUVIAL-FAN DEPOSITS IN WESTERN FRESNO COUNTY, CALIFORNIA[1]

WILLIAM B. BULL

U.S. Geological Survey, Sacramento, California

ABSTRACT

Deposition on alluvial fans is caused mainly by decrease in depth and velocity of flow, which results from increase in width as a flow spreads out on a fan. The alluvial-fan deposits of western Fresno County, California, are mudflows, water-laid sediments, and intermediate types. Common voids in the deposits are intergranular openings between grains that are held in place by a clay bond, bubble cavities, interlaminar openings in thinly laminated sediments, and buried, but unfilled, polygonal cracks.

INTRODUCTION

Alluvial fans are widespread topographic features in the arid and semiarid areas of the world, and about one-fifth of California is covered by alluvial-fan deposits. Ground water in many parts of the western United States is pumped from alluvial-fan deposits. Despite their importance, however, little detailed study has been made of these deposits. The purpose of this article is to provide basic information about the types of deposits of alluvial fans in western Fresno County, Calif.

The area discussed in this article includes about 1,400 square miles of the west side of the San Joaquin Valley and the adjacent Coast Ranges (fig. 1). Between the flood plains of the San Joaquin River and Fresno Slough and the foothills to the southwest is a strip of coalescing alluvial fans 12–19 miles wide. The slope ranges from about 10 feet per mile near the base of the larger fans to about 150 feet per mile on the upper parts of some smaller fans. The larger foothill groups have maximum altitudes of about

[1] Publication authorized by the Director, U.S. Geological Survey. Manuscript received September 18, 1961; revised manuscript received March 26, 1962.

2,700–3,400 feet, but the main Diablo Range has several peaks higher than 5,000 feet. Both the foothill belt and the Diablo Range have a rugged terrain cut by many steep canyons.

Little Panoche, Panoche, and Cantua Creeks are intermittent streams that flow along their entire length only during, or shortly after the winter rainy season. The streams heading in the foothill belt are ephemeral and flow only as a direct consequence of a particular storm. The characteristics of the flow range from clear water to viscous mud, and are determined by the intensity and duration of rainfall, amount of vegetative cover, rock type, and slopes of the drainage basin.

The average annual rainfall on most of the Diablo Range is 8–20 inches. The Diablo Range intercepts much of the rain thus creating a rain shadow on the foothill belt and the west side of the valley.

All the drainage basins studied except the basin of Little Panoche Creek are underlain chiefly by marine sedimentary rocks of Cretaceous and Tertiary age. The basins that head in the main Diablo Range are underlain in part by sandstone, shale, and siltstone of the Franciscan formation and serpentinized ultrabasic rocks. Cretaceous and Tertiary marine rocks are mainly mudstone, shale, and sandstone, but concretionary sandstone and conglomerate occur in some drainage basins.

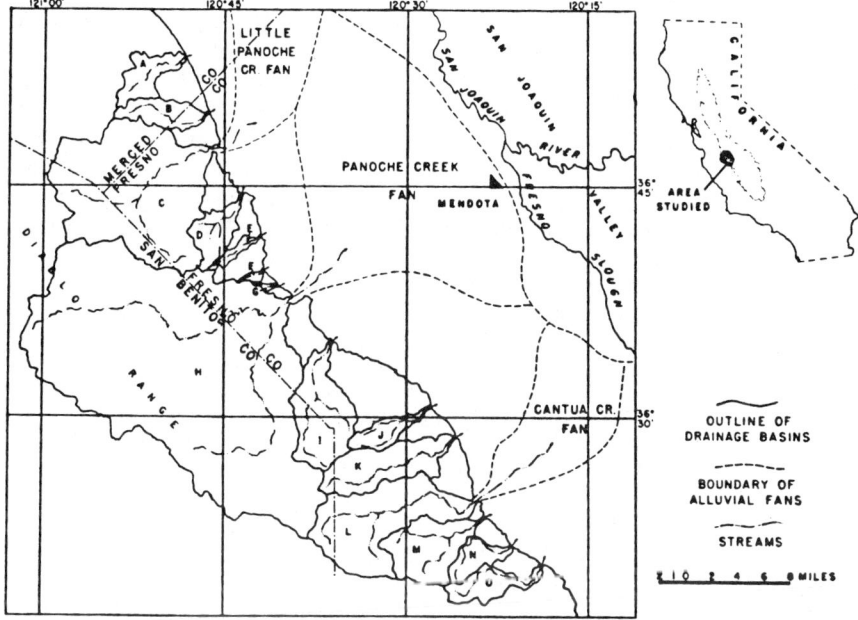

Fig. 1.—Larger alluvial fans and drainage basins studies in western Fresno County, California. Drainage basins: J, Arroyo Ciervo; K, Arroyo Hondo; L, Cantua Creek; E, Capita Canyon; O, Domengine Creek; F, Dosados Canyon; G, Escarpado Canyon; A, Laguna Seca Creek; C, Little Panoche Creek; N, Martinez Creek; D, Moreno Gulch; H, Panoche Creek; M, Salt Creek; L, Tumey Gulch; B, Wildcat Canyon.

ALLUVIAL-FAN DEPOSITS

Deposition on alluvial fans commonly is believed to be caused by a decrease in slope where a stream enters the apex of the fan (Longwell et al., 1948, p. 87–88; Blissenbach, 1954, p. 178). Such a change in stream gradient does not exist in western Fresno County; therefore deposition must be caused by other factors. Deposition on the fans is caused mainly by the decrease in depth and velocity of flow that results from the increase in width as the flow spreads out on the fan.

When water from the flow infiltrates into the ground, the decrease in volume of flow also causes deposition.

Flow in the stream channels of western Fresno County varies from clear water to viscous mud, and a usable classification must include deposits of mudflows, streamflows, and all the intermediate gradations. It should be applicable to both surface and subsurface deposits and should be adaptable to alluvial-fan deposits in other areas.

The sheetflood-streamflood-stream classification (McGee, 1897; Davis, 1938; Blissenbach, 1954, p. 178) of alluvial-fan deposits was not used because it is difficult to infer the shape of a deposit and the place of deposition from subsurface samples.

Mudflow deposits and water-laid sediments are distinct depositional types; however, a sharp classification line cannot be drawn between mud and water, and mudflow deposits and water-laid sediments are separated by deposits whose properties are intermediate between the two. A new name is not given to this third group, and they are referred to simply as deposits that are intermediate between mudflow deposits and water-laid sediments. The mudflow, intermediate, and water-laid classification is independent of the over-all shape of the deposit and the place of deposition, and the scheme can be used to classify subsurface samples, because parameters from grain-size analyses can be used to help describe each type.

MUDFLOW DEPOSITS

Blackwelder (1928, p. 466) defined a mudflow as a type of flow intermediate between a landslide and a waterflood, and it is generally recognized that mudflows follow definite stream channels. The mudflows in western Fresno County usually form during periods of intense rainfall.

The conditions favoring the formation of mudflows are summarized by Rickmers (1913, p. 195), Blackwelder (1928, p. 478–479), and Sharpe (1938, p. 56). These conditions include (1) unconsolidated material that contains enough clay to make it slippery when wet, (2) slopes that are steep enough to induce rapid erosion or sloughing of material, (3) short periods of abundant water, (4) insufficient vegetative protection. All these conditions are present at certain times in the foothill belt of western Fresno County.

Mudflows are a common and frequent type of flow in western Fresno County. Mudflows were deposited on some of the alluvial fans in every year between 1955 and 1962, and aerial photographs taken before 1955 show areas of deposition so recent that vegetation had not grown over the deposits.

Mudflows move in surges down a stream channel (Conway, 1893, p. 292; Blackwelder, 1928; and Sharp and Nobles, 1953). The surges may be so viscous that they come to a halt in the stream channel, or they may be ripples that catch up with the front of a flow. The surges can result from temporary damming of a mudflow by obstructions or constrictions in the channel, or they can be caused by the peak flow from tributaries entering the main channel at different times.

When a mudflow approaches the lower end of the stream channel it may overtop the banks, making lobate tongues of mud that are most common on the outside of the bends in the stream channel. Farther downstream the mudflow overtops both banks and spreads out as a sheetlike deposit on the fan.

The thickness of a mudflow decreases uniformly downslope from the point where it starts to spread out on the alluvial fan. A sequence of mudflows deposited during the 1958 season on the Arroyo Ciervo fan decreases in thickness from 1.6 feet to 0.3 foot in 0.7 mile.

Some idea of the viscosity of a mudflow can be obtained from the orientation and positions of the gravel-size material. A fluid mudflow shows graded bedding and horizontal orientation of the flat gravel fragments; but a viscous mudflow commonly has no graded bedding, and the larger rock fragments may be oriented in vertical planes as well as in other positions. Most of the mudflows studied were the fluid type.

A section of part of a polygonal block of

a mudflow is shown in plate 1, A. The mudflow has no graded bedding, and the disk-shaped gravel fragments are oriented mainly in vertical positions. The sample contained 6 per cent gravel, 40 per cent sand, and 54 per cent silt and clay.

One of the better criteria for identifying mudflows is the presence of abrupt, well-defined margins along the sides and downslope edges of the deposit. Most of the mudflows studied in western Fresno County were $\frac{1}{8}$ to 1 inch thick at their edges.

Most mudflows have been described as having considerable fine material mixed with coarser fragments. The mudflows in western Fresno County contain abundant clay derived from the mudstone and shale of the foothill belt. The clay occurs as a thick film around the sand and gravel grains or as a matrix that partly fills the intergranular voids. The mudflows studied are visibly poorly sorted, containing material that grades from clay to gravel. Mudflows incorporate very light material, such as fragments of vegetation and rabbit pellets, that would be carried away by waterfloods.

Very few grain-size analyses of mudflows have been published. The author made grain-size analyses of fifty mudflow samples collected from the deposits of the 1957 and 1958 seasons, and the cumulative curves of some of these mudflows are shown in figure 2, A. The slope of the curves is gentle for all the samples, which range from clay to predominantly gravel.

The grain-size analyses supply quantitative information for some of the visual observations. For example, the clay content of the fifty samples analyzed ranged from 12 to 76 per cent and averaged 31 per cent.

The sorting of mudflows is a parameter that can aid in the classification of subsurface deposits, such as core-hole samples. The three sorting indexes used were the Trask sorting coefficient, S_0; the quartile deviation, QD_ϕ; and the ϕ standard deviation, σ_ϕ (Inman, 1952). The sorting of the mudflow deposits is summarized in table 1.

Classification is more difficult if features such as lobate tongues and well-defined margins are covered by more recent deposits. The sorting indexes that were used to define the classification types were determined by the samples that had clear depositional characteristics. The lower limits for the sorting indexes of mudflows were set at 5.0 for S_0, 4.1 for σ_ϕ, and 2.3 for QD_ϕ. Quantitative properties such as clay content and sorting may be different for mudflows from different source rocks, but the principle of using sorting to help classify alluvial-fan deposits can be used in other areas. Most of the samples analyzed by the author contain shale fragments and a montmorillonite clay in addition to quartz, feldspar, and other rock and mineral fragments.

Seven samples from the first mudflow of the 1958 season on the Arroyo Ciervo fan were analyzed to evaluate downslope changes in grain size. In $\frac{3}{4}$ mile, the grain size of the coarsest percentile decreases from 18 to 3 mm., and the median grain size decreases from 0.22 to 0.12 mm. The amount of clay is almost the same for most of the samples. These relationships show that the mudflow represents mass flowage in which there is sorting of the coarsest fragments but no sorting of the finer-grained matrix.

WATER-LAID SEDIMENTS

Water-laid sediments are most common on the fans of the intermittent streams, but the deposits of some ephemeral streams consist chiefly of such sediments. Deposition is from flows that contain much less debris than do mudflows. The relative abundance of water allows the finer material to be winnowed from the sand, and the resulting de-

---

PLATE 1

A, Mudflow deposit. Scale 1/100-foot divisions.
B, Water-laid sediment.
C, Intermediate-type deposit. Scale 1/16-inch divisions.

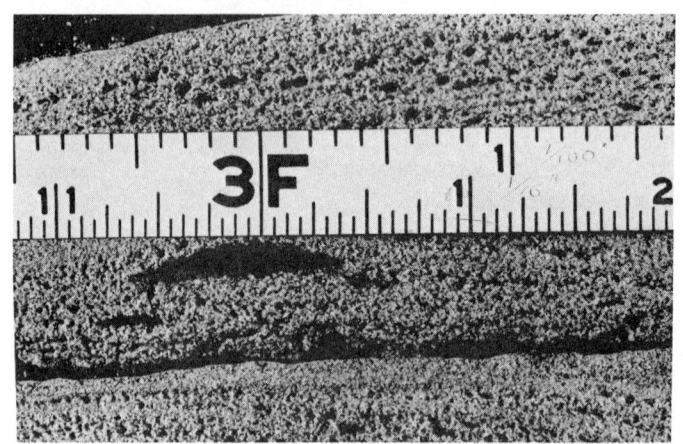

Types of alluvial-fan deposits

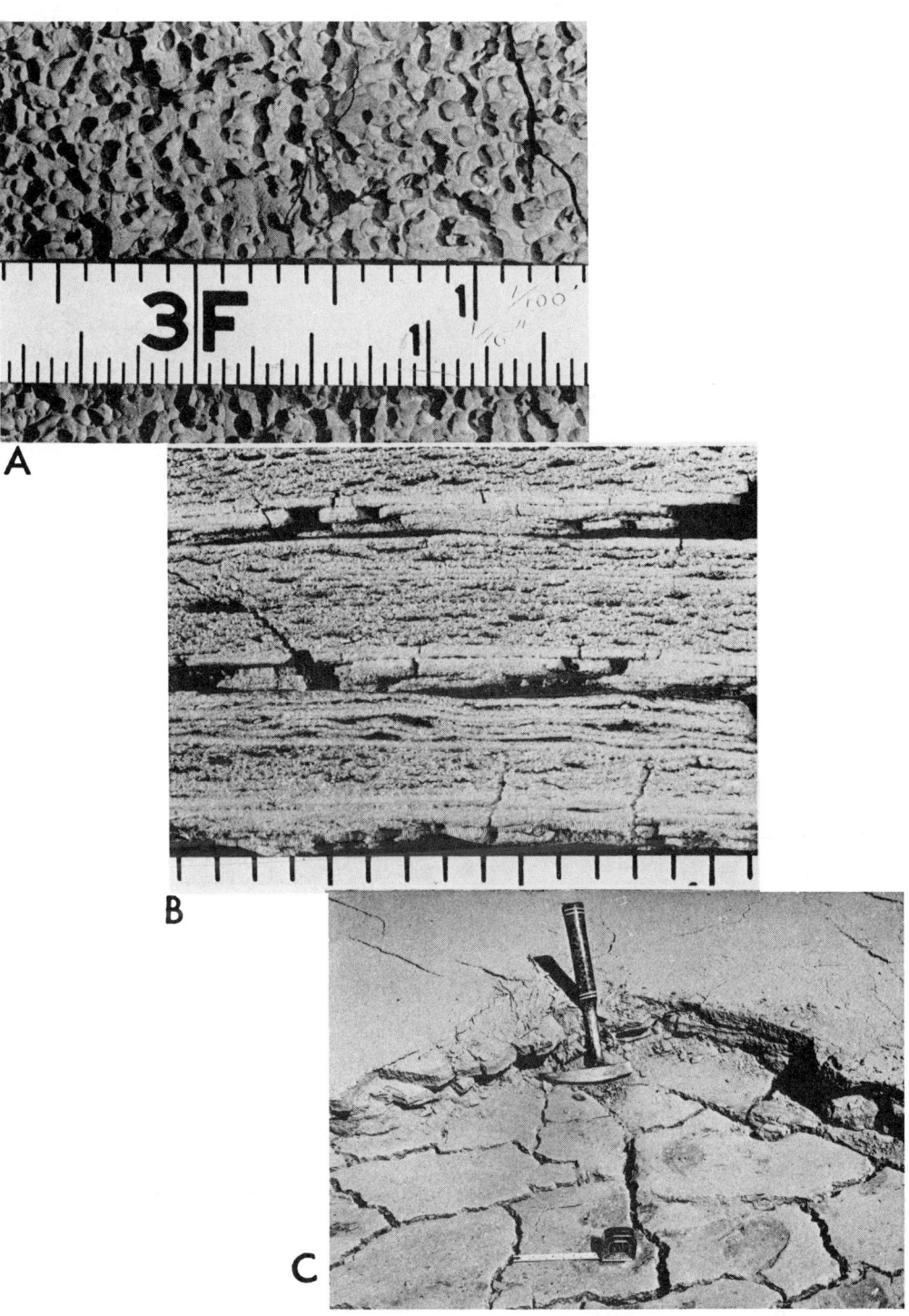

Textural and structural features of alluvial-fan deposits

posits usually are composed of well-sorted sand and clayey silt.

Two types of water-laid sediments occur on the alluvial fans. Most of the water-laid sediments consist of sheets of sand and silt deposited by a network of braided streams. The maximum reported depth of the sheets of water was estimated to be 6 inches. The shallow distributary channels were described as continually filling up with debris and then shifting a short distance to another location.

Such a flow forms a sheetlike deposit of

TABLE 1
DEPOSITIONAL CHARACTERISTICS AND SORTING OF ALLUVIAL-FAN DEPOSITS

DEPOSITIONAL CHARACTERISTICS

Water-laid Sediments	Intermediate Deposits	Mudflow Deposits
No discernible margins; usually clean sand or silt; cross-bedded, laminated, or massive.	No sharply defined margins; clay films around sand grains and lining voids; graded bedding and oriented fragments.	Abrupt, well-defined margins, lobate tongues; clay partly fills intergranular voids; may lack graded bedding and particle orientation.

SORTING INDEXES

	S_0	σ_ϕ	QD_ϕ
	Fifty Mudflow Samples		
Range................	5.0–25.	4.1 –6.2	2.3 –4.7
Average..............	9.7	4.7	3.1
	Sixteen Intermediate-Type Samples		
Range................	2.6– 5.0	3.1 –4.7	1.4 –2.3
Average..............	4.0	3.9	2.0
	Thirty-six Water-laid Samples		
Braided-stream deposits:			
Range................	1.1–2.7	0.48–2.4	0.15–1.4
Average..............	1.5	1.1	0.60
Stream-channel deposits:			
Range................	1.3–4.8	0.82–3.4	0.42–2.3
Average..............	2.3	2.0	1.1
All water-laid sediments:			
Range................	1.1–4.8	0.48–3.4	0.15–2.3
Average..............	1.8	1.4	0.79

The second type consists of sand and gravel deposited in the beds of the main stream channels.

Ranchers and pipeline-maintenance men have described surges of water spreading out over the fans in shallow sheets or bands. sand traversed by shallow channels, which repeatedly divide and join. Most of these braided stream channels are less than a foot deep, and many are less than 4 inches deep. They are separated by low bars or islands. Such a sheet of sand has no distinct margins

---

PLATE 2

A, Bubble cavities in clay.
B, Interlaminar openings in silt. Scale 1/100-foot divisions.
C, Buried, but unfilled, polygonal cracks.

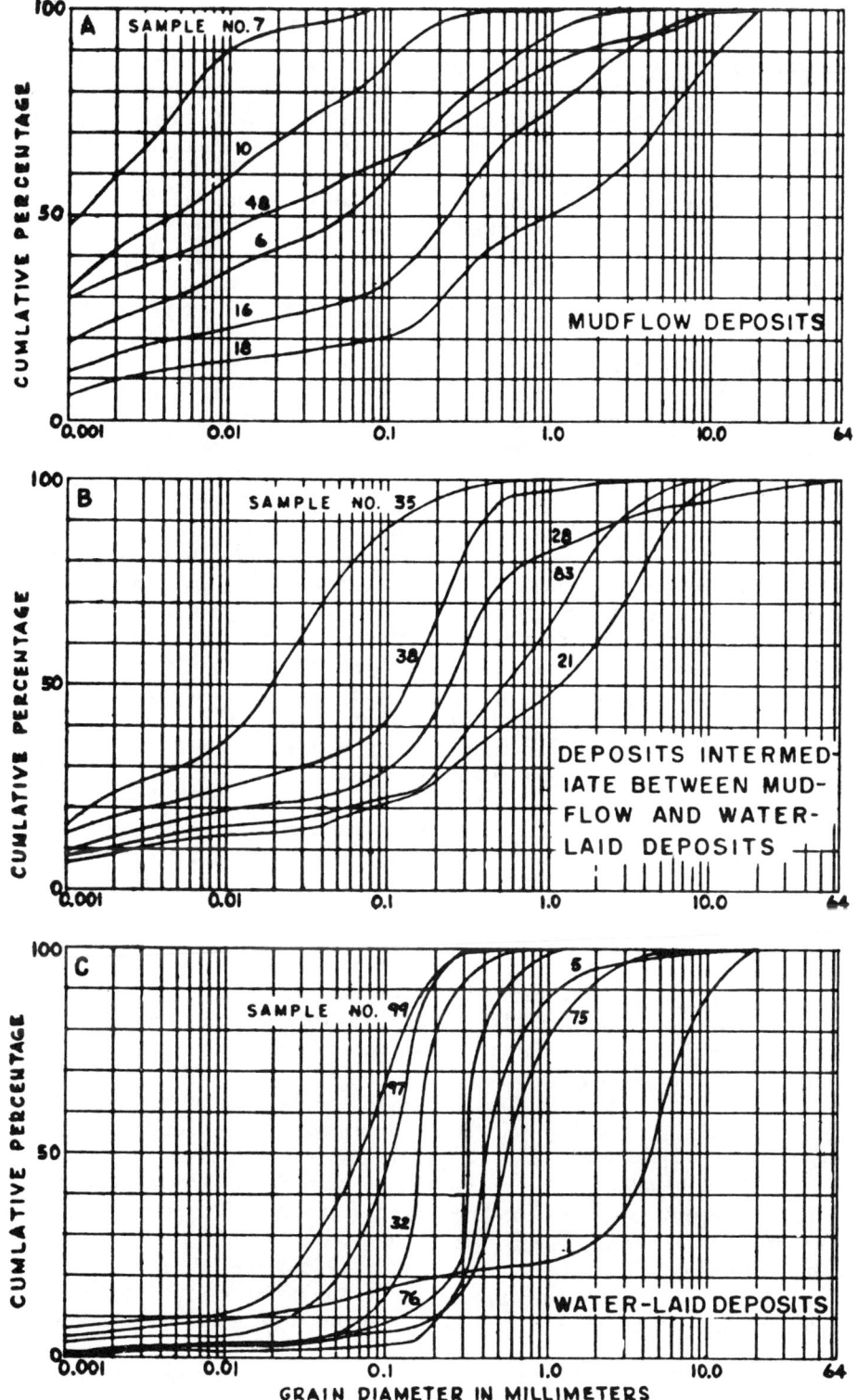

FIG. 2.—Cumulative curves of alluvial-fan deposits

and the silt or sand at the edges usually decreases in thickness until it is a thin film blending with the underlying soil. In general, the water-laid deposits are well sorted, and they may be crossbedded, laminated, or massive. Platy fragments, such as shale chips, have a definite horizontal preferred orientation, or imbrication.

Plate 1, B, shows a water-laid sand that is well sorted and has little clay between or around the grains. The cavities were formed by air entrapped at the time of deposition.

The water-laid sediments in the main stream channels usually are coarser grained and more poorly sorted than the sheetlike water-laid sediments.

The samples for the cumulative curves shown in figure 2, C, range from silty sand to silty gravel, but the central part of each curve has a steep slope. On either side of the central part, a flattening of the curve indicates a coarse or fine "tail" of grain-size distribution. The "S-shape curves" of the water-laid sediments are in marked contrast to the "straight-line" curves of the mudflow deposits. Some of the alluvial-fan sands, such as samples 32 and 76, are as well sorted as beach sands, and even the silty gravel (sample 1) shows remarkably good sorting in its coarsest two-thirds.

Grain-size analyses were made on thirty-six water-laid sediments. Twenty-three of these samples were from sheetlike braided-stream deposits and from sandbars in the main stream channels, and thirteen were from the beds of the main streams and larger distributary channels. The clay content of the water-laid sediments averaged 6 per cent regardless of the place of deposition. In general, the upper limits of sorting for the water-laid sediment classification were 3.0 for S_0, 2.3 for σ_ϕ, and 1.6 for QD_ϕ. Six of the stream-channel deposits have higher sorting indexes, but the poor sorting usually was caused by sand filling the interstices between particles of gravel size. The sorting of the stream deposits is summarized in table 1.

INTERMEDIATE DEPOSITS

The deposits of the intermediate group have characteristics between those of water-laid sediments and mudflow deposits. Most intermediate deposits have the following depositional characteristics. The deposits have no sharply defined margins, and the clayey sediment thins outward until it appears to blend with the soil. The material has a visibly poor degree of sorting, but it does not have the extremely poor sorting of mudflows. The clay occurs as films around the sand grains and as a partial filling in the intergranular voids. Most of the intermediate deposits have gravel-size fragments that are oriented horizontally and concentrated in the bottom part of the bed, causing graded bedding. A sample from an intermediate deposit is shown in plate 1, C.

The cumulative curves of some intermediate deposits are shown in figure 2, B. The general slope of the curves is intermediate between those of the other classifications.

The average clay content of sixteen samples is 17 per cent. The sorting indexes used for the intermediate group are as follows: S_0, 3.0–5.0; σ_ϕ, 2.9–4.1; and QD_ϕ, 1.6–2.3; the average sorting of the intermediate deposits is summarized in table 1.

TEXTURAL AND STRUCTURAL FEATURES

The alluvial-fan deposits in western Fresno County have a variety of textural and structural features. Some of the features already mentioned are graded bedding, laminations, crossbedding, and the orientation or imbrication of the larger fragments. Textural and structural voids, such as intergranular voids, bubble cavities, interlaminar voids, and polygonal cracks, will be discussed in this section. Many voids may be gradational between two or more types.

The most common voids are the intergranular openings. Their size is controlled by the particle size, sorting, and degree and type of packing. In some samples the voids constitute a "normal" porosity; however, much sand has large intergranular voids that result from poor packing of the sand grains. The grains may be held in place by a dry clay bond.

Bubble cavities are formed when air is trapped at the time of deposition, and they occur in all types of alluvial-fan deposits except water-laid gravelly sediments. Bubble cavities in mudflows and water-laid sediments have been mentioned by Sharp and Nobles (1953, p. 554), and Crandell and Waldron (1956, p. 352, 361) describe cavities that may have been formed by air entrained in a volcanic mudflow.

Mudflows pick up air in two ways. Air may be incorporated into mudflows as they move down both tributary ravines and main stream channels and this entrained air actually may decrease the viscosity of the mudflow. The other source of air is the deposits beneath the mudflow, as it spreads out on the alluvial fan. Part of the air in the soil moves upward and becomes trapped by the mudflow to form bubble cavities. Air may be trapped by water-laid sediments if they are deposited on soils that contain air which can move up into the saturated sediments. Bubble cavities are much more common in the water-laid sediments of ephemeral streams than of intermittent streams Probably more air is available for entrapment on the fans of ephemeral streams because the surface deposits tend to dry out more completely between periods of flow. The intermittent streams may flow continuously for several months, keeping the surface deposits saturated.

Bubble cavities are definitely more abundant in mudflow deposits than in water-laid sediments. The cavities in mudflows tend to be of irregular shape, and smaller than a sixteenth of an inch across. Bubble cavities are scattered throughout most mudflow deposits, but they are concentrated adjacent to shale fragments. If part of a mudflow is clay or silty clay, the cavities are larger and more spherical than the cavities in mudflows that contain large amounts of sand and gravel. An unusually large number of bubble cavities is shown in the mudflow sample in plate 2, A.

Bubble cavities in water-laid sediments are spherical or irregular. Most cavities are $\frac{1}{16}-\frac{1}{8}$ inch across, but some are larger than $\frac{1}{2}$ inch. The bubble cavities tend to be irregular in the coarser grained sediments, as is shown in plate 1, B.

Many water-laid sediments are thinly laminated. In general the laminations appear to represent fluctuations or repeated periods of deposition. Each lamina may result from a surge of water that eventually catches up with the main front of water spreading out over the fan. The deposit is crossbedded if the laminations are inclined to the main bedding planes.

Many laminated sediments have irregular disk-shaped openings between the laminations. Most of the interlaminar openings may be due to warping as the deposit dried. The warping probably was caused by an unequal distribution of clay, which caused differential shrinking. Some interlaminar openings are shown in plate 2, B.

Mud cracks that form a polygonal pattern are present in most dry clay-rich deposits. Water-laid clay is not common on the alluvial fans in western Fresno County, but polygonal cracks are a characteristic feature of the mudflow and most of the intermediate-type deposits. The spacing of the cracks increases with an increase in thickness of a deposit, and the width of the cracks increases with an increase in the clay content.

Polygonal cracks were found in many test pits. Some of these cracks were filled with sand or silt, but many were open although they were buried by several feet of deposits. Buried, but unfilled, polygonal cracks are shown in plate 2, C. The tape is on a mudflow that has well-developed polygonal cracks. Several inches of water-laid sand on top of the mudflow were removed.

Open cracks will be filled if sediment-bearing water flows over the cracks, and this may explain why some cracks are filled. The buried open cracks probably have been preserved by the following sequence of events. The mudflow was deposited first, and the sand was deposited before the mudflow dried enough to make a pattern of polygonal cracks. Then the whole deposit dried out after the rainy season. The sand at the surface became dry first, but prominent cracks did not develop because of the low clay content, which was estimated to be about 10 per cent. The mudflow dried next, and a pattern of shrinkage cracks developed as the clay (34 per cent) became drier. The sand did not fall into the open cracks because it had attained enough dry strength to bridge the gap between the polygonal blocks.

ACKNOWLEDGMENTS.—This paper has been prepared from an investigation of deposits susceptible to near-surface subsidence made by the U.S. Geological Survey in co-operation with the California Department of Water Resources. Stanley N. Davis and George A. Thompson, of Stanford University, gave advice on certain aspects of the field and laboratory work. The author wishes to thank Joseph F. Poland, of the U.S. Geological Survey, for critically reviewing this article.

REFERENCES CITED

BLACKWELDER, E., 1928, Mudflow as a geologic agent in semiarid mountains: Geol. Soc. America Bull., v. 39, p. 465–484.

BLISSENBACH, E., 1954, Geology of alluvial fans in semiarid regions: Geol. Soc. America Bull., v. 65, p. 175–189.

CONWAY, W. M., 1893, Exploration in the Mustagh Mountains: Jour. Geography, v. 2, p. 289–303.

CRANDELL, D. R., and WALDRON, H. H., 1956, A recent volcanic mudflow of exceptional dimensions from Mt. Rainier, Washington: Am. Jour. Sci., v. 254, p. 349–362.

DAVIS, W. M., 1938, Sheetfloods and streamfloods: Geol. Soc. America Bull., v. 49, p. 1337–1416.

INMAN, D. L., 1952, Measures for describing the size distribution of sediments: Jour. Sed. Petrology, v. 22, p. 125–145.

LONGWELL, C. R., KNOPF, A., and FLINT, R. F., 1948, Physical geology: New York, John Wiley & Sons, 602 p.

McGEE, W. J., 1897, Sheetflood erosion: Geol. Soc. America Bull., v. 8, p. 87–112.

RICKMERS, W. R., 1913, The Duab of Turkestan: New York, Cambridge Univ. Press, 563 p.

SHARP, R. P., and NOBLES, L. H., 1953, Mudflow of 1941 at Wrightwood, southern California: Geol. Soc. America Bull., v. 64, p. 547–560.

SHARPE, C. F. S., 1938, Landslides and related phenomena: New York, Columbia Univ. Press, 137p.

TRASK, P. D., 1930, Mechanical analyses of sediments by centrifuge: Econ. Geology, v. 25, p. 581–599.

SEDIMENTATION OF AN ALLUVIAL FAN IN SOUTHERN NEVADA[1]

BRIAN J. BLUCK
Department of Geology, University of Glasgow, Scotland

INTRODUCTION

The Arrow Canyon Range, some 60 miles north of Las Vegas, Nevada, has well developed alluvial fans on its eastern side (fig. 1). Although the nothern area is one of external drainage, the extreme southern part is at present one of internal drainage into the Dry Lake Playa. The whole region was considered by Longwell (1945, p. 109) to be one of external drainage during the Pleistocene.

The fan studied here (figs. 1 and 2) shows several stages of growth but with two distinct phases present. The earlier one is a series of coarse and fine conglomerates composed of limestone particles, all of which decline in size away from the source area. It has a well developed profile, is poorly sorted, and locally cemented with calcium carbonate. It was extensively developed at one time, filling its canyon to a variable depth. The surface of the fan is badly weathered in many places, and has secondary drainage developed on its lower part. It is clear from the sorting and lack of well developed bedding that it is chiefly a mudflow type of deposit.

The later phase is a well sorted deposit composed of limestone fragments, and has abundant evidence of it being of stream origin. It has a trench about 40 feet deep in the fan head, and has almost removed the earlier phase from the fan bay. It is one of a series of such deposits, each having its origin in one particular part of the source area. The last stage of deposition came from the canyon shown in figure 2 as A.

MAXIMUM PARTICLE SIZE

The deposits of each of the two phases of this fan were sampled for changes in the maximum

[1] Manuscript received April 25, 1963.

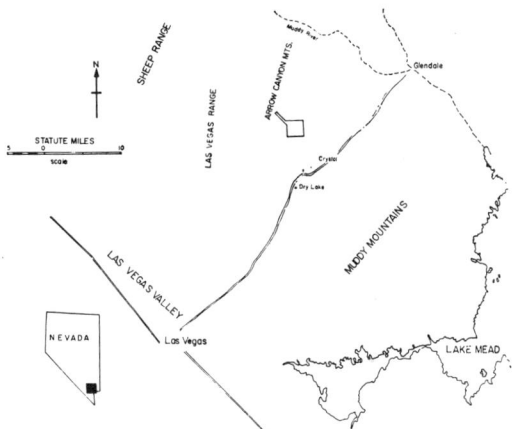

FIG. 1.—Location of area of study and position of the alluvial fan.

particle size. In both, the intermediate axes of the ten largest particles were measured from an area of 25 sq. yards and where possible at intervals of 100 yards. In the earlier phase (hereafter referred to as the mudflow phase) the least weathered part of the fan was chosen where the particles were in the same state and of the same size as originally deposited. Counts were made only on the coarse conglomerates. In the later phase (hereafter referred to as the stream phase) a check was made on the distribution of maximum sizes across the channel when deposits of this phase were found in the fan head area, but such variations in size were found to be very small.

The location of the samples together with the results are shown in figs. 2 and 3. On the mudflow phase there is an exponential decrease in the particle size with a high rate of decline and a low coefficient of correlation, whereas on the stream

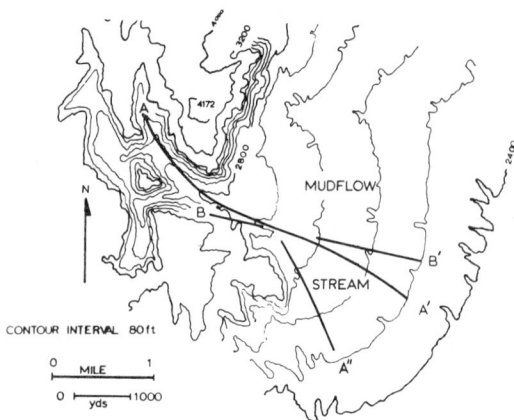

Fig. 2.—Topography of the fan surface and adjacent mountain areas. Along the line A-A' measurements were made on the maximum particle sizes of the stream deposit, and part way along A-A'-A" measurements of sphericity were made on the stream pebbles. Along the line B-B' measurements were made on the maximum particle sizes of the mudflow.

able to it during its reworking of the mudflow.

The mudflow phase consisting of coarse and fine conglomerate would be mixed in the stream during the reworking process and this would bring about a dilution of the larger sized particles in the stream. The amount of fine conglomerate is small, and the effect of this on the stream particle size is considered to be of little importance. An important factor effecting the reduction in particle size between these two deposits may be abrasion. The boulders of the stream phase have the following characteristics: fresh faces, cracking developed over the surface, (see Kuenen, 1956, p. 352) and chipping (Kuenen, 1956, p. 351) is evidenced by numerous conchoidal fracture planes on some particles. Splitting is not considered to be important in the size and weight reduction of pebbles (Kuenen, 1956, p. 364–365), but may be of much greater importance in particles of boulder size, where planes of weakness such as tectonically induced joints and fractures opened by desert weathering would make splitting easier under the impact of a larger weight. Weathering of the stream deposits between the periods of stream activity may also be an important factor, and has been discussed by Rougerie (1951) and Bouillet (1953). It is suggested that weathering of both the stream and the mudflow deposits coupled with splitting of the fragments during

phase there is an exponential decline in particle size, but with a lower rate of decline and a higher coefficient of correlation. The rate of change in particle size decline on the steam phase is lower at about 4000 yards (compare Krumbein, 1942, A, p. 1381–1382, and fig. 11). The particle sizes brought in by the tributary streams may be of different size from those of the main stream at the point of entry, and so change the maximum particle size of the main stream deposit. One such change was recorded where the tributary had a maximum particle size of 95 cm, and the main stream one of 71 cm. The profile of the main stream phase shown in fig. 3, is taken from the topographical map of the area, and is an exponential curve (see also Krumbein, 1937, p. 586–589). The close relationship between the rate of change in slope of an alluvial fan surface, and the rate of change in its maximum particle size has been demonstrated by Blissenbach (1952, 1954, p. 182), and is confirmed in the example studied here.

Although the stream phase has extensively reworked the mudflow, the size of the constituent particles in the stream deposit is much lower and the magnitude of this difference is greater towards the source area, (see fig. 3). There are a number of explanations either singly or in combination which could account for this difference. The surface of the mudflow has suffered notable chemical and physical weathering; boulders are often broken into numerous pieces or are chemically weathered on the undersides. Thus the stream deposit would have lower grain sizes avail-

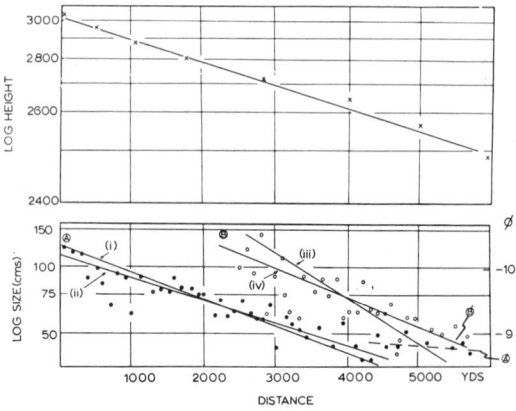

Fig. 3.—The top diagram shows the stream profile along A-A' in fig. 2. The bottom diagram shows the changes in maximum particle size for the stream along the line A-A' in fig. 2, and for the mudflow along the line B-B' in fig. 2. In the stream deposit, for the first 4,500 yds. the correlation coefficient = −0.88 and (i) is $D = 25885 + 2506\phi$ and (ii) is $S\phi = -10.146 + 0.000309 D$. In the mudflow for all data given, the correlation coefficient = −0.76 and (iii) is $D = 18089 + 1483\phi$ and (iv) is $S\phi = -11.057 + 0.00389 D$. The dashed line for the stream deposit denotes the change in the rate of particle size decline.

transportation in the stream phase are the two important causes of the particle size reduction between the mudflow and stream phase.

From these observations, it is concluded that it would be possible for an alluvial fan to undergo reduction in its particle size without moving any additional distance away from its source area.

SPHERICITY

In the collection of pebbles for the study of sphericity a unit grid was used at each locality shown in figure 4, taking roughly 15–20 pebbles from four positions within an area of nine square yards. An analysis of between sample variation was conducted, and is summerized in figure 4. Because sphericity is a function of size (Allen, 1948, p. 312, table 1, and Sneed and Folk, 1958, p. 141), the samples were confined to the 30–45 mm grade. Only the stream deposit was sampled.

Figure 4 shows the areal distributions of the mean sphericity values on the stream deposit. The results are variable and do not show any overall change (see Krumbein 1942, A, p. 1383, fig. 12 and Blissenbach 1954, p. 183, fig. 6). The variation in the standard deviation of the sphericity values are shown in figure 5. There is an increase in the standard deviation along the fan head trench, and a decline towards the base of the fan.

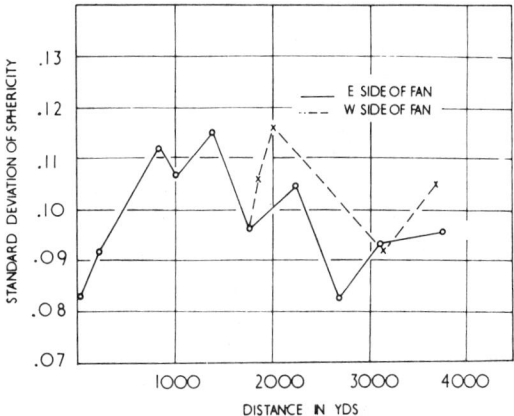

FIG. 5.—The distribution of the standard deviation of the sphericity values.

SHAPE

The same samples used for sphericity measurements were classified according to the Zingg (1935, in Krumbein, 1941) shape classification. The relative abundance of various shapes of pebbles is given in figures 6, 7, 8 and 9, together with graphs showing their rate of frequency change. Blade shaped pebbles show some systematic variations in frequency. There is a general increase in blades in the fan head channel, a decline on the fan proper, and a slight increase towards the base of the fan. These trends are shown in figure 7, and are found on each side of the fan.

Figure 9 shows that round pebbles decline in

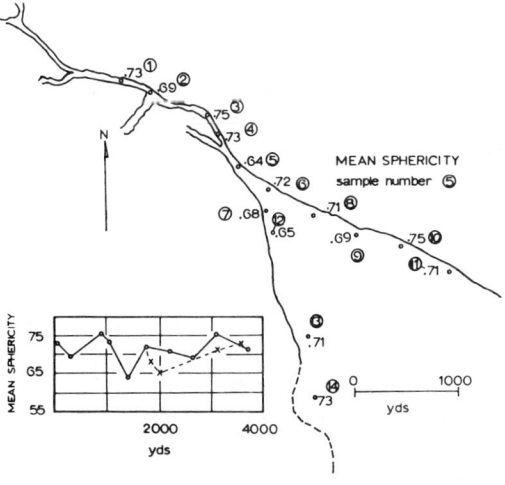

FIG. 4.—The distribution of mean sphericity values in the stream deposit. The dashed line is for the W and the solid for the E side of the fan. The differences of the mean sphericity value between samples, 1–2, 2–3, 3–4, 4–5, 5–6, 6–8, 8–9, 9–10, 10–11, 5–7, 12–13, are significant at the 0.1% level, and between samples 13–14 significant at the 0.5% level. Across the fan the differences of mean sphericity between samples 8–12, 9–13 are significant at the 0.1% level and between samples 10–11 significant at the 0.5% level.

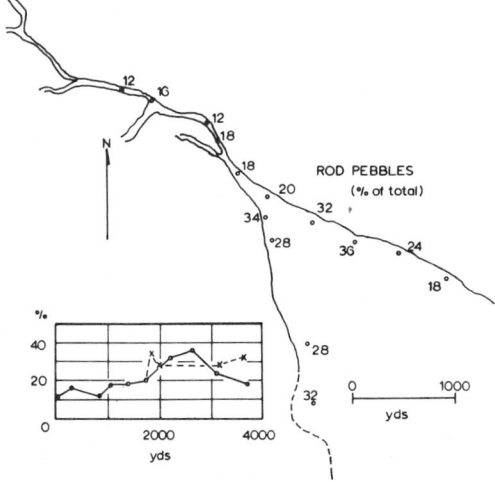

FIG. 6.—The variations in the percentage of rod shaped pebbles. The dashed line in the inset represents the W and the solid the E sides of the fan.

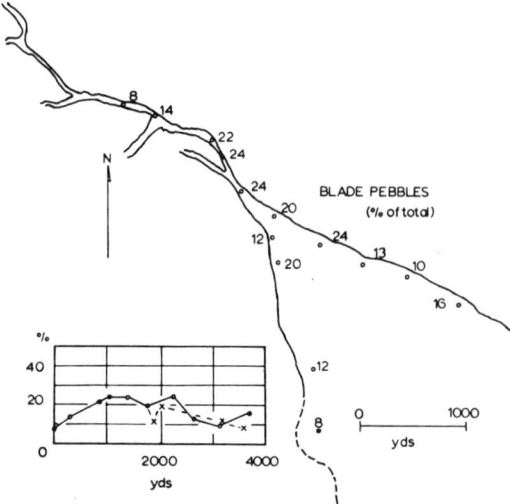

Fig. 7.—The variations in the percentage of blade shaped pebbles. The dashed line in the inset represents the W and the solid the E sides of the fan.

abundance downstream. The rate of decline is far less on the west side of the fan, than on the east side but the east side shows an increase in round pebbles at the fan base.

Discs are the most abundant pebbles in the size range studied here, and also the most variable in frequency. In the fan head trench, there is a rapid decline in their abundance, followed by a rapid increase in abundance that continues onto the fan. On the western side the increase is less rapid than on the east. This general increase takes place at the lower end of the fanhead trench.

There is an increase in rods downstream in the fan head trench, and this increase continues on into the fan proper, but it declines in localities near the fan head on the eastern side, and shows a slight increase on the west. There appears to be a small positive correlation between the frequency of rod and blade shaped pebbles, and between the standard deviation of sphericity and the abundance of round shaped pebbles. A suggestion of a negative correlation exists between the frequency of round and rod shapes.

In evaluating the downstream changes in form it is difficult to discern the effects of type of stream bed and the character of the many episodes of transportation of this deposit. However, certain characteristics are consistent with current ideas on pebble movement. The increase in rods downstream is believed to be due to selective tractive transport, and follows from the experimental investigations of Krumbein (1942, B). Sneed and Folk (1958) also found an increase in particles of this shape downstream in a fluvial environment. Deposition of the rods is most prominent in the area downstream of the fan head trench, where a decline in stream velocity took place. The character of the disc frequency distribution may be related to the selective transport in suspension, and this also follows from the work of Krumbein (1942, B). A similar increase in discs has been recorded by van Andel and others (1954, p. 114) and explained by them as being a result of transport in suspension.

It is probable that the deposition of these pebbles is related to several different phases of

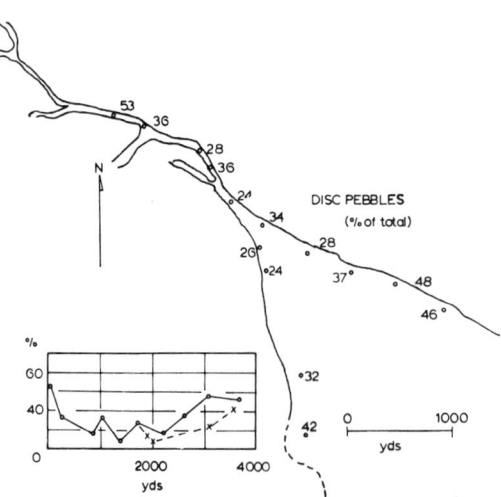

Fig. 8.—The variations in the percentage of disc shaped pebbles. The dashed line in the inset represents the W and the solid the E sides of the fan.

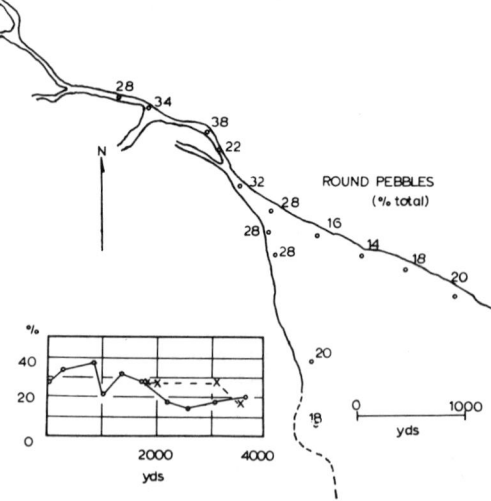

Fig. 9.—The variations in the percentage of round shaped pebbles. The dashed line in the inset represents the W and the solid the E sides of the fan.

transportation, each with its separate characteristics, and each selectively transporting a different group of pebble shapes. Thus the change in the frequency of rod and disc shapes may not be related to the same phase of transportation.

ORIGIN OF THE FAN HEAD TRENCH

To account for the development of the fan head entrenchment, a hypothesis similar to that of Eckis (1928, p. 236–238) is proposed. Figure 10 diagrammatically illustrates the profile of the mudflow as deduced from a study of remnants of its weathered surface in the fan bay and canyon. The gradient in the fan bay and canyon is gentle (the particle size decline was noted to be less rapid here, than on the surface below the fan head trench) and increases downstream. The stream deposit, however, has a profile as shown in figure 10, and is entrenched on the mudflow (fig. 10) with the greatest difference in profile levels being at the fan head. It is suggested that the mudflow, being of a torrential nature, was confined in the canyon area and deposited little sediment there, with the result that the profile was gentle. When emerging from its canyon, it began to deposit its load and build up a steeper gradient. The stream, however, had a lower sediment to water ratio, cut a sinuous type of channel on the mudflow in the fan bay, and was clearly not confined in the canyon in the way suggested by the mudflow. At the canyon mouth it began immediately to cut back a new profile, with the greatest difference in the levels of the two profiles being at this initial point of cutting back. Such conditions appear to have existed on many other fans in the vicinity of the Arrow Canyon Range, and the origin of their fan head trenches may well be similar to that proposed here.

CONCLUSIONS

An exponential rate of particle size decline in both stream and mudflow phases of this fan is brought about by selective sorting and abrasion. The rate of decline of the maximum particle size with distance is more rapid and its correlation coefficient is lower in the mudflow than in the stream deposit. Another particle size decline takes place through time, and is effected by the break down of these particles by weathering and

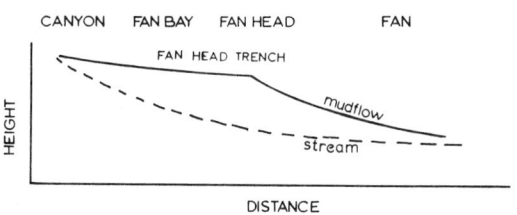

Fig. 10.—This is a diagramatic illustration of the profiles of the stream and mudflow and is based on observations made in the field and on areal photographs.

splitting. Although it is not possible to determine accurately the respective amounts of size reduction brought about by each of these factors, it is clear that size reduction by weathering and mechanical breakage is considerable. The effect of these latter processes declines downstream and in the smaller size grades.

Changes in sphericity are irregular, but a fairly uniform increase in rod shaped pebbles takes place downstream, and is caused by selective transportation in traction. An increase in discs takes place at the base of the fan head trench and is due to movement in suspension. This increase in discs is greatest where there is also a break in the rate of decline of maximum particle size. It is suggested that this change in the rate of particle decline is due to a change from an area where deposition comes predominantly from the tractive to one where it is predominantly from the suspended load.

The change from a phase of mudflows to one of streams effected a change in the profile of the alluvial fan, and resulted in the development of the fan head trench. Unlike Eckis (1928), this fan head entrenchment is only considered "normal" in the sense that a change in the type of depositional agent is normal during alluvial fan building.

ACKNOWLEDGMENTS

The writer wishes to thank Professor R. L Langenheim of the University of Illinois for help in choosing this particular fan for study, and Professor T. N. George of the University of Glasgow for reading the manuscript. The Graduate College of the University of Illinois provided a grant for field work and this is gratefully acknowledged.

REFERENCES

ALLEN, P., 1948, Wealden Petrology: the top Ashdown pebble bed and the top Ashdown sandstone: Quart. Jour. Geol. Soc. London, v. 104, p. 257–321.
ANDEL, TJ. H. VAN., WIGGERS, A. J., AND MAARLEVELD, G., 1954, Roundness and shape of marine gravels from Urk (Netherlands), a comparison of several methods of investigation: Jour. Sedimentary Petrology, v. 24, p. 100–116.
BLISSENBACH, E., 1952, Relation of surface angle distribution to particle size distribution on alluvial fans: Jour. Sedimentary Petrology, v. 22, p. 25–28.

———, 1954, Geology of alluvial fans in semi-arid regions: Geol. Soc. America Bull., v. 65, p. 175–190.
BOUILLET, G., 1953, Usure et fragmentation des calcaire jurassiques de la region de Bourges (Cher): Soc. Geol. France Bull., ser. 6 v. 3, p. 13–22.
ECKIS, R., 1928, Alluvial fans in the Cucamonga District, Southern California: Jour. Geology, v. 36, p. 224–247.
KRUMBEIN, W. C., 1937, Sediments and exponential curves: Jour. Geology, v. 45, p. 577–601.
———, 1941, Measurement and geological significance of shape and roundness of sedimentary particles: Jour. Sedimentary Petrology, v. 11, p. 64–72.
———, 1942A, Flood deposits of Arroyo Secto, Los Angeles County, California: Geol. Soc. America Bull. v. 53, p. 1355–1402.
———, 1942B, Settling velocity and flume behaviour of non-spherical particles: Am. Geophys. Union Trans., p. 621–633.
KUENEN, PH. H., 1956, Experimental abrasion of pebbles. 2. Rolling by currents: Jour. Geology, v. 64, p. 336–368.
LONGWELL, C. R., 1945, Low angle normal faults in the Basin and Range Province: Am. Geophys. Union Trans., v. 26, p. 107–118.
ROUGERIE, G., 1951, A propos de l'etude morphoscopique des galets equatoriaux: Soc. Geol. France Compte Rendu, p. 80–82.
SNEED, E. D., AND FOLK, R. L., 1958, Pebbles in the lower Colorado River, Texas. A study in particle morphogenesis: Jour. Geology, v. 66, p. 114–150.

ERRATUM

Page 397, line 15 in the lefthand column should read: "was conducted, and is summarized in . . . ."

Part II

MODERN ALLUVIAL FAN DEPOSITS

Editor's Comments
on Papers 6 Through 12

6 **WINDER**
  Alluvial Cone Construction by Alpine Mudflow in a Humid Temperate Region

7 **LEGGETT, BROWN, and JOHNSTON**
  Alluvial Fan Formation near Aklavik, Northwest Territories, Canada

8 **DENNY**
  Fans and Pediments

9 **HOOKE**
  Processes on Arid-Region Alluvial Fans

10 **WASSON**
   Intersection Point Deposition on Alluvial Fans: An Australian Example

11 **WASSON**
   Catchment Processes and the Evolution of Alluvial Fans in the Lower Derwent Valley, Tasmania

12 **HOOKE and ROHRER**
   Geometry of Alluvial Fans: Effect of Discharge and Sediment Size

The number of studies of modern alluvial fans has increased greatly in the 1960s, 1970s, and 1980s. There are many significant studies of alluvial fans from many parts of the world, as well as some experimental studies of manmade alluvial fans in the laboratory. I have selected seven papers that provide examples of the variety of modern alluvial fans and the depositional processes that act on them in different climatic settings.

Paper 6 is a detailed study of a debris flow (referred to in the paper as a mudflow) that deposited sediment very rapidly on a small alluvial fan or cone in Alberta. Although Sharpe (1938) had previously divided

Editor's Comments on Papers 6 Through 12

mudflows into semiarid, alpine, and volcanic types, most workers considered debris flows and mudflows to be characteristic of arid and semiarid alluvial fans. Paper 6 is significant because it demonstrates that debris flows are also important features of fans in humid climates. The careful, detailed description of the debris flow and its characteristics provided great impetus to subsequent studies of alluvial fans. Studies of ancient alluvial fan deposits, in particular, had been and continue to be strongly biased toward interpretations of arid-climate conditions, because so many significant studies of modern fans were in the western United States. Winder presently is in the Department of Geology, University of Western Ontario.

Paper 7 is a study of some alluvial fans located north of the Arctic Circle on the northeast flank of the Richardson Mountains in the Northwest Territories of Canada. This paper was selected because it is one of the first important papers on high-latitude alluvial fans. Using surface and subsurface data, Leggett and his colleagues demonstrate some of the remarkable variability of modern fans from different environments. These fans are composed mostly of organic silt derived from Cretaceous sandstone, siltstone, and shale. They are also frozen much of the year and underlain by permafrost, certainly a wholly different situation from that which prevails in arid- and humid-region alluvial fans. Robert Leggett, the senior author of this paper, has retired from the National Research Council and is presently a consulting geologist in Ottawa.

Paper 8 returns us to the arid southwestern United States with a study of alluvial fan and pediment development in the Death Valley and Rio Grande Valley areas. Denny focuses on the effects of fanhead trenching, capture and piracy of stream channels on fans, and the sequential growth and development of alluvial fans, as beautifully demonstrated in some of his figures. He also discusses the development of desert pavement and varnish on abandoned segments of arid-region fans, the relationship between size of fans to size of drainage area, the effects of climatic changes on alluvial fan growth, and the development of pediments. Denny's paper integrates many years of previous study in the southwestern United States (Denny, 1941, 1965; Denny and Drewes, 1965). He has recently retired from the U.S. Geological Survey and lives in New London, New Hampshire.

Paper 9 summarizes detailed studies of three alluvial fans east of the Owens Valley and some laboratory experiments in which small scale-model alluvial fans were built. Many aspects of alluvial fan sedimentation are covered in Hooke's paper, including the concept of fan segmentation, fan geometry, weathering processes, the concept of midfan intersection point deposition, debris-flow and streamflow sedimentation, sieve deposition, fanhead entrenchment, and channel vs. overbank sedimentation. Hooke's detailed maps of the three fans clearly show the spatial distribution of the different types of deposits. The laboratory studies provided the opportunity to study at another scale the various depositional processes and proximal-to-distal changes in various sedimentary parameters.

Paper 10 and Paper 11 provide excellent examples of depositional processes on alluvial fans from two different parts of Australia. Paper 10 discusses intersection-point deposition of sieve lobes during a major rainstorm adjacent to Lake George in New South Wales. The gravelly sieve lobes were deposited downfan from incised channels in a relatively humid climate, a setting quite different from that described in Paper 9. Wasson's careful mapping and detailed study of these deposits contributes much new information about the processes of deposition on small alluvial fans.

Paper 11 discusses alluvial fans constructed during the last glacial period in the Derwent Valley of Tasmania. Both debris-flow and streamflow processes deposited sediment on these fans, which were subsequently covered by eolian sand, silt, and clay. Wasson discusses in some detail the effects of glaciation in the drainage basin on fan sedimentation, and the means of distinguishing glacially deposited sediments from talus accumulations, landslide deposits, and fan deposits. His stratigraphic cross-sections beautifully show the interfingering of streamflow and debris-flow deposits on the fans, followed by incision and dissection of the fans. He was also able to determine a hiatus in fan sedimentation, marked by widespread deposition of a 1-m-thick sandy loam of eolian origin. Wasson completed his Ph.D. thesis on alluvial fans in southeastern Australia and has continued to work on alluvial fan deposits in Australia (Wasson, 1975a, 1975b; 1977, 1979). He is presently at the Australian National University in Canberra.

The final paper on modern alluvial fans, Paper 12 combines field and laboratory studies in an attempt to understand the relation of sediment size and discharge to the slope and shape of alluvial fans. Hooke and Rohrer produced a succession of 17 fans in a box in the laboratory; they varied discharge and sediment size, and built fans of varying shape and size. They compared their results with field data from natural fans in the Death Valley region, and generated some models that characterize the geometry, surface shape, and slope of fans. Hooke has continued to work on various aspects of alluvial fan morphology and sedimentation since completing his thesis and presently teaches at the University of Minnesota (Hooke, 1968a, 1968b, 1972; Hooke and Rohrer, 1977).

REFERENCES

Denny, C. S., 1941, Quaternary Geology of the San Acacia Area, New Mexico, *Jour. Geology* **49:**225-260.

Denny, C. S., 1965, Alluvial Fans in the Death Valley Region, California and Nevada, *U.S. Geol. Survey Prof. Paper 466,* 62p.

Denny, C. S., and H. Drewes, 1965, Geology of the Ash Meadows Quadrangle, Nevada-California, *U.S. Geological Survey Bull. 1181-L,* p. L1-L56.

Hooke, R. LeB., 1968a, Steady-State Relationships on Arid-Region Alluvial Fans in Closed Basins, *Am. Jour. Sci.* **266:**609-629.

Hooke, R. LeB., 1968b, Slopes of Alluvial Fans (abs.), *Geol. Soc. America Spec. Paper 101,* p. 97-98.

Hooke, R. LeB., 1972, Geomorphic Evidence for Late-Wisconsin and Holocene Tectonic Deformation, Death Valley, California, *Geol. Soc. America Bull.* **83:**2073-2098.

Hooke, R. LeB., and W. L. Rohrer, 1977, Relative Erodibility of Source-Rock Types, as Determined from Second-Order Variations in Alluvial-Fan Size, *Geol. Soc. America Bull.* **88:**1177-1182.

Sharpe, C. F. S., 1938, *Landslides and Related Phenomena,* Columbia University Press, New York, 137p.

Wasson, R. J., 1975a, Evolution of Alluvial Fans in Two Areas of Southeastern Australia, Ph.D. thesis, Macquarie University, Sydney, Australia.

Wasson, R. J., 1975b, Alluvial fan, *Australian Geog.* **13**(2):157-159.

Wasson, R. J., 1977, Late-Glacial Alluvial Fan Sedimentation in the Lower Derwent Valley, Tasmania, *Sedimentology* **24:**781-799.

Wasson, R. J., 1979, Sedimentation History of the Mundi Mundi Alluvial Fans, Western New South Wales, *Sed. Geology* **22:**21-51.

6

Copyright © 1965 by the National Research Council of Canada
Reprinted from *Canadian Jour. Earth Sci.* **2**:270–277 (1965)

ALLUVIAL CONE CONSTRUCTION BY ALPINE MUDFLOW IN A HUMID TEMPERATE REGION

C. G. WINDER

Department of Geology, University of Western Ontario, London, Ontario

Received February 25, 1965

ABSTRACT

In a humid, temperate area of central-western Alberta, a mudflow occurred along Hell's Creek, a short tributary of the Smoky River in July, 1962, during a heavy rainstorm. Flows had not occurred during the previous 3 years, even though rains of comparable intensity had fallen. The source area is a steep-sided rock amphitheater, cut in deformed shales and siltstones, and almost devoid of vegetation. The mud moved over a mile in a narrow stream valley and spread across the lower part of an alluvial cone. The exposed lithology of the cone suggests that mudflow is the principal agent in construction of this and other cones of the region. Deposition of the cone has diverted the course of a large river.

INTRODUCTION

Mudflow deposition on alluvial cones in arid and semiarid regions has been described by several authors (Rickmers 1913; Blackwelder 1928; Buwalda 1951; Blissenbach 1954; Beaty 1961; Bull 1962, 1963). Trowbridge (1911, p. 740) described such a deposit in the Owens valley, California, but attributed the movement of huge boulders to running water. Mudflow levees were described by Sharp (1942) in the St. Elias range of the Yukon, an area with aridity of a special type. Sharpe (1938, p. 57) classified mudflows as semiarid, alpine, and volcanic and wrote (p. 59) that the alpine type is less common than the semiarid. The conclusion might be drawn that mudflow deposition is characteristic of semiarid and arid regions. However, Fryxell and Horberg (1941) described mudflows in the Grand Tetons of Wyoming where the precipitation averages 27 in., exceeding the arbitrary 20-in. limit for a semiarid region. They point out that between 8 000 and 10 000 ft, the rainfall increases at the rate of 1 in. per 1 000 ft and cloudiness tends to inhibit the drying of mountains. Rapp (1960) has described in great detail mudflow and other earth-slide phenomena in an area of high precipitation at latitude 68° in northern Norway. The suggestion is made by this author that alpine mudflow in alluvial cones in humid regions is not considered common, largely because the particular type of deposition has not been recognized.

The Smoky River in west central Alberta (Fig. 1) is one of the major drainage channels flowing northward to the Peace River. Near Grande Cache Lakes (Irish and Thorsteinsson 1957) in township 57, range 8 (lat. 53°55′ N., long. 119°09′ W.) the river flows up to 5 mi per hour during flood stage through a steep-sided valley averaging 4 000 ft wide. Glaciation of the valley is indicated by a U-shape, erratics, and small sand and gravel deposits on the valley sides at low elevation. Numerous small streams from 1 to 6 mi in length drain the adjacent hillsides. Most of these tributaries have little effect on the course of the river but some of the *shortest*, such as Hell's Creek, Barret Creek, and

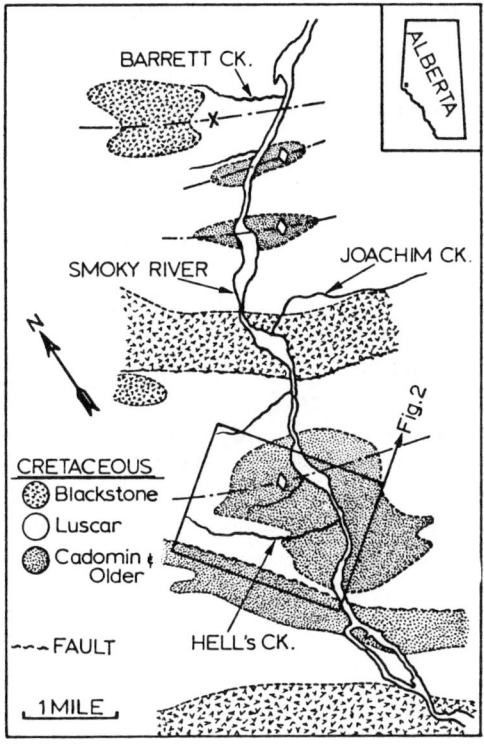

FIG. 1. Index map and general geology along the Smoky River, Township 57-58, Range 8-9, Alberta (Irish 1954 and Irish and Thorsteinsson 1957).

Joachim Creek have an alluvial cone of sufficient size to divert the river against the rock wall on the opposite valley side. During runoff subsequent to a heavy rain, creek water has a high mud content, and small pebbles and cobbles can be heard rumbling along the bottom.

The Smoky River area has a humid temperate climate (Table I) although

TABLE I

Ranges of regional climatic factors covering the Smoky River area
(Atlas of Canada, 1957)

Total mean precipitation	18–20 in. } variability 20%
Total mean snowfall	75–100 in.
Number of days snowfall	60
Number of days of measurable precipitation	120–140
Mean dates of snow cover	October 27 – April 25
Maximum depth of snow	20–30 in.
Mean annual total bright sunshine	1 800–2 000 hours
Mean daily temperatures	January 10–15°
	July 55–60°
Range of minimum temperatures	January −30 to −40°
	July 55–60°
Lowest recorded temperature	−60°
Highest recorded temperature	96°

precipitation and temperature data for a full year are not known for the local area. The total mean precipitation is about 20 in. and measurable precipitation, either rain or snow, falls on the average every third day. The average daily temperature during the summer is below 60 °F and during the winter, well below freezing.

Temperature and rainfall records were kept during the summer months of 1960 and 1961 at Gustav Flats, elevation 3 000 ft, approximately 1 mi south of Hell's Creek. No records were kept in 1962 when the mud flow occurred. Twice daily readings were made on a Taylor maximum–minimum thermometer and daily readings in a standard copper rain gauge. The data (Table II) are within the general ranges given above but it is notable that the temperature is close to or below freezing during 1 day every month. Furthermore, the total rainfall comes in short periods of high precipitation, and commonly with one-half of the total for 1 month falling during 1 day. The local topography has considerable influence on the weather and heavy rain can be concentrated in a relatively small area. The local relief is 4 000 ft and lower temperatures occur at higher elevations. The highest point is 6 986 ft on Mount Hamell just above the headwaters of Hell's Creek. The physiographic and climatic conditions are similar to the Kärkevagge area, Norway (Rapp 1960) but there, average temperatures are 10 to 15° lower and annual precipitation is almost twice as large.

PHYSIOGRAPHY AND GEOLOGY

Hell's Creek is a southeasterly flowing tributary of the Smoky River, draining a large rock amphitheater on the southeast side of Mount Hamell (Fig. 2). The stream course is about 2 mi in length and has an average gradient of a thousand feet per mile. Observations during three summers were that the normal flow of water is small, and shortly after a heavy rain the quantity and velocity of water would not prevent a truck from crossing. The width of the water in the channel is 4 to 6 ft but often less. The creek is usually dry before the end of the summer.

The stream course of Hell's Creek can be divided into three sections. The lower section crosses an alluvial cone 3 000 ft in width and 6 000 ft along the valley with the highest point about 450 ft above river level. The stream flows with a gradient of about 800 ft per mile through a sinuous steep-sided V-shaped valley cut into alluvial material, to an average depth of 25 ft except in the last 400 ft where the sides are low. Numerous abandoned channels up to 4 ft in width and depth radiate in a serpentine fashion from the apex of the cone. A dense cover of conifers and ferns grows on the surface but the "stony" nature of the cone is evident. Both walls of the present stream gulley (Fig. 3) expose a heterogeneous mixture of clay to angular boulders up to 3 ft without any obvious sorting. The largest observed boulder was rounded, about 5 ft in diameter, and in the stream gulley about half way across the cone.

The middle section of the stream flows through a highly sinuous, deep gulch cut in steps and waterfalls through the lower part of the Luscar Formation,

PLATE I

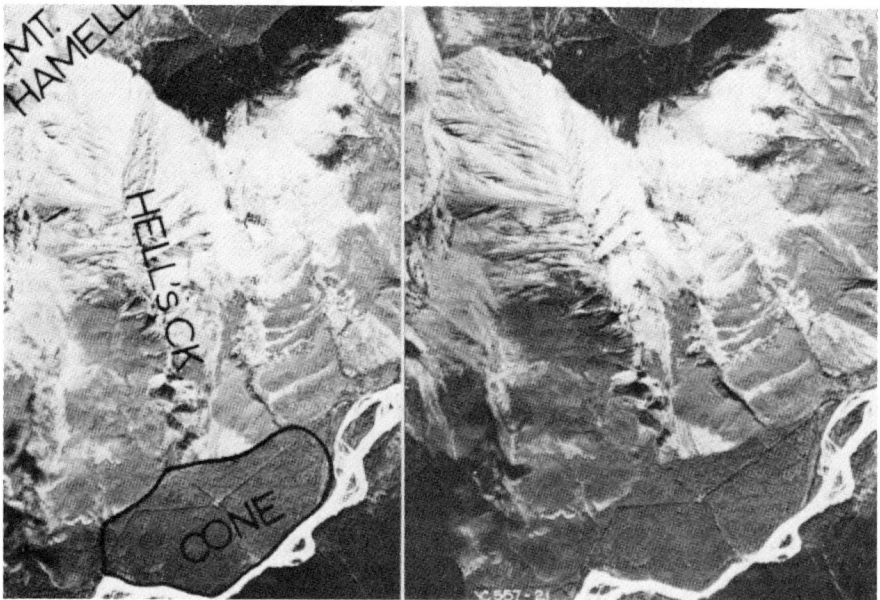

FIG. 2. Stereoscopic air photos of the Smoky River, flowing to the right almost due north, and Hell's Creek flowing from a rock amphitheater on the southeast side of Mount Hamell. The length of the cone from apex to base along Hell's Creek is about 3 000 ft. Height of the cone is about 450 ft. Total maximum visible relief is about 3 000 ft. Photograph taken September 1961.

PLATE II

FIG. 3. Consolidated bouldery clay near the apex of the alluvial cone in the gully of Hell's Creek. Total exposure at this point is about 35 ft.
FIG. 4. Edge of the mud flow across lower road on the Hell's Creek alluvial cone showing the edge concentration of boulders and logs, July 18, 1962.

TABLE II

Temperature and rainfall data, Gustav Flats, Smoky River, Alberta

	June 10 – September 5, 1960				May 26 – September 7, 1961				
	June	July	August	September	May	June	July	August	September
Temperature									
Highest	77	92	93	76	87	90	96	91	60
Lowest	29	28	29	24	29	29	35	34	27
Mean max.	72.6	80	60	64.4	77.2	76.5	75.5	76.2	55.2
Mean min.	38.9	41	40	31.4	36.5	41.3	45.1	42	35.5
Mean	55.8	60.5	54	47.9	56.9	58.9	60.3	59.1	45.4
Rainfall (in.)	3.51	0.28	1.47	0.01	0.29	1.91	2.62	1.91	0.49
Total rainfall for above period	5.27						7.22		

Days of notable rainfall

1960	1961
June 20–0.9	June 12–0.76
June 21–1.59	Aug. 17–0.93
Aug. 1–1.0	Aug. 25–0.76

Lower Cretaceous, consisting of thin-bedded siltstones, shales, some thin sandstones, and coal. The lowest waterfall is over the Cadomin Conglomerate, which forms a razor-back ridge on the north side of this rugged gorge. The upper slopes of the stream valley, particularly on the south side, are covered with conifers.

The upper section of the stream is a deep V-shaped amphitheater of continuous rock exposure without vegetation to impede erosion. The maximum dimensions of the amphitheater are 6 000 ft in length, 3 000 ft across the top, and 2 000 ft deep at the lower end. The volume of material excavated to form the amphitheater is at least five times the volume of the existing alluvial cone at the lower end of the stream. Observations on the quantity of eroded debris remaining temporarily in the amphitheater were not made but the quantity in the canyon, and in the Barrett Creek amphitheater, is considered insignificant. The north wall and the lower part of the south wall are composed of the lower part of the Luscar Formation. The Grand Mountain thrust, cutting through the lower third of the south wall, has raised a continuous exposure of thin bedded siltstones and shales of the Upper Jurassic Nikinassin Formation which, in a general way, have an appearance similar to the Luscar. Structurally, the north wall consists of the south limb of a sharp anticline on the south side of the Campbell or Sterne Creek anticlinorium, a fold with an inverted U-shaped cross section. (Dahlstrom (1960, Fig. 6) has an oblique airphoto of this unique fold and the Hell's Creek amphitheater, but formation labels on the picture are not correct. The Luscar Formation includes the section marked Nikanassin, which is below the upper dotted line tracing the Cadomin Conglomerate.) The south wall is a monoclinal structure, cut by a fault resulting in numerous flexures and distortion of the bedding, which are quite evident in Fig. 2.

OCCURRENCE AND CHARACTER OF MUDFLOW

During the summer of 1959, two roads were built across the fan, a straight one at high level for a seismic profile, and another following the sinuous horse trail near the river. The upper road crossed the 25 ft stream gulley by a fill with a "culvert" of bulldozed trees. The valley above was soon filled with alluvium to the level of the road. For the 10 day period prior to July 17, 1962, some rain fell every day, varying from brief showers to thundershowers. At 7.30 p.m. that evening, two men travelled by truck down the valley along the lower road. At 9.00 p.m. an exceptionally heavy rainfall began and the water level in the tributary streams rose rapidly to a "roaring" level. Intense claps of thunder caused noticeable local vibration. At 10.00 p.m., the two men returned and found both lower and upper roads at the Hell's Creek crossing blocked with a wide expanse of water-saturated conglomeratic mud. The amount of water flowing in Hell's Creek did not prevent their crossing on foot but they sank to their knees in mud in several places.

The mud covered a triangular area with a 200 ft front along the river,

400 ft up the creek channel to an estimated depth of 2 ft (Fig. 4). The mud did not form a continuous sheet but flowed down numerous gullies on the surface of the cone in the vicinity of the stream. The total quantity of mud in this flow was insignificant in comparison to the total volume of the cone. The boulders are angular blocks and slabs of siltstone with rounded edges and corners, presumably as a result of abrasion by transport in the highly viscous medium. Rapp (1960, p. 156) observed that pebbles and boulders in mudflows are commonly oriented transverse to the direction of flow. The cone was extended into the river on a curved front a distance of at least 5 ft. As can be seen in Fig. 2, the narrowest part of the Smoky River in the general area exists at this point because of encroachment by the cone. The creek channel immediately above the point of mud spread is a steep-sided valley rapidly increasing to a depth of 25 ft. Mud was found as high as 12 ft on the valley sides. In the area of the spread, mud coated the trunks of trees up to 5 ft above the stream channel level. The flow could be zoned with a central area of mud along the stream channel and a border area standing about 6 in. above the central zone with a concentration of mud and coarse clastics up to boulder size, and trees. The trees must have been uprooted from growth position as they were green and most of the branches had been ripped off. The next morning, the mud was still in a highly fluid state as velocities resulting from bulldozing were comparable to those of flowing water. The boulders were carried along at a slightly lower velocity. More specific data on the character of such mud are given by Sharp and Nobles (1953). The similarity of the fluid mud (Fig. 4) and the consolidated material of the cone (Fig. 3) seems obvious. Bull (1962) has described the texture of similar material in an alluvial cone.

At the upper road, the fill for the road, which temporarily determined the upstream gradient, remained intact except for some loose material which had been eroded from the downstream side. The steep-sided channel above and below the fill was clear of mud except for that plastered on the side of the valley. A small quantity had spread across the area where material had been excavated for the road fill. The amount of water flowing in the stream next morning was normal for a period of heavy rain.

It is interesting to note that during the same night a mudflow blocked the road across a huge cone at the mouth of Barrett Creek. This small tributary, about 4 mi down river from Hell's Creek, flows out of a barren amphitheater of rocks which have the same stratigraphic position as those along Hell's Creek, and are tightly folded in an asymmetrical anticline.

INTERPRETATION

Over a period of time, quantities of mechanically derived coarse and fine clastic must accumulate in the rock amphitheater of Hell's Creek. Saturation by a heavy rain would not necessarily cause movement and some triggering event would seem required. Short periods of heavy rain had not resulted in mudflow during the three summers previous to 1962. The intensity of local rainfall in the drainage basin of Hell's Creek may have been greater than during the 3 previous years although mudflow in Barrett Gulch, 4 mi away, simultaneous with that at Hell's Creek, would suggest this factor has secondary importance. Snow is not considered a factor as the amphitheater is clear by mid-July. A rock fall or local slumping, possibly initiated by vibrations from a clap of thunder, could have caused a large quantity of wet material in an unstable topographic position to acquire instantaneous fluidity. Rapp (1960, pp. 162, 163) reported that earth slides occurred during a thunderstorm and suggested that 2 in. of diurnal rainfall is sufficient to release earth slides on unstable slopes. The high gradient in the narrow, steep-sided channel would maintain or possibly increase the initial velocity. The speed was sufficiently high at the upper road to "squirt" the mud across the road fill into the channel below with little or no spread. Where the channel is not confined in the vicinity of the lower road, the area of the mud spread was determined by the topography and the volume of mud. The consistency of the mud was sufficient to resist immediate erosion and the cone was extended into the Smoky River, thus restricting its channel further.

The period of flow was short, but sporadic occurrences at unknown intervals over a long period have built a large sediment mass of heterogeneous texture. The stream of insignificant size that normally flows across the surface of the cone could not have transported the numerous boulders and deposited them with a matrix of clay-sized material.

CONCLUSIONS

Mudflows are probably common in humid temperate climates. In areas of high relief, a rugged, steep-sided rock amphitheater devoid of vegetation and cut in fine clastics fractured by folding and faulting can supply mud and boulders to build a large alluvial cone and divert the course of a major river. High rainfall and possibly melt water from snow is necessary to saturate the sediment but special circumstances, for example a local clap of thunder, probably initiate the flow of mud. Construction activities in the vicinity of such a physiographic setting should be protected from mud inundation.

This author has recollections of many cones of this type in the mountainous country of western Alberta and northeastern British Columbia. One or two could be recognized as resulting from landslides by a fairly constant width along the side of the mountain, and the highly irregular, hummocky surface of the slide material. But numerous cones have a moderate regular slope, a relatively small stream on the surface, and a bouldery surface with relatively even relief. Construction by surface stream transport of material was a "conventional" thought. Only a small part of the total thickness of the Hell's Creek cone is exposed but the heterogeneous mixture of clay to boulder size material and the actual occurrence of a mudflow are suggestive of the depositional process for the whole mass. Formation by mudflow might possibly be the rule and not the exception for cones with the particular physiographic setting in alpine regions.

ACKNOWLEDGMENTS

Thoughtful criticism has been given by G. G. Suffel, University of Western Ontario; K. K. Landes, University of Michigan; M. J. Frarey, Geological Survey of Canada; and W. B. Bull, United States Geological Survey. N. S. Armstrong of Sparton Air Service, Calgary, supplied the air photos.

REFERENCES

ATLAS OF CANADA. 1957. Can. Dept. Mines Tech. Surv., Ottawa.
BEATY, C. B. 1961. Boulder deposit in the Flint Creek valley, western Montana. Geol. Soc. Am. Bull. **72**, 1015.
BLACKWELDER, E. 1928. Mudflow as a geological agent in some arid mountains. Geol. Soc. Am. Bull. **39**, 465.
BLISSENBACH, E. 1954. Geology of alluvial fans in semiarid regions. Geol. Soc. Am. Bull. **65**, 175.
BULL, W. B. 1962. Relationship of textural (CM) patterns to depositional environment of alluvial fan deposits. J. Sediment. Petrol. **32**, 211.
―――― 1963. Alluvial-fan deposits in western Fresno county, California. J. Geol. **71**, 244.
BUWALDA, J. P. 1951. Transportation of coarse material on alluvial fans. (Abs.) Geol. Soc. Am. Bull. **62**, 1497.
DAHLSTROM, C. D. A. 1960. Concentric folding. Edmonton Geol. Soc. Field Guide Book, Rock Lake. p. 82.
FRYXELL, F. M. and HORBERG, L. 1941. Alpine mudflows in Grand Teton National Park, Wyoming. Geol. Soc. Am. Bull. **54**, 457.
IRISH, E. J. W. 1954. Copton Creek, Canada. Geol. Surv. Can. Map 1041A.
IRISH, E. J. W. and THORSTEINSSON, R. 1957. Grande Cache map-area. Geol. Surv. Can. Map 1049A.
RAPP, A. 1960. Recent development of mountain slopes in Kärkevagge and surroundings, Northern Scandinavia. Chap. 7. Geograph. Annaler, **42**, 148.
RICKMERS, W. R. 1913. The Duab of Turkestan. Cambridge Univ. Press, London.
SHARP, R. P. 1942. Mudflow levees. J. Geomorph. **5**, 222.
SHARP, R. P. and NOBLES, L. H. 1953. Mudflow of 1941 at Wrightwood, southern California. Geol. Soc. Am. Bull. **64**, 547.
SHARPE, C. F. S. 1937. Landslides and related phenomena. Columbia Univ. Press, New York.
TROWBRIDGE, A. C. 1911. The terrestrial deposits of Owens valley, California. J. Geol. **19**, 706.

7

Copyright © 1966 by R. F. Leggett, R. J. E. Brown, and G. H. Johnston
Reprinted from *Geol. Soc. America Bull.* **77**:15-30 (1966), courtesy of the
Geological Society of America

Alluvial Fan Formation Near Aklavik, Northwest Territories, Canada

R. F. LEGGET *Division of Building Research, National Research Council, Ottawa, Ontario, Canada*
R. J. E. BROWN } *Soil Mechanics Section, Division of Building Research, National Research Council,*
G. H. JOHNSTON } *Ottawa, Ontario, Canada*

Abstract: When the Canadian Government decided to relocate the town of Aklavik in the delta of the Mackenzie River, Northwest Territories, the search for a new site involved detailed site investigations of four prospective locations. One was on the gently sloping surface of an alluvial fan between the Richardson Mountains and the western rim of the Mackenzie Delta. Predictions from surficial evidence and from a detailed study of aerial photographs were invalidated when drilling in the perennially frozen ground disclosed at least 40 feet of organic silt and no sand or gravel as had been expected. The adjacent mountains of Cretaceous age supply materials eroded from sandstone and shales. The action of frost probably induces mechanical disintegration. Turbulent stream flow continues the erosion process, which appears to be still further aided by the annual growth of fast-growing sedges and grasses on the surface of the stream meanders. The short but warm growing season leads to rapid decay of this grass cover; this combined with the annual layers of stream-bed material results in the fans being composed predominantly of organic silt with only minor quantities of coarse-grained material.

INTRODUCTION

Late in 1953 the Government of Canada decided that a new townsite should be found for the settlement of Aklavik, Northwest Territories, situated on the Peel Channel of the Mackenzie Delta. This decision was prompted by the ice content of the deltaic soils underlying the townsite that makes the soils unsuitable for building foundations, proper sanitation facilities, and airstrip construction. Danger from river flooding and the limited size of the site added to the local problems (Merrill and others, 1960).

A survey team under the auspices of the

Dept. of Northern Affairs and National Resources (Canada) was organized during the winter of 1953–1954 for the purpose of selecting a new site and was led by C. L. Merrill of that department. Personnel consisted of engineers and specialists in geography and geology from the staffs of the following Canadian federal departments: Northern Affairs and National Resources, Mines and Technical Surveys, Public Works, Transport, and National Health and Welfare, and from the Division of Building Research of the National Research Council.

Uncontrolled mosaics of aerial photographs covering the entire Mackenzie Delta and the adjacent uplands were assembled and studied in Ottawa during the winter of 1953–1954. Intensive study revealed that several potential sites existed on each side of the delta that merited investigation in the field.

Therefore, during the summer of 1954 detailed site investigations were conducted at each of the potential townsite areas. One of these areas consisted of a series of coalescing alluvial fans on the west side of the Mackenzie Delta, 12 miles southwest of Aklavik. Personnel of the relocation survey team conducted detailed soils and permafrost investigations on two of the alluvial fans.

The results obtained from this test drilling were unexpected in that the anticipated predominance of sand and gravel usually associated with alluvial fans was not found. The character of the soil formation thus revealed eliminated this site as a location for the new town, but at the same time raised questions regarding the formation of alluvial fans, which the writers have studied since the original field work was completed and upon which they now report.

DESCRIPTION OF ALLUVIAL FANS

Location

The coalescing alluvial fans form a plain tilted slightly downward to the east between the Richardson Mountains and the Husky Channel, a secondary channel that flows down the extreme west side of the Mackenzie Delta in northwestern Canada (Fig. 1; Pl. 1, fig. 1). The southern extremity of the Mackenzie Delta lies 90 miles north of the Arctic Circle. The delta is a low flat area approximately 40 miles wide and 130 miles long, covering an area of about 4700 square miles. It consists of silts, with some fine sand and clay, and is interlaced by many river channels and spotted with thousands of stagnant lakes (Mackay 1956; 1963). The Richardson Mountains rise to approximately 2000 feet above sea level on the western flank of the delta overlooking the alluvial fans. Thirteen miles to the southeast, at the southern end of the alluvial plain, the mountains rise to more than 2800 feet. To the west of this flank they rise to more than 5000 feet (Pl. 1, fig. 1).

The fan selected as the most suitable for a townsite area is the most northerly of the coalescing series and is 1 mile west of the junction of the Peel and Husky channels. Most of the observations were made on this fan (hereafter referred to as the "willow fan"; see description of fan vegetation) which is 1.75 miles long from the apex to the toe and 2.5 miles wide. Six miles to the south observations were made on another fan (hereafter referred to as the "spruce fan"; see description of fan vegetation) of about the same dimensions.

Climate

For more than 30 years there has been a weather station located in the delta at Aklavik. The climate in the region of the alluvial fans may vary slightly from that of Aklavik because of their location at the base of the Richardson Mountains but it can reasonably be assumed to be very similar to that of Aklavik.

Although the Mackenzie Delta reaches the arctic coast of Canada, its climate is essentially continental in character, with long cold winters and short, warm summers. Combined with the low elevation of the Mackenzie River valley and the protection afforded by the mountains to the west, the local climate is one of the factors associated with the extension of the tree line in the delta to within 20 miles of the arctic coast.

Study of the records (1931–1960) gives the following temperature information:

	°F
Mean January daily temperature	−19.9
Mean January daily minimum temperature	−26.9
Mean January daily maximum temperature	−12.9
Mean July daily temperature	56.5
Mean July daily minimum temperature	48.0
Mean July daily maximum temperature	64.9
Mean annual temperature	15.6
Mean annual minimum temperature	8.1
Mean annual maximum temperature	23.1

Extreme lowest recorded temperature since 1921	−62
Extreme highest recorded temperature since 1921	93

Particularly notable are the low mean annual temperature (in contrast with the mean July above freezing, giving an average total of 105 days without frost. On the average, however, only 66 of these would occur in succession.

The oscillation of the air temperature around the freezing point (32°F) during the year is relevant to the subject of this paper. According to

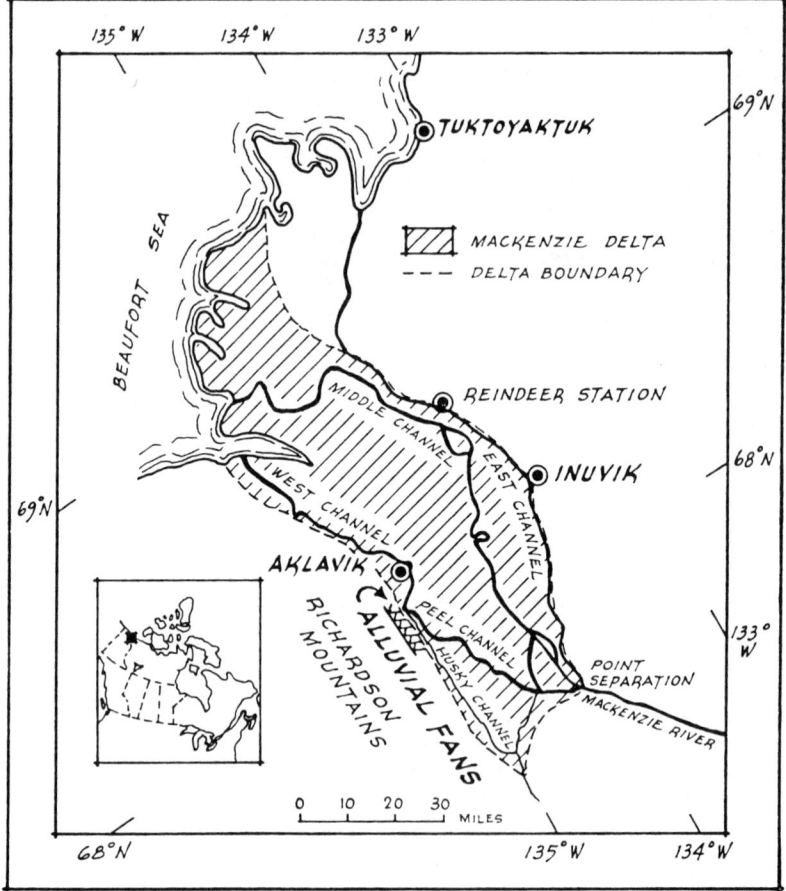

Figure 1. Map of Mackenzie Delta, Canada, showing location of alluvial fans

daily maximum of 64.9°F), showing clearly the relatively short duration of high summer and the long duration of the steadily cold winter.

At Aklavik, the average frost-free period each summer is only 66 days, from June 15 to August 20. It has been as short as 32 days and as long as 102 days. The dates of the last frost in spring and the first frost in autumn vary by about 1 month on each side of the average dates given. During most of June, July, August, and early September the temperature remains

Fraser (1959), who suggested that a freeze-thaw cycle is represented by a rise to 34°F following a drop to 28°F, the average number of freeze-thaw cycles occurring annually at Aklavik is 25. Examination of temperature records throughout Canada shows that the average annual frequency of freeze-thaw cycles decreases steadily with increasing latitude through central Canada. For example, Regina, Saskatchewan (lat. 51°N.), has an annual average of 60 cycles; Fort Smith, Northwest Territories (lat. 60°N.), has 48; and Eureka, Northwest Territories (lat. 80°N.), has 12. Although the annual frequency of freeze-thaw cycles decreases toward the north, the duration of the freezing part of the cycle increases, and the minimum below-freezing temperatures become lower.

average less than 5 miles per hour. For almost half the time, the wind comes from the north or northwest, blowing inland from the Beaufort Sea, and for about one-third of the time it is from the south or southeast.

The hythergraph of the climate at Aklavik is

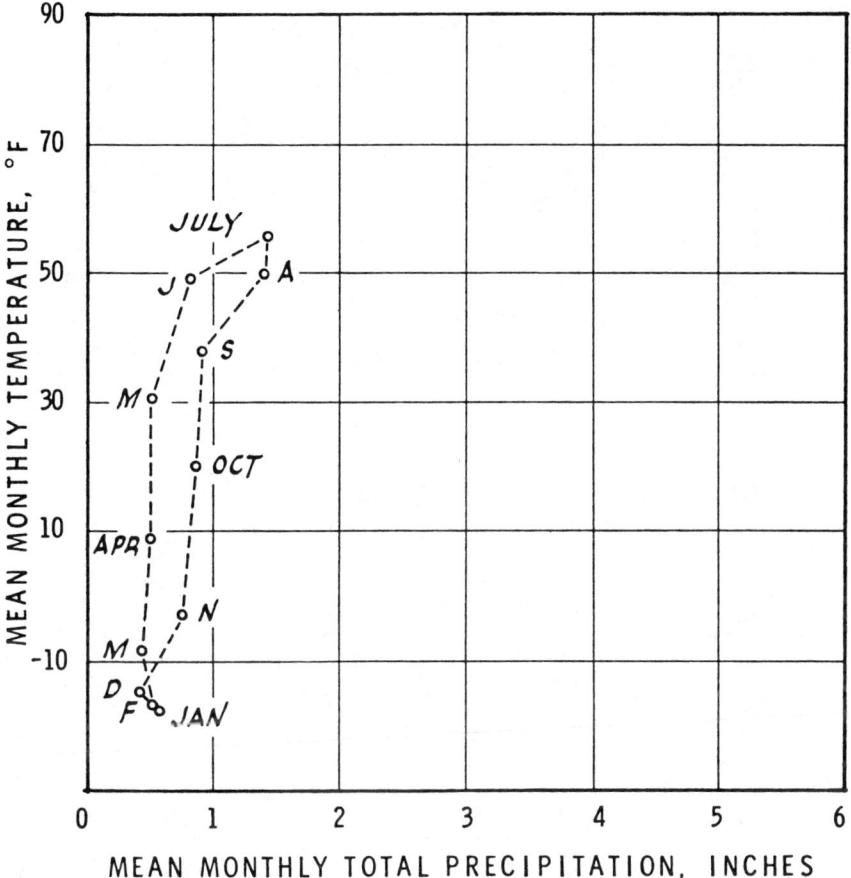

Figure 2. Hythergraph of Aklavik, Northwest Territories, Canada

The mean annual precipitation at Aklavik is 8.84 inches, the mean annual rainfall is 3.90 inches, and the mean annual snowfall is 49.4 inches. Rain has never been reported in the 4 months from November to February and is rare in March and April. Snow has fallen in all months of the year, although it rarely falls in July.

Aklavik is not very windy compared to the country as a whole. The strongest winds occur in June, with an average of only 8 miles per hour. In November and December the winds

shown in Figure 2. The main features of the climate are the short, warm summers; the long, cold winters; the low number of annual freeze-thaw cycles; the low annual precipitation; and the low wind speeds.

Surface Features

PHYSICAL GEOLOGY AND RELIEF: The western limit of the Keewatin Ice Sheet appears to have been in the delta area, the ice being blocked on the west by the barrier of the Richardson Mountains. Moving north along

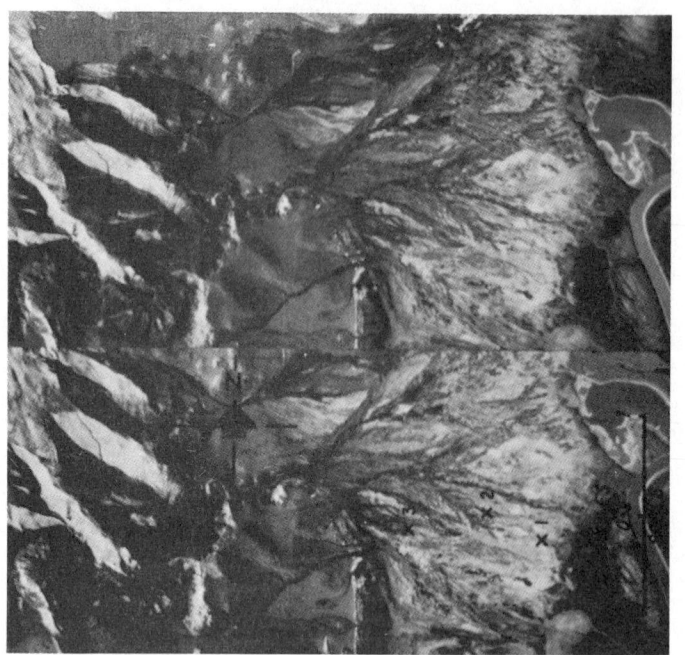

Figure 2. Stereo pair of willow fan situated between Mackenzie Delta (bottom of photographs) and Richardson Mountains (top of photographs). (RCAF A 12861-182 and 183) x, borehole location; 1, Hole no. 1; 2, Hole no. 2; 3, Hole no. 3

Figure 1. Aerial view of alluvial fans looking south from an altitude of 20,900 feet. Mackenzie Delta on left with meandering Husky Channel at toe of fans. Note V-shaped valleys in Richardson Mountains on right. Borehole location shown by x. (RCAF air photo T5-10L, August 1944)

AIR PHOTOGRAPHS OF ALLUVIAL FANS, NORTHWEST TERRITORIES, CANADA

Figure 1. Aerial view of willow fan looking northwest (June 1954)

Figure 2. Ground view of willow fan (August 1954)

ALLUVIAL FANS NEAR AKLAVIK, NORTHWEST TERRITORIES, CANADA

Figure 1. Willow fan gully (April 1954)

Figure 2. Sandstone bluffs at entrance of willow fan gully (August 1960)

Figure 3. Weathered shale blocks in Richardson Mountains (April 1954)

Figure 1. Stream flowing on willow fan, 2000 feet southeast of Hole no. 3 (May 1954)

Figure 2. Freshly deposited material on willow fan, 2000 feet southeast of Hole no. 3 (May 1954)

SURFACE OF ALLUVIAL FANS NEAR AKLAVIK, NORTHWEST TERRITORIES, CANADA

the trough now occupied by the Mackenzie Delta, the ice straightened and smoothed off the eastern scarp of the mountains. Deposits and land forms associated with glaciation, such as till, kames, moraines, and melt-water channels, abound on the eastern slopes, and erratics occur to elevations of 3000 feet above the fans. West of the eastern scarp the relief is that of a rolling plateau dissected by many V-shaped valleys (Mackay, 1963).

The alluvial fans have been formed from disintegrated and decomposed products from the Richardson Mountains. Both the fans and the surface deposits of the Mackenzie Delta are believed to be predominantly postglacial. Two distinct grades were present on each of the fans investigated (Riccio, 1962). The first, relatively steep, extends from the apex of the fan downward for a distance of from one-third to one-half the length of the fan; on the willow fan this grade is 1.9 per cent and on the spruce fan, 6.1 per cent. The second grade, to the deltaic deposits at the toe of the fan, is longer and lower; at the willow fan it is 1.3 per cent and at the spruce fan, 2.3 per cent.

VEGETATION: The low elevation of the Mackenzie River valley, which is flanked by protecting mountains over much of its course, favors a northward extension of the great northern coniferous forest into the delta to about 20 miles south of the arctic coast. In the forested part of the delta spruce is dominant; poplar, birch, and thickets of alder and willow are also present. In many places, however, there are open patches of meadows unable to support any tree growth. North of the tree line the vegetation consists of mosses, grasses, sedges, and heaths, with widely scattered scrub willow and alder.

The tree growth on the willow fan, consisting of willow and alder thickets up to 15 feet high, is typical of the north half of the alluvial plain (Pl. 2, fig. 2). These thickets are interspersed with open meadow-like areas in which sedge (predominantly *Carex aquatilis Wahl*) and grass grow to a height of about 15 inches in only a few weeks every summer. There are some areas of bare (new) soil and an increasing number of sedge and grass tussocks toward the toe of the fan. Other ground plants include Sphagnum and other mosses, lichen (*Cladonia sp.*, *Cetraria sp.*), Labrador tea, juniper, blueberry, and wintergreen. The over-all pattern of the vegetation appears to trace the past movements of the discharge from the creek and thus intensifies the braided appearance of the fan when viewed from the air or by means of aerial photographs (Pl. 1, fig. 2; Pl. 2, fig. 1).

The tree growth on the spruce fan is typical of the southern half of the alluvial plain. Its vegetation is similar to the willow fan previously described with the addition of spruce trees, which reach a maximum height of 25 feet. The average height and density of the tree growth exceeds that on the willow fan, and hence the braided effect seen from the air is more pronounced.

The boundary between the fans and delta is fairly well delineated by vegetation. This is true even on the spruce-covered fans where the height and density of trees are somewhat less than those on the adjacent delta. The factors that allow these fans in the south half of the alluvial plain to support spruce growth in contrast to the scrub willow and alder growth of the northern fans present a problem beyond the scope of this paper.

FROST PHENOMENA: The ground surface of all fans is characterized by the following microrelief forms: polygons in local depressions surrounded, or partly surrounded, by raised rims up to 5 feet in height; complete and partial polygons about 30 feet in diameter delineated by troughs of depths measured in inches; and frost mounds up to 5 feet in diameter and 1 foot high. There are numerous signs of intensive frost action throughout the entire fan area. Large areas contain frost mounds, and polygonal cracks are found everywhere. Permafrost is continuous and extends to depths of several hundred feet.

Soils

BOREHOLE LOCATIONS: Subsurface investigations were carried out by drilling and sampling methods (Pihlainen and Johnston, 1954) to obtain undisturbed cores of the perennially frozen soils. Three holes were drilled in the willow fan and one on the spruce fan (Pl. 1, fig. 1; Pl. 1, fig. 2; Fig. 3; Pihlainen and others, 1956).

Hole no. 1, located at the toe of the willow fan, was advanced to a depth of 32 feet with core recovery of more than 90 per cent. Drilling difficulties were experienced at Hole no. 2, located part way up the fan. Two core barrels froze in at depths of 5 and 12 feet; samples were obtained only to a depth of 5 feet, and the hole had to be abandoned in order to recover the core barrels. Hole no. 3, located near the apex of the fan, was advanced to a depth of 38 feet with core recovery averaging 50 per cent. Hole no. 4, located near the middle of the spruce fan, was drilled to a depth of 29 feet, with core recovery averaging 60 per cent. (Detailed soil profiles are shown in Fig. 4.)

TEST RESULTS: Undisturbed samples were examined for ice segregation characteristics, and laboratory tests were carried out to identify the various soils found. Although the observations reported in this paper pertain mainly to the willow fan, some results of investigations on the spruce fan are included as supplementary material.

The soils consist predominantly of a light-brown to light-gray silt-sized soil, with varying amounts of organic material, thin layers (1 inch to 1 foot) of friable mudstones, and some more resistant sandstone pebbles. Twenty-three samples were analyzed for grain-size distribution. Figure 5 shows envelopes of the graphical records of the test results; samples from the toe of the willow-covered fan are all smaller than fine sand (0.2 mm). Grain-size distribution limits for samples from the top of this fan show larger particle sizes grading down from approximately the coarse-sand size (less than 2.0 mm). Samples from the middle of the spruce fan (with a higher surface gradient) show a grain-size distribution envelope similar

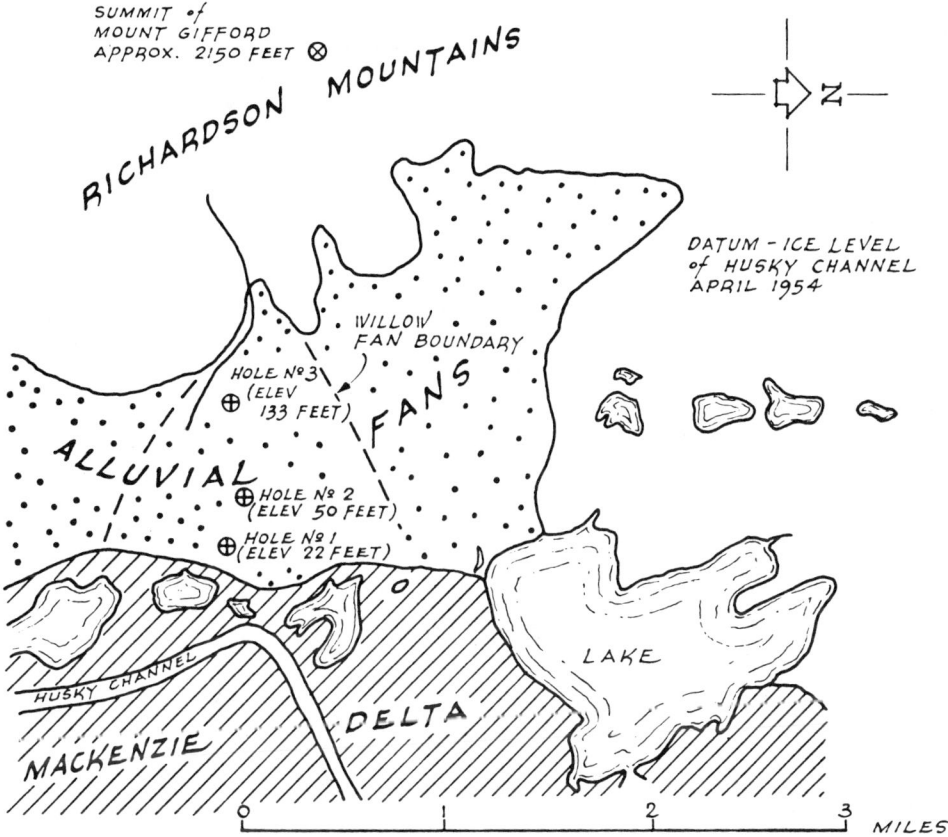

Figure 3. Map of willow fan area near Aklavik, Northwest Territories, Canada

to that at the top of the willow-covered fan except that more silt- and clay-size particles are present.

In general, the soils found were nonplastic. Atterberg limit tests, by which the plasticity of a soil is determined, were conducted on a number of samples, but plasticity-index values for only nine were obtained. The test results are listed in Table 1, and a plot of plasticity index vs. liquid limit is shown in Figure 6. The graph shows that the points lie along the A-line[1] and

---

[1] A-line represents the empirical boundary between typical inorganic clays (which are generally above the A-line) and plastic materials containing organic colloids.

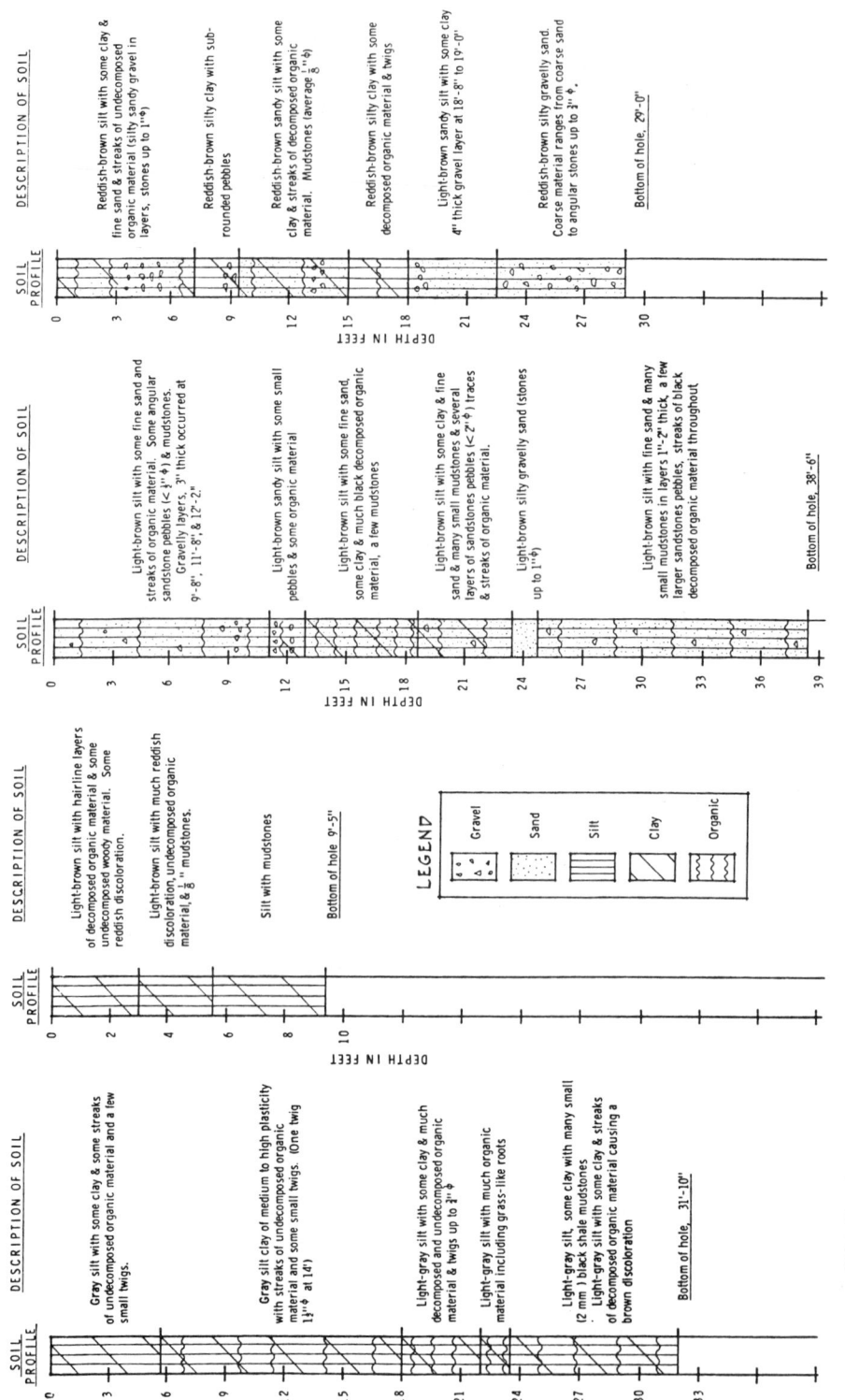

Figure 4. Detailed soil profiles from alluvial fan boreholes near Aklavik, Northwest Territories, Canada

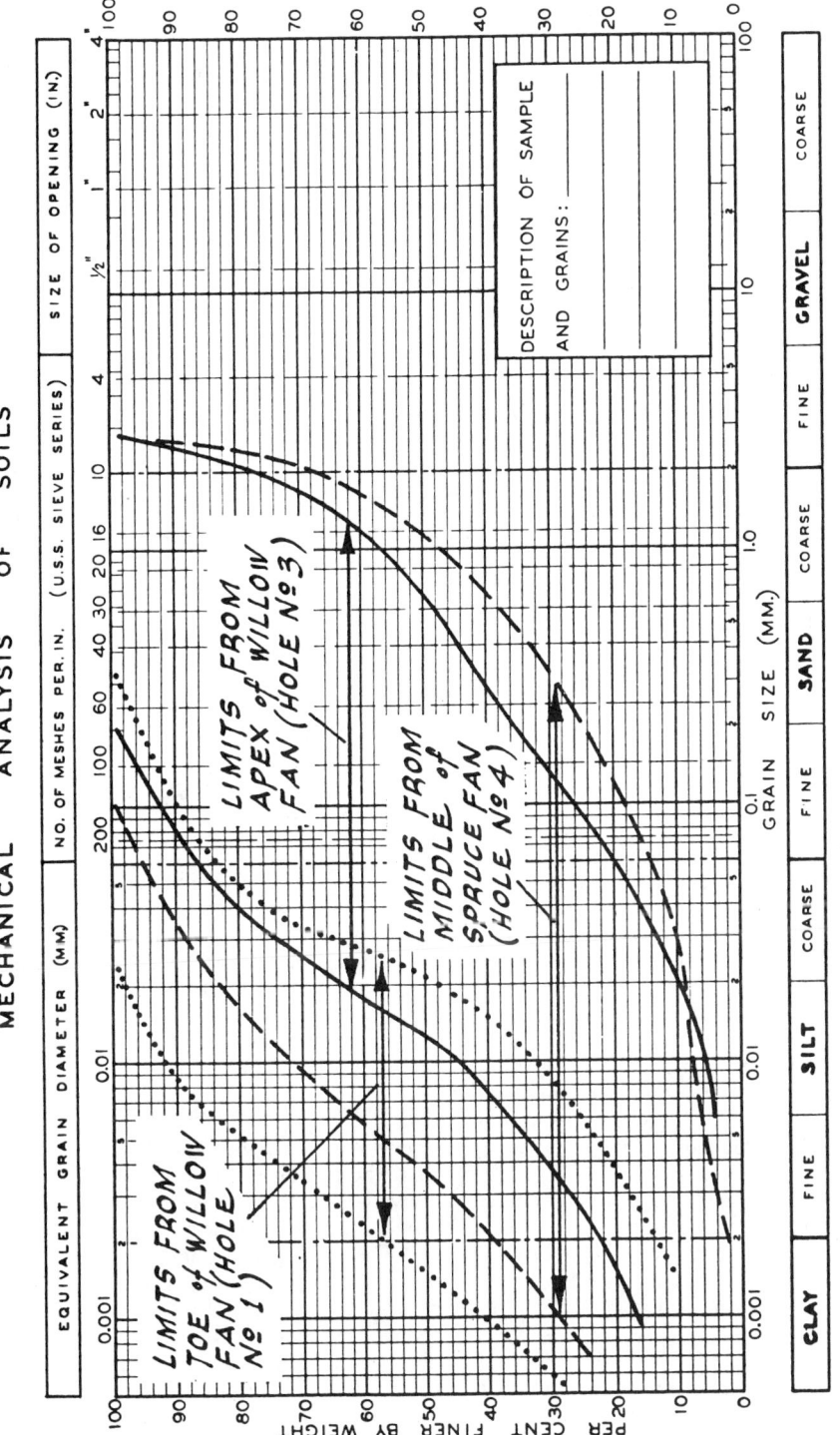

Figure 5. Envelopes of grain-size distribution curves for fan soil samples from near Aklavik, Northwest Territories, Canada

that the soils are predominantly organic and inorganic silts and silt clays (Casagrande, 1948).

A number of soil samples were examined under the microscope. Many angular quartz particles were observed in the soils from the upper part of the fan, indicating a sandstone origin. A reddish discoloration caused by the ferruginous deposits was quite prominent in these soils, but was not noticeable in those from the toe of the fan.

Thirty samples were taken from depths between 2 and 31 feet in Hole no. 1 for determining moisture (ice) content; this is expressed as a percentage of the weight of dry soil in the sample. Moisture contents averaged 85 per cent and ranged from 27 to 154 per cent. Ice segregation consisted of horizontal layers ranging in thickness from hairline (less than 1/32 inch) to 3/4 inch, averaging 1/16 inch.

POLLEN ANALYSIS: In an attempt to determine the past vegetation associated with the fan deposits, samples from the four boreholes were examined by Dr. N. W. Radforth of the Department of Biology, McMaster University, Hamilton, Ontario. A pollen analysis showed that all major groups of plants were represented in the microfossils. Those commonly occurring in existing northern peats were present in all four boreholes, including microfossils of the three

TABLE 1. PLASTICITY AND ACTIVITY OF SOIL SAMPLES FROM ALLUVIAL FANS NEAR AKLAVIK, NORTHWEST TERRITORIES, CANADA

Location	Sample depth (in feet)	Atterberg limits			Clay fraction per cent. < 0.002 mm	Activity PI/per cent clay
		Liquid limit	Plastic limit	Plasticity index		
Willow fan Hole no. 1	8	49.8	31.1	18.7	53	0.353
	14.5	55.1	32.3	22.8	48	0.475
	17	47.0	26.3	20.7	24	0.863
	22	47.3	28.8	18.5	43	0.430
Hole no. 3	21	40.1	30.9	9.2	16	0.575
Spruce fan Hole no. 4	7.5	43.7	28.3	15.4	39	0.395
	8.5	37.1	23.4	13.7	39	0.352
	16	40.3	25.5	14.8	35	0.423
	21	51.5	30.4	21.1	40	0.527

inch, and averaging 1/16 inch. In ten samples taken from depths of 1.5 to 5 feet, in Hole no. 2 the moisture contents averaged 74 per cent, ranging from 31.2 to 193.9 per cent. Horizontal ice layers varied in thickness from hairline to 1/2 inch. Moisture content samples were taken about every 2 feet in Hole no. 3. In 25 samples from depths of 7 to 38 feet moisture contents averaged 72 per cent, ranging from 17.1 to 299.1 per cent. Ice segregation consisted predominantly of horizontal layers ranging in thickness from hairline (less than 1/32 inch) to 3/4 inch, and averaging 1/16 inch. In Hole no. 4 moisture content samples were taken at least every 2 feet of penetration. In 28 samples from depths of 3 to 23.5 feet moisture contents averaged 83 per cent, ranging from 20.0 to 253.3 per cent. Ice segregation consisted of horizontal main peat types—amorphous-granular, fine-fibrous, and coarse-fibrous. It is impossible to reconstruct the distribution of the original plant cover, but variations in microfossil type frequency, indicating a predominance of shrubs, sedges, grasses, and lichens, suggest that the original plant cover was similar to that of the present.

SOURCE OF FAN MATERIAL

The willow fan has evidently been formed from the deposition of disintegrated materials derived from the Richardson Mountains and transported through the gully at the apex of the fan. The rocks cut by the gully are mainly interbedded buff to brown fine- and coarse-grained sandstones, light- to dark-gray shales, and light-brown to dark-gray silty and sandy siltstones of Cretaceous age (Jeletzky, 1958) (Pl. 3, fig. 1). All are subject to severe frost action, as is indicated by the shattered appearance of many of the outcrops. The fine-grained sandstones are the more resistant, and sandstone pebbles found in the boreholes on the fan are predominantly of this material. The coarse-grained sandstones are friable and appear to disintegrate much more rapidly, particularly during the snow-melt period when they are easily eroded by the run-off (Pl. 3, fig. 2). In general, they are completely disintegrated by the time they are deposited on the fan by the gully stream. The siltstones and shales, also fine- and coarse-grained, are somewhat more resistant but do break down fairly quickly under frost and water action, resulting in deposition as individual particles or as small "mudstones," the latter being found even in Hole no. 1 at the toe of the fan. Many of the layers of siltstone have been cemented by a ferruginous coating up to 1/8 inch thick that has penetrated joints and fractures in the rock.

Sandstone boulders up to several feet in diameter can be seen on the high talus slopes rising above the narrow gorge of the gully. Some have clearly been disintegrated by natural weathering into flat subangular fragments. Disintegration appears to be more intense on the north-facing slope than on the south-facing slope, the gully being oriented approximately east-west. Resistant shale outcrops form distinct ridges on a saddle above the fan (elevation 1180 feet); these form steps, each about 2 feet high. The surface of the shale is severely weathered and cracked by frost, giving a series of small, friable round-topped blocks (Pl. 3, fig. 3). Between the ridges the shale is darker in color and has weathered to flakes approximately 1/8 inch in size. Distinct polygonal cracking producing polygons about 8 feet in diameter is evident in this black shale.

There are soil slumps at an elevation of 800 feet above sea level on a north-facing ridge. The almost vertical slip faces are spaced about 10 feet apart and range in height from 2 to 6 feet. The exposed soil is a light-brown silt with friable mudstones in the top 2 feet and streaks and flakes of organic material throughout the face of the exposure.

The gully from which the fan-forming stream emerges is about 150 feet deep close to its mouth and is V-shaped, with side slopes very steeply graded and in places at angles exceeding 45 degrees (Pl. 3, fig. 2). Material in the bed of the gully varies in size from silt-sized fines to

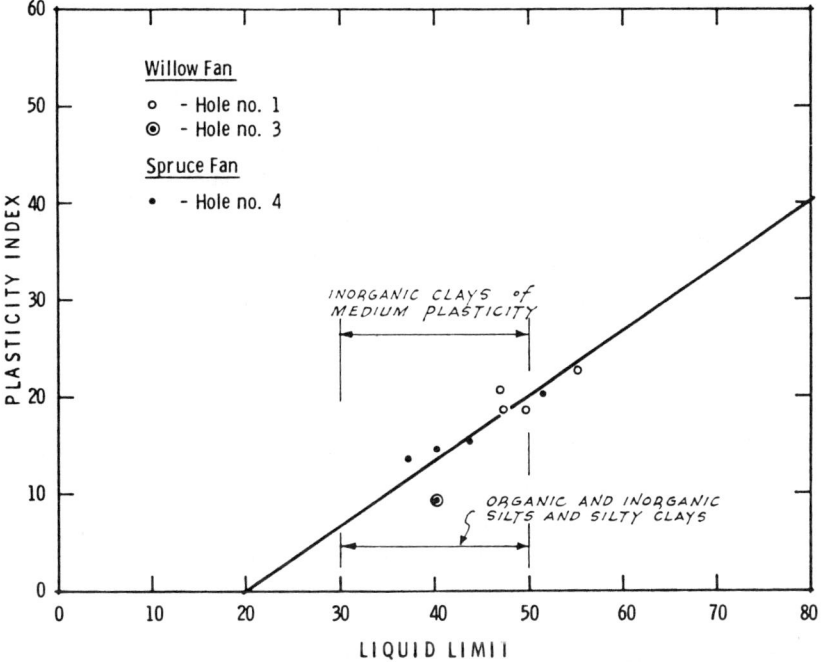

Figure 6. Plasticity chart for alluvial fan soils from near Aklavik, Northwest Territories, Canada

boulders several feet in diameter. Many outcrops of the buff to dark-brown sandstone, weathered to white, and of red-stained siltstone are evident. Rock outcrops and boulders all exhibit severe surface weathering characteristics, which is particularly noticeable in the boulders of shale that clearly come from the uppermost beds above the gully.

FORMATION OF FANS

Introduction

As an immediate result of the disclosure in all three boreholes that the fans were underlain by silt with high ice content similar to that at the existing site of Aklavik, the site had to be abandoned as a prospective location for the new town. The existence throughout the fans of fine-grained soils derived from weathering of solid rock in the adjacent mountains suggests active disintegration of the parent rock material.

A literature search revealed that specific references to alluvial fan deposits or formation appear to be limited. The only paper on alluvial fans in polar regions describes fan development in an adjacent area in Alaska (Anderson and Hussey, 1962). It deals only with surficial features, however, for the authors have apparently not undertaken subsurface investigations. (A listing of papers on alluvial fans is given in the Bibliography.) No publication has been found that is relevant to the problem reviewed in this paper, a fact that explains the limited list of references.

What factors contribute to the breakdown of parent rock into the materials forming the fans? The nature of the parent rocks is clearly a first determinant, but the local climatic conditions and weathering to which the rocks are subjected are equally important. The process of transportation and deposition must be considered, and vegetation on the fans may be of some significance. Detailed study of the properties of the soil samples obtained from the boreholes could assist in the elucidation of the problem.

Freeze-Thaw Tests

The role of alternate freezing and thawing in the disintegration of rock material in northern regions is somewhat difficult to assess or to separate from the role of other physical processes. Although the concept of strongly predominant mechanical weathering in northern regions has been widely accepted, it is suggested that the fact that shattered rock is more evident in northern than in southern Canada is not solely the result of lower temperature or of high frequency of freeze-thaw cycles. (In fact, the frequency of freeze-thaw cycles decreases with latitude, as has already been mentioned.) The evident abundance of shattered rock may derive to an important degree from the absence of a concealing and insulating mantle of snow and vegetation in the north and may therefore be a secondary effect of the climatic factors of low annual temperature and precipitation (Fraser, 1959).

Determination of the number of freeze-thaw cycles that actually occur in nature for comparison with laboratory freeze-thaw tests is not an easy or convenient climatic analysis. The total amount of shattering appears to depend not only on the number of cycles, but also on the length and intensity of each cycle and on the removal of the debris so that fresh rock becomes available (Tricart, 1956). Experiments carried out with many rapid cycles to simulate the influence of thousands of years produced only a small thickness of frost-shattered material. Nonetheless, in an effort to obtain some idea of the role of alternate freezing and thawing on the formation of the fine-grained fan soils, laboratory tests were conducted on sandstone and shale fragments from the gully above the willow fan. Rock fragments were subjected to two daily freeze-thaw cycles (dry during freezing part of cycle), one cycle consisting of 6 hours of freezing down to 28°F and 6 hours of thawing up to 34°F. Between October 1956 and July 1957 the fragments were subjected to 400 cycles. This would be roughly equivalent to the number of cycles naturally occurring over a 16-year period (25 per year at Aklavik, *according to* Fraser, 1959). Grain-size distribution before and after the test are shown in Figure 7; in the range from about 0.1 to 10 mm the material finer by weight increased by about 10 to 20 per cent. In the range from about 1.0 to 3.5 mm a similar increase was noted.

On the basis of this testing alone it is virtually impossible to assess the relative importance of freeze-thaw cycles in nature on the disintegration of the source rock material. In view of the disintegration observed in the laboratory tests, however, it appears that the process is significant; it is possibly of paramount importance in the initial stages of mechanical breakdown.

Process of Deposition

Flow from the postglacial gully onto the fan is intermittent. For most of the year winter conditions result in complete freezing, and it is only

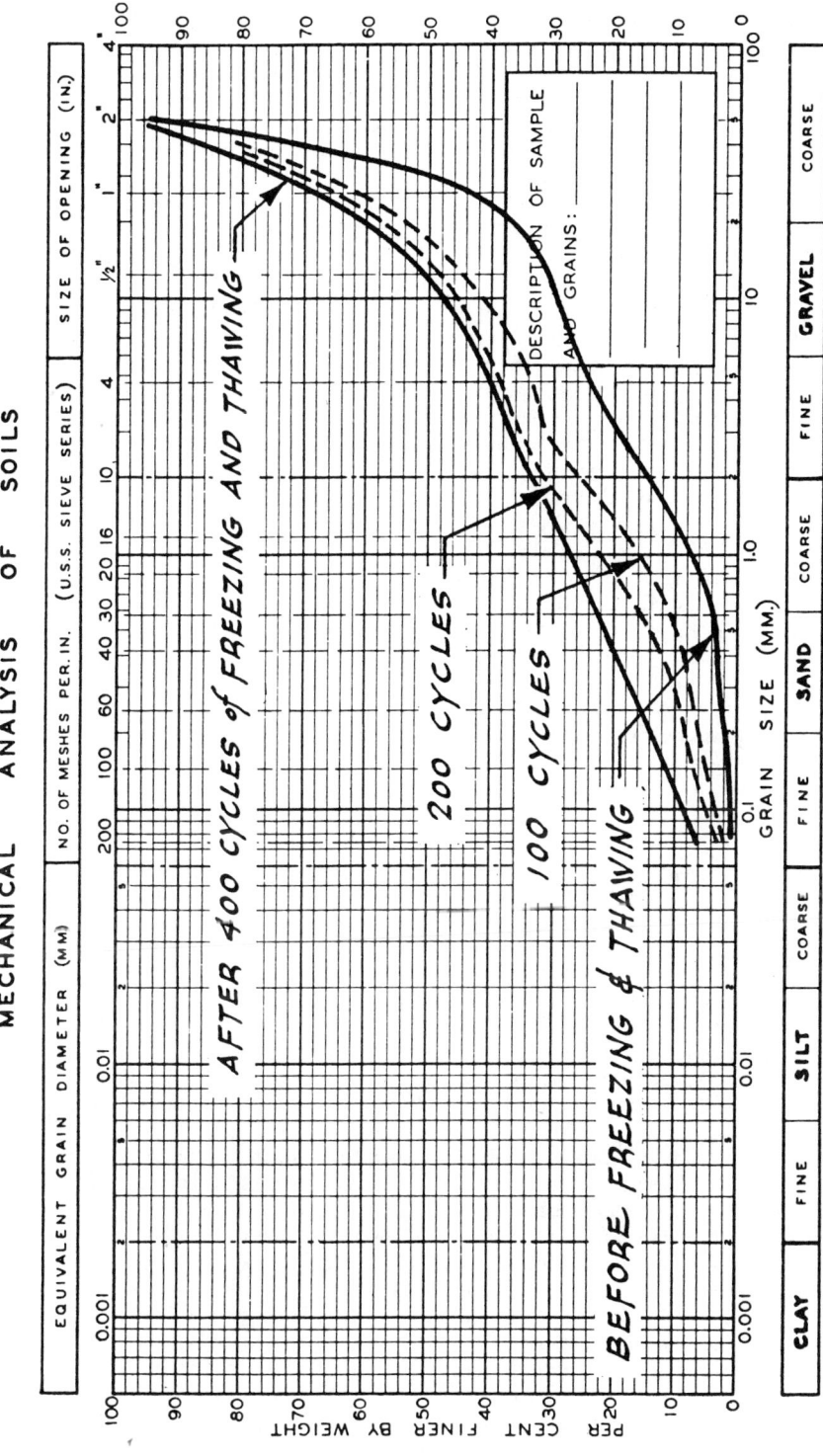

Figure 7. Disintegration of rock sample from willow fan gully near Aklavik, Northwest Territories, Canada, by freeze-thaw cycling in laboratory (Test no. 1)

with the advent of the very short summer season that the snow cover begins to melt and provide any stream flow. Drifting of snow in the gully, due to winter winds, results in an appreciable local storage of water despite the low annual precipitation for the region. Relatively slow melting of the snow drifts, especially those on the north-facing slopes, leads to a moderate regulation of run-off, with the result that there is usually some flow from the gully as late as August.

Emerging from the gully in the spring, and carrying a heavy bed-load of freshly disintegrated rock fragments with some silt, the stream has cut a trench about 6 feet wide and up to 6 feet deep in the loose scree deposits. Within half a mile of the end of the gully, however, the stream bed begins to flatten out and the stream itself takes on a braided pattern. As the entire region is underlain by permafrost, it is not until the end of August that the very shallow active layer has thawed even to a depth of about 18 inches. By that time, stream flow has almost ceased. Practically all the flow from the gully, therefore, debouches onto a fan of solidly frozen material. Accordingly, once the apex of the fan is passed, the stream flow spreads out rapidly, its velocity falls quickly, and its load of solid material is deposited in thin sheets on the particular area over which the stream happens to be flowing.

So complete is the control exercised by the frozen character of the material onto which the stream discharges that for no greater distance than half a mile from the end of the gully is there a properly defined stream bed; the course followed by the stream is diverted from side to side of the fan area by any unusual (and quite random) accumulation of deposited bed-load material. It is this irregular process of deposition and change of route that gives the upper part of the fan surface its characteristic braided appearance.

The intermittent character of the flow even in what may appear to be regular stream beds is indicated by the dense growth of scrub willow over the upper part of the fan (Pl. 4, fig. 1), and by the occurrence of small open sedge and grass meadows interspersed between the willow thickets.

Examination of the exposed mud-flats (Pl. 4, fig. 2) shows that the deposited material consists mainly of light-gray silt-sized particles, with many black streaks close to the surface suggestive of decomposing organic material. Shale and sandstone pebbles, subrounded to subangular and rarely more than 1 inch in diameter, lie scattered over the surface. Where there is any accumulation of these coarser particles, random counts showed less than one-third to consist of the more resistant materials.

Along the south side of the fan, in its upper reaches, is a levee-like ridge which generally follows the most southerly line of stream deposition. It is about 3 feet high and from 10 to 30 feet wide. There is a similar land form on the north side of the fan, but it is not nearly so well defined. The character of these ridges and the distribution of willow and meadow vegetation, described previously, suggest great variation in the stream flow from the gully from year to year, with occasional major changes in the pattern of flow over the fan, probably in years when there is unusually heavy run-off.

Disintegration of Material

Disintegration of the fan material appears to begin with weathering of the sandstone and shale beds into which the gully has been cut. Frost action and alternate freezing and thawing combined with water action are the initial weathering agents contributing to the breakdown of the material. The melting snow provides erratic stream flow from the gully onto the fan, concentrated in the early part of the short summer when the fan material onto which the flow discharges is still completely frozen. The stream flow spreads out over the frozen surface of the fan, resulting in rapid deposition of its bed and suspended loads. The upper part of the fan is composed of relatively coarser material resulting from the deposition of heavier sand particles and more resistant fine-grained sandstone pebbles. The soils in the lower third of the fan are made up of predominantly silt-sized particles with few, if any, pebbles.

After deposition on the fan, further alteration of the material is possible, and if so, it appears to be associated with organic material. Examination of the character of the near-surface deposits in areas of bare, fresh soil and in some of the grass and sedge meadows on the fan, as well as discussions with botanists who have visited the Mackenzie Delta, suggest that only sedges and grasses can be considered as potential factors in any further breakdown of the material, if such occurs. (The presence of undecomposed twigs of alder and willow in the boreholes to considerable depth tends to confirm the negligible effect of these larger organic features.) The sedges, predominantly *Carex aquatilis Wahl*, and grasses grow rapidly in the short but warm summer and provide annually an appreciable source of organic material. Possibly the addition of this decomposed material to the freshly deposited mineral fan material contributes to the final transformation to the predominating organic silt.

CONCLUSION

Probably the most significant feature of this study is the evidence it yields of the unreliability of detailed standard photo interpretation of land forms such as alluvial fans in arctic regions. Prior to field investigations, it was assumed by the authors and others who examined the original photographs of Plate 1 that the fans must consist predominantly of granular material in view of their proximity to the source material and from all the surficial evidence presented by the stereoscopic view. Even on the site, first impressions tended to confirm this conclusion, although the presence of the microrelief features described previously indicated the possibility of fine-grained materials. More detailed inspection of the process of stream formation raised doubts in the minds of those accustomed to studying arctic soils. The results of the test borings and soil tests were sufficiently unexpected, however, that the preparation of this paper appeared to be desirable, even though it is necessarily incomplete.

More field observations and further laboratory tests will be necessary in order to assess with certainty the relative importance of the various factors contributing to the rapid breakdown of the initial products of rock fragmentation into the organic silt revealed in the boreholes. The concluding hypothesis, that the influence of decomposed sedges and grasses contributes to the final transformation of the parent rock to the predominating organic silt in the fans, is perhaps unusual, but it is regarded by the authors as a possibility. Positive proof would require prolonged laboratory experimentation and extensive field observations; but it is improbable that such work can be undertaken in the foreseeable future, because of the isolation of the location and the greater importance of so many other terrain problems that must be faced in the slow development of arctic regions.

REFERENCES CITED

Anderson, G. S., and Hussey, K. M., 1962, Alluvial fan development at Franklin Bluffs, Alaska: Iowa Acad. Sci., Proc., v. 69, p. 310–322

Casagrande, A., 1948, Classification and identification of soils: Am. Soc. Civil Engineers, Trans., v. 113, p. 901–991

Fraser, J. K., 1959, Freeze-thaw frequencies and mechanical weathering in Canada: Arctic, v. 12, p. 40–53

Jeletzky, J. A., 1958, Uppermost Jurassic and Cretaceous rocks of Aklavik Range, northeastern Richardson Mountains, N.W.T.: Geol. Survey Canada, Paper 58-2, 84 p.

Mackay, J. R., 1956, Mackenzie deltas—A progress report: Canadian Geographer, no. 7, p. 1–12

—— 1963, The Mackenzie Delta area, N.W.T: Ottawa, Geog. Branch, Memoir 8, 202 p.

Merrill, C. L., Pihlainen, J. A., and Legget, R. F., 1960, The new Aklavik: search for the site: Eng. Jour., v. 43, p. 52–57

Pihlainen, J. A., and Johnston, G. H., 1954, Permafrost investigations at Aklavik, 1953: Natl. Research Council Canada, Div. Bldg. Research, Tech. Paper no. 16 (NRC 3393), 16 p.

Pihlainen, J. A., Brown, R. J. E., and Johnston, G. H., 1956, Soils in some areas of the Mackenzie River Delta region: Natl. Research Council Canada, Div. Bldg. Research, Tech. Paper no. 43 (NRC 4096), 26 p.

Riccio, J. F., 1962, A geological and geographical appraisal of alluvial fans: The compass of sigma gamma epsilon, v. 39, no. 2, p. 87–95

Tricart, J., 1956, Étude expérimentale du problème de la gélivation: Biuletyn Peryglacjalny, no. 4, p. 285–317

BIBLIOGRAPHY

Beaty, C. B., 1963, Origin of alluvial fans, White Mountains, California and Nevada: Assoc. Am. Geographers Annals, v. 53, no. 4, p. 516–535

Blackwelder, E., 1928, Mudflow as a geologic agent in semiarid mountains: Geol. Soc. America Bull., v. 39, p. 465–484

BIBLIOGRAPHY

Blissenbach, E., 1954, Geology of alluvial fans in semiarid regions: Geol. Soc. America Bull., v. 65, p. 175–189

Bull, W. B., 1963, Alluvial fan deposits in western Fresno County, California: Jour. Geology, v. 71, p. 243–251

—— 1964, Geomorphology of segmented alluvial fans in western Fresno County, California: Geol. Survey Prof. Paper 352-E, p. 89–128

Conway, W. M., 1893, Exploration in the Mustagh Mountains: Jour. Geology, v. 2, p. 289–303

Davis, W. M., 1938, Sheetfloods and streamfloods: Geol. Soc. America Bull., v. 49, p. 1337–1416

Eckis, R., 1928, Alluvial fans in the Cucamonga District, southern California: Jour. Geology, v. 36, p. 224–247

McGee, W. J., 1897, Sheetflood erosion: Geol. Soc. America Bull., v. 8, p. 87–112

Sharp, R. P., 1942, Mudflow levees: Jour. Geomorphology, v. 5, p. 222–227

Sharp, R. P., and Nobles, L. H., 1953, Mudflow of 1941 at Wrightwood, southern California: Geol. Soc. America Bull., v. 64, p. 547–560

Wiman, S., 1963, A preliminary study of experimental frost weathering: Geog. Annaler, v. 45, p. 113–121

Manuscript Received by the Society September 16, 1964

8

Copyright © 1967 by the American Journal of Science
Reprinted from Am. Jour. Sci. **265**:81-105 (1967)

FANS AND PEDIMENTS*

CHARLES S. DENNY
U. S. Geological Survey, Washington, D. C.

ABSTRACT. Alluvial fans and pediments are discussed within the framework of an open system. Most desert piedmonts include both fans and pediments and have systems of stream channels with contrasting drainage area, discharge, gradient, and bedload. The drainage net on a piedmont includes streams that head in the adjacent highlands and others that rise on the piedmont and are tributary to the streams from the highlands. The reason for these contrasts is geologic; it is the presence of a highland that sheds coarse detritus to an adjacent piedmont where the debris is stored for a time until it is weathered and then removed both in solution and by stream erosion.

Washes in desert areas probably increase in discharge downstream only in mountains; on piedmonts, discharge may decrease rapidly downfan. Using the open-system concept, a piedmont may be considered as approaching a steady state when the rate of movement of detritus from mountain to piedmont is nearly the same as the rate at which material moves from piedmont to playa or flood plain. To maintain such a steady state requires broad areas of piedmont where erosion is the dominant process. Whether these areas of erosion are areas of more or less bare rock—pediments— or whether they are abandoned segments of fans depends upon the geometry of highland and lowland. Large highlands and small lowlands favor complex fans where half the surface area may no longer receive sediment but be subject to erosion, whereas broad lowlands and small highlands favor extensive pediments.

An open-system approach to the origin of fans and pediments encourages the study of present-day processes and opposes the current tendency to rely solely on climatic change as the explanation for features whose mode of formation may be in doubt.

INTRODUCTION

This paper is concerned with the application of the principle of dynamic equilibrium to the formation of alluvial fans and pediments (Strahler, 1952; Hack and Goodlett, 1960; Hack, 1965; Chorley, 1962). In this discussion, piedmont refers to the plain with locally complex topography that slopes from highland to playa or flood plain. The piedmont is an important part of the arid landscape of the Basin and Range province. Part of a piedmont may be a more or less bare rock surface, a pediment. Part is alluvium, and if its surface has the form of a section of a very low cone it is a fan. In the Death Valley region most piedmonts are largely coalescing alluvial fans; in parts of the Rio Grande Valley the pediment is the dominant form. Piedmonts with discontinuous gravel aprons and areas of bare rock occur also in humid regions.

The interpretation of piedmont formation under the concept of dynamic equilibrium assumes that a piedmont acquires its specific shape or morphology because it is acted on by certain processes and will maintain this morphology as long as these processes are in operation. If these processes change their character or intensity, then the piedmont changes

* Publication authorized by the Director, U. S. Geological Survey.

its morphology. Process and form are indivisible. Neither can exist without the other. In piedmont formation the energy of the system depends primarily upon the height of the mountains above the surrounding piedmont. This energy performs work through such agents as running water, weathering, and mass wasting to maintain the piedmont in its proper state as regards its size, shape, character of surface, and so on. A steady state may be maintained at different levels, depending especially upon the ratio between the rates of its "gains" and "losses". For piedmonts this point refers to the rates of movement of material from the upland to the piedmont and from the piedmont to the adjacent playa or flood plain.

In the Death Valley region, many piedmonts have large segments that do not receive detritus from the mountains. Material is being removed from these segments by local runoff. A certain amount of detritus is added to the piedmont each year, and a certain proportion of the total amount of fan debris is removed each year. The fan (part of the piedmont) grows larger year by year, but eventually the rates of supply and removal must be equal. The piedmont has then attained equilibrium; the volume of detritus on the piedmont is constant and will remain the same as long as the environment does not change. Nikiforoff (1942, 1959) devised a simple mathematical formula to illustrate the equilibrium principle as applied to soil formation, and Hack (1965, p. 7) has described how the formula can be applied to talus accumulation at the base of a cliff. It can be shown to apply also to piedmont formation when the rates of supply and removal are equal.

Desert piedmonts have two systems of stream channels, those that head in the adjacent highlands and others that rise on the piedmont and are tributary to the streams from the highlands. The reason for these contrasts is geologic. The highland sheds coarse detritus to an adjacent piedmont where the debris is stored for a time until it is weathered and then removed both in solution and by stream erosion. The highland owes its existence either to tectonic movements or to resistant bedrock.

Washes in desert areas probably increase in discharge downstream only in mountains; on piedmonts, discharge may decrease rapidly downfan. For this reason the amount of debris supplied to a desert piedmont will greatly exceed the amount removed from it unless the area of piedmont that is being eroded is large. To maintain a steady state between form of piedmont and rates of supply and removal requires broad areas of piedmont where erosion is the dominant process. Whether these areas of erosion on the piedmont include broad sloping plains or pediments or whether the areas of erosion are deeply gullied into badlands depends primarily on geology of mountain and basin.

This picture of piedmont development in the desert is an outgrowth of a study of alluvial fans in the Death Valley region made in 1956-58 (Denny, 1965; Denny and Drewes, 1965) and of earlier studies in New

Mexico (Denny, 1941). In the preparation of this paper I am indebted to C. C. Nikiforoff for advice and encouragement. I am also particularly grateful to J. T. Hack and R. P. Sharp for their suggestions and critical discussion.

A THEORY OF ALLUVIAL-FAN FORMATION

A fan is "an accumulation of debris brought down by a stream descending through a steep ravine and debouching in the plain beneath, where the detrital material spreads out in the shape of a fan, forming a section of a very low cone" (Howell and others, 1960, p. 105). The prime requisite of fan formation, the setting of highland and lowland side by side, can be explained in various ways. In Virginia, it is produced by the differential erosion of adjacent bodies of rock of varying resistance; in the Death Valley region it is the result of tectonic movements. In common with stream channels in humid regions, it is probable that the average channel slope of a desert wash in the mountains is inversely proportional to its discharge, but for a given discharge it is directly proportional to a function of the material it transports and the resistance of the bedrock or material that encloses the channel (Hack, 1957). If the mountain is composed dominantly of resistant rock, the streams in the mountain will have much steeper gradients than those in the adjacent lowland. Where a stream reaches the mountain front it encounters these gentle slopes and deposits the part of its load that is too coarse to be moved on such subdued gradients. The wash builds a fan in the lowland and alluviates its channel in the mountain. It thus has a smooth slightly concave-upward profile from valley floor well up into the adjacent highlands.

Deposition is by no means the only process that operates on an alluvial fan. Although a small fan at the base of a Recent fault scarp may be entirely covered by water during floods, many large fans include areas that have not received sediment for a long time. The broad apron of coalescing fans that form the piedmont in Death Valley east of the Panamint Range is a mosaic of areas of desert pavement and of wash (Noble and Wright, 1954, fig. 2; Hunt and Mabey, 1966). The local relief on the fan surface near the mountain front is more than 100 feet. More than half the surface of the fan is desert pavement and gully—segments where weathering and erosion have been the dominant processes for a long time. Sedimentation is now limited to less than half the surface of the fan.

In the arid cycle of Davis (1905), fans are characteristic of the youthful stage and grow larger and larger until the mountains are considerably reduced, broad alluvial aprons slope toward extensive playas or flood plans, and the several originally enclosed basins have become integrated. This picture of the long-continued expansion of alluvial fans is not entirely consistent with the existence of large fans where more than half the surface is being eroded and deposition is limited to only a small segment. If we consider fan formation in the framework of a dynamic system, a more consistent hypothesis can be devised.

Initially, let us suppose that a graben begins to be lowered across an area of low relief. At once fans begin to form adjacent to the initial fault scarp on either side of the sinking block. If the rate of depression were exceedingly slow, perhaps the rate of erosion of the bordering fans might soon approach their rate of growth. Such fans might indeed be so small as to pass almost unnoticed.

If the rate of depression of the graben is rapid, and if the mountains are large and the lowlands are broad, fans may grow and coalesce. Assume that the volume of material supplied to these fans per year is constant. Actually, of course, the amount of detritus brought to a fan varies from year to year, and the rate can be regarded as constant only as an average over many years. The torrential summer cloudbursts in desert regions have a highly erratic distribution both in time and in space. Floods that come from the mountains and spread water only near a fan's apex are a more frequent occurrence than those due to precipitation that falls directly on a fan. Mountains induce precipitation, concentrate flow, and have channels that do not lose as much water by percolation as those on fans. On the ideal fan debris is commonly deposited near the fan's geometrical apex. Movement of debris downfan from the apex is erratic, taking place in short periods of flow separated by long intervals of dryness. An occasional flood may carry debris from mountain to playa, but floods that do not pass far beyond the apex are probably more frequent.

Many large fans are dissected near the mountain front so that their upper part no longer receives sediment from the mountain. The loci of deposition is shifted away from the mountain front, away from the fan's geometrical apex. This incision is an important step in the history of a complex fan and probably may occur for several different reasons, depending on the local geology (Eckis, 1928, p. 236-241).

Faulting may cause fan-head trenching. Most active fans in the Death Valley region have a smooth, slightly concave upward profile from the central playa or flood plain well up into the adjacent highlands. A relative sinking of the valley floor, expressed by a fault scarp on the fan a short distance below the apex, like some of those that cross fans in Panamint and Death Valleys (Jennings, 1958), may cause a steepening of the slope of the fan-building wash for a short distance upfan from the scarp, perhaps to a point inside the mountain front. Thus the depositional apex of the fan is shifted downfan to the scarp. Probably the stream channel above this new apex will soon be regraded to another smooth slightly concave upward profile similar to that present before faulting. Regrading would probably involve a slight downcutting along the entire length of the stream channel in the mountains and the deposition of this debris downfan from the new apex.

Fan-head trenching may be the normal consequence of large variations in flood discharge. That is, during a long period of time a fan will have a tendency to grow at its apex, but occasionally an exceptionally large flood will trench the fan near its apex and deposit near its toe.

Consider the hypothetical example shown in figure 1. Map A shows several fans at the base of a fault scarp. Map B (fig. 1) shows the hypothetical fans after fan-head trenching has taken place. Detritus from the mountain bypasses segment (1) by way of a narrow trench and is deposited as fan segment (2) which laps up over a part of the outer edge of segment (1). Weathering and erosion by local runoff and by wind become the dominant processes on abandoned segment (1). Gullies are carved in it by local runoff—one is shown in map B—and desert pavement develops on the interfluves. These gullies that head on the fan—the fan washes—erode the fan and deposit their load either on the adjacent playa or flood plain or further downfan. Thus, fan washes are to be contrasted with those from the highland—the mountain washes—that are primarily agents of deposition. Near its headwaters at point x, the floor of the gully may be below the level of the adjacent large wash at y. If the mountain wash erodes its left bank at y, the wash may cut through its bank and spill down into the meandering gully at x. This piracy leads to the abandonment of segment (2) and the enlargement of (3) (map C).

The area of the fan that no longer receives detritus from the mountain, or the area where weathering and erosion are dominant, is now greater than before piracy took place. This increase in the area being eroded causes an increase in the rate of removal of material from the fan. Because we have assumed in this hypothetical analysis that the volume of material deposited on the fan per year is constant, any increase in the area of erosion requires a decrease in rate of growth of the fan. Other drainage diversions will take place; one such event is suggested in map C. These changes will cause further shifts in areas of deposition and corresponding increases in total area being eroded. This ideal fan is approaching a time when the amount supplied to it per year will equal that removed by erosion. The fan will then be in a steady state of fan formation. Thereafter the fan will not grow larger either in area or in volume, but the geographic position of active and abandoned segments will be continually changing. Mountain washes deposit, and fan washes erode; an equilibrium between them is maintained by piracies.

Requirements of the theory.—The interpretation of piedmonts as the steady state of an open system requires a shift from time to time in the loci of deposition and of erosion. Otherwise part of the piedmont's surface would grow higher and higher while the remainder would be reduced to lower and lower elevations. The transformation of the surface of an abandoned segment of a fan (part of the piedmont) into

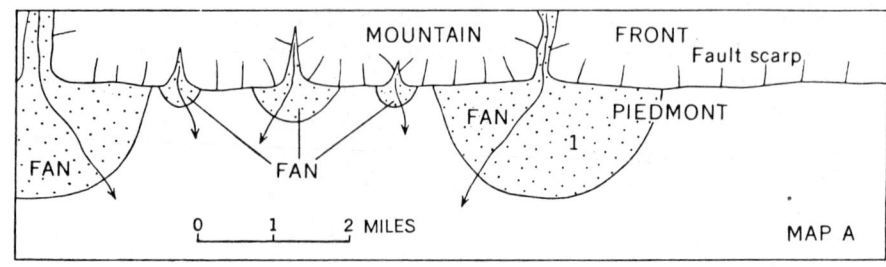

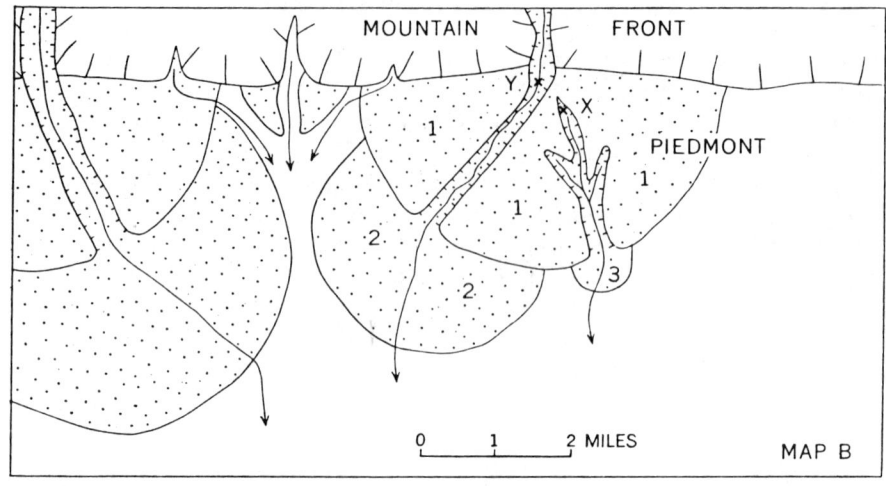

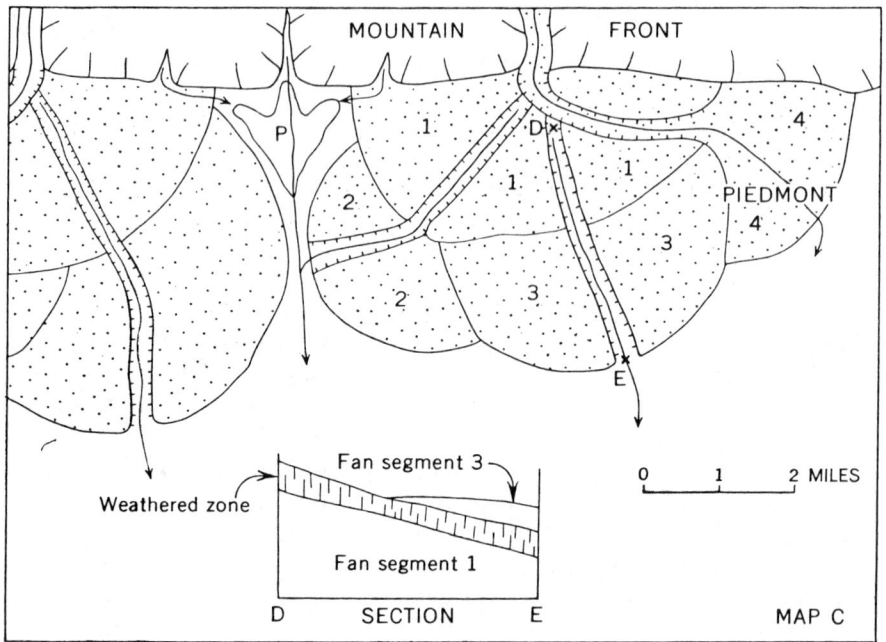

desert pavement is a part of the process of removal and is discussed later. Changes in area of deposition take place because washes with differing gradient and load occur side by side and lead to drainage diversion or piracy. The juxtaposition of contrasting washes is caused by a variation in geology and hydrology between mountain and fan. The mountain may be made of resistant rock that yields coarse detritus which is carried on a steep gradient out onto the piedmont and deposited as a fan. The small, commonly meandering gully that heads on the piedmont has a fine-grained bedload because it comes from unconsolidated material, in part from weathered debris—the pavement and the weathered zone beneath it. The slope of the gully is less than that of the fan-building wash to which it is tributary. The difference in profile is commonly so great that the wash from the mountain crosses the fan at higher elevations than its own downstream tributary, the meandering gully. The gradient of the gully can be low because its bedload is fine grained and because rapid runoff on adjacent pavements tends to increase gully discharge. It is the erosion by meandering gullies, caused by local runoff on the piedmont, that is continually removing debris from the fan.

Piracy on piedmonts, related to lithologic contrasts between highland and lowland, has been noted by several authors (Rich, 1935; Mackin, 1936; Hunt, Averitt, and Miller, 1953; Knechtel, 1953; Hack, 1960). The Shadow Mountain fan in the Amargosa Valley, California, is described below to illustrate the part played by piracy in the maintenance of the steady state of fan or of piedmont formation.

Shadow Mountain fan as an illustration.—A piedmont extends from the west base of Shadow Mountain, east of Death Valley Junction, California, for 3 to 6 miles to the flood plain of Carson Slough, a tributary of the Amargosa River (fig. 2), which lies about 3000 feet below the mountain top. (See map of Shadow Mountain fan, pl. 1 in Denny, 1965). The piedmont is largely coalescing alluvial fans; there are but two small areas of pediment, K and L, that cut across deformed Tertiary(?) rocks (Denny and Drewes, 1965, pl. 1).

The Shadow Mountain fan heads in a reentrant on the north slope of Shadow Mountain. The fan is concave upward in an east-west direction, the average slope being about 700 feet per mile near the mountain front, decreasing to less than 100 feet per mile near Carson Slough. The fan is built of fragments of quartzite and other resistant rock.

Fig. 1. Maps showing development of alluvial fans. Scales approximate. Map A, small fan at base of recently elevated mountain. Map B, wash from mountain has dissected original fan, segment (1), and is building a new fan, segment (2). Another wash, heading in abandoned segment (1), is building a new fan, segment (3), at its mouth. Map C, two drainage diversions have taken place causing wash from mountain to abandon segments (2) and (3) and to build new segment (4) at its mouth. Section D-E, stratigraphic relations betweens fan segments (1) and (3) as they are exposed in wall of gully D-E.

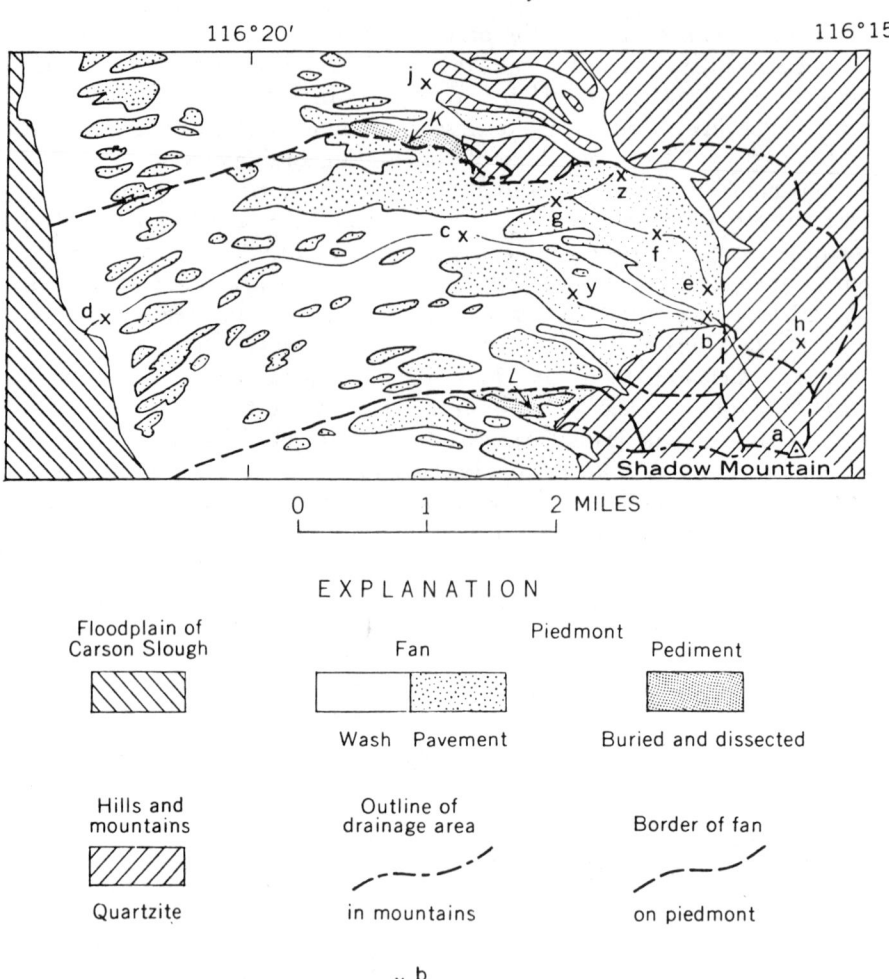

Fig. 2. Map of Shadow Mountain piedmont on east side of Amargosa Valley near Death Valley Junction, California.

Near the mountain front the material is pebble and cobble gravel; near the toe the gravel fraction is finer grained, and more than half of the fan debris is sand and silt.

The fan is a complex mosaic of desert pavements and of washes with a total surface area of about 9 square miles. The desert pavements that cover about one-third of the fan are smooth, gently sloping surfaces composed of closely packed angular fragments of rock. All the large areas of pavement are dissected by steep-sided gullies (most are not shown on fig. 2). Pavements are described later. Washes that cover the rest of the fan consist of braided channels and gravel bars. About one-third of the area of washes is floored with unweathered gravel that is moved by present-day floods; the remaining washes have their surface material coated with desert varnish and have been dry for a long time, perhaps for 2000 years (Hunt and Mabey, 1966). The local relief between pavement and wash ranges from about 40 feet near the fan's apex to only 1 or 2 feet near the toe.

On Shadow Mountain fan, the juxtaposition of washes with contrasting gradient and bedload has led to piracy. One of the principal fan-building washes heads at the mountain summit and runs northward and westward to Carson Slough. In figure 2 the course is roughly a-b-c-d. A second wash heading in the pavement on the north side of the fan at point e runs through the pavement for about 1½ miles (e-f-g) to join the large wash at c. At points approximately a mile up each wash from where they join, the gradient and size of bedload of the large wash are about twice that of the small one. Throughout much of its length the bed of the large wash lies below the adjacent pavement, but at some points, such as y, the wash's bed is even with the pavement, and unweathered detritus laps onto a floor of varnished stones.

Another large wash skirts the northeast edge of the fan, draining a large area on the north slope of Shadow Mountain; the headwaters are near h. This wash passes north of the fan by way of several channels that reach the piedmont near j. In actuality, runoff from the north slope does not reach j because the upper part of the large wash has been diverted westward at z, in the northeast part of the map area, and now drains to the Shadow Mountain fan (z-g-c). Before piracy took place, the bed of a small west-flowing wash heading on the pavement near z lay about 10 feet below that of the large wash to the northeast. Apparently the large wash cut through its south bank near z to empty into the adjacent, small, low-lying channel.

On the bed of the large wash near the elbow of piracy the new west-flowing channel is separated from the older northwest-trending one (trending toward j) by a gravel bank only 2 or 3 feet high. Clearly this drainage diversion was initiated but a short time ago. A high flow of water down the large wash could still supply the water to both channels at z. Because of this piracy, the amount of water and detritus reaching the Shadow Mountain fan in time of flood will increase whilst that supplied the piedmont to the north will decrease. This shift in the equilibrium between form and process will probably cause alluviation of the narrow gully in the pavement (z-g) and the gradual building of a low cone of detritus near the mouth (c).

This piracy is but the most recent one. The gravel beneath the broad area of pavement on the fan's north edge was derived in part from the north slope of the mountain (Denny, 1965, fig. 16). A later diversion of water away from the fan established the northwesterly drainage (z-j) that has only just now been beheaded.

Piracy is what maintains the balance on the piedmont between the processes of erosion, deposition, and weathering. The shifting from time to time of the loci of these processes places a constraint on long-continued fan-building. That is why the surface area of many fans, regardless of their past history, bears a simple functional relation to the area of their highland sources. Weathering and erosion have been for a long time the dominant processes on the large pavement near the northeast edge of the fan, while deposition has been dominant on the piedmont to the north, near j. As a result of the recent piracy, a reduced amount of debris now reaches the piedmont northwest of the fan where I would now expect stream channels to be incised and the material on the interfluves to weather. Meanwhile the rate and area of sedimentation on the fan itself will increase because its source area has been doubled. Some of the small areas of pavement on the lower half of the fan are only a foot or two above adjacent washes and may in time be buried under detritus.

It is interesting to note that these examples of localized deposition or dissection on the Shadow Mountain fan and in the adjacent mountains are the same sort of features that in Deep Springs Valley, California, are believed by Lustig to be indicative of climatic change (Lustig, 1965). Shifts in loci of deposition from within the mountains to points far down the fan, fan-head trenching, abandoned channels (channels no longer connected to mountains), and hanging tributary fans in the mountains are all features of Deep Springs Valley that have their counterparts on the Shadow Mountain fan.

Desert pavement, an example of a steady state.—The abandoned segments of many fans that have not received sediment from the mountain for a long time are traversed by gullies that head in the abandoned segment (map A, fig. 3). The gullies, many of which have meandering courses, range in depth from a few feet to many tens of feet. Near its headwaters the floor of a gully lies well below the bed of the adjacent wash (section B-B') although the gully empties onto the wash farther downfan. On most fans, except those composed of granitic detritus (R. P. Sharp, written communications, 1963), the gullies are separated by flattish areas of desert pavement.

On most abandoned segments, the area of pavement is considerably greater than the area of meandering gully, suggesting, at first glance, that a once more extensive pavement is now being dissected. This suggestion implies that pavement-forming processes were dominant during some interval in the past, but that now dissection is the sovereign process. The absence of ungullied pavements, the ubiquity of meandering gullies, and the presence of pavements on the walls of many gullies leads me to wonder if pavements on fans are not always dissected, that the processes of pavement formation and dissection go hand in hand (Sharon,

Fans and pediments 91

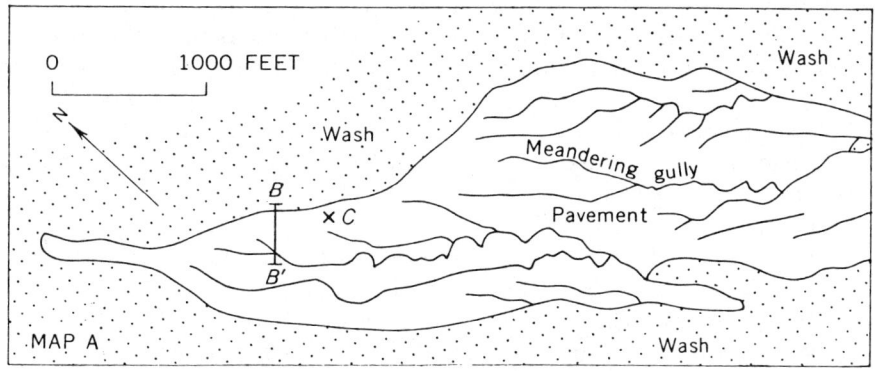

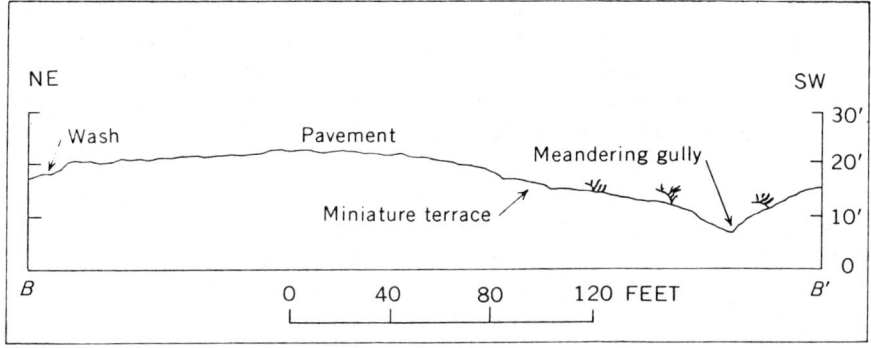

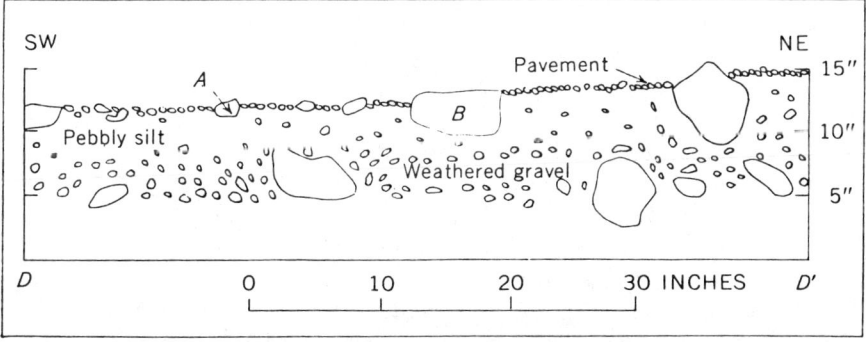

Fig. 3. Map, profile, and section of pavements on fans in Amargosa Valley, near Death Valley Junction, California (Denny, 1965, pl. 2, figs. 8 and 11). Map A, map of pavement and surrounding washes on Bat Mountain fan about 4 miles northwest of Death Valley Junction. Meandering gullies head in pavement. B-B′, profile across parts of pavement and wash showing miniature terraces. Floor of gully lies nearly 10 feet below bed of adjacent wash. For location see map A. D-D′, section of pavement on Shadow Mountain fan (near g, fig. 2). Armor of stones rests on pebbly silt that is transitional downward into weathered gravel. Large stones are risers of miniature terraces.

1962, p. 133). Doubtless there are places where pavement dissection reflects some tectonic or climatic event. Nevertheless I believe that many "dissected" pavements are approaching the steady state of an open system within the broader system of the piedmont as a whole. Pavement dissection tends to balance pavement formation. The pavement on a fan is dependent for its existence on the shifts in the locus of deposition caused by piracy.

Desert pavements are smooth, gently sloping surfaces composed of closely packed rock fragments commonly coated with desert varnish or faceted by solution. The stones range in diameter from a fraction of an inch to several feet and are slightly rounded to angular. Many of the latter are broken fragments of an abraded pebble, cobble, or boulder. The pavement is an armor that promotes runoff and protects the underlying material from removal by water or wind. Pavements rest on and in a layer of silty material ranging from about one to several inches thick that is transitional downward into weathered gravel (sec. D-D', fig. 3). R. P. Sharp (written communication, 1963) states that in many places the silty layer is less than an inch thick and rests with sharp contact on underlying material.

The silty layer beneath the armor of stones appears to be a concentrate formed in part by weathering and in part as a consequence of the processes that produce the stone armor, processes that transform an initial surface of channels and gravel bars into a smoth pavement. Some of the silty material may be the result of the weathering in place of gravel similar to that which underlies it. Under some pavements, the gravel is strongly weathered to depths of several feet, and some of the stones crumble in the fingers. However, in many places the gravel beneath pavements appears to be little weathered, and the silt between gravel and pavement must have been introduced. Perhaps in such places the silt was deposited by the wind, and the stones became set on and in it by lateral movements when the silt was wetted. Lateral movement of a saturated surface layer, as explained below, might also move stones upward to the pavement from beneath the silt.

The overall slope of a pavement is downfan; in map A of figure 3 this would be to the southeast. The local slope is generally toward the adjacent wash or meandering gully (sec. B-B') and may be nearly at right angles to the overall slope. Most pavements are broken by miniature terraces with risers less than an inch high which trend at right angles to the slope of the pavement for tens of feet. The treads of some of these terraces are composed of small fragments and coarse sand forming discontinuous bands across the pavement. In many cases the treads are indistinguishable from the adjacent pavement in surface form or in varnish coating, and it is only the riser that can be distinguished.

These miniature terraces apparently form by downslope flow when the underlying silty material is wetted to depths of 2 to 3 inches. The risers are miniature scarps formed at right angles to the pavement slope. It is this downslope movement, aided by surface wash and by wind action, that gradually transforms the channels and gravel bars of an abandoned wash into a smooth pavement.

Observations over several years led Hunt and Washburn (1960, 1966) to postulate that miniature terraces in Death Valley are not forming at present but are largely relics of a moister climate. In the Amargosa Valley, miniature terraces are forming at the present time. I observed a pavement after a heavy rain and found that the upper 2 inches of the silt layer was wet, whereas the material below was dry. When I removed a stone such as A in section D-D' of figure 3, the surrounding silt and its stone shell tended to flow down into the hole formerly occupied by the stone. In section D-D' the pavement is broken by two rock-defended terraces with risers about an inch high. When the surface layer of saturated silt moves downslope (to the left in D-D'), it piles up behind large stones such as B that are anchored in firm dry ground beneath. Smaller stones in the pavement, such as A, rest on the wet silt and are carried along with it.

Apparently pavements are born dissected. Once a segment of a fan ceases to be an area of sedimentation it begins to be gullied by local runoff while the interfluves between gullies are gradually smoothed. The smoothing of the interfluves involves the movement of fine debris down into an adjacent gully, both from the pavement itself and from the weathered zone beneath the pavement. This movement, which tends to fill up the gully, is opposed by the transporting power of local runoff down the gully. The stream in the gully deposits its load where it emerges downfan onto a broad wash. Pavement formation and gully erosion may tend to balance each other, to approach a steady state. Thus a pavement once formed may persist for some time. Although the actual stones forming it are gradually comminuted by weathering, they are replaced by others as the surface is gradually lowered by lateral movement of a surface layer, by rainwash, and by deflation. The appearance of a pavement may thus remain much the same. Similar pavements may be of diverse ages.

The deeper that gullies are cut below adjacent pavements the smaller the areas of pavement become, until they are finally completely consumed. On the Shadow Mountain fan the areas of pavement near the apex have the form of long fingers separated by narrow gullies. Some of the fingers have rounded tops, all areas of essentially smooth pavement having been destroyed. An area of pavement will ultimately be destroyed by erosion or buried beneath new increments of fan debris.

Discussion.—Thus far I have discussed fan formation only in terms of a steady state, but I do not mean to imply that all features of fans or of the piedmonts of which they are a part are in or approaching a steady state. Some features are probably relics of past conditions. The

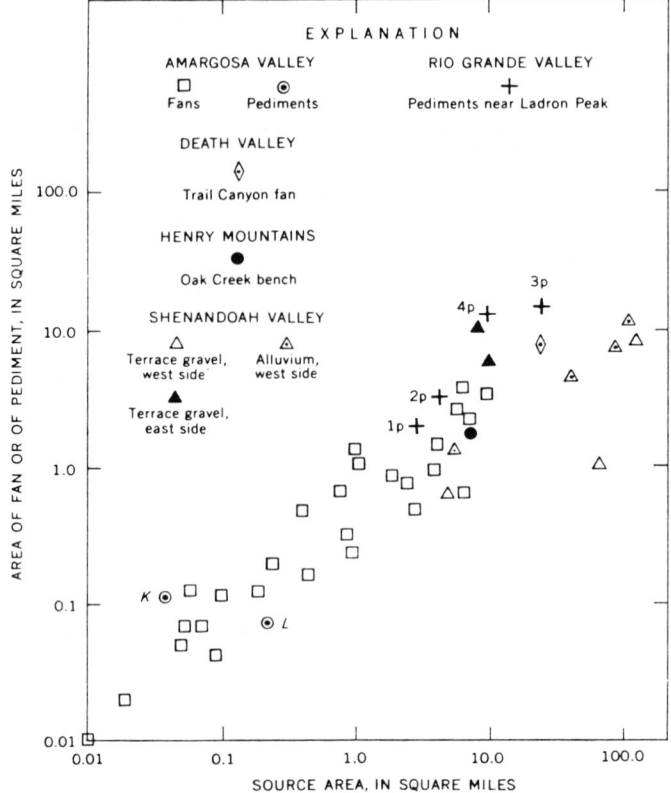

Fig. 4. Logarithmic graph showing the relation between the area of a fan or pediment and its source area in adjacent mountains. Data for Amargosa Valley from Denny (1965), for Shenandoah Valley from Hack (1965).

matter hinges on the rapidity with which landscape features adjust themselves to changes in the rates of the processes that act on them (Chorley, 1962, p. 6). Lustig studied alluvial fans in Deep Springs Valley, California, and concluded that climatic changes best account for their characteristics, their form tending toward an equilibrium with each successive climatic episode (Lustig, 1965). As already noted, many of the features cited by Lustig from Deep Springs Valley as evidence of climate change can, on the Shadow Mountain fan, be explained as the result of piracy. Without doubt, climatic changes have occurred and have influenced fan morphology, but the changes involve only the rates at which the processes operate. Changes in rates cause changes in the steady state, not a change in the nature of the system as a whole.

The example of a steady state described earlier involved equal rates of sediment supply and removal from a fan or piedmont. We have no data on what the actual rates involved may be, but we can compare the area of a fan with that of its source. Figure 4 shows that the size of fans in the Amargosa Valley, California, is an exponential function of the size of the source area. As we are dealing with fans of contrasting size, lithology, and geologic history, this relationship indicates a steady state between form and process.

Lustig believes that the surface areas of the alluvial fans in Deep Springs Valley, California, do not show a relation to their catchment areas in the adjacent mountains. He attributes this discrepancy to the fact that planimetric measurement of surface area may not accurately reflect the volumes of sediment that have been either removed from the catchment area or deposited in the fan (Lustig, 1965, p. 134). However a study of the topographic map of Deep Springs Valley indicates a fan area-source area relation in Deep Springs Valley similar to that in the Amargosa Valley if most of the southwest-sloping floor of the valley northeast of the playa is included as part of the fan with the largest catchment basin. It seems clear that the areas of fans will soon adjust to geologic changes caused by faulting or tilting and to changes in stream regimen due to piracy or climatic change.

The contrast in surface area between the small fans on the east side of Death Valley and the large fans on the west side does not reflect the volume of sediment the adjacent mountains have contributed to this basin. The source areas to west and to east are about the same size. The mountains to the west (Panamint Range) are about 3000 feet higher than those to the east (Black Mountains). This difference doubtless causes sedimentation on the Panamint Range fans to be faster than on the Black Mountain fans and will, in itself, cause the fans to be larger on the west side of the valley. However, the small surface area of the eastern fans is due also to eastward tilting of the valley floor, which causes more rapid deposition on the playa at the toes of the Black Mountain fans than at the toes of those to the west. To maintain the surface of the playa in a horizontal position, the eastern fans are being buried more rapidly than those to the west.

Another difficult problem is the exact nature of climatic change. What is the range in variability in the rates at which various processes operate from time to time within a given time interval? Could the features cited by Lustig as indicative of climatic change be merely the result of variations in magnitude and frequency of floods during the last few hundred years? I know of no actual measurements that might have a bearing on this point and offer only an analogy with features in a humid region, specifically the valley of Little River in the Appalachian Mountains in Virginia, described by Hack and Goodlett (1960). There, over a 7-year period, floods of various intensities produced terraces and dissected alluvial fans analogous in many respects to features in desert mountains. The features in Virginia do not record a climatic change.

Implications for stratigraphy.—The interpretation of fan stratigraphy under an open system differs from one based on a cyclic approach. Under Davisian theory, a dissected fan, such as most of those in Death Valley region, is the product of one or more depositional episodes separated by erosional intervals. The historical interpretation suggests that such a fan was built largely during glacial epochs and was weathered and dissected during parts of interglacial intervals. Under dynamic theory, the processes of erosion, weathering, and deposition are all in operation at the same time. It is only the relative intensity of these processes from place to place that has changed since the fan first came into existence. The concept of time independence of a fan requires the continued maintenance or renewal of the factors that brought it into being, such as relative position of mountain and basin, rate of sedimentation, and so on. In a tectonically active region such as Death Valley these factors may have obtained for a long time.

The difference in stratigraphic interpretation can be illustrated by an example based on the ideal fan described earlier. In figure 1 each segment is about the same size, but such perfect equilibrium is unlikely. One segment probably grows larger than another and partly buries the older one. In map C of figure 1, for example, segment (3) buries the lower edge of segment (1). An exposure along the bank of the gully that dissects both segments, section D-E, records first deposition of segment (1), then weathering of segment (1), and finally deposition of segment (3).

Under Davisian theory, this cross section could be the result of a series of episodes that affected all fans in the area. Perhaps deposition took place in a fluvial episode followed by weathering at a time when flooding was infrequent. Under dynamic theory, this section is interpreted as indicating that two episodes during which sedimentation was the dominant process at this place were interrupted by an interval during which weathering and erosion were the dominant processes at this same point in space. While segment (1) was being weathered, other segments of the fan were being eroded, and still others were receiving sediment.

The concept of simultaneous deposition and weathering on the surface of a fan was clearly expressed by Rollin Eckis (1934, p. 101):

> A consideration of the processes at work upon the present alluvial surfaces suggests the mode by which the complex series of alternating residual clays and unweathered deposits has accumulated on the piedmont slopes. Portions of the alluvial cones (fans) that are undergoing active deposition are being covered by unweathered gray deposits. Simultaneously, on the portions of these same cones remote from active deposition the gravels at the surface are breaking down to form red-brown soil clays. During the long and complex history of alluvial deposition in this area, accumulation of detritus has not been continuous over the whole cone, consequently red soil clays have developed on different parts of the cones at different periods, later to be buried by fresh deposits, and thus alternate with unweathered or partly weathered deposits in vertical section.

It may well be true that climatic change alters the rates of the processes that control alluvial-fan formation, but stratigraphic sections such as described above do not prove that this is so. Under dynamic theory, erosion, deposition, and weathering operate concurrently; they vary in relative intensity from place to place, not from time to time. In the absence of independent evidence, the presence of similar stratigraphic sequences on adjacent fans is not proof that the fans are contemporaneous, nor does it demonstrate that the sequences are climatically controlled. Perhaps they are, but this kind of evidence does not prove it.

ORIGIN OF PEDIMENTS

As used in this paper, the term pediment (Bryan, 1922, p. 88) refers to the part of the piedmont that is a more or less bare rock surface. Although pediments are commonly veneered with alluvium, the underlying surface on the bedrock—the pediment— is visible in many places. Where the veneer of alluvium is smooth and not gullied, it is sometimes difficult to decide whether the landform is pediment or fan.

Many geologists have speculated about the manner in which erosion fashions a pediment and about the origin of the sharp break in slope between mountain and pediment (see summary in Tuan, 1959). I will consider only the geometry of fan and pediment in areas where the mountains are of resistant rock and the valleys are of weak rock, a situation that leads to stream piracy and changes in areas of erosion and deposition. The mountains may be the result of faulting or of differential erosion, and the climate may be humid or arid. Examples are drawn from the arid Amargosa Valley, from the semiarid Henry Mountains and Rio Grande Valley, and from the humid Shenandoah Valley.

Shadow Mountain piedmont.—The pediments on the piedmont near Shadow Mountain in the Amargosa Valley illustrate the development of such surfaces in areas of high relief and active deformation. In this valley the areas of pediment are small. The climate is arid; precipitation is about 3 to 4 inches per year. The mountains are of resistant quartzite and limestone and furnish a coarse detritus to the piedmont.

On either side of the Shadow Mountain fan, weakly consolidated and deformed beds of sandstone, claystone, and fanglomerate are exposed in shallow gullies. These Tertiary rocks are overlain unconformably by a few feet of gravel. The unconformity at the base of the gravel is a buried pediment. These pediments (areas K and L, fig. 2) are in reentrants between large alluvial fans. They were eroded, buried, and reexposed by washes tributary to the large fans which thus act as a local base level.

The area of pediment at K is larger and that at L is smaller than the size of fans with comparable source areas (fig. 4). This difference is perhaps related to differences in the washes that formed the pediments. The wash at K has a low gradient because its source is an area of weak Tertiary rocks that yields a fine detritus. The wash at L has a coarse bedload and steep gradient because its source is in resistant limestone. The gradient and bedload of the wash at L more nearly resemble those of the adjacent fans than those of the wash at K.

The washes in which these pediments are exposed head in small segments of the adjacent hills, commonly meander, and resemble the meandering gullies in areas of pavement. Both types of gully have small drainage areas, low gradients, and fine bedloads. The only distinction between them is that some have cut down through the unconsolidated fan deposits into the underlying deformed bedrock. Because these gullies and the large washes from the mountain join downfan, adjacent segments of the two systems are at different levels and invite piracy. The gullies in pediment area L (fig. 2) lie 10 to 20 feet below the wash to the north and are separated from it by a low narrow ridge of gravel capped by desert pavement. If the wash should erode its south bank, it will spill down into the gullies and bury the pediment under fan gravel.

From the point of view of process the areas of dissected fan and gullied pediment are the same. They are both areas where gully erosion is the dominant process. The only real difference between the gullies is that those of the pediment, as commonly recognized, are eroded in the rocks of the mountain block or in deformed sediments of the adjacent depositional basin, whereas the gullies in the abandoned segments of a fan are flowing in alluvial materials that have a fan-shaped surface and have undergone little or no deformation.

The areas of pediment in the Amargosa Valley, controlled by the level of the adjacent fans, will remain small as long as the high relief between mountain and basin maintains large fans. Piracy may cause changes in location of areas of pediment but not in their relative proportion to the area of piedmont as a whole.

Henry Mountains piedmont.—The piedmont around the Henry Mountains, Utah, shows how the relative size of mountain and basin may influence the development of pediment and fan. The mountains are small and the surrounding lowlands are large. The piedmont north of the mountains, shown in figure 5, is about twice as long, from Table Mountain to Fremont River, as that west of Shadow Mountain. This contrast in size is one of the reasons why pediments are more extensive than fans on the Henry Mountains piedmont. The mountains are a group of isolated peaks produced by differential erosion of small bodies of resistant porphyry and indurated sedimentary rocks surrounded by broad areas underlain by shale. The climate is semiarid; precipitation ranges from about 7 inches on the piedmont to perhaps as much as 15 inches on the mountains.

On the piedmont north of the mountains (fig. 5), gravel (Qg_1 and Qg_2) composed chiefly of fragments of porphyry and indurated sedimentary rocks is deposited by streams heading on or near Table Mountain, while washes heading on the piedmont are continually removing gravel and carving valleys bordered by bare rock surfaces or pediments. The loci of erosion and deposition shift from time to time because of piracy.

Fig. 5. Map showing gravel deposits, areas of pediment, and the loci of stream piracies on a small segment of piedmont north of Henry Mountains, Utah (Hunt, Averitt, and Miller, 1953, pls. 18 and 20).

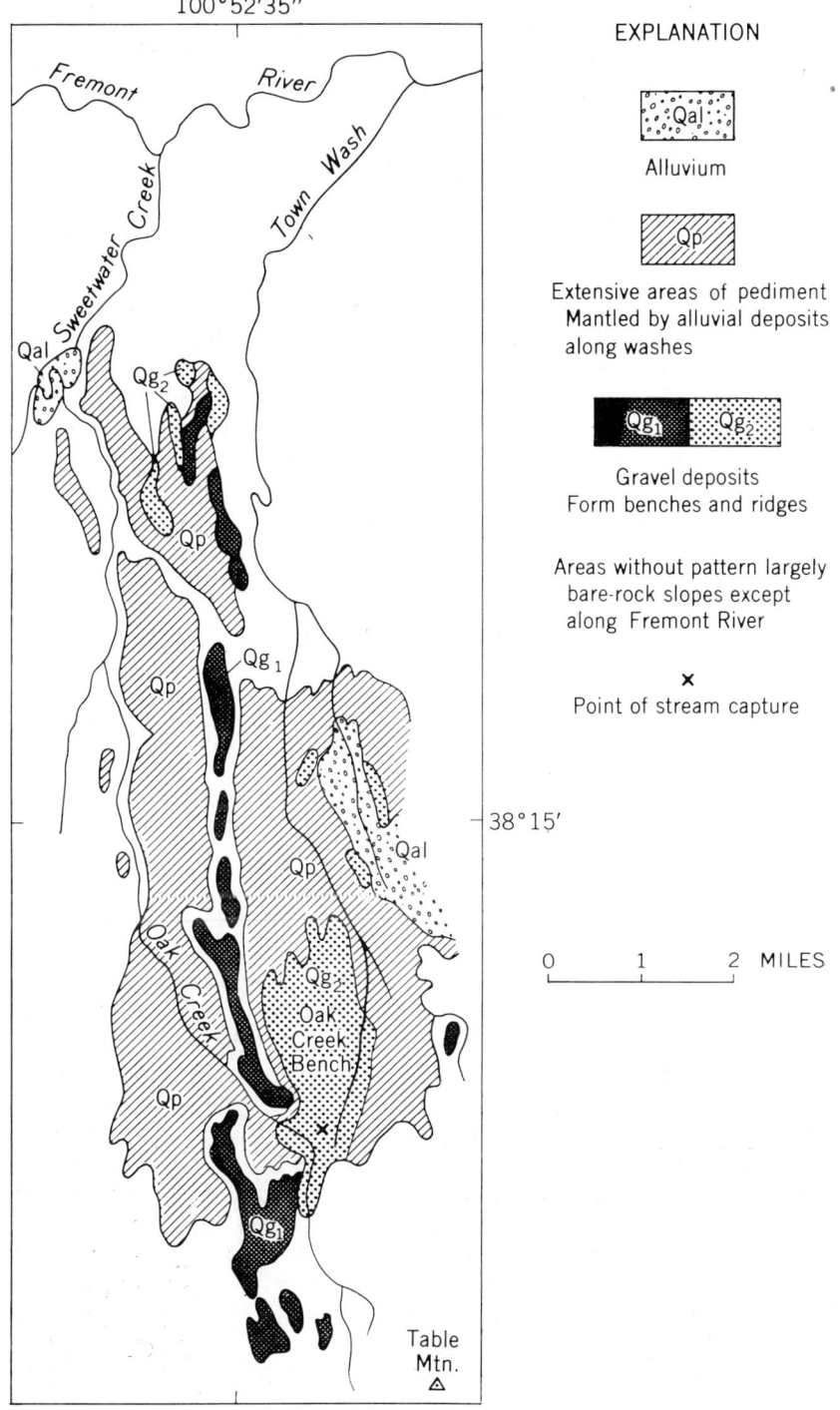

The gravel (Qg_2) of Oak Creek bench was deposited by the ancestral Oak Creek on a surface, probably a pediment, cut on the Blue Gate Shale. At present, Oak Creek flows in a shallow channel across the head of the bench to point X (fig. 5) where it turns sharply west and flows down into a broad valley. This broad valley was eroded by the stream that captured the ancestral Oak Creek at point X. Oak Creek is now aggrading the broad valley but is threatened with imminent capture by a tributary of Town Wash which has its headwaters close to point X. The divide between Oak Creek and Town Wash is only 10 feet high (Hunt, Averitt, and Miller, 1953, p. 195).

The piedmont north of the Henry Mountains is twice as long as that west of Shadow Mountain. If the systems of which these two piedmonts are a part are nearly in equilibrium, the present rate of supply of detritus to the larger piedmont is nowhere near enough to cover it. The size of Oak Creek bench relative to its source area on Table Mountain falls at the bottom of the field of scatter of the points in figure 4 representing fans in Amargosa Valley. In semiarid southern Utah one might expect a greater area of sedimentation for a given source area than in arid southern California (Langbein and Schumm, 1958). The discrepancy may be due to erosion that is removing the gravel that underlies Oak Creek bench.

Rio Grande Valley.—The following discussion of the Rio Grande Valley illustrates piedmont formation under a semiarid climate where the piedmont is graded to a through-flowing river. Changes in the master stream can affect the piedmont in various ways, depending upon the geologic setting of the valley.

The piedmonts of the Rio Grande Valley in central New Mexico are largely pediments cut across weak rocks and are veneered with gravel from the adjacent highlands of resistant rock. The valley is semiarid although slightly more humid than the Henry Mountains. Precipitation ranges from 10 inches on the piedmont to 20 inches in the highlands. The pediments are extensive relative to the size of the adjacent mountains, and their gravel veneers are in places thick enough to obscure the underlying bedrock. The pediments rise steplike above the river and were described by Bryan and others (Bryan, 1932; Bryan and McCann, 1938; Denny, 1941; Wright, 1946) as primarily stream-eroded surfaces graded to a successively lowered base level, the Rio Grande.

North of Ladron Peak, which lies about 15 miles west of the junction of the Rio Puerco and the Rio Grande, are the remnants of three pediments that slope northward and eastward toward the Rio Puerco about 5 miles east of the eastern limit of the area shown in figure 6. (Denny, 1941, fig. 2). The small remnant of the Tio Bartolo pediment is a steeply sloping, fan-shaped mass of bouldery gravel resting in part on pre-Tertiary rocks. The profile on top of the gravel was thought (Denny, 1941) to be graded to a level about 400 feet above the Rio Grande. The Valle de Parida pediment, a gravel-covered and slightly dissected plain,

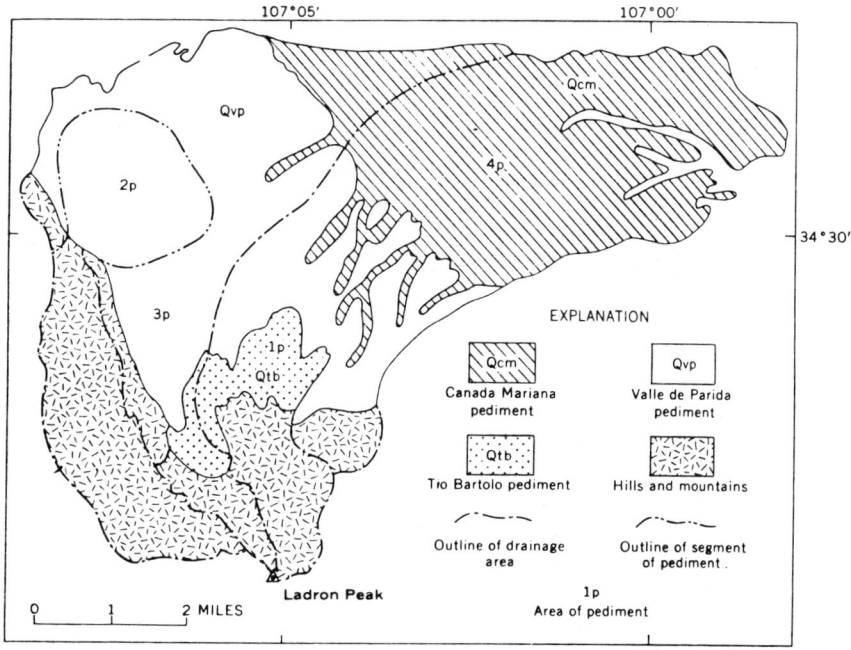

Fig. 6. Map of pediments north of Ladron Peak, Socorro County, New Mexico. Map generalized from figure 2 in Denny (1941) and from Riley and Mesa Aparejo 15' quadrangles, U. S. Geological Survey.

slopes northeastward to a level about 150 feet above the Rio Grande. At a lower level and separated from the higher pediment by a gently sloping dissected escarpment lies a third pediment, the Cañada Mariana, graded to a level about 50 feet above the Rio Grande.

On figure 6, I have measured the area of these dissected pediments, or parts of them, and also the areas that drained to them and supplied their gravel cover. Plotted on figure 4, the points that represent these pediment remnants fall on the upper side of those for the Amargosa Valley fans but show a similar exponential relation to the size of their source areas.

In 1941, I stated, under Davisian theory, that these three pediments were primarily the result of lateral planation by streams from Ladron Peak. The grade of these streams was successively lowered because the Rio Grande, the local base level, periodically lowered its channel. The scarps between the pediments were said to have retreated headward parallel to themselves.

Under dynamic theory the formation of these pediments may be thought of as a continuous process, not dependent solely on successive lowerings of the Rio Grande separated by long intervals when the river's grade was stabilized. It would be surprising to have a relation between size of pediment and its source if the lower pediment is enlarging itself at the expense of the higher one. Rather it is probable that these three pediments near Ladron Peak, although their initiation may have been related to changes in the level of the Rio Grande, are now in equilibrium with present processes. Their variation in size and shape may be related to the local geology.

The Tio Bartolo pediment is cut across resistant pre-Tertiary rocks and shows evidence of local piracies. It resembles in form the areas of pavement on the upper part of the Shadow Mountain fan. This pediment is now bypassed by streams from Ladron Peak, and the gravel that caps it is old but has no necessary relation to any pause in the downcutting of the Rio Grande. It may well stand above the Valle de Parida pediment because of differences in the underlying bedrock that cause differences in the bed load of the streams which traverse these pediments. Differences in bed load lead to differences in gradient between the two pediments. That the size of the two lower pediments is also related to the local geology rather than to changes in the level of the Rio Grande is suggested by their area-source relations as shown in figure 4. Perhaps the northwest-trending break in slope between Valle de Parida and Cañada Mariana pediments reflects a change in the underlying bedrock from more resistant older Cenozoic or pre-Tertiary rocks on the west to less resistant later Cenozoic continental sediments on the east. The more resistant rocks may act as a local base level and retard the dissection of the Valle de Parida pediment.

The Rio Grande Valley pediments illustrate, as have those in Utah and California, the overall importance of the local geology in pediment formation. Isolated areas of pediment such as those near Ladron Peak do not necessarily have historical significance. They can be related to bedrock differences or to piracy. This does not mean that an extensive pediment in the Rio Grande Valley, such as the Ortiz pediment (Wright, 1946, pl. 10), does not record a long period of stability for the Rio Grande. Probably it does.

A second characteristic of many of the broad areas of pediment in the Rio Grande Valley is that such surfaces are smooth. This is due at least in part to their extensive gravel covers which in places are so thick that parts of these pediments should be classed as fans. The fan-shaped area of the Valle de Parida pediment, area 2p, is an example. The area-source relations (fig. 4) show that this fan is as much a part of the equilibrium of the piedmont north of Ladron Peak as are the associated pediments where slightly deformed older rocks are visible beneath a gravel cap. In places outside the area shown in figure 6 where the pediment surface beneath the gravel cap is well exposed, the pediment has a much greater local relief than the top of its gravel cap.

The smooth gently sloping gravel surfaces of the Rio Grande Valley contrast with the steeper and more irregular gravel surfaces in the Amargosa Valley. Although the major differences between these arid and semiarid piedmonts are due to geologic factors, it is probable that the greater rates of erosion and deposition on the semiarid piedmont favor gentle gradients and extensive gravel blankets.

Shenandoah Valley.—The gravel-covered footslopes of the Shenandoah Valley, Virginia, are an example of piedmont development in a humid region where the relief between mountain and basin is due to differential erosion of rocks of different resistances, rather than to faulting. The only tectonic requirement is that the region be maintained at an altitude sufficient to cause vigorous erosion for a long time. In the Shenandoah Valley solution is an important process, not only as a major factor in basin formation but also in the degradation of fans and pediments.

The rocks of the Shenandoah Valley region include a thick sequence of limestones, dolomites, and shales that form lowlands, and resistant quartzites and igneous rocks that form highlands. Alluvial terraces occur in the lowlands where a large stream enters from the adjacent mountains. These terraces and the adjacent flood plains have the form of fan-shaped, gravel-covered, and dissected pediments (Hack, 1960, 1965). Piracies take place where streams heading in the carbonate rocks are tributary to fan-building streams and have lower gradients than the main streams.

The area-source relationship of the gravel-covered piedmonts of the Shenandoah Valley is similar to that of the Amargosa Valley fans. In both regions the juxtaposition of hard and soft rocks leads to the formation of piedmonts alike in form in spite of great differences in climate. The Shenandoah Valley piedmonts are comparable in size with those on the west side of Death Valley in California. For comparison I show the area-source relation of the Trail Canyon fan in figure 4.

Although the overall extent of the gravels in the Shenandoah Valley is roughly proportional to the area that supplies the coarse detritus, in detail it is the modern, gravel-covered flood plains that have the most orderly area-source relation. Points representing the flood plains of four streams on the west side of the valley are shown in figure 4. Gravel-covered terraces that are remnants of abandoned flood plains of these same streams are smaller than their modern counterparts (fig. 4) and are no longer in equilibrium with the streams that deposited the gravel. The gravel of the terraces is being eroded by local runoff and by weathering. Gravel aprons east of the valley are larger and perhaps thicker than those on the west (fig. 4). The eastern aprons are now subject to both deposition and erosion and probably are in a steady state.

CONCLUSIONS

Many piedmonts are storage areas for coarse detritus in transit from highland to adjacent basin. If the basin is small the detritus builds alluvial fans that may completely cover the adjacent piedmont. Where a large basin adjoins highlands of resistant rock, fans may form only a part of the neighboring piedmont; on the remainder of the piedmont streams will carve broad pediments veneered with gravel. Piracies occur because of contrasting gradients between mountain and piedmont streams and cause shifts in loci of erosion and deposition, leading to the establishment of a steady state. Such systems exist in both arid and humid regions in spite of differences in the rates of erosion, of deposition, and of weathering.

A Davisian approach to the origin of piedmonts fosters a search for features having historical significance. It leads to an overreliance on climatic change as an explanation for features whose actual mode of origin may be in doubt. A buried soil, for example, under the historical approach suggests a past climatic change, whereas in the equilibrium concept this feature is regarded as part of a single system in equilibrium.

An analysis of the origin of piedmonts as an open system emphasizes both the geometry of a piedmont and the processes that act upon it. This does not exclude the recognition of features having historical significance, but it does force one to investigate present-day processes. Only from an understanding of the results of the forces at work today can we hope to recognize and explain relict forms.

REFERENCES

Bryan, Kirk, 1922, Erosion and sedimentation in the Pagago country, Arizona, with a sketch of the geology: U. S. Geol. Survey Bull. 730, p. 19-90.

───────, 1932, Pediments developed in basins with through drainage as illustrated by the Socorro area, New Mexico [abs.]: Geol. Soc. America Bull., v. 43, p. 128-129.

Bryan, Kirk, and McCann, F. T., 1938, The Ceja del Rio Puerco: A border feature of the Basin and Range province in New Mexico: Pt. 2, Geomorphology: Jour. Geology, v. 46, p. 1-16.

Chorley, R. J., 1962, Geomorphology and general systems theory: U. S. Geol. Survey Prof. Paper 500-B, 10 p.

Davis, W. M., 1905, The geographical cycle in an arid climate: Jour. Geology, v. 13, p. 381-407.

Denny, C. S., 1941, Quaternary geology of the San Acacia area, New Mexico: Jour. Geology, v. 49, p. 225-260.

───────, 1965, Alluvial fans in the Death Valley region, California and Nevada: U. S. Geol. Survey Prof. Paper 466, 62 p.

Denny, C. S., and Drewes, Harald, 1965, Geology of the Ash Meadows quadrangle, Nevada-California: U. S. Geol. Survey Bull. 1181-L, 56 p.

Eckis, Rollin, 1928, Alluvial fans of the Cucamonga district, southern California: Jour. Geology, v. 36, p. 225-247.

───────, 1934, South Coastal-basin investigation; geology and ground water storage capacity of valley fill: California Dept. Public Works, Water Resources Div. Bull. 45, 279 p.

Hack, J. T., 1957, Studies of longitudinal stream profiles in Virginia and Maryland: U. S. Geol. Survey Prof. Paper 294-B, p. 45-97.

───────, 1960, Interpretation of erosional topography in humid temperate regions: Am. Jour. Sci., v. 258-A, Bradley Vol., p. 80-97.

───────, 1965, Geomorphology of the Shenandoah Valley, Virginia and West Virginia, and origin of the residual ore deposits: U. S. Geol. Survey Prof. Paper 484, 84 p.

Hack, J. T., and Goodlett, J. C., 1960, Geomorphology and forest ecology of a mountain region in the central Appalachians: U. S. Geol. Survey Prof. Paper 347, 66 p.

Howell, J. V., and others, 1960, Glossary of geology and related sciences, with supplement, 2d ed.: Washington, D. C., Am. Geol. Inst., 325 and 72 p.

Hunt, C. B., Averitt, Paul, and Miller, R. L., 1953, Geology and geography of the Henry Mountains region, Utah: U. S. Geol. Survey Prof. Paper 228, 234 p.

Hunt, C. B., and Mabey, D. R., 1966, Stratigraphy and structure, Death Valley, California: U. S. Geol. Survey Prof. Paper 494-A, 162 p.

Hunt, C. B., and Washburn, A. L., 1960, Salt features that simulate ground patterns formed in cold climates: U. S. Geol. Survey Prof. Paper 400-B, p. 403.

——————, 1966, Patterned ground, in Hunt, C. B., Robinson, T. W., Bowles, W. A., and Washburn, A. L., Hydrologic basin, Death Valley, California: U. S. Geol. Survey Prof. Paper 494-B, p. B104-B133.

Jennings, C. W., compiler, 1958, Geologic map of California, Death Valley Sheet: California Div. Mines.

Knechtel, M. M., 1953, Pediments of the Little Rocky Mountains, north-central Montana [abs.]: Geol. Soc. America Bull., v. 64, p. 1445.

Langbein, W. B., and Schumm, S. A., 1958, Yield of sediment in relation to mean annual precipitation: Am. Geophys. Union Trans., v. 39, p. 1076-1084.

Lustig, L. K., 1965, Clastic sedimentation in Deep Springs Valley, California: U. S. Geol. Survey Prof. Paper 352-F; p. 131-192.

Mackin, J. H., 1936, The capture of the Greybull River: Am. Jour. Sci., 5th ser., v. 31, p. 373-385.

Nikiforoff, C. C., 1942, Fundamental formula of soil formation: Am. Jour. Sci., v. 240, p. 847-866.

——————, 1959, Reappraisal of the soil: Science, v. 129, no. 3343, p. 186-196.

Noble, L. F., and Wright, L. A., 1954, Geology of the central and southern Death Valley region, California, in Jahns, R. H., ed., Geology of southern California: California Dept. Nat. Resources Div. Mines Bull. 170, p. 143-160.

Rich, J. L., 1935, Origin and evolution of rock fans and pediments: Geol. Soc. America Bull., v. 46, p. 999-1024.

Sharon, David, 1962, On the nature of hamadas in Israel: Zeitschr. Geomorphologie, neue folge, v. 6, p. 129-147.

Strahler, A. N., 1952, Dynamic basis of geomorphology: Geol. Soc. America Bull., v. 63, p. 923-938.

Tuan, Yi-Fu, 1959, Pediments in southeastern Arizona: California Univ. Pubs. Geography, v. 13, 163 p.

Wright, H. E., Jr., 1946, Tertiary and Quaternary geology of the lower Rio Puerco area, New Mexico: Geol. Soc. America Bull., v. 57, p. 383-456.

ERRATUM

Page 83, the fourth line from the bottom should read: "flood plains, and the . . . ."

9

Copyright © 1967 by The University of Chicago
Reprinted from *Jour. Geology* **75**:438–460 (1967), by permission of The University of Chicago Press

PROCESSES ON ARID-REGION ALLUVIAL FANS[1]

ROGER LeB. HOOKE

Department of Geology and Geophysics, University of Minnesota

ABSTRACT

Alluvial fans were studied in the field, largely in the desert regions of California, and in the laboratory. Field study consisted of detailed mapping of parts of four fans and reconnaissance work on over one hundred additional fans. Features mapped included the nature and age of deposits, material size, and channel pattern. In the laboratory small alluvial fans were built of mud and sand transported through a channel into a 5-foot by 5-foot box under controlled conditions.

Material is transported to fans by debris flows or water flows that follow a main channel. This channel is generally incised at the fanhead, because there water is able to transport on a lower slope the material deposited earlier by debris flows. The main channel emerges onto the surface near a midfan point, herein called the "intersection point." On laboratory fans most deposition above the intersection point is by debris flows that exceed the depth of the incised channel. Fluvial deposition dominates below the intersection point. This depositional relation probably also occurs on natural fans.

On fans deficient in fine material large discharges may infiltrate completely before reaching the toe of the fan. Coarse debris is then deposited as lobate masses, herein called "sieve deposits." In many respects sieve deposits resemble debris-flow deposits, but they lack primary fine material, and fresh lobes are highly permeable.

INTRODUCTION

The object of this study was to understand processes acting on fans and the features produced by these processes. The conclusions reached are based on reconnaissance and detailed fieldwork and on a concurrent laboratory study. The reconnaissance work included qualitative observations of features such as debris size and lithology, and channel form on more than one hundred fans in the deserts of California. Measurements of fan area and slope were made in some instances.

Four fans were selected for detailed fieldwork; three of them are discussed in this paper (fig. 1). The detailed study involved geomorphic mapping by plane-table methods or on large-scale aerial photographs, measurement of channel cross-sections, and study of weathering phenomena.

The laboratory study focused attention on aspects of fan morphology that might otherwise have been overlooked or misidentified in the field, and also led to qualitative relationships among fan slope, water discharge, debris-flow behavior, and depth of fanhead incision. However, exact modeling of a specific fan cannot, at present,

[1] California Institute of Technology contribution no. 1393. Manuscript received October 18, 1966.

be used to obtain quantitative data, because scaling relationships for sediment transport by streams and debris flows are not sufficiently well known. Consequently fans built in the laboratory were treated as small fans in their own right, not as scale models of natural fans.

NATURAL FANS STUDIED IN DETAIL

Each of the three fans studied in detail (figs. 2, 3, and 4) had six distinguishable ages of alluvium (table 1). The youngest unit on each consisted of material in the presently active main channel. Abrasion during transport has given this debris a light gray color. A slightly older channel on Trollheim Fan also contains abraded gravels but is separated from the main channel by a 2-foot high levee.

The second unit also consists of abraded gravels either in abandoned channels or in terraces 2–10 feet above the main channel. However, the abrasion coating is perceptibly duller, and channel banks have slumped owing to the absence of frequent cutting. The beds of these channels are higher than that of the main channel where the two diverge; thus they will carry flow only if exceptionally high floods occur or if the main channel aggrades.

In the third unit is placed the oldest material clearly related to channeling. The abrasion coating has been removed by weathering, and in its place there is a brown weathering rind or, on Shadow Rock Fan, a light-brown desert varnish. On Gorak Shep Fan this unit consists of terraces 1–4 feet above the younger channel deposits. On the other two fans it is an old channel that widens down fan and diverges from the main channel at an elevation of 4 feet or more above the latter's bed. Topography in these channels is much more irregular than that on still older creep-smoothed surfaces.

Once channel deposits have been mapped, the remainder of the fan can usually be separated into two or three additional units. An important characteristic for distinguishing these units from each other and from channel deposits is the extent to which open pore space, commonly found in fresh gravels, has been filled by secondary fines during post-depositional weathering. Owing to lithologic differences in the source areas, pore-space filling proceeds at different rates on different fans, and the youngest unit in which the pore space appears to be filled will vary in relative age from fan to fan. For instance on Trollheim Fan the oldest

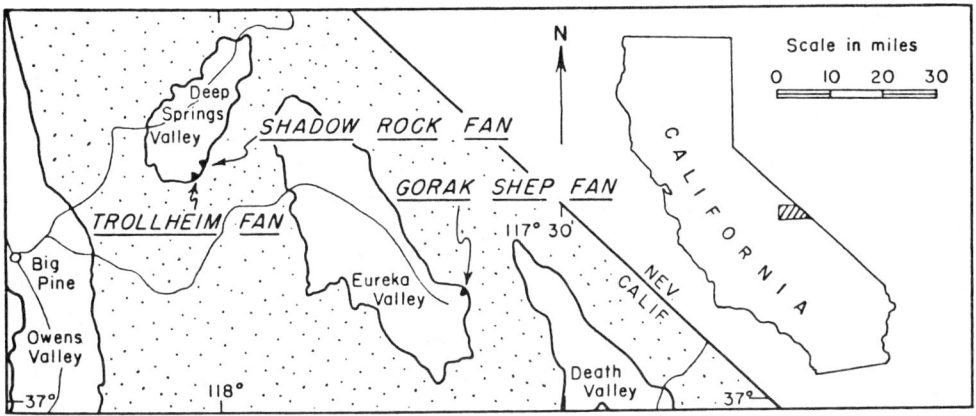

Fig. 1.—Index map showing locations of fans studied in detail. Exact locations are given in the Appendix

TABLE 1

UNITS MAPPED ON NATURAL FANS

Relative Age*	Gorak Shep Fan	Shadow Rock Fan	Trollheim Fan
1......	Recent channel deposits	Present channel	Present channel Overflow channel
2......	Overflow channel deposits	Terrace and/or older channel	Older channel material
3......	Oldest channel deposits	Oldest channel	Oldest channel material
4......	Youngest fan surface	Youngest fan surface	Younger fan surface
5......	Older fan surface	Older fan surface	Older fan surface
6......	Oldest fan material	Oldest fan surface	

* Correlations shown do not necessarily reflect absolute age.

channel material (table 1, unit 3) has been filled, but on Gorak Shep Fan it still has open pore space. The extreme is represented by Shadow Rock Fan, where open pore space occurs on the oldest fan surface. Because filling of the pore space promotes creep, older surfaces are generally smoother than channel bottoms, and desert pavements are commonly developed on them.

Additional weathering characteristics and topographic position were also used for distinguishing older units. For instance, on Gorak Shep Fan three units are identified by their positions in successively higher terraces near the fanhead. Terracing results from intermittent uplift on a fault (fig. 2). In contrast the three older units on Shadow Rock Fan are distinguished primarily by weathering phenomena. The first two of these units were identified by the relative darkness of desert varnish. In some places the contact was well defined, both on air photographs and on the ground, but elsewhere the age assignments were very subjective. Rocks in the oldest unit have an even darker desert-varnish coating. Furthermore, the unit occurs as low terraces and has a "softer texture" on air photographs. This textural difference probably results from vegetative differences, which in turn reflect the presence of additional secondary fines and thus greater weathering.

The relative age assignments of the three oldest units on Shadow Rock Fan are supported by the areal ratio of patches of fine detritus to areas of coarse material (fig. 3). Because the fines are secondary, older units should have proportionally more fine material.

Weathering characteristics also provided a basis for distinguishing the two oldest units on Trollheim Fan. Quartzite boulders in the oldest unit have a distinctly darker desert-varnish coating than those on the younger fan surface, and many carbonate boulders on the older surface have been reduced to crumbly fragments while those in the younger surface retain their original form. The older surface is also smoother and supports a denser vegetation cover.

SEGMENTATION

Gorak Shep and Trollheim fans have segmented profiles (fig. 5). Bull (1964b) has shown that segmentation may result from tectonic disturbance of the fan-source system. For instance, uplift of the source area, or a decrease in fan slope by tilting, may cause a new fan to be built out over the head of the old fan.

On both Gorak Shep and Trollheim fans the segmentation is of this type; that is, the youngest segment has developed near the fanhead, probably as a result of one of the above-mentioned causes. The conclusion that the upper segment is youngest in these two cases is based on: (1) apparent overlapping of material in the lower segment by that at the toe of the upper segment (pl. 1E), (2) weathering characteristics suggesting that the upper segment contains fresher material (esp. fig. 4), and (3) grading of the main channel to the slope of the upper segment. In cases where a lower segment is

---

PLATE 1

Laboratory apparatus and natural fans.

A, Water source for laboratory apparatus.

B, Working area of laboratory apparatus. Letters and dashed lines are referred to in text.

C, Exposure of gravels on Gorak Shep Fan in southeast bank of main channel upstream from fault scarp. Stratification is well defined. Void space in a lens of gravel (*arrow*) near the top of the bank has not yet been filled with secondary fines. There is no evidence of debris-flow action.

D, Sieve deposition on Shadow Rock Fan. In this instance the hummocky topography results from sieve deposition, but similar topography can be formed by debris-flow action. Note lighter color of more recent deposits to left of center.

E, Sieve deposition on Gorak Shep Fan. Note the bank in the foreground (*outlined*) emerging from beneath the sieve lobes. This bank is the "youngest fan surface material" in the vicinity of P on figure 2. *Arrow*, fault scarp near fanhead (see fig. 2).

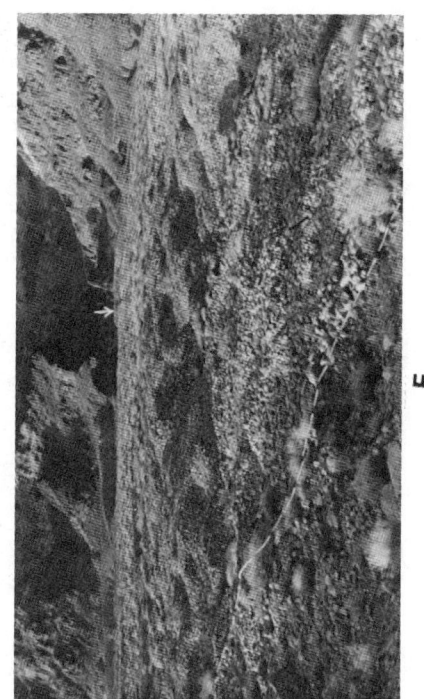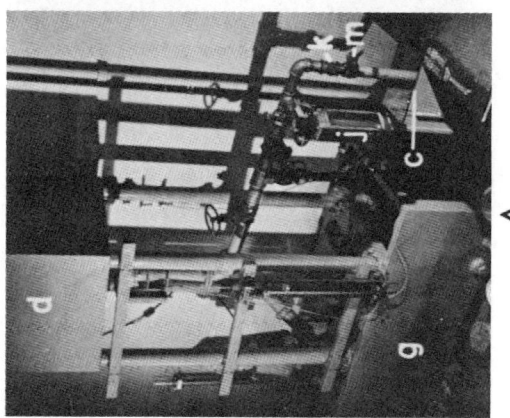

youngest, the upper segment is usually deeply incised, the channel being graded to a lower one (Bull, 1964b, p. 102 and fig. 66).

Although Trollheim Fan has four segments (fig. 5), Shadow Rock Fan, about a mile northeast of Trollheim, is apparently unsegmented. Both fans should have experienced the same tectonic history. The absence of segmentation on Shadow Rock Fan is attributed to the coarseness of material and importance of infiltration (sieve deposition) on this fan. Under these conditions fan slope appears to be less sensitive to tectonic influences.

MEASUREMENTS OF WEATHERING PHENOMENA

On Gorak Shep and Shadow Rock fans measurements were made that substantiate the age distinctions discussed above. On the latter, gradients of fronts of many sieve lobes (fig. 9) were measured, and steep slopes were found on a higher percentage of lobes in younger units (table 2). Because these lobes were probably deposited at or very near the angle of repose, and because the slopes would be expected to decrease with age due to creep, the data agree with age distinctions based on desert-varnish development.

On Gorak Shep Fan carbonate cobbles and boulders commonly contain quartz inclusions. Because quartz is more resistant to weathering than carbonate, quartz bands and knobs stand out in relief on more weathered rocks. Rocks showing such differential weathering were sought and the *maximum* relief measured. With the exception of one area where only ten such rocks could be found, between seventy and seventy-seven (usually seventy-five) measurements were made in each of the nine areas studied (table 3). Actual relief was not recorded, but measurements were put in class intervals, and a mean was calculated

TABLE 2

SLOPES OF SIEVE LOBES ON SHADOW ROCK FAN

Unit (See Fig. 6)	% of Slopes Greater than Indicated No. of Degrees		
	32°	30°	28°
Present channel..............	9	30	39
Terrace above present channel and older channel.........	7	22	34
Oldest channel..............	0	13	25
Youngest fan surface........	3	5	19
Older fan surface...........	0	2	5
Oldest fan surface...........	*	*	*

* Data inconclusive.

PLATE 2

Laboratory fans. Scale in *B* is graduated in inches; those in *E–H* are in centimeters.

A, Sieve deposition on fan A-1 after eighth episode.

B, Fan B-1, built entirely with debris flows.

C and *D*, Fan B-2, channel after episode 17 (*C*) and debris flow of episode 18 that followed this channel (*D*). Note lobe of debris that left channel at bend.

E and *F*, Fan B-2, debris flow of episode 38 (*E*) is being eroded by flow in *F*. Note debris-flow levees in *E* and distributary flow pattern below intersection point in *F*.

G and *H*, Fan B-3 at or near end of episodes 24 and 30, respectively. Owing to the absence of debris flows, existing channels at the fanhead are not cut down below bankfull stage as shown by the water level in *H*, and both fans are extensively braided, even near the fanhead.

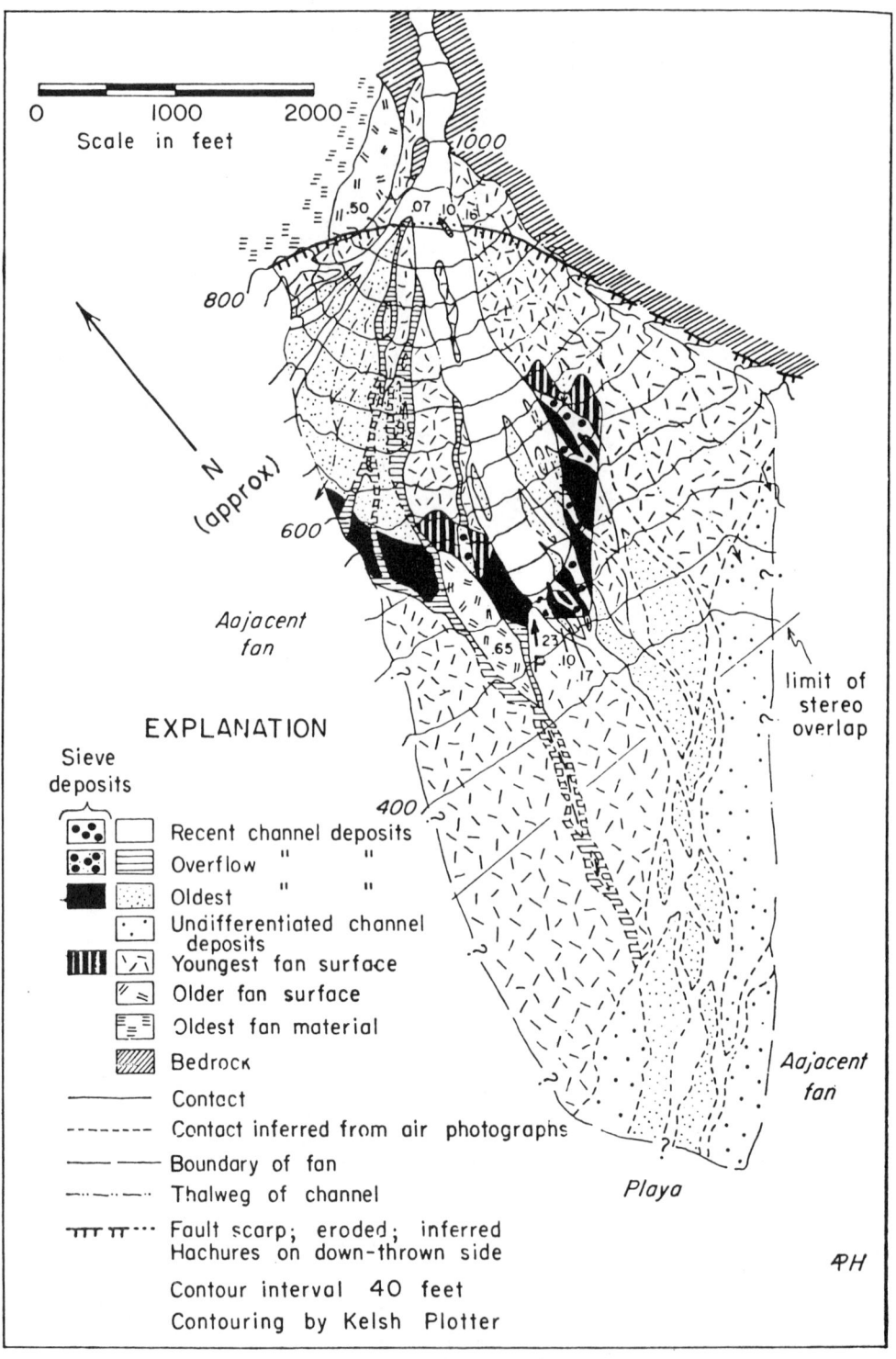

FIG. 2.—Map of Gorak Shep Fan constructed in the field on hazy, high-altitude air photographs. *P*, location of photograph in plate 1*E*. Numbers are discussed in text and table 3.

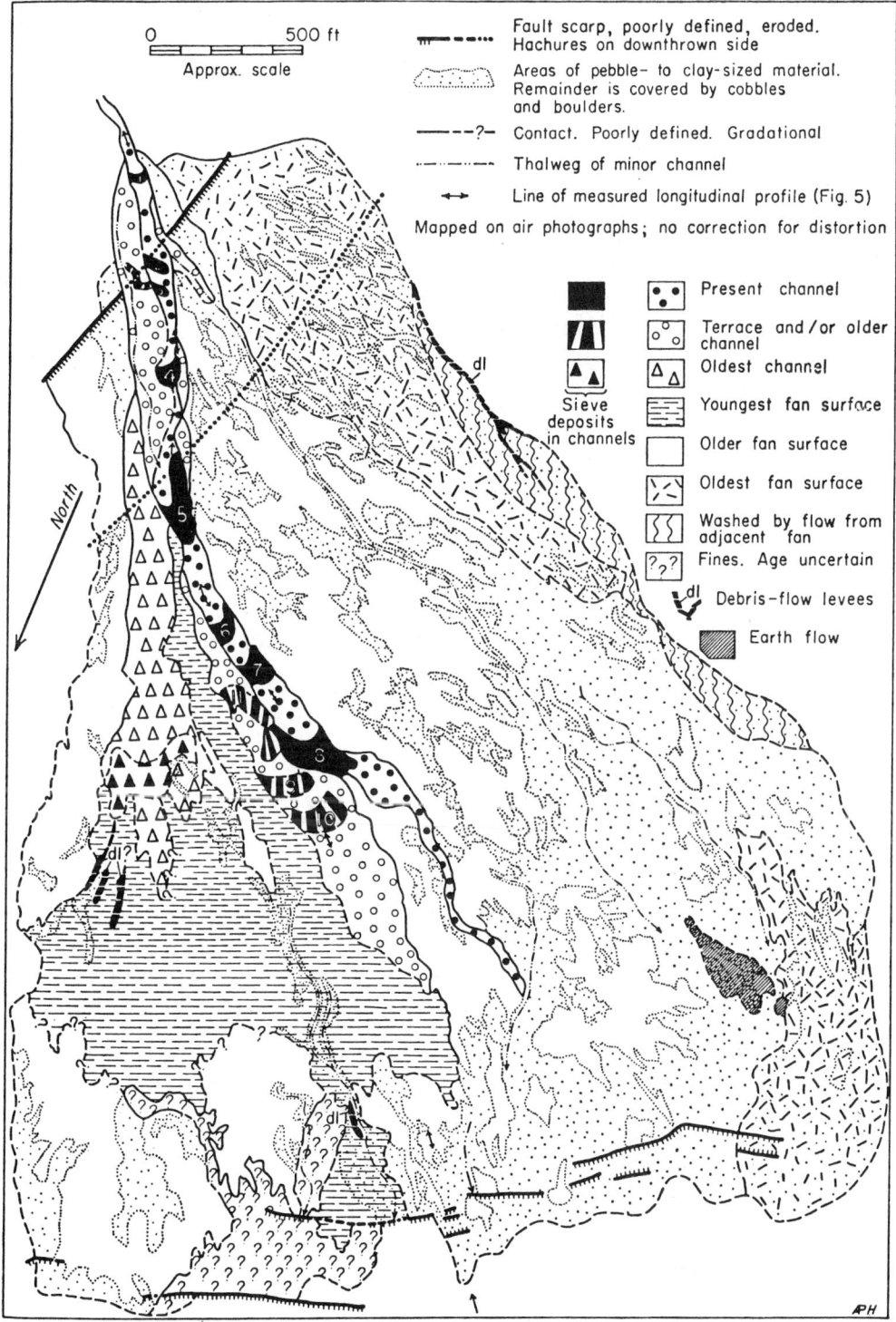

Fig. 3.—Geomorphic map of Shadow Rock Fan

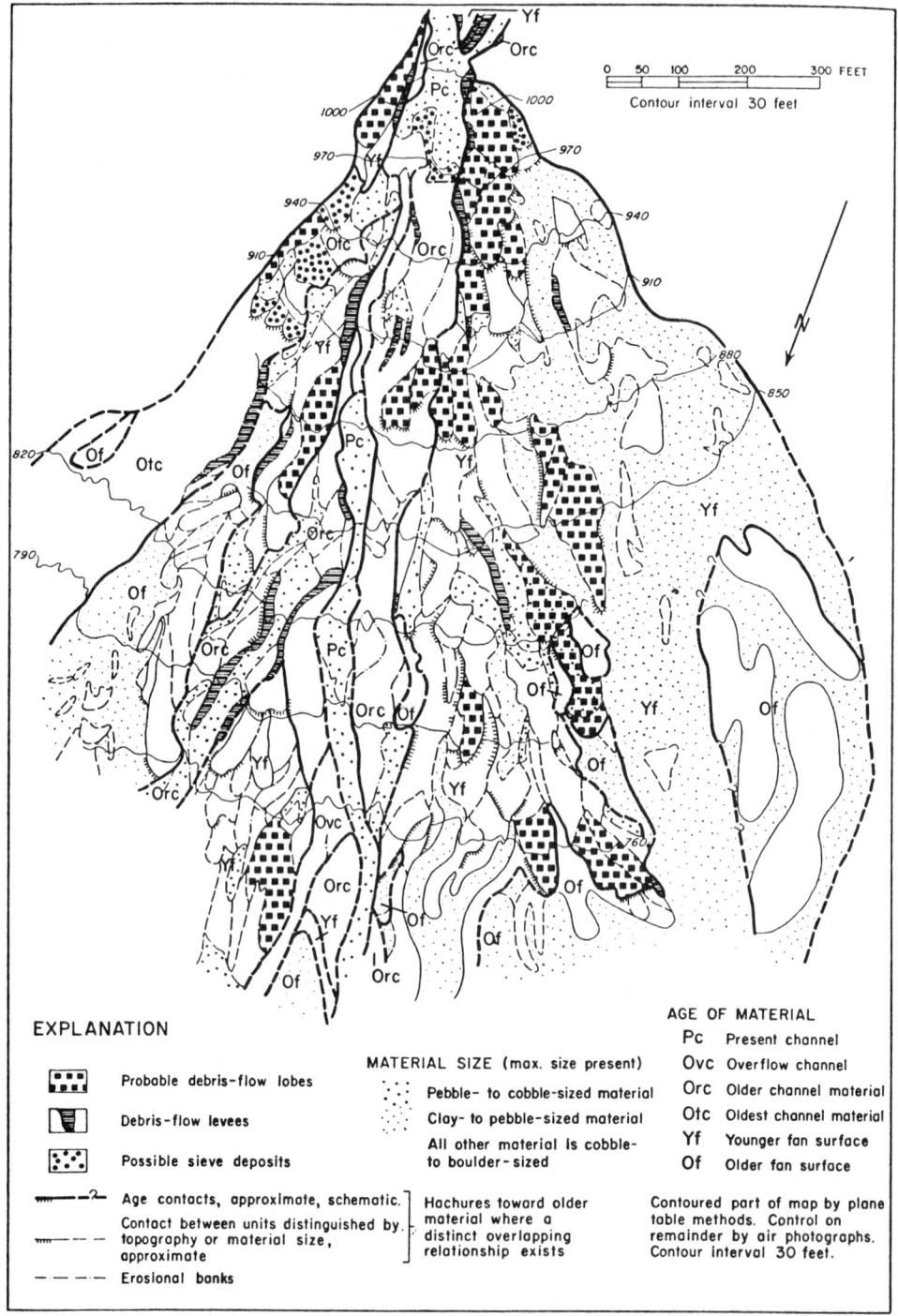

FIG. 4.—Geomorphic map of the upper (most recent) segment of Trollheim Fan. Relationship between younger and older fan surfaces reflects segmentation (fig. 5). Other contacts are based on material size and topographic (frequently lobate) form. Debris-flow and sieve deposits are tentatively identified using criteria described in the text. Identification of sieve lobes is much more tenuous than identification of debris-flow lobes. Where no distinction is made, deposits are either fluvial or are of much less certain origin.

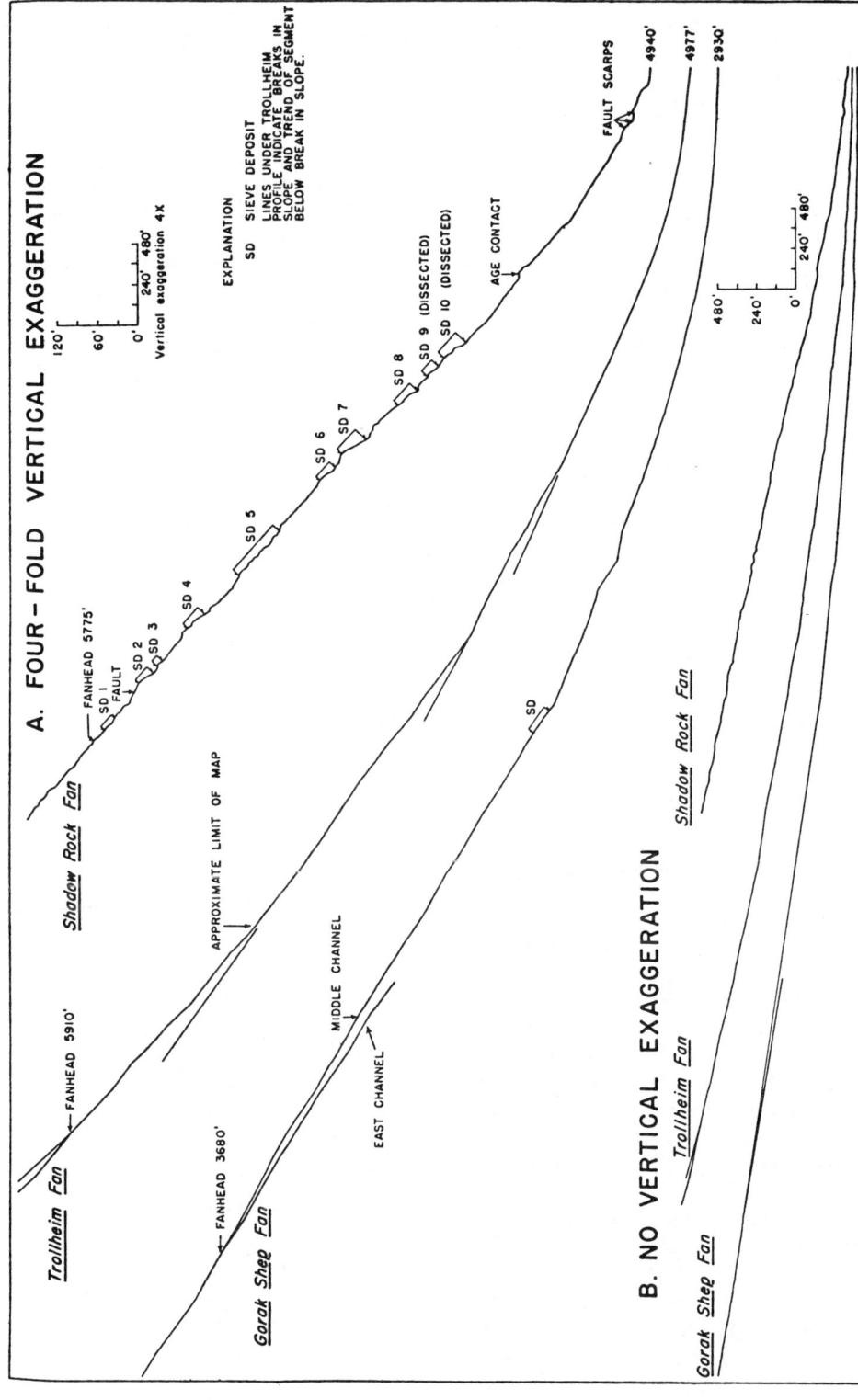

FIG. 5.—Longitudinal profiles of fans. Profiles were measured in the field with alidade and plane table. Lower limit of map of Trollheim Fan (fig. 4) is shown on Trollheim profile.

(table 3). Because the maximum relief was measured on each rock, the calculated mean of these values is referred to as a *mean-maximum* value. In general only rocks greater than 6 inches in diameter were measured, but this requirement was relaxed in small or relatively unweathered areas where suitable rocks were difficult to find.

Measurements on the upper segment were consistent with age relationships based on terracing and pore-space filling, the depth of differential weathering being progressively greater on successively older units. The good agreement between the two measurements on the youngest fan surface in this segment (0.16 and 0.17 inches) suggests that other differences, though small, are significant.

Measurements on the lower segment are also consistent with mapped age relationships in this segment, but these units all appear to be older than those with which they are correlated on the upper segment. Because the upper segment is the youngest, this greater differential weathering is not surprising. Water crossing the upper segment has reworked and abraded material in the lower segment, but existing weathering relief on rocks was not completely smoothed during the reworking. Thus this older material retains some evidence of its greater age.

Correlation of the two areas of "older fan surface," one in the upper segment and the other in the lower, is based largely upon the data in table 3. The material of this age in the lower segment may be part of a presegmentation fan surface.

CONCLUSION

From this discussion of age relationships two points should be emphasized. First, because fans are surfaces of continuous deposition on a geologic time scale, any age distinction is arbitrary, and gradational relationships are common. Second, the "age" of an area is not so much a measure of the time since deposition as it is of the time since the material was last subjected to abrasion by water.

TABLE 3

MEAN-MAXIMUM DEPTH OF DIFFERENTIAL WEATHERING ON GORAK SHEP FAN

Map Unit (See Fig. 2)	Mean-Maximum Depth of Weathering (Inches)	
	Upper Segment	Lower Segment
Recent channel deposits	0.07	0.10
Oldest channel deposits	0.10*	0.17
Youngest fan surface	0.16 0.17†	0.23
Older fan surface	0.50	0.65

* Only ten rocks measured.
† Measurements on opposing members of a paired terrace.

LABORATORY APPARATUS AND PROCEDURE

Small alluvial fans were built of mud and sand in the laboratory apparatus shown in plate 1. Water was pumped into the constant-head tank, d, (pl. 1A), whence it flowed through a series of pipes and valves into the inlet box c. The channel b (pl. 1B), leading from the inlet box to the working area, a, was 4 inches wide and 50 inches long and had a slope of 0.16. Debris was picked up in the channel and deposited as a fan in the 5-foot × 5-foot working area. Water left the working area through the outlet gate e, and sediment trap, f, and returned to the reservoir, g, below the constant-head source by way of the return pipe, h.

The working area had a slope of 0.0076 toward the outlet. However, the instrument carriage, i, carrying a point gage, rode on horizontal rails. Thus all elevation measurements were made with respect to a horizontal datum plane. Discharge was measured with a Fischer and Porter flowrater meter, j, and was regulated with the plug valve, k. Discharges on the order of 0.002 and 0.02 cfs were used. The fast-action valve, m, was used to turn the flow on and off, so the setting of the plug valve could be left unchanged during a run.

SERIES A

The first series of experiments comprised twelve *runs*. Each run consisted of eight to twelve or thirteen *episodes*. Each episode involved packing the channel, b, with a fixed amount of material, and then opening the fast-action valve for a fixed length of time, D, with the discharge, Q, determined by the setting of the plug valve. During each run Q and D were held constant. Following the eighth episode in each run a contour map of the resulting fan was made, and morphological features were mapped (fig. 6).

The quantity of material used in each episode was held constant for the entire series of runs and amounted to roughly 17 lb. dry weight. This debris was in the coarse-sand to granule range on the Wentworth scale and was poorly sorted (fig. 7).

SERIES B

Finer sediment (fig. 7) and lower discharges were used in Series B experiments. The inlet channel, b, was narrowed to 2 inches by placing a partition down the middle of the previous channel, and the bulkhead, n, was extended as shown by dashed lines on plate 1B. This eliminated undesirable effects caused by asymmetry of the earlier arrangement.

Series B consisted of three runs, each involving thirty-five or more episodes. An episode usually consisted of a debris flow followed by a water flow, but in several episodes only one type of flow was used (table 4). A small amount of dry sediment was added to the channel before most water flows. This provided material for transport by the water and for backfilling of the channel to keep it graded to the level of the rising fanhead.

Debris flows were made by mixing a slurry of water and sediment in a graduated can, from which it was easily poured into the channel just above the bulkhead. Flow density was determined by weighing the can containing the known volume of slurry. This method of determining the density was used during the last half of run B-2 and during run B-3. Earlier density measurements, made on volumes 5–10 cm.[3], are less reliable. The range in density of flows is given in table 4.

Viscosity measurements were made on two samples with a Stormer Viscometer (table 5). Owing to a finite yield strength, flowage did not occur at low applied stresses. After the yield strength of the material was exceeded, the viscosity decreased, at first slowly and then more rapidly, finally approaching a constant value at an applied stress slightly less than twice the yield strength.

SIMILARITY OF PROCESS ON LABORATORY AND NATURAL FANS

Laboratory fans were not intended to be scale models of a natural fan but are treated as small fans in their own right. However, gross scaling relationships between debris size and discharge should be met, and the general equivalence of slopes on laboratory and natural fans indicates that they have been. Laboratory-fan slopes were usually between 4° and 8°, values that are about average for natural fans with a high percentage of cobbles and boulders. Under conditions of sieve deposition, slopes ranged from 7° to 13°, in good agreement with average slopes on Gorak Shep (9°) and Shadow Rock (13.5°) fans in regions of sieve deposition (fig. 5).

The use of observations on laboratory fans as a basis for conclusions regarding natural fans requires further justification.

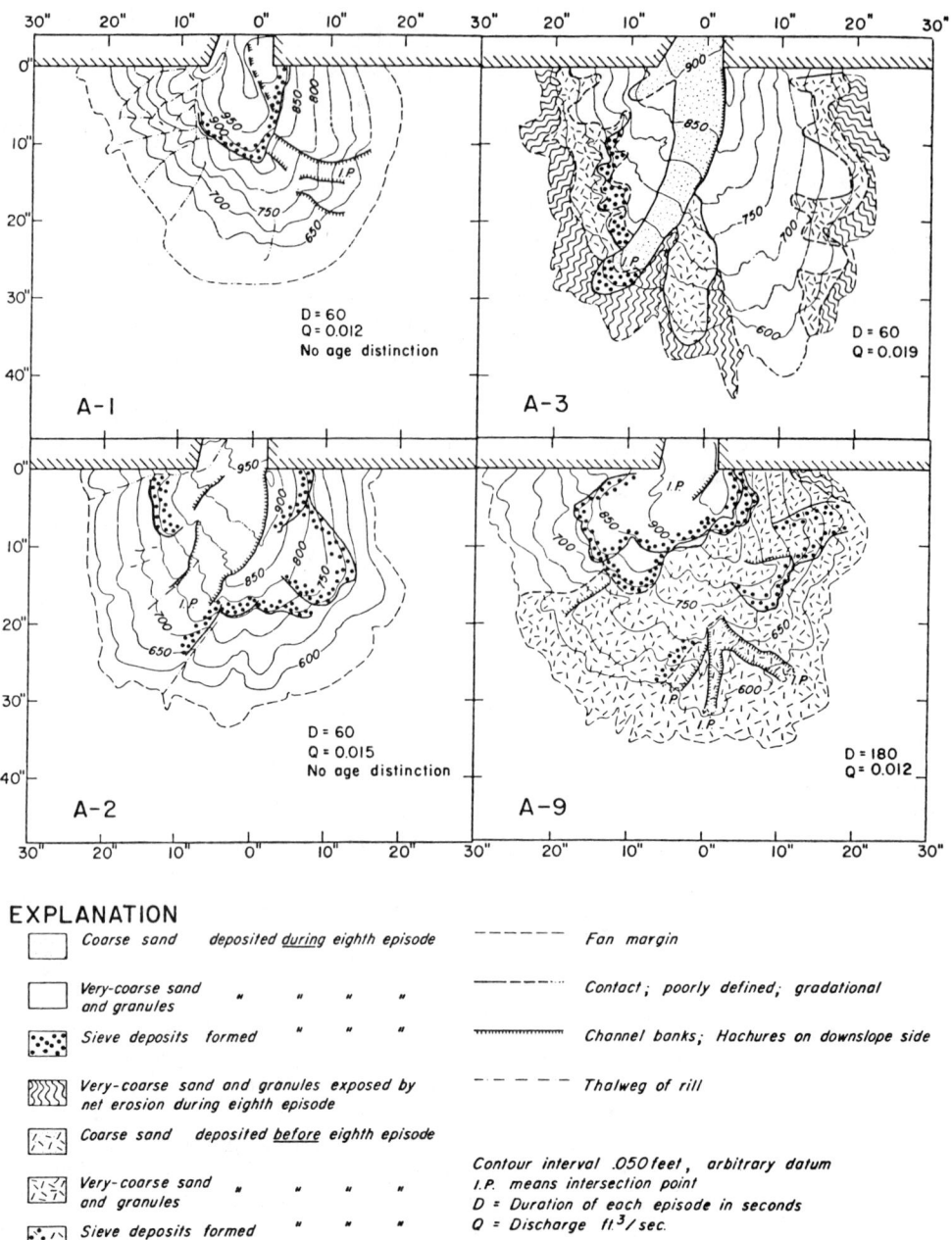

FIG. 6.—Maps of laboratory fans, Series A. Each map depicts a separate fan composed of material accumulated during eight episodes of deposition. For any one fan each of the eight episodes had a fixed discharge, Q, and lasted for a fixed length of time, D. For example, after the map of fan A-2 was drawn, the working area (pl. 1B) was completely cleaned. A new fan, A-3, was then built, using eight episodes of 60 sec. duration each and a discharge of 0.019 feet^3/sec. The fan surface prior to the eighth episode was identified by a thin coat of spray paint. Erosion around the edge of the fan (e.g., A-3) is, in part, the result of floating of grains attached to a film of paint. In runs labeled "no age distinction" the surface was not painted prior to the eighth episode.

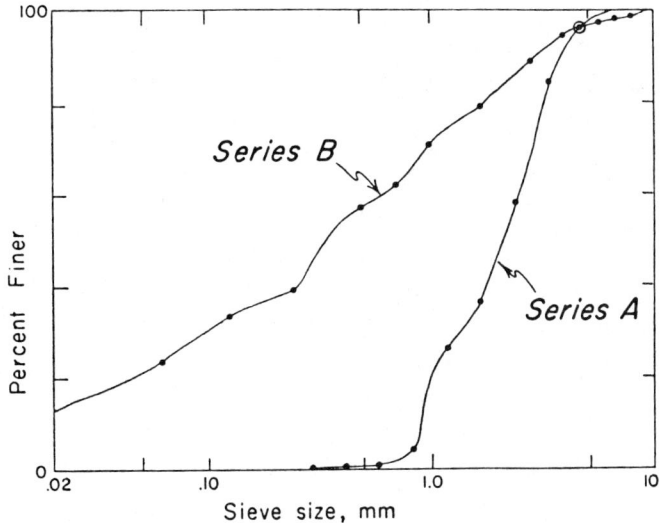

FIG. 7.—Size distribution of material used in the laboratory experiments

TABLE 4

RANGE OF VARIATION OF PARAMETERS BETWEEN EPISODES IN SERIES B

No. of Episodes	Debris-Flow Data		Water-Flow Data		
	Volume (Liters)	Density (Gm/Cm³)	Discharge (Cfs)	Duration (Minutes)	Amount of Dry Debris Added (Pounds)
Run B-1					
35........	0.5–4.0*	1.49–1.85	No water flow	No water flow	No water flow
Run B-2					
43........	0.5–1.5	1.52–1.69	0.002	2	1.5
Run B-3					
66:					
1–34.....	No debris flow	No debris flow	0.002–0.004	Usually 2 or 4; rarely 6 or 12	2.0
35–66....	0.7–1.5	1.54–1.73	0.002–0.004	Usually 2 or 4; rarely 6 or 12	0.5

* Average 1.5 liters.

This procedure is based upon the postulate that processes in the laboratory are similar to those in nature and that the morphologic effects of these processes are the same on both laboratory and natural fans. The following laboratory examples of such processes and effects are followed, in parentheses, by a reference to a paper describing the same feature or characteristic in nature.

Runoff events during Series B usually began with a debris flow and ended with a water flow of substantially longer duration from the channel when the flow became deep enough to overtop the banks (pl. 2D), and wide lobes formed below the intersection point (Beaty, 1963, p. 520). When laboratory flows overtopped the channel banks without forming diverging lobes, levees were left as the flow receded (pl. 2E). Contacts between fresh laboratory flows and the older fan surface were always sharp (pl. 2B, D and E), as in natural flows unmodified by flood waters (Blackwelder, 1928, p. 470; Jahns, 1949, p. 13).

TABLE 5
RESULTS OF VISCOMETER TESTS

	Test 1	Test 2
Density, gm/cm^3	1.42	1.48
Viscosity at applied stress ≈ twice the yield strength, poises	0.42	0.92, 1.02*

* Viscosity increased during the test for unknown reason.

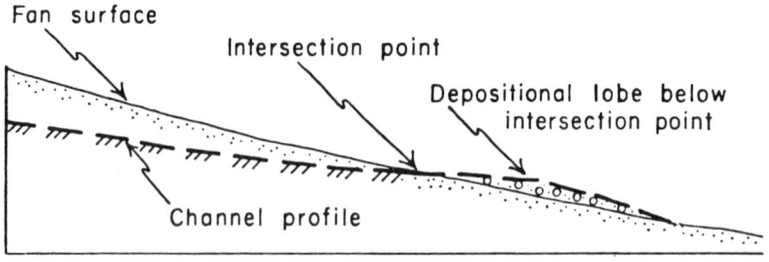

FIG. 8.—Idealized sketch of intersection-point relationships

(Pack, 1923, p. 355; Beaty, 1963, p. 520–521). The debris flows had steep, rounded fronts; deeper parts moved faster than shallow parts, and surges moved more rapidly than the front (Jahns, 1949, p. 12–13). Fresh flows were remobilized by subsequent surges if insufficient drying time was allowed (Sharp and Nobles, 1953, p. 551). If the front was not remobilized, the surges caused small pressure ridges to develop parallel to the flow borders (Jahns, 1949, p. 13). The main body of laboratory flows followed the existing main channel (pl. 2C and D), but small lobes diverged

The main channel on laboratory fans had a slope less than the slope of the adjacent fan surface and thus merged with the surface at a point commonly near midfan and herein called the *intersection point* (fig. 8). Material deposited here is generally coarser than the average material found in the channel and in some instances forms a small secondary fan on the surface below the intersection point. Sometimes several small channels head in this deposit (pl. 2F). Diversion of the main channel was often caused by plugging with debris-flow material (pl. 2C–E), and channels disappear beneath such deposits upstream (fig. 6, runs A-1 and A-9) (Beaty, 1963, p. 527–528). Abandoned main channels were also observed to end upstream at the banks of deeper channels. These channel patterns are all common in nature.

In several instances laboratory study has been used as a basis for interpretation of field observations. The agreement of laboratory and field data in these cases is further support for similarity of process. The best examples of such agreement are found in the ensuing discussions of sieve deposition and fanhead incision and in the dependence of fan slope upon discharge, debris caliber, and depositional process (Roger LeB. Hooke, manuscript in preparation).

DEBRIS FLOWS AND FLUVIAL PROCESSES

MECHANICS OF SEDIMENT TRANSPORT

Early workers (Gilbert, 1882, p. 183–184; Trowbridge, 1911, p. 738–744) assumed that fans were built by running water alone. The importance of debris flows was not widely recognized prior to Blackwelder's (1928) discussion of this process. Since then knowledge of sediment transport has increased to the point where some differences between water and debris flows can be discussed.

Natural debris flows may have viscosities of over 1,000 poise and densities of 2.0 to 2.4 (Sharp and Nobles, 1953, p. 552–553). In contrast, the viscosity of water is about 0.01 poise, and the density of a stream carrying suspended sediment is only slightly greater than unity. Furthermore, the response of water to an applied stress, τ, can be represented by

$$\tau = (\mu + \eta) \frac{du}{dy},$$

where du/dy is the rate of shear in turbulent flow and μ and η are the molecular and eddy viscosities, respectively. The eddy viscosity generally *increases* with rate of shear, but the molecular viscosity is constant (Rouse, 1950, p. 88). In contrast, debris flows behave as quasi-plastic substances, and their rate of deformation under stress is represented by

$$(\tau - \tau_0) = \mu(\tau) \frac{du}{dy}, \qquad \tau \geq \tau_0,$$

where τ_0 is the yield strength, and $\mu(\tau)$ indicates that the apparent viscosity, μ, is a function of the applied stress (Leopold, Wolman, and Miller, 1964, p. 31). No flow occurs until the yield strength is exceeded, but then the viscosity *decreases* gradually with increasing applied stress until a constant value is reached. The value of τ_0 and the form of the function $\mu(\tau)$ differ among flows, owing to differences in flow density and in mineralogy of the silt-clay fraction. This is the behavior observed in laboratory debris flows, but natural flows are probably not fundamentally different.

Owing to the high density of mud, forces on rocks in debris flows are substantially different from those on comparable rocks in a stream. The submerged weight of a rock is reduced, perhaps by more than 60 per cent, relative to its submerged weight in water, and bed shear forces are increased. Furthermore, the combination of the low-density contrast between rocks and mud and the high density of mud reduces the settling velocity (Vanoni, 1962, p. 84). Thus gravity forces tending to prevent motion are reduced, and drag forces are increased.

The ability of a stream to transport sediment in suspension is closely governed by a balance between the settling velocity of the sediment and net upward transport of material by turbulent eddies (Vanoni, 1963). With the reduced settling velocity in debris flows, coarser material can be maintained in suspension. Dispersive pressure forces (Bagnold, 1954, 1955) may also keep material in suspension in debris flows with high concentrations of pebbles to boulders.

These considerations suggest one fundamental difference between water flows and debris flows. Whereas streams vary their sediment load readily by deposition or erosion and will continue to flow as long as a slope exists, debris flows cannot selectively deposit any but the coarsest fragments. This means that a debris flow cannot turn into a stream by deposition. Both types of flow are formed by water moving over and entraining loose sediment, but at some point sediment entrainment becomes irreversible. Perhaps this is the best distinction between streams and debris flows.

FIELD CRITERIA USED TO RECOGNIZE
DEBRIS-FLOW DEPOSITS

Debris-flow deposits consist of cobbles and boulders imbedded in a matrix of fine material. They are generally poorly sorted, and individual flows are unstratified. Recent debris flows have flat tops, steep sides, and a lobate form. The base makes an abrupt contact with underlying material. Rain and rill erosion, creep, and weathering eventually modify this distinctive morphology and internal character. Cobble and boulder accumulations with low relief are commonly the only remaining evidence, but their position as levees along channels or as lobes diverging from channels usually leaves little doubt as to their origin, especially after debris-flow behavior is seen in the laboratory.

EFFECTS OF DEBRIS FLOWS ON FAN MORPHOLOGY

Transportation of coarse material and levee formation.—One characteristic feature of many alluvial fans is a large number of cobbles and boulders. The occurrence of this material in levees and lobes and its association with material of recognizable debris-flow origin suggest that much of it is transported to the fan by debris flows.

Coarse material frequently accumulates at the front of a debris flow and is shoved aside by the advancing snout, forming levees that confine the remainder of the flow (Sharp, 1942, p. 225). The bouldery, sharp-crested levees common on many fans were probably formed in this way.

Levees also form when a debris flow at peak discharge overflows channel banks. Levees of this type were observed on laboratory fans (pl. 1E) and on debris flows on Surprise Canyon Fan in Panamint Valley, California. Such levees are generally wider and have more rounded crests than those described by Sharp.

A distinct sorting of stones, with pebbles on the inside (toward the flow) and coarser material on the outside, was observed on several natural debris-flow levees. This sorting probably does not result from selective deposition of coarser material by slower flow because the size difference is not large and the coarser material is probably not heavy enough to be dropped independently of the main body of the flow. Instead, coarser material may selectively migrate to the surface. Either the higher surface velocity or the sliding and rolling processes described by Sharp (1942, p. 225) then move it to the front and edges of the flow, where it is deposited when the edges stop moving.

Radial variations in alluvial-fan stratigraphy.—Because mud has a finite yield strength, debris flows stop when the shear stress on the bed, τ_0, no longer exceeds the yield strength of the mud, τ_c, or

$$\tau_0 = \rho g d S < \tau_c,$$

where g is the gravitational acceleration and ρ, d, and S are the density, depth, and hydraulic gradient of the flow, respectively. Laboratory observations suggest that this condition may result from a loss of water to the underlying dry fan material, thus increasing the yield strength, or from a decrease in either the hydraulic gradient ($\approx$ fan slope), or flow depth, as the flow moves down fan and spreads out. Thus the areal extent of a debris flow is limited by its volume and yield strength and by the slope of the fan surface. Furthermore, its down-fan extent is determined in large part by the degree to which existing channels prevent lateral spreading at the fanhead.

Consequently, most deposition near the toe of fans is probably not by debris flows but by running water, although the material deposited may have been eroded from debris-flow deposits higher on the fan. In contrast, much of the deposition near the fanhead is probably caused by debris flows overtopping the channel banks. Thus the stratigraphy of fans on which debris-flow deposition has been important should be inhomogeneous; debris-flow deposits nearer the fanhead should interfinger in the midfan region with water-sorted material deposited nearer the toe. This variation was observed on laboratory fans B-2 and B-3.

Natural radial inhomogeneity was observed in the Pliocene Ridge Basin Group (Crowell, 1954), a dissected alluvial-fan basin filling about 75 miles north of Los Angeles. The unit nearest the source area, the Violin Breccia (Crowell, 1954), is an unbedded to poorly bedded cobble and boulder conglomerate with a matrix of sand- to clay-sized material. This unit is probably of debris-flow origin. Outward from the source area it grades into fluvially bedded conglomerate with smaller clasts, thence into well-bedded, rarely cross-bedded, sandstone with interbedded shale, and finally into shale. The conglomeratic rocks interfinger with the sandstone, and the sandstone with the shale, as observed on laboratory fans. The shale is near the center of the basin, about 4 miles from the source, and presumably represents playa deposition.

RELATIVE SIGNIFICANCE OF DEBRIS FLOWS AND WATER FLOWS

As Blackwelder (1928, p. 473–474) observed, the proportion of recognizable debris-flow material in and on fans varies widely. Shadow Rock and Gorak Shep fans (figs. 2 and 3) have practically no recognizable debris-flow deposits, but Trollheim Fan (fig. 4), about a mile south of Shadow Rock Fan, consists predominantly of debris-flow material. These differences appear to be lithologically controlled. The source area of Gorak Shep Fan is underlain by resistant carbonate rocks, and valley-side slopes are steep. Thus material is quickly transported to the fan with little pretransport weathering. As a result there is little fine material in the source area, and debris-flow formation is inhibited. Similarly, the drainage area of Shadow Rock Fan is underlain predominantly by resistant quartzite of the Campito Formation (Nelson, 1966), and fine material is again lacking. In contrast, exposures of readily weathered sandy dolomite of the Reed Formation (Nelson, 1966) in the source area of Trollheim Fan contribute substantial amounts of fine material and are thought to be responsible for debris-flow deposition on this fan. Climatic and topographic factors do not vary significantly between Trollheim and Shadow Rock fans and hence are unlikely to be responsible for the differences in mode of deposition.

Volumetric estimates of the role of debris flows in transporting material to a fan are difficult to make. Characteristically heterogeneous debris-flow deposits are commonly sorted and stratified by subsequent water flows. This may be partially responsible for the lower percentage of debris-flow material recognized by Blissenbach (1954, p. 179) on fans in areas of higher precipitation. Furthermore, the best exposures of fan stratigraphy are usually in the main channel above the intersection point. Because debris-flow deposition is likely to be more common near the fanhead, estimates may be biased in its favor.

INFILTRATION AND SIEVE DEPOSITION

Earlier students of alluvial fans have remarked upon the significance of infiltration in reducing the sediment-carrying capacity of fluvial flows and instigating deposition (Trowbridge, 1911, p. 738; Eckis, 1928, p. 237; Blissenbach, 1954, p. 178; Bull, 1964a, p. 17). However, Bull (1964b, p. 104) has shown that water on some fans composed of fine material does not percolate below the root zone of vegetation.

Laboratory observations suggest that permanent features attributable to infiltration will not be found on fans unless moderately large water flows infiltrate completely before reaching the toe of the fan. Surficial wetting of dry fan material may cause deposition during small flows or in the initial stages of large flows, but the amount of such deposition is small, and any features produced will be minor and will be destroyed by subsequent higher discharges.

However, if the fan material is sufficiently

coarse and permeable, as in laboratory fans of Series A (fig. 7), the entire flow may infiltrate before reaching the toe of the fan. Under these conditions a lobe of debris is deposited at the point where water is unable to effect further transport. Because water passes through rather than over such deposits, they act as strainers or sieves by permitting water to pass while holding back the coarse material in transport. I call the lobate masses thus formed "sieve lobes" or "sieve deposits," and the mode of formation is sieve deposition.

Extensive deposits inferred to have been formed in this way were found on seven natural fans. Four are in the southeast material to the front of the lobe, and a new barrier is formed just up fan from the preceding one.

Deposition of coarser material along the lateral edges of a lobe, where competence is reduced, may confine the flow temporarily. However, lateral shifting of the flow, a decrease in infiltration as fine material is deposited upstream, or a slight increase in discharge can result in diversion of the flow over the lobe's flanks or front. In such cases rapid erosion ensues, and a fluidized debris mass shoots down fan a few centimeters and stops. Backfilling behind the mass proceeds until this part of the lobe is built up to the elevation of the remainder.

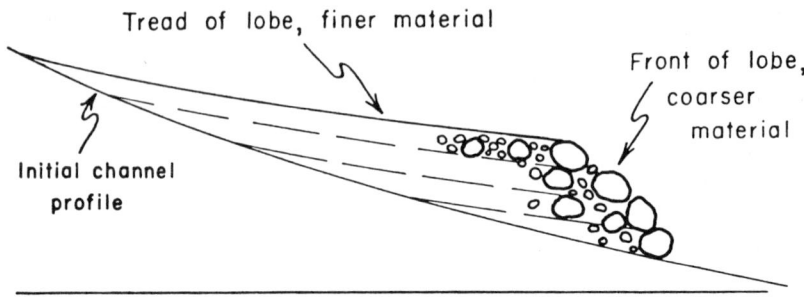

Fig. 9.—Schematic sketch of growth of a sieve lobe

corner of Deep Springs Valley, California, one of which, Shadow Rock Fan (fig. 3; pl. 1D), has been mapped in detail. Two others, including Gorak Shep Fan (fig. 2; pl. 1E), are in the southeast corner of Eureka Valley, California. The last is about 2,800 feet N. 16° E. of Badwater in Death Valley.

LABORATORY OBSERVATIONS

On laboratory fans, sieve deposition is initiated by deposition of granules which form an initial debris barrier. The channel slope immediately upstream from this barrier is reduced by backfilling (fig. 9), and further deposition ensues. Deposition does not actually occur in layers as represented in figure 9, but schematically this provides an easy way to visualize the process. When regrading of the channel above one barrier is complete, water is again able to transport

Sieve deposition occurred both near midfan and at the toe of laboratory fans. In the first case, initial deposition resulted from infiltration, and the infiltrating water continued as subsurface flow until it encountered the impermeable bed of the laboratory apparatus and emerged around the toe of the fan. In the second case deposition was instigated by the break in slope at the toe of the fan, and water passing through a lobe emerged immediately below it.

The radial position of sieve lobes on laboratory fans was controlled by a balance between discharge and infiltration rate. For instance, with decreasing discharge and increasing fan thickness (increasing infiltration rate) the deposits on fans in figure 6 are found progressively nearer the fanhead. Note also on fan A-3 that the sieve lobe at the end of the main channel is farther down fan than the lobes deposited earlier during the same episode. This is because the infiltration rate was decreased by deposition of fines near the end of the episode.

In summary, sieve deposition on laboratory fans is a consequence of the coarseness of material and resulting high infiltration rate. Sieve deposition may be initiated either by complete loss of discharge through infiltration, as is the case for sieve deposition near midfan, or by a break in slope, as at the toe of laboratory fans. Sieve deposition on natural fans is a result of the same factors, as the subsequent discussion of two examples will demonstrate.

SIEVE DEPOSITION ON SHADOW ROCK FAN

The drainage basin of Shadow Rock Fan is underlain predominantly by quartzite of the Lower Cambrian or Precambrian Campito Formation (Nelson, 1966). This rock is resistant to weathering as evidenced by a dark desert varnish on rocks on older parts of the fan and by the paucity of fine material in both recent and older deposits (fig. 3). Consequently infiltration rates are high in the older deposits as well as in recent sieve lobes, and sieve deposition can occur anywhere on the fan (fig. 5). The location of a given sieve lobe is determined by the size of discharge; larger discharges continue farther down fan before loss of water is sufficient to cause deposition. Once the flow has infiltrated it apparently does not return to the fan surface. Hence freshly abraded gravel does not extend to the toe of the fan (fig. 3), as it does below the sieve deposits on Gorak Shep Fan (fig. 2).

In this situation sieve lobes formed by low discharges are commonly modified by subsequent higher flows. Modification involves dissection of the original deposit, which is usually as wide as the channel that formed it, leaving several small disconnected lobes. For instance, deposits 5 and 6 on Shadow Rock Fan (fig. 3) apparently were dissected during construction of deposit 7, whereas lobes 1–4 are relatively unaltered and thus postdate 7. Modification of lobes is an important process, as unmodified deposits are rare on older parts of the fan surface.

SIEVE DEPOSITION ON GORAK SHEP FAN

Sieve deposition on Gorak Shep Fan occurs at the down-fan limit of the upper segment (figs. 2 and 5). Since the upper segment is youngest, its slope is the steady-state slope of the fan under present conditions. The decrease in competence at the break in slope between the two segments presumably causes deposition and is believed to be responsible for the existing sieve deposits.

The source area of Gorak Shep Fan is underlain by a thick section of resistant early Paleozoic (?) carbonate rocks, primarily dolomite. Slopes are steep, have virtually no soil cover, and supply the fan with pebble- and cobble-sized debris, with a minimum of fines. On the other hand, older deposits on the fan have a matrix of sand- to clay-sized particles produced by postdepositional weathering. The individual pebbles appear to be in contact with their neighbors (pl. 1C) and would not need to readjust and settle if the fines were removed. Furthermore, a few layers of coarse-pebble gravel with unfilled pore space were observed in vertical exposures of the "youngest fan surface" material (pl. 1C). In at least three instances (pl. 1C) such layers are overlain by granule-sized material in which the void space is filled. These relationships suggest that the fines are secondary and that they were trapped by the granule layer during illuviation.

Because recent channel deposits rest on older material in which the void space is filled, infiltration alone is not sufficient to instigate sieve deposition. A break in slope, as at the toe of laboratory fans, is also necessary. Water passing through sieve deposits on Gorak Shep Fan emerges below them and continues down fan as surface runoff, as the extensive channel development on the lower segment indicates (fig. 2). Thus sieve deposition on Gorak Shep Fan occurs in a restricted zone, in contrast to the ubiquitous occurrence of such deposits on Shadow Rock Fan. This contrast reflects the coarser debris and much greater resistance to weathering of the quartzite on Shadow Rock Fan, where older deposits still have open pore space.

DISTINGUISHING BETWEEN DEBRIS-FLOW AND SIEVE DEPOSITS

Both sieve and debris-flow deposition may take place on the same fan. However, for debris flows to form, substantial amounts of fine material must be present in the source area. Conversely sieve deposition cannot occur if too much fine material is available. Therefore, one process usually predominates, and conclusive evidence that one is significant usually implies that the other is much less important. For instance, Shadow Rock Fan is predominantly composed of sieve deposits, but four linear ridges (fig. 3) are probably debris-flow levees, possibly built by flows originating on the fan. A fifth levee, the long, branched one on the south side of the fan, was apparently built by a debris flow originating in the dolomitic terrane between Shadow Rock Fan and the fan to the southwest. Similarly there are lobes of material among the debris flows of Trollheim Fan that have been mapped as possible sieve deposits (fig. 4).

Distinguishing between materials deposited by these two processes after they have been modified by erosion and downslope movements is not easy. Only a few of the units mapped on Trollheim Fan could be identified with any confidence. The following criteria were developed through comparison of the deposits on Trollheim Fan with each other and with those on Shadow Rock Fan:

1. Recent sieve deposits are composed of pebble- to boulder-sized material without fines and thus have a high infiltration rate. In contrast, recent debris-flow deposits have a matrix of fine material.

2. Sieve deposits rarely contain the especially large boulders (> 3-foot diameter) found in many debris flows.

3. Relatively unmodified debris flows on Trollheim Fan are 2–10 feet thick and several times as wide. In contrast, the fronts of sieve deposits on Shadow Rock Fan are commonly 10–30 feet high, but, owing to dissection, their treads are usually narrow and rounded in cross-section.

4. Contacts between debris flows and the underlying material are usually sharp and well defined, and flows appear to overlap the older deposits. Contacts between sieve deposits and underlying material are generally gradational and rarely give the impression of an overlapping relationship.

5. Debris flows can generally be traced some distance up fan and have a slope approximately equal to the fan slope, whereas sieve lobes typically have short treads with slopes less than the fan slope (figs. 5 and 9).

6. Debris-flow levees are distinctive and are indicative of debris-flow action.

7. Fresh sieve deposits are clearly related to a channel from the fanhead, but fresh debris flows may not be related to any visible channel.

CHANNELS AND DEPOSITION ON ALLUVIAL FANS

Over short periods of time, deposition on both laboratory and natural fans is localized (figs. 2–4 and 6). However, shifting of the locus of deposition on laboratory fans results in relatively uniform deposition over the entire fan surface, and deposition on unsegmented natural fans is probably also uniform if the time scale involved is sufficiently long. On segmented fans, deposition can be considered uniform only on the active segment.

If long-term deposition occurs uniformly over the entire fan, whereas short-term deposition is localized, then periodically debris must be transported across upper parts of the fan and deposited at points more distant from the source area. Such transportation is accomplished primarily by flows in channels. This interpretation explains the apparent contradiction of a depositional fan whose most obvious surficial features are channels, a fact also recognized by Tolman (1909, p. 157), and provides a framework within which to discuss deposition.

POSITION OF THE INTERSECTION POINT

The intersection point on laboratory fans is commonly near midfan. This appears to be because fluvial deposition predominates near the toe and occurs without down-fan migration of the intersection point, while overbank debris-flow deposition predominates near the fanhead. Thus the average radial position of the intersection point should be related to the relative importance of debris flows and fluvial processes in transporting material to a fan. Measurements on laboratory fan B-3 support this conclusion. The average distance of the intersection point from the fanhead during the first thirty-four episodes, during which water flows alone were used, was 19 cm., and the average for the remaining thirty-two episodes, each of which began with a debris flow, was 41 cm. The fan radius was about 80 cm. This difference is obvious in plate 2C and H.

The situation on natural fans is complicated by segmentation, and no general statement is possible without more detailed study of the relationship between segment boundaries and the intersection point. On fans on which the most recent segment includes the fanhead (figs. 2 and 4, and east side of southern Death Valley), the intersection point is often nearer the lower boundary of the segment than might be expected from laboratory observations. This apparently reflects the ability of water flows to transport substantial volumes of debris across the gentler slopes of lower segments. Thus lower segments must be aggrading, but more slowly than the upper segment.

The intersection point on laboratory fans shifted gradually due to debris-flow and fluvial deposition. The intersection point would migrate up fan as low banks of the main channel were buried. Subsequent water flows then eroded a new channel offset laterally from the previous course. Similar processes have been described by Eckis (1928, p. 234–235) and Bull (1964a, p. 27–28) from observations during waterfloods on natural fans. Such shifting is necessary for uniform deposition.

Diversion of the main channel near the fanhead helps distribute debris-flow material over upper parts of the fan and also results in a rapid shift of the intersection point. Remnants of older channels from which the flow has been diverted (figs. 3 and 4) testify to the importance of this process. Beaty (1963, p. 527–528) has discussed some common causes of diversion.

FANHEAD INCISION

In this report a fanhead is considered incised if it seems unreasonable to expect overbank flooding by water flows at least once in a few decades. The discharge required to produce overbank flooding may be approximated by means of the Manning equation if the width, depth, and slope of the channel are measured and the hydraulic roughness estimated. In the areas investigated the discharges so calculated are totally unreasonable in terms of the area and hydrologic characteristics of the drainage basins. For example, a discharge of approximately 33,000 cfs would be required for overbank flooding on Gorak Shep Fan, which is incised about 4 feet and has a drainage area of less than 1 sq. mile. Using this definition, a wide or steep channel, such as that on Gorak Shep Fan, need not be especially deep to be incised.

Most geologists have assumed that material at the fanhead was deposited before fanhead incision occurred and that some fundamental change in regime was responsible for incision (e.g., Lustig, 1965, p. 171). Observations on laboratory fans suggest that incision can be the natural result of an alternation of debris flows and water flows (pl. 2E and F) and that fanheads are, so to speak, "born incised." This idea was first suggested to the author by David Schleicher. Owing to the high viscosity and finite yield strength of debris flows, the hydraulic gradient required for their movement is greater than that required by water for transport of much of the finer material in them. Consequently water flows tend to erode channels in debris-flow deposits. This has been observed in nature by Pack (1923, p. 355) and Blackwelder (1928, p. 470) on a debris flow in Utah. Pack (1923, p. 355) reports that "in places the debris deposited by the preceding mudflows was incised sufficiently to permit the water to flow ... in fairly well-defined channels. Well-washed and well-sorted boulders and gravels were strewn along the water channels forming a bold contrast with the heterogeneous masses deposited by the mudflows." Even though a fanhead is incised, debris flows may exceed the channel depth and deposit broad sheets of material on the fan surface above the intersection point. This occurred repeatedly on laboratory fans. Such overbank deposition need not result in diversion of subsequent water flows, as once the peak discharge has passed, the level of flow commonly recedes, and water tends to follow the original course (pl. 2E and F).

Fanhead incision also occurs on laboratory fans when the locus of deposition shifts to a place that has not received sediment for several episodes. The slope toward such topographic lows is greater than the steady-state slope of the fan under prevailing discharge and sediment-caliber conditions. Incision due to this process alone lasts only until the low area is built to the level of the rest of the fan. On natural fans headward erosion and capture of the main channel by channels heading on the fan (Denny, 1965, p. 16, 38) may result in this kind of incision.

The depth of incision produced in these ways, and uninfluenced by any other factors, may be called the depth of *normal* fanhead incision. In many instances fanhead incision is so great that overbank deposition on the adjoining fan surface is impossible. It is inferred that in these instances the fan-source system has been disturbed, resulting in segmentation. Climatic change and tectonic movement are two common causes of such disturbance (Bull, 1964b, p. 100–113). Abnormally deep fanhead incision is not so widespread as would be expected if climatic change were the principal factor, so tectonic movement is probably the more common cause. Fans on the west side of southern Death Valley, for example, have been deeply incised as a result of eastward tilting of the valley.

Depth of normal fanhead incision.—Laboratory run B-3 was designed to study factors affecting the depth of normal fanhead incision. It was already clear from the results of run B-1 (pl. 2B) that no incision would occur on laboratory fans built entirely by debris flows with no intervening or subsequent water flows. During the first half of run B-3 only water flows were used. The main channel was incised (i.e., deeper than bank-full flow) at the end of only six out of the first twenty-six episodes (fig. 10; pl. 2G and H). In five of these episodes (17, 18, 20–22) the flow had just shifted to a part of the fan that had not received debris for several episodes, and incision was due to the steeper slope toward these areas, as described above. The same explanation probably holds for incision following episode 26, but the cause was not as clear in this instance.

The duration of water flow was increased three times during these first twenty-six episodes, but simultaneous changes in the depth of the channel were not observed. Furthermore, the flow at the end of episodes 19 and 25 was clear, rather than muddy, and sediment transport was small. This suggests that the flow was unable to transport the material armoring the channel bed and thus could not erode the channel to a lower slope. Thus the absence of incision cannot be attributed to lack of sufficient time for erosion.

Following the twenty-sixth episode the discharge was doubled. As a result, the slope of transportation was decreased and the fan regraded. This accounts for the incision observed at the end of episodes 27–29, 32, and 33 (fig. 10) and for the relatively constant elevation of the fanhead during these episodes. During episodes 30, 31, and 34 the fan was already regraded in the direction of flow, and the channel was no longer incised. Shifting of the flow to a steeper part of the fan accounts for the incision during episodes 32 and 33.

Starting with 35, each episode began with

a debris flow and was followed by a water flow that was carrying sediment. The rapid increase in depth of incision and in elevation of the fanhead are clearly shown in figure 10. During episodes 35–47 the steady-state depth of incision was approximately 0.090 feet. When deposition in the channel caused the depth of incision to drop below this value, overbank deposition by debris

The depth of incision is decreased until overbank deposition near the fanhead again proceeds at the same rate as deposition elsewhere on the fan. Alternatively, if overbank deposition at the fanhead occurs frequently and the rate of deposition is faster here then elsewhere on the fan, the depth of incision should increase until depositional rates are again equal.

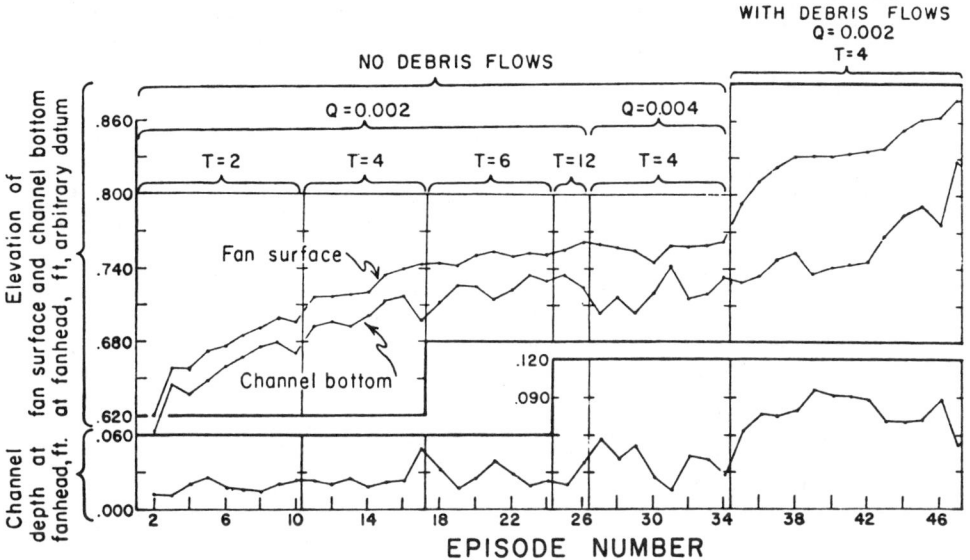

FIG. 10.—Changes in fanhead elevation and depth of incision during laboratory run B-3. T is the duration of water flows in minutes, and Q is the water discharge in cubic feet per second.

flows near the fanhead resulted in an increase in the fan-surface elevation (episodes 35–38 and 43–45, fig. 10).

As a result of this laboratory study, it is inferred that the primary factor affecting the depth of *normal* incision on natural fans with a significant amount of debris-flow deposition is the magnitude of debris flows. For uniform deposition over the entire fan surface, it appears that debris flows must occasionally exceed the channel depth and deposit material near the fanhead. If the depth of incision is so great that overbank deposition does not occur for a geologically long period of time, deposition farther down fan should eventually result in backfilling in the channel above the intersection point.

ACKNOWLEDGMENTS.—I am indebted to Robert P. Sharp for my interest in alluvial fans and arid-region geomorphology and for many helpful suggestions and stimulating discussions in the field and during preparation of the manuscript. Vito A. Vanoni and Norman H. Brooks read a draft of the manuscript and willingly discussed hydraulic and hydrologic problems that arose. They also made available the facilities of W. M. Keck Laboratory at the California Institute of Technology for the laboratory study. The paper has also benefited from critical review by W. B. Bull, C. S. Denny, and H. E. Wright.

R. DeVault and my wife rendered cheerful assistance during a hot summer in the field, and my wife drafted the figures. F. Cammack,

B. Carter, R. Greenway, S. Savin, and especially E. Daly assisted in certain aspects of the field- and laboratory work.

The research was supported by a grant from the Arthur L. Day Fund of the Geological Society of America and by research funds and a fellowship from the National Science Foundation. This paper is the first of two based upon the author's doctoral thesis at the California Institute of Technology.

APPENDIX: LOCATION OF FANS

Gorak Shep Fan is located in the vicinity of 37°07'45" N., 117°38'27" W. in southern Eureka Valley, California, and is on the Last Chance Range 15-min. Quadrangle.

Shadow Rock Fan is located in the NW $\frac{1}{4}$ sec. 16, T. 8 S., R. 36 E., on the Blanco Mountain 15-min. Quadrangle, California, and is in southeastern Deep Springs Valley.

Trollheim Fan is located in the E $\frac{1}{2}$ NW $\frac{1}{4}$ sec. 20, T. 8 S., R. 36 E., on the Waucoba Mountain 15-min. Quadrangle, California, and is in southeastern Deep Springs Valley.

REFERENCES CITED

BAGNOLD, R. A., 1954, Experiments on a gravity-free dispersion of large solid spheres in a Newtonian fluid under shear: Roy. Soc. (London) Proc., ser. A, v. 225, p. 49–63.

——— 1955, Some flume experiments on large grains but little denser than the transporting fluid, and their implications: Inst. Civil Engineers Proc., pt. 3, v. 4, p. 174–205.

BEATY, C. B., 1963, Origin of alluvial fans, White Mountains, California and Nevada: Assoc. Am. Geographers Annals, v. 53, p. 516–535.

BLACKWELDER, E., 1928, Mudflow as a geologic agent in semi-arid mountains: Geol. Soc. America Bull., v. 39, p. 465–480.

BLISSENBACH, E., 1954, Geology of alluvial fans in semi-arid regions: Geol. Soc. America Bull., v. 65, p. 175–189.

BULL, W. B., 1964a, Alluvial fans and near-surface subsidence in western Fresno County, California: U.S. Geol. Survey Prof. Paper 437-A, p. A1–A71.

——— 1964b, Geomorphology of segmented alluvial fans in western Fresno County, California: U.S. Geol. Survey Prof. Paper 352-E, p. 79–129.

CROWELL, J. C., 1954, Geology of the Ridge Basin area: California Div. Mines Bull. 170, Map Sheet 7.

DENNY, C. S., 1965, Alluvial fans of the Death Valley region, California and Nevada: U.S. Geol. Survey Prof. Paper 466, 62 p.

ECKIS, R., 1928, Alluvial fans of the Cucamonga District, southern California: Jour. Geology, v. 36, p. 224–247.

GILBERT, G. K., 1882, Contributions to the history of Lake Bonneville: U.S. Geol. Survey 2d Ann. Rept., p. 167–200.

JAHNS, R. H., 1949, Desert floods: Eng. and Sci. Jour., v. 12, p. 10–14.

LEOPOLD, L. B., WOLMAN, M. G., and MILLER, J. P., 1964, Fluvial processes in geomorphology: San Francisco, W. H. Freeman & Co., 552 p.

LUSTIG, L. K., 1965, Clastic sedimentation in Deep Springs Valley, California: U.S. Geol. Survey Prof. Paper. 352-F, p. 131–192.

NELSON, C. A., 1966, Geologic map of the Blanco Mountain Quadrangle, Inyo and Mono counties, California: U.S. Geol. Survey Map GQ-529.

PACK, F. J., 1923, Torrential potential of desert waters: Pan-American Geologist, v. 40, p. 349–356.

ROUSE, H., ed., 1950, Engineering hydraulics: New York, John Wiley & Sons, 1,039 p.

SHARP, R. P., 1942, Mudflow levees: Jour. Geomorphology, v. 5, p. 222–227.

——— and NOBLES, L. H., 1953, Mudflow of 1941 at Wrightwood, southern California: Geol. Soc. America Bull., v. 64, p. 547–560.

TOLMAN, C. F., 1909, Erosion and deposition in the southern Arizona bolson region: Jour. Geology, v. 17, p. 136–163.

TROWBRIDGE, A. C., 1911, The terrestrial deposits of Owens Valley, Calif.: Jour. Geology, v. 19, p. 706–747.

VANONI, V. A., ed., 1962, Sediment transportation mechanics: introduction and properties of sediment: Am. Soc. Civil Engineers Hydraulics Div. Jour., July, 1962, p. 77–107.

——— 1963, Sediment transportation mechanics: suspension of sediment: *ibid.*, Proc., September, 1963, p. 45–76.

ERRATUM

Page 458, line 21 in the lefthand column should read: "commonly recedes, and . . . ."

INTERSECTION POINT DEPOSITION ON ALLUVIAL FANS: AN AUSTRALIAN EXAMPLE

BY R. J. WASSON

Introduction

Many alluvial fans are trenched and the slope of the channel bed is generally less than the slope of the fan surface. Hooke (1967) has called the point where the channel emerges onto the fan surface "the intersection point", and he noted that material deposited at this point sometimes "... forms a small secondary fan on the surface below the intersection point" (p. 450).

Hooke argued that gullies on fans can result in either even deposition over the surface of the fan, or incision. In the former case, intersection point deposition occurs as a gully shifts the locus of deposition, and even deposition is maintained. In the latter case, disturbance of this even deposition will result in incision of the gully so that overbank deposition can no longer occur at the fanhead. Hooke (1967) suggested that the disturbance is commonly induced by either climatic change or tectonic movement (cf. Bull 1964, Lustig 1965). Intersection point deposition will also occur during incision.

It is the aim of this paper to provide an account of some contemporary and older examples of intersection point deposits on fans near Lake George in SE Australia. To anticipate later conclusions, the deposits described here are associated with gullies which are actively incising the fans.

The Lake George scarp

Lake George lies in an internal drainage basin in the southern tablelands of New South Wales.

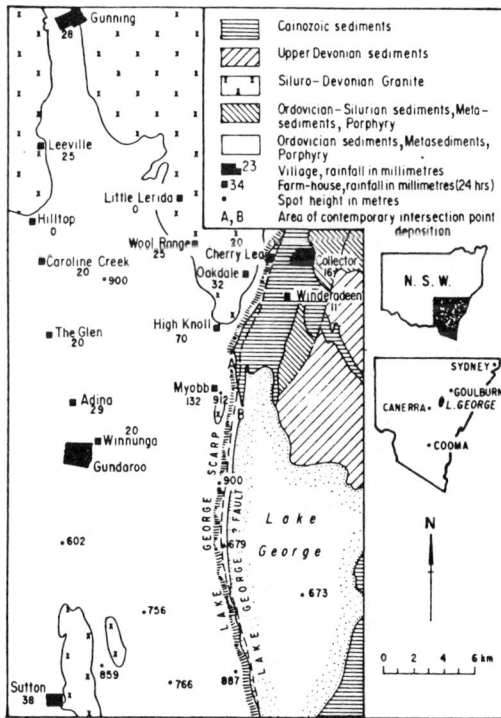

Fig. 1. Location and geological map showing the distribution of rainfall on 18 January 1972. The area of contemporary intersection point deposition lies along the Lake George Scarp between A and B.

187

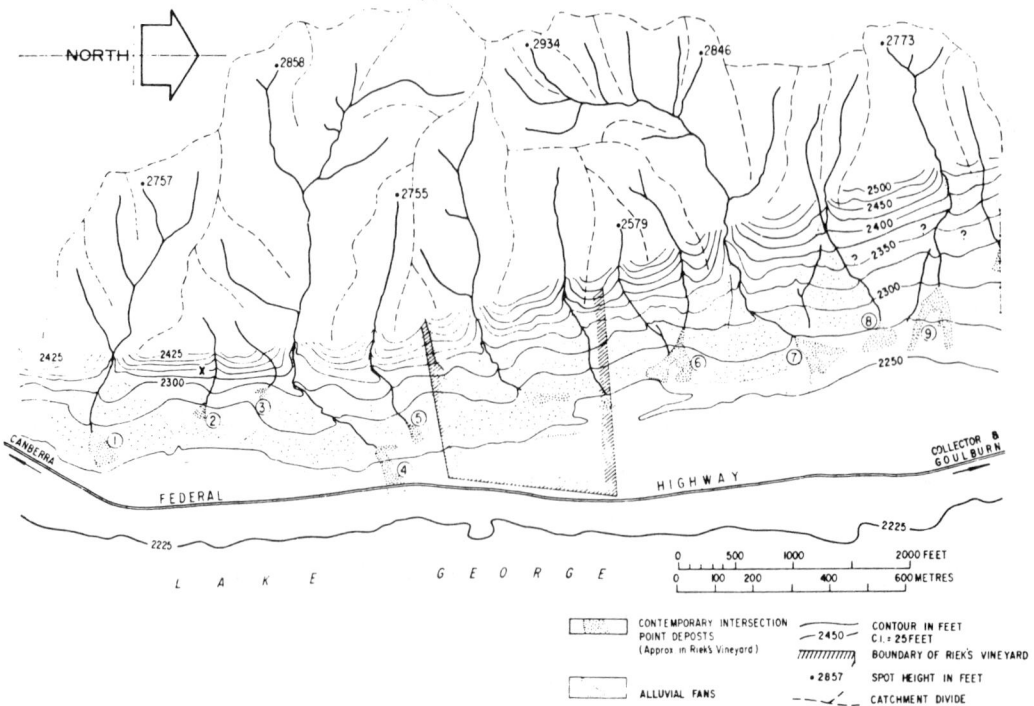

Fig. 2. Topo-planimetric map showing the locations of alluvial fans and contemporary intersection point deposits. The base map from which this figure was prepared is part of a photogrammetric survey between Geary's Gap and Collector carried out by the N.S.W. Department of Main Roads in 1965—1966 (Plan 3C 120 Sheet 3).

The Lake George scarp is a bold topographic feature which forms the western boundary of the basin. The rocks of the scarp are primarily phyllite, schist, acid volcanics, tuff and gneissic granite. Along the foot of the scarp (or Cullarin Range) are a number of alluvial fans (Figs. 1, 2, 3) which coalesce laterally and in many areas both overlie and are interbedded with lacustrine beach deposits of rounded and sub-rounded pebble gravel. The toe of one of the largest fans was cliffed by waves when the lake stood slightly higher than its present level.

From the scarp front most of the gullies run laterally towards the fan margins. Judged from the observations of local residents and property owners, the gullies appear to have been relatively inactive for at least 24 years prior to 18 January 1972. On that date a high intensity rainstorm fell on the scarp producing landslides within the catchments of the alluvial fans and on the scarp front, and formed intersection point deposits with a total volume of approximately 4700 m^3.

Meteorological conditions

The severe rainstorm of 18 January, 1972 was a highly localised event of short duration. This conclusion is based upon an eyewitness account by Dr. E. F. Riek, a part-time vigneron who is cultivating land at the base of the scarp. Riek's land lies roughly in the centre of the area of contemporary intersection point deposition. An account of the storm's effects on the top of the scarp, ca. 1.5 km W of the vineyard, and valuable precipitation records, were provided by Mrs. May of the property known as Myobb.

The rainstorm appears to have been most intense in the vicinity of Myobb (Fig. 1), for at the house 132 mm was recorded in a little less than 2 hours. The storm consisted of two periods of very high intensity rainfall which were separated by about 30 minutes, an observation of both of the eye-witnesses.

At High Knoll, 4 km N of Myobb, 70 mm of rainfall was recorded during the storm, but

Fig. 3. The Lake George scarp showing part of an alluvial fan and contemporary intersection point deposit 6. The *Stipa variabilis* in the foreground grows in a depression behind a beach ridge, from which the photograph was taken.

only 32 mm fell at Oakdale, 7.5 km NNE of Myobb (Fig. 1). Independently, Dr. Riek estimated a rainfall of between 130 and 150 mm on his vineyard. Based on farmers' observations, it would appear that the violent part of the storm was restricted to the eastern edge of the plateau which is bounded on its eastern side by the Lake George scarp. The bulk of the rain fell between Myobb homestead and the edge of Lake George.

Precipitation records at Myobb provide only 14 years of measurements (1958—1971), during which time the average annual precipitation (which here includes the water equivalent of snow) was 720 mm. This figure compares with the official average recording at Collector village of 697 mm which is based on a 75 year record (1891—1969) (Records of the Commonwealth Bureau of Meteorology). In the 14 year Myobb record, daily precipitation has never exceeded the 132 mm of 18 January, 1972, and the highest daily total prior to the storm was 88 mm recorded in 1959. Daily measurements of 25 mm or more have been made on 77 occasions between 1958 and 1971, whereas recordings of 50 mm or more have been made on only 12 occasions during the same period. Unfortunately, these figures provide no information about intensity as expressed in mm/hour, because they are all 24 hour totals, and they do not differentiate between individual storms which may occur within a 24 hour period.

All property owners who either live, or consistently work, near Myobb are unanimous in asserting that a storm of the intensity of that of 18 January, 1972 has not occurred during the time that they have resided in the area; periods ranging from 24 to 40 years.

Contemporary intersection point deposits

The contemporary intersection point deposits were examined 5 months after their development, and, as a consequence of stock trampling, some superficial disturbance had occurred. Very little vegetation had grown on the deposits and their gross morphology seemed unaltered. Emphasis was placed on

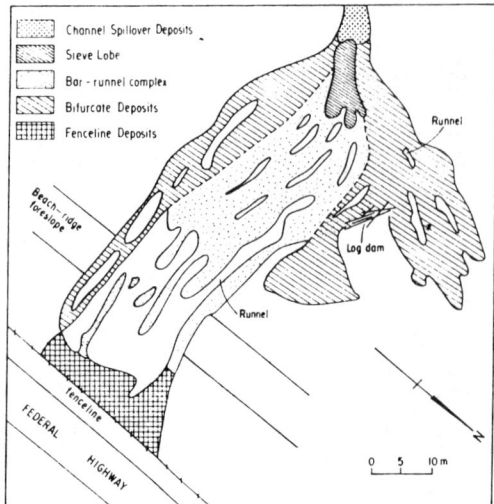

Fig. 4. Morphological plan of contemporary intersection point deposit 4.

intersection point deposit 4 (see Fig. 2) where a particularly fine array of morphological features is displayed. The significance of morphological features differentiated on deposit 4 was evaluated on the remaining eight intersection point deposits, and some general conclusions drawn.

Intersection point deposit 4 lies at the end of an erosional gully which has trenched the northern side of an alluvial fan. The deposit is complex, but some zones of similar morphology can be distinguished and are represented in Fig. 4. At the upstream end of the deposit the channel walls have been overtopped and buried by debris which has spread laterally. This debris will be called the "channel spillover deposits". About four metres down the deposit from the end of the gully lies a body of gravel which is broadly lobate (Fig. 5). "Fingers" of gravel extend a short distance downslope from the lobe which is about 11 m long and 5 m wide. The largest particle in the lobe is 60 cm long on its longest axis, and has a volume of about 0.08 m^3. This lobate deposit strongly resembles features which Hooke (1967) called "sieve lobes".

The distinguishing characteristic of the sieve lobe is that it is built of the bulk of the largest gravel particles which occur in the intersection point deposit. The boulders are partially overlain by pebbles and cobbles, and the largest particles have dammed smaller ones, producing a localised and rude imbrication with upslope dip. No overall preferred orientation of the a-axes of the gravel could be discerned, and particles of varying shapes and sizes are jumbled together. The voids between the large gravel particles are partially filled by fine-grained deposits which vary markedly in texture from point to point, and which consist of various proportions of granules, sand and silt. Nowhere are the largest particles submerged by these finegrained deposits.

Downslope of the sieve lobe lies a complex area of gravel (to be known as the "bar-runnel complex") which spreads laterally only slightly (Fig. 4). Approximately 11 m downslope of the distal end of the sieve lobe there are runnels within the gravel deposits. The grass within the runnels was intact but flattened, only small quantities of pebble gravel lying on the grass. The runnels are ca. 1 m wide at their proximal ends, and are flanked by particles of all sizes. The areas of gravel which lie between the runnels are often spool-shaped (cf. Krigström 1962) and adopt both a tabular (to slightly convex) and distinctly ridged cross-profile. Both the tabular and ridged bars are similar in form to the embryo spool-bars described by Krigström (see his Fig. 2) in Iceland. The bars have an average thickness of ca. 30 cm.

The area which is free of sediment in the bar-runnel complex increases downslope until a break of slope occurs across the foreslope of an old lakeshore ridge. Above the foreslope the mean slope angle is 3°. Some of the bars terminate above the foreslope, while others continue over it, spreading out towards the base of the foreslope. The area of the foreslope covered by debris is small relative to the rest of deposit 4.

Below the foreslope, deposits occur adjacent to a reconstructed fence; the former fence having been destroyed by the transportation of gravel. Much of the deposit upslope of the fence consists of granule and pebble gravel, cobbles occurring locally. The surface of this deposit consists of tabular bars of gravel which partially overlie one another, producing smooth boundaries, in plan, between deposits of markedly different texture. Smallscale braiding is also in evidence on the surfaces of the bars.

Spreads of gravel on either side of the sieve

Fig. 5. Part of the sieve lobe of contemporary intersection point deposit 4. Part of the southern bifurcate deposits can be seen in the middle distance. Flow was from right to left. The notebook is 12 cm long.

lobe are distinct on deposit 4. The characteristics of these "bifurcate deposits" which distinguish them from the deposits in the bar-runnel complex are: much finer texture, only occasional ridged bars, more tabular bars, and more runnels further upslope. The southerly arm of the bifurcate deposit gradually merges with the bar-runnel complex at the downslope end of the complex. The northern arm has itself been bifurcated by a log-dam (see Fig. 4), and the area immediately east of the dam is characterised by linear ridges. The finer bifurcate deposits overlie the north-western edge of the bar-runnel deposits, and it seems that the log-dam occurred during the deposition of the bar-runnel gravels, and before the deposition of the northern bifurcate gravels.

The two parts of the northern arm of the bifurcate deposit are texturally distinct, the more westerly being finer than the easterly deposit which fans out downslope of the log-dam.

Observation of flow conditions

Dr. E. F. Riek observed flow conditions within the gullies and on the contemporary intersection point deposits in the immediate vicinity of his vineyard, and on the slopes of the scarp face. His observations were qualitative, and largely incidental to his main concern, that of saving various earthworks on his property. It should also be borne in mind that Dr. Riek's observations were recorded by this author some five months after the storm. Nevertheless, some valuable indications of the flow conditions have been obtained from the eyewitness, evidence which has assisted in the interpretation of some of the depositional forms observed in the contemporary deposits. Three of the intersection point deposits which Dr. Riek directly observed were subsequently ploughed, prior to this author's observations.

The first part of the storm, which began at about 5 p.m., produced clearwater flows in the gullies. Rick believed that the bulk of the gravel which had been lying in the previously dry channels was removed by the clearwater flow, including material which he had gathered from the surrounding pastureland. The largest particles moved by the clearwater flow had a maximum diameter of about 60 cm, and they were always completely immersed in water during their transportation.

The clearwater flow was followed by more turbid flow in which fine sediments became increasingly important. This change occurred towards the end of the first part of the storm. The clearwater flows appear to have both loosened and saturated channel wall sediments. Moreover, hillslopes had been saturated and landslides occurred in the steep catchments of the fans. Only one landslide occurred on the scarp front (at X on Fig. 2). Most landslides occurred on the lower parts of slopes which are being undercut by streams, and a few landslides occurred high on valley side slopes. Debris flow levees lead from the landslides to the channels below.

It would appear that the landsliding occurred, at the earliest, towards the end of the clearwater flow. The more turbid flow was accompanied by large quantities of green timber which is thought to have been dislodged from the forested catchments by the landsliding. The largest of the trees transported in the gullies was about 6 m long. The combination of landsliding and bank caving contributed a large quantity of fine sediment which produced, according to Dr. Riek, mud slurries within the gullies and localised slurry flows on the intersection point deposits. Riek's observations suggest that the slurries were essentially fluviatile deposits with a high sediment concentration, rather than debris flows in the sense of Johnson (1970).

A particularly valuable observation made by Dr. Riek was that the water level, and the quantity and calibre of sediment carried in the channels, oscillated considerably during the duration of flow. Riek also noted that at no time were the channel banks overtopped, except, of course, at the intersection point. The oscillation in flow conditions was also assisted by log-dams which occurred within the gullies, and which generally trapped first boulders and then smaller particles. The dams periodically burst, producing a sudden but relatively minor increase in flow depth downstream. A similar phenomenon has been described by Sharp and Nobles (1953), Blackwelder (1928), and Singewald (1928) in debris flows.

The turbid slurries were followed by clearwater flows which in some areas were of short duration, but on deposit 4 lasted until about 9 p.m., 4 hours after the beginning of the storm. The clearwater carried some sediment, but generally the water only re-sorted some of the slurry deposits, removing the surface fines and redistributing the surface gravel into small-scale braid bars.

The formation of contemporary intersection point deposit 4

The morphological description of deposit 4, in combination with the reconstruction of flow conditions afforded by the observations of Dr. Riek, allows some discussion of the genesis of the deposit.

The formation of deposit 4 began with a clearwater flow which moved particles up to cobble size, the debris spreading laterally only slightly. Then began the construction of the bar-runnel complex (Fig. 4). The construction of the log-dam probably occurred at an early stage in deposition, resulting from the initial scouring of the dead timber which was already in the channel. The log-dam diverted part of the bar-runnel complex gravel, so limiting the lateral spread of the gravel. At sometime near to the change from clearwater flow to turbid flow, boulders began to accumulate in a sieve lobe just downstream of the intersection point.

The position of the accumulation of coarse particles in the sieve lobe in deposit 4 is not a chance occurrence, but is related to the rapid change in the hydraulic conditions of flow at the intersection point. Upstream of the intersection point, water and sediment were confined, thereby maintaining sufficient depth of water to immerse the largest particles. Past the intersection point the confining channels were absent, water depth decreased as the fluid spread laterally, so exposing the largest particles. The concomitant reduction of tractive force lead to deposition of the largest particles.

The sieve lobe appears to have developed by both a lee-shadow effect and damming. The largest boulders appear to have been dammed behind low bars and ridges of cobble gravel. This accumulation of coarse particles assisted the development of the lobe by producing a lee-shadow effect, so adding sediment to the downslope sides of the developing lobe. The boulders also dammed material on their upstream sides.

The construction of the sieve lobe caused subsequent loads of coarse sediment to bifurcate on either side of the bar-runnel complex. This later deposition partially overran earlier deposits, including the upstream end of the sieve lobe, and much of the lobe was flooded by fine-grained sediments ranging from silt to granules. The disposition of these finegrained sediments between the boulders and cobbles, but never fully immersing the large blocks, suggests that the deposition of the sieve lobe was followed by a mud and sand slurry which flowed between the blocks rather than with them or over them (cf. Hooke 1967).

The bar-runnel complex developed by a number of bifurcations of single channels attendant upon localised boulder and cobble accumulations. Runnels are distinguished by the dearth of sediment within them, abrupt boundaries formed by bars or ridges on either side, and a general flattening of the grass sward within them. A few of the runnels have been partially infilled. Some of the unfilled and infilled runnels anastomose (in the sense of Leopold and Wolman 1957, p. 40), that is, they both bifurcate (or fork) from a common trunk, and rejoin. However, bifurcation is far more common than is rejoining.

Krigström (1962, p. 331) reported that in braided sandur streams, dry channels and bars are almost always crossed by younger channels. Leopold, Wolman and Miller (1964, p. 291) noted that large braided rivers are characterised by continuous shifting of their channels, a feature which rapidly produces abandoned channels (Fahnestock 1963). However, in the contemporary intersection point deposits there is little evidence that the various runnels have shifted their courses. Within the bar-runnel complex, only one example of diagonal trenching of a bar was seen. Here, the younger channel had been partially filled with gravel. Evidence of trenching of bars may well be completely obscured by later deposits. While the morphological patterns of bars and runnels are similar to some braided streams, there is little evidence of shifting of runnels, distinguishing this deposit from streams where the term "braided" would be unreservedly applied.

Parts of the bifurcate deposits display tabular bars, (generally not spool-shaped) which have migrated by deposition at their distal faces. These bars are ca. 10 cm thick and their average length in the direction of maximum slope is ca. 1 m. The bar faces are sinous in plan, and represent very sharp textural breaks between the bars and the underlying sediment. Internally, the tabular bars display horizontal bedding of the platy metasediment pebbles and granules. Very few of the maximum projection planes of particles are lying at an angle which deviates from the slope of the surface beneath the bar. This observation can be generalised to the great bulk of the contemporary intersection point sediments.

The significance of the sieve lobes

The characteristics described for deposit 4 are not all common to all of the remaining eight deposits that were examined. The critical morphological feature within the deposits is the sieve lobe, the presence or absence of which significantly effects the rest of the deposit. A sieve lobe developed in six of the deposits, and the size, shape, and position of the lobe is a function of particle size, the slope of the surface of the alluvial fan immediately downslope of the intersection point, and hydraulic conditions.

The significance of the sieve lobe can be demonstrated on deposit 9. This deposit has been markedly bifurcated (Fig. 2) either side of a well developed sieve lobe. Bifurcation probably occurred early in the construction of the deposit, for there is no equivalent of the bar-runnel complex of deposit 4. A minor log-dam occurred at the downslope margin of the sieve lobe, and this dam diverted some material to the northern side of the southern limb of the deposit. There is some evidence of re-working of bars by channels which were braided.

The absence of a sieve lobe has a dramatic effect on the morphology of an intersection point deposit. Deposit 8 is almost perfectly fan-shaped (Fig. 2), and the reason for this remarkable lack of bifurcation is the absence of gravel larger than cobble size, and the very small proportion of cobbles incorporated in the deposit. There are no bars of high local relief, no sieve lobe, and a very few fragments of wood which could congregate as a log-dam. Bars have been reworked by braided channels which have been unrestricted in their movement, and tabular bars have developed in the body of the deposit. The absence of both wood and gravel larger than cobble size can be explained by a low point in the northern wall of the gully, some 300 m upstream of the intersection point. The channel immediately upstream of the low point is oriented 045°, whereas the channel immediately downstream of the low point is oriented 073°. Water, wood, and coarse sediment moving to 045° failed to round the bend in the channel and spilled out of the gully onto the fan surface. The fine-grained sediment was carried past the bend and deposited at the intersection point.

The effect of the absence of a sieve lobe can again be seen on deposit 5, where the surface of the alluvial fan immediately downslope of the intersection point is the steepest of all such slopes examined, with a gradient of 7°. The largest gravel fragments are only cobble size and did not congregate into a sieve lobe but spread in irregular longitudinal ridges. The absence of boulders within this deposit, in combination with a steep transporting surface, has meant that material has moved considerably further downslope from the intersection point than was possible on deposit 4. This increased mobility of the sediment has produced a diffuse, scattered deposit in the upper parts, with a complex bar-runnel and tabular bar deposit in the lower parts.

Older intersection point deposits

The discussion so far has established a qualitative relationship between the contemporary intersection point deposits and certain flow conditions within the gullies. During the life of the gullies there have been several events similar to that of January, 1972. For example, upstream of the intersection point of deposit 4 there are three bouldery, grass covered lobes, the axes of which deviate from the thalweg of the gully. Surrounding the lobes are indistinct fan-shaped spreads of vegetated sediment which lie on the alluvial fan surface. These features are depicted on Fig. 4 where they are seen in association with the gully which ends in deposit 4. The bouldery lobes and surrounding fan-shaped deposits occur in association with the gullies on a number of the alluvial fans. The lobes and surrounding deposits are simply sieve lobes within intersection point deposits which are older than the contemporary deposits discussed in the first half of this paper.

The intersection point deposits, both old and contemporary, and the gullies are related in two ways. The first is shown in Fig. 6 where the deposits are distributed along the length of the gully. The second is depicted in Fig. 7 where the contemporary deposits lie on the old deposits. The first situation occurs where the gully has changed course and extended itself downfan between floods. The second situation occurs where the gully has neither significantly changed course, nor extended itself downfan between floods.

The explanation of the first situation (Fig. 6) is simply that each flood produces an intersection point deposit in which a sieve lobe develops, and subsequent, but considerably less violent flows lengthen the gully. Where a gully can freely change course the lengthening occurs in a direction away from the sieve lobe because the lobe offers resistance to the between-flood flows. As the gully is lengthened the bed is lowered, a characteristic of discontinuous gullies in which the gradient of the bed

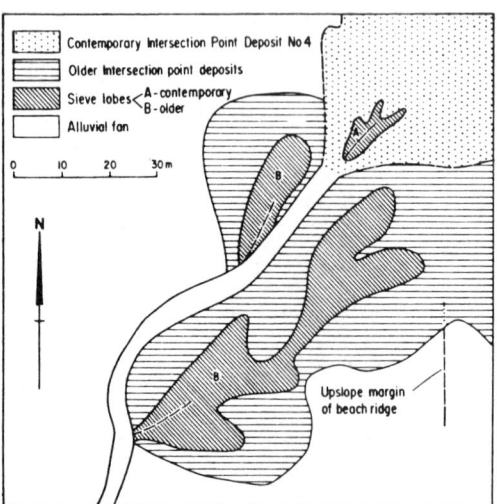

Fig. 6. Contemporary intersection point deposit 4, showing the relationship between the older intersection point deposits, the channel, and the sieve lobes.

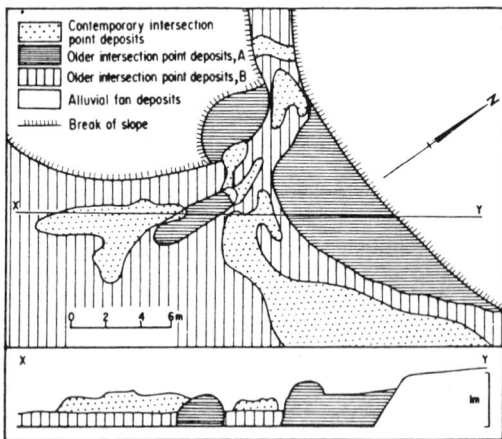

Fig. 7. Contemporary intersection point deposit 5, showing the relationship between the contemporary and older intersection point deposits.

is held constant (Leopold, Wolman and Miller 1964). Lowering of the bed effectively protects the older lobes from erosion.

The exact mechanism of gully lengthening in the first situation is obscure. Scour occurred just below the intersection point on contemporary deposit 5, and very small discontinuous gullies have developed on the alluvial fan surfaces underneath the margins of other contemporary deposits. These small scour features could develop into larger discontinuous gullies by headward retreat (Blong 1966, Schumm and Hadley 1957, Brush and Wolman 1960), so lengthening the main gully by the familiar mechanism of gully coalescense.

The second situation (Fig. 7) is rather more complex, for between-flood flows appear to accomplish some erosion of intersection point deposits. Each intersection point deposit is complicated by the erosional remnants of earlier deposits, and complex sedimentation patterns have resulted:

1. In Fig. 7 there are intersection point deposits of three different ages. Deposit A is the oldest and apparently the most dissected. Deposit B is inset to A and appears to be relatively intact. The contemporary deposit lies on deposit B but is patchy. All of this intersection point deposition has occurred in a shallow trench in the alluvial fan.
2. Contemporary deposit 3 is an extremely small feature (Fig. 2), in which bifurcation has been promoted by a low and cobbly mound; part of an older intersection point deposit.
3. Contemporary deposit 6 displays a very complex depositional pattern, which is the result of a mosaic of older bouldery deposits and a network of rills. There was some between-flood lengthening of the gully but in general the topographic pattern, which pre-dates the contemporary deposition, has been reinforced.

Discussion and conclusions

Contemporary intersection point deposition on alluvial fans at Lake George was brought about by a high intensity rainstorm of relatively low frequency. The nature of the flow in the gullies which trench the fans changed from clearwater with very low sediment concentration, to mud slurries, followed by a final phase of clearwater.

The most important feature of the intersection point deposits is the sieve lobe, the presence or absence of which dictates the form of the remainder of the deposit. If the lobe develops fairly early, but not at the beginning of the deposition, then a bar-runnel complex will develop and the remainder of the sediment will be bifurcated. If the lobe does not develop then the deposit will be almost perfecty fan-shaped.

The construction of the sieve lobe depends upon a sediment supply which is very poorly sorted; in the case described at Lake George the sizes represented range from clay to boulders. It is most unlikely that sieve lobes would develop in well sorted sediment.

Intersection point deposition at Lake George is a periodic phenomenon, and the event of January 1972 was the latest in a series of unknown duration. The gullies in the large alluvial fans have been lengthening, and as a result the intersection points have been moving down the fans. Each time a significant flood occurs, intersection point deposition is accomplished. Subsequent flows lengthen the gullies by routes which abandon the intersection point deposits.

The gullies on the small alluvial fans have not been lengthening, and as a result each significant flood produced an intersection point deposit which was eroded by between-flood flows. The deposits lie on the erosional remnants of older deposits, producing complex sediment patterns.

Contemporary deposition on the alluvial fans at Lake George is restricted to intersection points. Overbank deposits were only seen at low points on gully walls, for example, upstream of contemporary intersection point deposit 8. In general, however, there is no overbank deposition from the gullies.

Hooke (1967) argued that a fanhead is incised when "... it seems unreasonable to expect overbank flooding by water flows at least once in a few decades" (p. 457). As we have seen, there is no evidence (with one minor exception) of overbank deposition from the gullies at either the fanheads or closer to the intersection points. Therefore, applying Hooke's criterion, the gullies in the Lake George fans are incised.

Intersection point deposition on the Lake George fans is the consequence of gully incision. Once a gully is developed in a fan, then lengthening of the gully will result in regrading of the bed, to maintain a constant longitudinal gradient, thereby providing sediment for intersection point deposition further downfan. Regrading of the bed will, of course, deepen the incision.

The gully lengthening on the Lake George fans is moving gullies on adjacent fans closer to one another (Fig. 2, 3). If the lengthening continues then gullies from adjacent fans will join and perhaps ultimately reach the lake, in an integrated drainage network.

Acknowledgements

I would like to thank Dr. E. F. Riek (C.S.I.R.O. Division of Entomology) for providing me with his acute observations. Also, I am grateful to Professor J. L. Davies, Dr. R. J. Blong and Dr. M. Selby for their comments on earlier drafts of this paper. The work reported here was carried out during the tenure of a Commonwealth Postgraduate Research Scholarship.

Mr. R. J. Wasson, School of Earth Sciences, Macquarie University, New South Wales 2113, Australia

References

Blackwelder, E., 1928: Mudflow as a geologic agent in semi-arid mountains. *Geol. Soc. Amer., Bull.*, 39, 465—480.

Blong, R. J., 1966: Discontinuous Gullies on the Volcanic Plateau. *J. Hydrology* (N.Z.), 5 (2), 87—99.

Brush, L. M. and Wolman, M. G., 1960: Knickpoint behaviour in noncohesive materials: a laboratory study. *Bull. Geol. Soc. Am.*, 71, 59—74.

Fahnestock, R. K., 1963: Morphology and hydrology of a glacial stream. *U. S. Geol. Surv., Prof. Paper* 422-A, 70 pp.

Hooke, R. LeB., 1967: Processes on arid-region alluvial fans. *J. Geol.*, 75, 438—60.

Johnson, A. M., 1970: *Physical Processes in Geology.* Freeman. Cooper and Co., San Francisco, 577 pp.

Krigström, A., 1962: Geomorphological studies of sandur plains and their braided rivers in Iceland. *Geogr. Ann.* 44, 328—46.

Leopold, L. B. and Wolman, M. G., 1957: River channel patterns: braided, meandering and straight. *U.S. Geol. Surv., Prof. Pap.* 282-B.

Leopold, L. B., Wolman, M. G. and Miller, J. P., 1964: *Fluvial Processes in Geomorphology*, Freeman, San Francisco, 522 pp.

Schumm, S. A. and Hadley, R. F., 1957: Arroyos and the semi-arid cycle of erosion. *Am. J. Sci.*, 255, 161—174.

Singewald, J. T., 1928: Discussion of 'Mudflow as a geologic agent in semi-arid mountains by Eliot Blackwelder'. *Geol. Soc. Am., Bull.*, 39, 480—483.

ERRATUM

Page 89, line 41 in the lefthand column should read: "slope is ca 1 m. The bar faces are sinuous. . . ."

11

Copyright © 1977 by Gebrueder Borntraeger
Reprinted from *Zeitschr. Geomorphologie* 21:147–168 (1977)

Catchment processes and the evolution of alluvial fans in the lower Derwent valley, Tasmania

by

R. J. Wasson, Auckland

with 7 figures

Zusammenfassung. Um ein allgemeines Modell der Schwemmfächer-Bildung aufzustellen, ist es notwendig, eine große Anzahl von aktiven und relikten Schwemmfächern aus physiogeographisch verschiedenen Räumen zu untersuchen. Eine Studie an einer Reihe von relikten Schwemmfächern in SE-Tasmanien zeigt, daß sie durch die Mobilisierung von Material unter den Periglazialbedingungen der letzten Kaltzeit Tasmaniens entstanden sind. Die Schwemmfächer-Akkumulation erfolgte durch Schuttströme und sedimentreichen Abfluß aus Einzugsgebieten, die in den Höhen Periglazial- und Nivationsprozesse aufwiesen. Es wird angenommen, daß die Zerschneidung der Schwemmfächer erfolgte, als das Klima sich verbesserte und die Verdichtung der Vegetation im Einzugsgebiet die Sedimentbelastung des Abflusses reduzierte. Die Schwemmfächer wurden beim Meeresspiegelanstieg im Derwent Ästuar unterschnitten und kleine Einfüllungen entwickelten sich in den Schwemmfächer-Rinnen während des Spätholozäns.

Summary. The construction of a general model of alluvial fan evolution requires a large number of studies of both active and relict alluvial fans in diverse environments. A study of a suite of relict alluvial fans in southeastern Tasmania shows them to be the result of sediment mobilisation under cold climatic conditions during the last-glacial period of Tasmania. Fan accumulation proceeded by debris flows and sediment-charged water flows derived from catchments which at the highest altitudes were periglacial and nivational. Fan incision is thought to have occurred as climatic amelioration and catchment revegetation reduced the sediment yield/water discharge ratio. The fans were cliffed by rising sea level in the Derwent estuary, and small inset fills developed within the fan incisions during the late-Holocene.

Résumé. La construction d'un modèle général concernant l'évolution des cônes alluviaux requiert un grand nombre d'études de cônes alluviaux tant actifs qu'anciens dans des environnements divers. L'étude d'une succession de restes de cônes alluviaux dans le S.E. de la Tasmanie montre qu'ils résultent du déplacement de sédiments en climat froid pendant la dernière période glaciaire de la Tasmanie. L'accumulation des matériaux sur les cônes est due à des écoulements de débris et à des rivières chargées d'alluvions provenant d'endroits où, à cette altitude, ils étaient d'origine périglaciaire et nivale. L'incision des cônes de débris s'est vraisemblablement produite lorsque l'amélioration du climat et le retour du couvert végétal a amené la réduction du quotient fourniture de sédiment/débit liquide. Les cônes ont été attaqués en falaises par l'élévation du niveau marin dans l'estuaire du Derwent, et de petits dépôts intérieurs se sont développées dans les échancrures des cônes alluviaux à la fin de l'Holocène.

General introduction

The factors which control the accumulation and dissection of alluvial fans are numerous, and the relative effects of each of the factors is commonly in dispute. BEATY (1970) has drawn attention to the conflicting conclusions regarding contemporary conditions on alluvial fans in the area between the White Mountains and Death Valley in the southwestern United States, and considered that "... conflicts may some day be accomodated by a more general theory of fan formation" (p. 76). COOKE & WARREN (1973) confirmed BEATY's observation, and added: "At present the chronological evidence necessary to resolve some points in this controversy is rarely available" (p. 187).

A general model of fan evolution is not available, and it is the contention of this author that such a model will need to be constructed from a combination of studies of modern processes on alluvial fans in various environments, and from analyses of the history of the evolution of fans which are now active or relict. The historical approach is required to answer the questions: (1) Are all environments of fan accumulation and dissection represented among modern, active fans?; and (2) Have the processes of fan evolution remained constant throughout time in a particular area?

With the proposed method of constructing a general model of fan development in mind, it is the aim of this paper to present an historical account of the evolution of a suite of alluvial fans in the lower Derwent Valley of southeastern Tasmania (Fig. 1 and 2), the most southerly state of Australia.

Alluvial fan dissection

The reasons for fan dissection, and intersection point deposition (HOOKE 1967) are contentious and there are two groups of arguments; 1, dissection is induced by processes within the general fan accumulation regime; 2, dissection is the result of some fundamental change in regime. The proponents of the first argument invoke three causes: 1, extreme events of intense rainfall which carve out temporary trenches (BULL 1964; BEATY 1974; and cf. SCHICK 1974); 2, alternation of debris flows and water flows (HOOKE 1967); 3, continued downcutting within the fan catchment resulting in inevitable dissection of the fan to maintain an equilibrium profile (ECKIS 1928). This third cause must operate on a much longer time scale than either of the other two.

The proponents of the second argument invoke two causes: climate change (LUSTIG 1965, has suggested that a change of climate which increases the prevalence of debris flows will deepen any pre-existing channel because of an increase in tractive force) and diastrophic activity, particularly tilting and faulting

(HOOKE 1972; BULL 1964). The two arguments are usually offered as alternatives (e. g. LUSTIG 1965; HOOKE 1967; COOKE & WARREN 1973), but there is no reason to believe that all of the various causes proposed are mutually exclusive.

Fan dissection, as the result of a fundamental change in the regime of a fan, has been recognised by a number of writers (RYDER 1971 a, b; CHURCH & RYDER 1972; WILLIAMS 1970; 1973; HOPPE & EKMAN 1964), but no criteria are provided to distinguish between a trench cut by a change in regime and a temporary trench cut as part of the accumulation of the fan.

The difficulties of many of the arguments can be avoided by concentrating on the critical factor of fan dissection, namely, whether or not a trench cut into a fan provides sediment to the fan surface, or removes sediment and deposits it beyond the boundaries of a fan. In the former case the fan is considered to be active. The following scheme and terminology is proposed by the present writer:

Fan Entrenchment – Downcutting into a fan surface by a channel which contributes sediment to the fan surface. This contribution can be of any volume. Entrenchment usually occurs during fan construction.
Fan Incision – Downcutting into a fan surface by a channel which debouches beyond the fan margin. No sediment is contributed to the fan surface. Incision is usually associated with fan destruction.
Fan Dissection – A general term to include both entrenchment and incision.

Entrenchment must also be distinguished from channel formation which often accomplishes most sedimentation. Many fans are built largely by braided or distributing streams, and channels producing this accumulation are not necessarily entrenched. The entrenchment of a fan moves the locus of deposition away from the apex, and the new area of deposition often consists of shallow braided channels. The entrenchment is a form of dissection of the upper part of the fan (often called fanhead incision or trenching), and removes the fan apex from active deposition.

General setting of the Derwent Fans

Eroded remnants of alluvial fans occur along both sides of the lower reaches of the Derwent River estuary between the towns of New Norfolk and Bridgewater (Fig. 1). The fans are low gradient and generally concave in longitudinal profile. Surface gradients are less than 10° and the toes of the larger fans have gradients of less than 4°. The fans lie below generally steep catchments with often rectilinear slopes at angles of about 30°. The catchments are cut in a suite of Permian and Triassic mudstone, conglomerate, shale, siltstone, sandstone, arkose, limestone, coal, and intrusive Jurassic tholeiitic dolerite sills, dykes, and sheets. Dolerite caps topographic highs, and small patches of Tertiary basalt occur in the Derwent valley bottom. The gross landscape pattern of the Hobart district has been determined by Tertiary faulting. The section of the Derwent river which concerns us cuts across the Mt Dromedary Horst (SPRY & BANKS 1962) so that the orientation of the valley between the entry of the Lachlan River

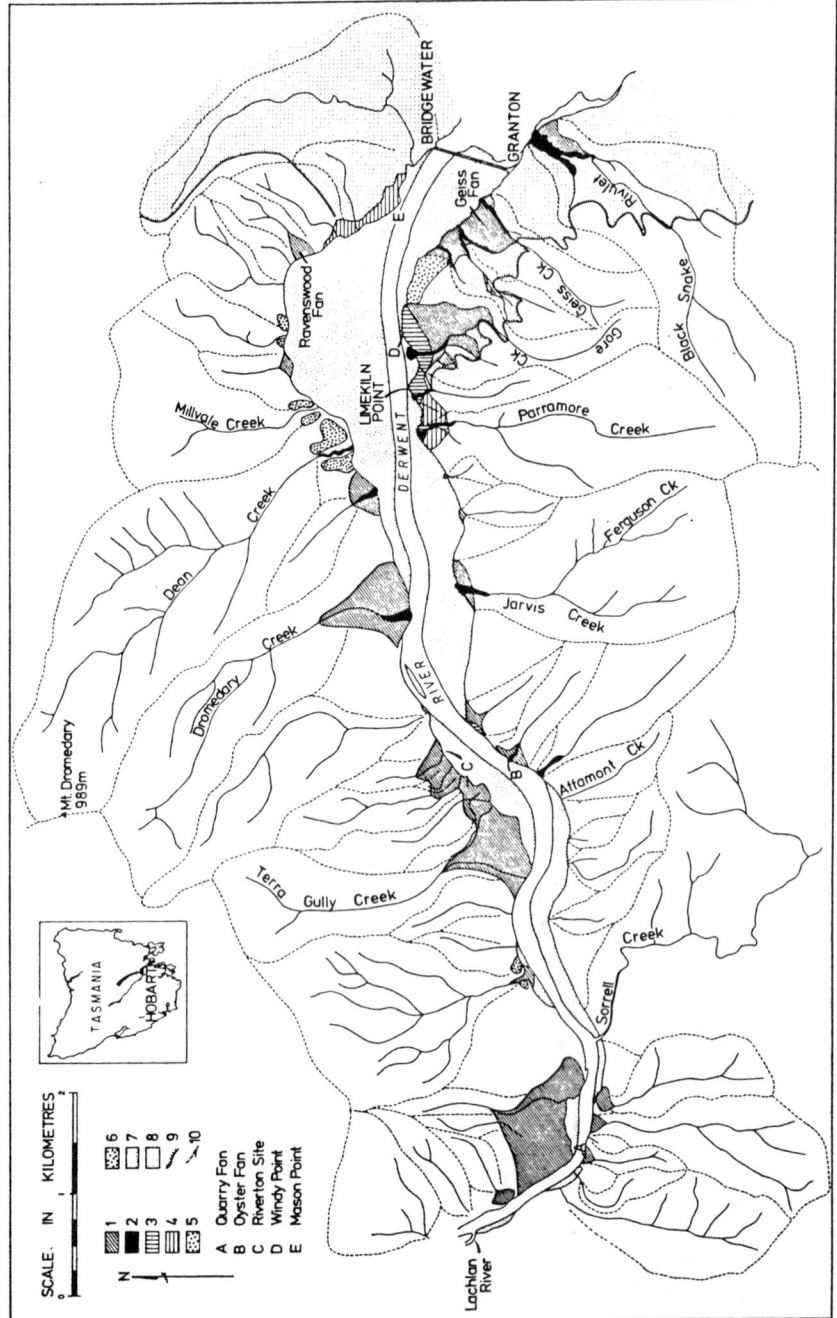

Fig. 2. Geomorphic map of the alluvial fans and their catchments. 1 = Alluvial Fan Remnants (Last Glacial); 2 = Inset Fills (Holocene); 3 = Aeolian Coversands on Fan Remnants; 4 = Aeolian Coversands not on Fan Remnants; 5 = Fluviatile Gravels; 6 = Strath Terraces; 7 = Estuarine Fill (Holocene); 8 = Low, Undulating Hills – Thin Gravel-rich Regolith; 9 = Boundary between Piedmont and High Hills; 10 = Catchment Divides.

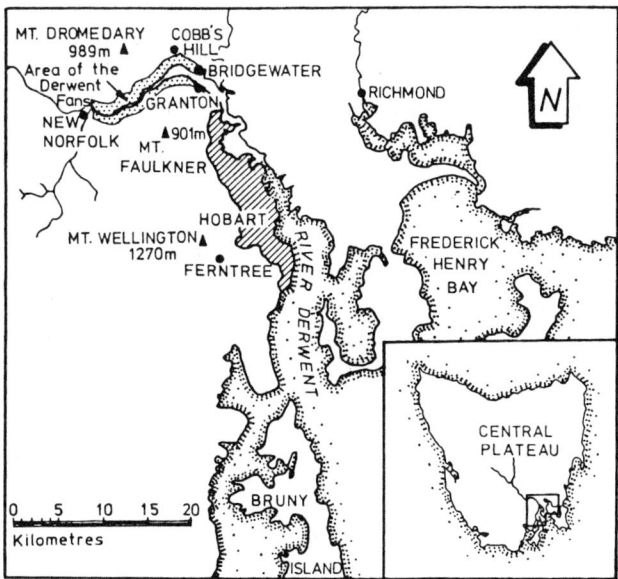

Fig. 1. Location map, south-eastern Tasmania.

and Limekiln Point (Fig. 2) is not fault-controlled, but the orientation of the valley below Limekiln Point is controlled by faults of the Derwent Trough. There is no evidence that the faults were active either during or since fan formation.

The area of the Derwent Fans lies in the Thornthwaite subhumid precipitation province (DAVIES 1967). Hobart has an average annual precipitation of 620 mm (91 year record, Australian Bureau of Meteorology, Records). The area lies on the boundary between the Thornthwaite warm and cool temperature provinces, and the mean annual temperature at Hobart is 12.4° C, ranging from a mean maximum of 16.7° C to a mean minimum of 8.1° C (88 year record). The lowest ground level temperature recorded in Hobart is −7.7° C (83 year record), and the average number of frosty nights (screen minimum less than or equal to 2° C) is 25 (13 year record). The mean annual temperature on Mt Wellington (1,270 m ASL), the highest point in the area, is 4.1° C, with a mean maximum of 7.2° C and a mean minimum of 1.0° C (8 year record).

The vegetation of the area is broadly zoned according to altitude with a *Eucalyptus* spp. woodland (classification after SPECHT 1972) on the lowest slopes, a low open forest of eucalypts on higher slopes, and a sub-alpine open forest dominated by *Eucalyptus delegatensis* at about 700–800 m above sea level. On the summits there is a sub-alpine woodland of *E. coccifera*. The bulk of the vegetation is either dry or moist sclerophyll woodland or forest, but there are some very small patches of mixed forest (wet sclerophyll and temperate rainforest) in sheltered areas.

The age of the alluvial fans

Dateable organic material has not been found in either the fans or in most associated deposits. Therefore, stratigraphic methods have been used to determine the age of the fans.

The alluvial fans overlie and, in places, surround alluvial gravels which make up terrace remnants and thinly veneer strath terraces. The terraces lie up to about 25 m above mean sea level. The alluvial gravels have been transported down the Derwent valley, as shown by bedding and imbrication directions. The fans are younger than these gravels which WASSON (1975) suggested are of last interglacial age.

The fans quite clearly dip beneath estuarine sediments which lie along the sides of the Derwent River estuary (Fig. 3). These estuarine sediments are the product of the most recent rise in sea level (i. e. post Wisconsin). Drilling information is available (Tasmania Public Works Department, files and plan 1501-3) for two cross-sections of the Derwent valley: one at Bridgewater, and the other near Attamont Creek (Fig. 2). At Bridgewater the base of the estuarine sediments lies directly on bedrock, and is between 18 and 20 m below present mean sea level. THOM & CHAPPELL (1975) have compiled a sea level curve for southeastern mainland Australia, showing that the sea was about 20 m below the present level between 8,500 and 9,500 radiocarbon year B. P. This result accords with the sea level curve of MILLIMAN & EMERY (1968), and BLOOM's (1971) ice volume curve.

Aeolian deposits mantle many of the alluvial fan surfaces (Fig. 2), forming sheets up to 1.5 m thick. These coversands (FLINT 1971) are poorly sorted mixtures of sand, silt, and clay, but fine sand constitutes the modal fraction. The surfaces of some of the sheets consist of a subdued dune relief, usually short longitudinal and parabolic dunes. The distribution of the coversands requires that the sediment was derived from the Derwent River when a source was available; that is, when sea level was lower than − 20 m. NICOLLS (1958) argued for a last glacial age for the coversands.

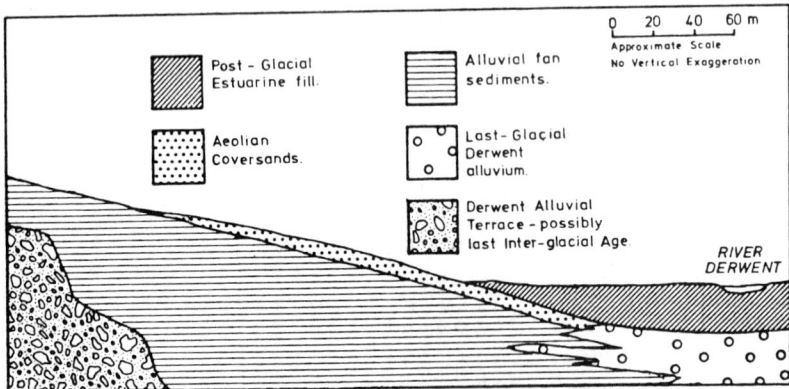

Fig. 3. Schematic representation of the stratigraphic relationships.

The schematic section in Fig. 3 is a synthesis of field observations and drilling information. It is evident that the alluvial fans predate both the post-Wisconsin estuarine fill, and the aeolian coversands.

All that now remains of the fans is a number of eroded remnants, but sufficient to estimate roughly their former extent. Assuming that either the gradient of the surfaces of the fan remnants (avoiding the eroded and often bevelled edges of the remnants) remained constant from fan apex to toe, or that the gradient lessened (cf. BLISSENBACH 1954), it is clear that the fans were built at a time of lower sea level. This conclusion was earlier reached by DAVIES (1967) in support of a last-glacial age for the fans.

A synthesis of the available evidence shows that a last-glacial age for the fans is most likely. The fans are therefore thought to have formed along the sides of the Derwent River, at this time graded to a lower sea level. At the end of fan accumulation, aeolian coversands developed as the result of deflation from the floodplain, and perhaps seasonally exposed channel sediments of the Derwent River. Between 8,500 and 9,500 years B. P. the sea began to encroach on the Derwent River, and estuarine sediments progressively covered fluviatile and alluvial fan sediments alike. The history of the fans, and associated deposits, will be returned to below.

The alluvial fan catchments

The fans are considered to be of last-glacial age, and DAVIES (1967) argued that they were fed with sediment from periglacial solifluction mantles which formed in high parts of the fan catchments. Slope mantles produced under cold climatic conditions, and believed to be of last-glacial age, occur in the catchments. Therefore, DAVIES' association is supported by this writer, and an attempt is made here to extend his work and detail the sedimentary conditions which prevailed in the catchments during fan accumulation.

The evidence for cold climatic conditions in the fan catchments will be considered below in some detail, but first it is necessary to comment on terminology. The term "periglacial" is here used in its broad sense, following DAVIES (1969), TRICART (1970), and WASHBURN (1973), and refers to a region in which climate promotes significant freeze-thaw activity, but does not necessarily maintain permafrost. The term "periglacial solifluction" is used for solifluction (as defined by ANDERSSON 1906) in a periglacial environment. "Nivation" refers to a variety of processes associated with semi-permanent snowdrifts (EMBLETON & KING 1968; DAVIES 1969; WASHBURN 1973).

The most recent overview of late-Quaternary periglacial and glacial environments in Tasmania is by DAVIES (1974) who considered that the lower limit of periglacial activity at the height of the last-glacial was about 450 m above present sea level. This limit was mapped at roughly the same altitude in an east-west direction across Tasmania, suggesting that it was dependent upon temperature, rather than precipitation. Precipitation had a decided east-west gradient as reflected in the last-glacial snowline. Modern periglacial processes are active above the present treeline. Some sorted polygons (WASHBURN 1973) are active on the summit of Mt Wellington (Fig. 1) at about 1270 m (COSTIN 1967), and sorted polygons are active at about 1500 m on the Ben Lomond plateau in north-eastern Tasmania (DAVIES 1969). DERBYSHIRE (1973) has recorded modern periglacial activity at 1371 m on Mt Rufus, at 1250 on Mt Campbell, and as low as 1005 m on the Murchison-Forth interfluve. All of these sites are on the Central Plateau (Fig. 1).

The catchments of the alluvial fans were investigated for evidence of periglacial solifluction, but it was necessary to extend fieldwork beyond the catchments to adjacent areas so that a complete picture of last-glacial conditions could be obtained. The deposits to be discussed below are believed to be of last-glacial age, a conclusion supported by LOVEDAY (1955), NICOLLS (1958), and DAVIES (1967, 1974). This conclusion is based on the evidence of the freshness of both the deposits and their topographic forms, and often the lack of overlying deposits of any thickness. The very few radiocarbon dates available support the interpretation of age.

The sediment mobilised on slopes in the fan catchments during the last-glacial can be divided into: block glacis (defined by CAINE 1968) and periglacial solifluction slope mantles at the highest points of the fan catchments; grèzes litées on lower slopes between about 600 and 140 m above sea level; rudely bedded and unbedded gravel-rich rubble deposits filling valleys below 300 m above sea level, and debris flows on the lower slopes, building footslope aprons at the same elevation as the alluvial fans.

On the undulating, narrow plateau surface which lies just below the summit of Mt Faulkner (901 m above sea level), there is a block glacis which is now relict beneath a *Eucalyptus delegatensis* sub-alpine open forest with a patchy but dense, shrubby understorey. Blocks are commonly covered by both foliose and crustose lichen. Parts of the block glacis have a bright brown (7.5 YR 5/8) sandy loam matrix, which can be examined in exposures along vehicular tracks. The matrix is obviously part of the glacis for it underlies blocks and surrounds others. It has not been washed in. The block glacis was derived from low dolerite ridges and moved into a valley on the western side of the summit plateau. The source of the blocks is low cliffs no more than 5 m high, and clearly rockfall was of no significance (cf. WASHBURN 1973, p. 63).

CAINE (1968), DAVIES (1958), CAINE & JENNINGS (1968) have argued convincingly that both blockstreams and block glacis of the type on Mt Faulkner are the result of periglacial conditions. The Mt Faulkner glacis was derived by frost-wedging of joint-bounded dolerite blocks, and, at the head of the glacis, the blocks are columns which have come directly out of the low ridge. Here there appears to have been no fine-grained matrix, and, following the argument of CAINE & JENNINGS (1968), the most reasonable explanation for the movement of blocks at the head of the glacis is that ice rather than sediment acted as the lubricant. Such a hypothesis demands a mean annual temperature below $0°$ C if the ice was permanent, but, if the ice was seasonal, then this palaeotemperature calculation is in error.

The brown and yellow-brown matrix which occurs within the lower slopes of the block glacis can be traced both upslope and downslope on the flanks of Mt Faulkner. Away from the glacis the brown sandy loam includes a variable proportion of clasts which have a strong downslope orientation of long axes. The fabric and texture of these deposits are not in themselves characteristic of any particular environment, as noted for similar deposits by GALLOWAY (1970), and WASHBURN (1973). But the distribution of this material, and its direct genetic association with a block glacis of undoubted periglacial origin, strongly suggests that the slope deposits were produced by periglacial solifluction (cf. GALLOWAY 1970). The slope mantle can be traced down the slopes of Mt Faulkner onto sandstone bedrock where the texture of the regolith changes but the mantle is the equivalent of the higher material and is believed to be of periglacial origins. The periglacial mantle can be traced to 610 m above sea level, where the mantle overlies a truncated palaeosol, the B horizon of which has many of the characteristics of a red podsolic as described by STACE et al. (1968).

The technique used to identify periglacial slope mantles on Mt Faulkner was used throughout the field area. No deposits below about 600 m above sea level could be convincingly identified as being periglacial, possibly the result of generally poor exposure below 600 m. No evidence was found to support recent suggestions (CHICK 1972) that periglacial conditions extended to sea level in Tasmania during the last glacial. Therefore, DAVIES' lower periglacial limit of about 450 m is not contradicted by evidence from the fan catchments, and immediately surrounding areas.

Grèzes litées ("... bedded slope deposits of angular, usually pebble-size rock chips and interstitial finer material...", WASHBURN 1973) occur between about 600 and 140 m above sea level in the catchments and adjacent areas. At Ferntree (on the flanks of Mt Wellington, Fig. 1) at an altitude of about 480 m above sea level, there is a grèzes litées deposit (derived from siltstone) overlying an unbedded slope mantle of unknown environmental affinities. The Ferntree grèzes litées are no longer actively accumulating. The slope is well vegetated, and the slope mantle at the present-day surface has none of the characteristics of the bedded sediments beneath (see Fig. 4).

Large pieces of charcoal within the basal mantle yielded a ^{14}C date greater than 40,000 years B. P. (I-8155). The grèzes litées display considerable variability in the ratio of clasts to matrix: clasts are here considered to be particles greater than $-1.0\ \phi$ (2 mm) on their intermediate axis. Most of the matrix is silt (0.06–0.002 mm), and its proportion varies between 1 % and 40 %. Where the matrix proportion is high, the clasts show moderate to poor downslope orientation, and beds with matrix-supported clast fabrics (i. e., the clasts "float" in a finer-grainer matrix) are considered to be the result of debris flow (in the sense of JOHNSON 1970) only a few centimetres thick. Beds which have a low proportion of matrix show moderate to good orientation of clast long axes parallel to bed boundaries, and some upslope dip occurs as well.

The beds with low matrix proportions are dominantly the result of running water, showing bedding characteristics of alluvium (see ALLEN 1970; CONYBEARE & CROOK 1968), including inclined (foreset) bedding and cut-and-fill structures, but on a mean slope of 12°. Some of these beds are almost openwork, that is, the clasts support one another and many of the voids are not filled with matrix. These partially openwork beds show poor and often non-existent preferred orientation of clasts, and they could have been the result of rockfall from bedrock

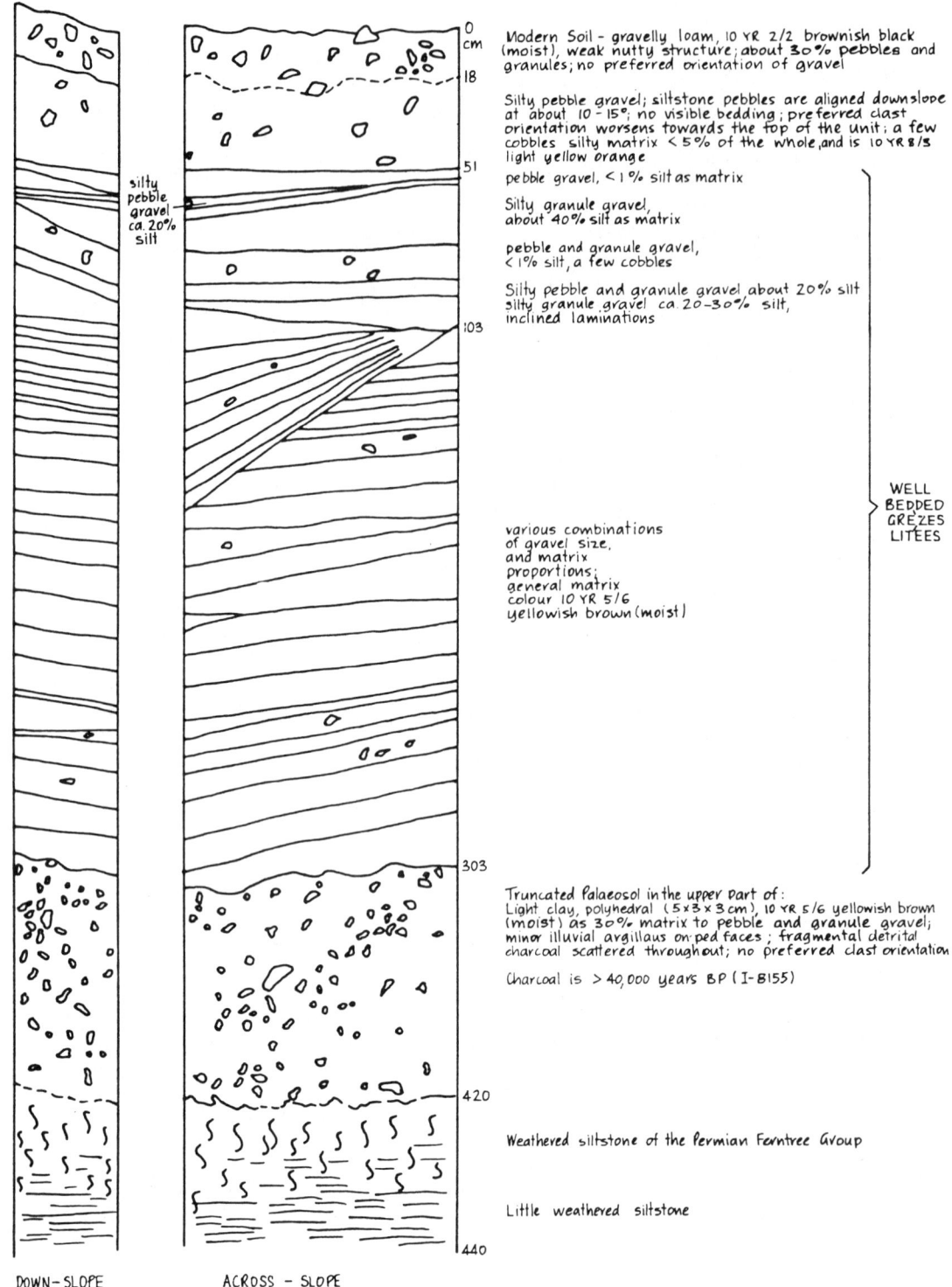

Fig. 4. Ferntree Grèzés Litees with underlying and overlying deposits.

upslope. This suggestion is based on recent work by the author on debris slopes in the Pakistan Hindu Kush (to be published elsewhere).

Accounts of grèzes litées in the process of formation are few (MALAURIE & GUILLIEN 1953; RAYNAL in ALEXANDRE & MACAR 1960; DUTKIEWICZ 1967; PISSART 1967; JAHN 1960, 1961), and most of the literature is concerned with the interpretation of relict deposits (e. g. GUILLIEN 1973 and references; BOUT 1953; WATSON 1970 and references; SOONS 1962; SOUCHEZ 1964). BASTIN & GUILLIEN (1971) presented pollen evidence of the climatic conditions which prevailed during grèzes litées formation in Charente. A review of this literature (WASSON 1975) shows that grèzes litées are always associated with snowpatches, that is, with nivational conditions. As noted above, an interpretation of at least some types of beds within grèzes litées of the Hindu Kush does not require the presence of snow. However, the Ferntree (and other) slopes must have experienced conditions of slopewash more efficient than those usually produced by rainfall, but slopes essentially devoid of vegetation to allow the often very fine bedding to form.

At Cobbs Hill (Fig. 1) a grèzes litées deposit is made up of very sharp fragments of siltstone, the shape of which bears no relationship to the pattern of joints and bedding planes within the bedrock. There seems no alternative to a frost-shattering explanation of the sediment at this site, and, in combination with the evidence from the Ferntree site (and others in the fan catchments and adjacent areas), the grèzes litées deposit is thought to have formed under nivational conditions, and perhaps periglacial conditions at Cobbs Hill. This last site is at 180 m above sea level, below DAVIES' general lower limit of periglacial activity. But the Cobbs Hill site would have been a cold air sump during grèzes litées formation, as shown by exposures within the quarry. The peculiar topographic features of the site may explain the anomalous altitude of the deposit.

At altitudes generally below about 300 m, small valleys are infilled with rudely bedded and unbedded deposits of pebble and cobble gravel with variable proportions of finer-grained sediment. The processes of formation of these fills are not known, but combinations of mass-movement by creep and possibly by rapid flow, and sheetwash are likely to have been most important. These fills have none of the characteristics of the higher elevation slope mantles attributed to periglacial solifluction, but they are considered to have been broadly coeval with the periglacial instability. The onset of cold and probably dry climatic conditions in southeastern Tasmania lowered the treeline (DAVIES 1974; MACPHAIL 1975), and presumably disturbed biological and geomorphological equilibria at altitudes both above and below the treeline. The deglacial climatic amelioration probably had a similar effect, and it is in these changing equilibria that the cause of the low level slope instability might be found.

The low altitude slope mantles not only consist of unbedded and rudely bedded rubble, but also include well-bedded, gravelly, poorly sorted deposits adjacent to and between the alluvial fans. The beds consist of sheets (in three-dimensions) lying at angles between 10° and 20°, which formed footslope aprons. These aprons, judging from their dip, must have extended below present sea level onto the Derwent valley floor. The deposits formed by successive debris flows. This view is supported by the similarity between these deposits and beds

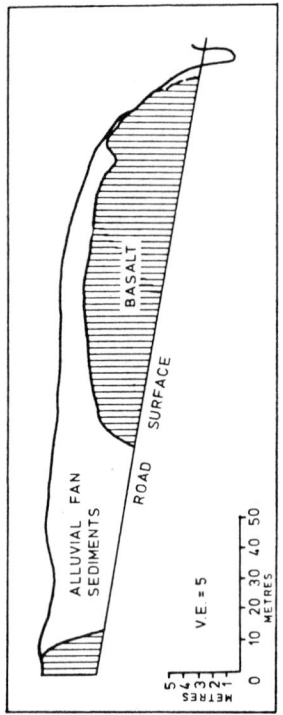

Fig. 5. Cross-section across Quarry Fan. The section is parallel to the Derwent River and is exposed in a cutting beside the Lyell Highway.

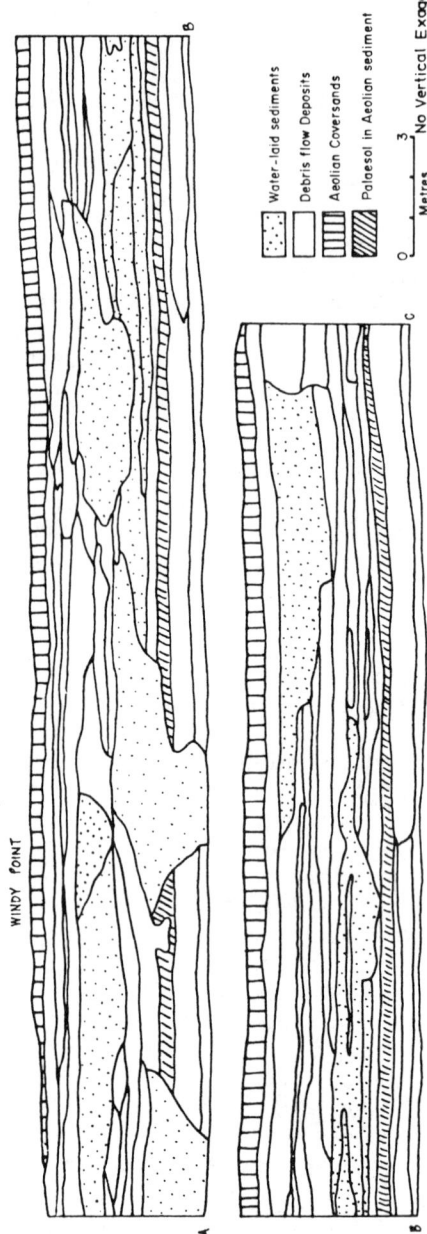

Fig. 6. Cross-section of Windy Point Fan at Windy Point.

within the alluvial fans which, it will be shown, are the result of debris flows. The slope mantles also conform to the characteristics of debris flows as described by JOHNSON (1970), FISHER (1971), and BULL (1972).

As judged from topography and the rock types represented within the debris flows, the deposits flowed down from heights generally no greater than 160 m above sea level, but at Wendy's Quarry (near Jarvis Creek, Fig. 2) the flows could have come from slopes 360 m above sea level.

The various low altitude slope deposits were clearly synchronous with alluvial fan deposition. At three sites on the north bank of the Derwent River, there are exposures of interdigitating slope and fan deposits. The view that the low altitude and higher altitude slope mantles are of the same age demands that the fans are of last-glacial age, a conclusion reached earlier by consideration of the fans themselves.

The accumulation of the alluvial fans

The form of the alluvial fans has been profoundly influenced by the pre-fan landscape. The strath and alluvial terraces formed a small part of a landscape dominated by long ridges which extended away from Mt Faulkner and Mt Dromedary towards the Derwent River. Valleys between the ridges became the loci of deposition as slopes became unstable during the last glacial. Fig. 5 shows a cross-section, parallel to the Derwent River, of Quarry Fan (Fig. 2). It is clear that the fan began as a valley fill between two ridges of basalt, and, as deposition progressed, one of the ridges was overtopped and the fan form became recognisable. Bedrock is close to the surface at a number of exposures in the fans on both banks of the Derwent River, and the sub-fan topography is undulating, and locally steep.

Extensive road cuttings and creek bank exposures provide an excellent opportunity to examine the fan sediments and the ways in which the fans were constructed. A detailed analysis of the sediments, and the processes of their accumulation, are the subject of another paper, and will be treated here in summary only. The sediments are bedded, and each of the beds has been placed into one of two major categories: water-laid deposits, and debris flow deposits (after BULL 1972). The distinction between these two categories rests upon sorting, sedimentary structures, and the shape of the deposits. In a poorly sorted deposit, such as is common in the Derwent Fans, the processes operating to produce water-laid deposits should produce sedimentary structures including horizontal and inclined laminations in sand-rich material, and imbrication and commonly lenticular beds in gravel-rich deposits. The absence of characteristically fluvial sedimentary structures in a poorly sorted deposit in which gravel is set in an unsorted finer-grained matrix, strongly suggests that the deposit is the result of a debris flow (cf. HOOKE 1967; JOHNSON 1970). This method of classification places BULL's (1964) intermediate deposits in the water-laid group, for if a deposit shows only rudimentary signs of fluvial sedimentary structures, it is classified as water-laid. It must be noted that the term debris flow is here used to include mudflows, for there is no compelling reason to separate the two phenomena.

The apical and middle parts of the fans are relatively coarse-grained, and consist of both debris flow and water-laid deposits. The lower, or distal, deposits are predominantly water-laid. This arrangement can be found on any radius of a fan, and has been described previously by Hooke (1967) and Bull (1972), both of whom argued that water-laid deposits predominate in the distal parts because fluvial action is capable of moving sediment further than debris flows. Price (1974) has reproduced the same pattern by mathematical simulation. Erosional features such as channels are important in the upper parts of the fans, but are scarce towards the toes of the fans. These channels are most commonly filled with water-laid sediments, and less commonly by debris flow. A typical association of the various deposits is shown in Fig. 6, a section exposed in a road cutting near mid-fan in the Windy Point Fan.

The debris flow deposits mainly occur as thin, wide sheets, which are generally parallel and sub-parallel (in three dimensions) both with each other, and with the fan surfaces. The flows lie at angles between 1° and 7°. The flows spread across the surfaces of the fans, but a few moved down channels cut within the fan surfaces. The U-shaped channels occasionally seen may have been cut by debris flows (see Johnson 1970) but the bulk of the sheet flows did not disturb the sediments over which they moved.

Most of the water-laid deposits in the apical to mid-fan locations occur in channels, but some occur in sheets which are either isolated from a channel, or are the result of a channel filling up and spilling over its banks. The infilling of the channels was almost certainly the result of back filling, a process described by Hooke (1967) and other authors. When the channel infilled, sediment was spread over the former channel by intersection point deposition (Hooke 1967, and Wasson 1974). The sheets of water-laid sediments which are unassociated with channels are the sheetflood deposits of Blissenbach (1954), and Bull (1964, 1972) and were probably formed as intersection point deposits.

The distal sediments are characteristically fine-grained, dominated by clay, silt, and fine sand. Minor beds of pebble and granule gravel occur, but otherwise depositional bedding is generally absent. These sediments are believed to be the suspended load of discontinuous channels which spread their load in thin sheets across the fan toes. In a few locations beds less than 10 cm thick can be seen within the clay-rich distal sediments. Therefore, these sediments are seen as water-laid, and deposited by sheetflood. Channels rarely reached the fan toes, as judged from the few infilled channels seen in distal sections. The distal channels are filled with pebble gravel lenses within sandy deposits, the fills being generally finer-grained than their proximal equivalents.

Rodine (1974) investigated the conditions under which debris flows are mobilised. Most debris flows begin as landslides from hillslopes, but can also develop by mobilisation of material which has accumulated from successive landslides (Johnson & Rahn 1970; Vinogradov 1969) in the lower parts of the landscape. Intermittent landsliding on slopes will contribute sediment to a channel, but flow may not occur because there is insufficient water and/or debris. Gradual accumulation of debris in channels over many years may provide the conditions necessary for debris flow, once water is made available.

The debris flows in the Derwent Fans are envisaged as forming in the way just described, but the argument comes from a need to explain the debris flows rather than from evidence of relict landslides within the catchments. The relict slope aprons, which are built of debris flow deposits and which lie at sea level, are also thought to have resulted from landslides. Deposits of slow mass-movement and slope-wash have been documented at all altitudes within the catchments, but the relationship between these deposits and the postulated landslides is not at all clear. Both snowmelt and rainfall are likely to have contributed moisture to the slope mantles, and thereby provide conditions suitable for failure and rapid mass-movement. Sediment initially mobilised by periglacial processes at high elevations, and by non-periglacial processes on lower slopes, was fed into channels by landsliding and probably by slower mass transfer directly into valley bottoms. Individual landslides of sufficient mass and lubrication may have continued directly from slope to fan, while others stored sediment in valley bottoms. The periodicity of the Derwent debris flows cannot be determined, but following the ideas of BEATY (1974), flows may have occurred only once in a few centuries. They can also occur with a return period of a few days, or hours, once the right conditions prevail (RODINE 1974). It is clear that debris flows do not occur in the catchments at the present time.

The water-laid deposits must have been derived from the same sediment sources as the debris flows. HOOKE (167) showed that water flows commonly follow debris flows on fans, redistributing and commonly dissecting, the debris. Such a combination of processes would produce many of the channels seen in the Derwent Fans, and provide some of the sediment which infilled the channels. Rainfall and snowmelt between debris flows presumably reworked debris accumulated in both the catchment channels, and channels cut into the fans.

The small channels seen in sections such as that depicted in Fig. 6 represent fan entrenchment, and were probably the result of one or all of the following causes: 1, high intensity water flows resulting from snowmelt or rainfall (as described by BULL 1964 and WASSON 1974); 2, alternations of debris flows and water flows (as described by HOOKE 1967); 3, toe trimming – when a fan's toe lies close to a stream which is flowing roughly at right-angles to the axis of the fan, a lateral shift of the position of the stream might erode the toe. A cliff in the toe is then a locus for gully initiation, and the gully will retreat headwards. Sediment is removed from the fan, and is either deposited as a secondary fan or is dumped directly into the stream which lies at the bottom of the fan. When sediment is removed from the fan, then the fan is described as entrenched, but if the supply of sediment from the catchment is sufficient to aggrade the distal part of the gully and the fan surface once again begins to receive sediment, the fan is said to be incised. Toe trimming, and its effects, is particularly well illustrated on a large number of fans in relatively confined valleys on the east coast of the North Island of New Zealand. Toe trimming can also be accomplished by wave attack during rising sea level, or by any other process which removes the toe of a fan.

Some of the channels cut into the surfaces of the Riverton coalesced fans (Fig. 2) are both wider and deeper than any of the fan sections. These Riverton channels have been only partly filled with sediments, and these sediments are identical to those which make up the body of the fans. These channels must have formed towards the end of the period of fan accumulation. They were probably the result of toe trimming by the Derwent River, and almost certainly supplied sediment directly to the river.

A hiatus in fan accumulation

Evidence of a hiatus in fan accumulation can be seen in both the Windy Point and Limekiln Point sections (Fig. 2), and at other sites. The evidence at Windy Point and Limekiln Point consists of a yellowish (10 YR 6/8) fine sandy clay loam which occurs as a sheet about 1 m thick within the body of the fans (see Fig. 6 and 7), and is quite unlike other sediments in the fans. Within the upper part of the sheet is a zone of polyhedral mottled sandy clay with clay skins on ped faces. This zone gives way to a weakly prismatic mottled sandy clay loam. The sequence is thought to be the truncated B horizon of a palaeosol developed within a parent material of sandy clay loam. The only evidence of an old soil is the illuvial clay on ped faces, and the slight rubification and induration of the putative B horizon.

The sediment has a clay fraction which is mineralogically different from that of the fan sediments. The $> 10.5 \phi$ fraction of the fan sediments (both water-laid and debris flow) consists of quartz, dolomite, calcite, feldspar, kaolinite, palygorskite, cristobalite, and mixed-layer minerals. The $> 10.5 \phi$ fraction of the yellow sediment does not contain dolomite, calcite, palygorskite, or cristobalite, all of which are derived from the calcareous marine rocks of the fan catchments. The yellow deposit does not appear to have the same provenance as the fan sediments, and therefore could be either marine or aeolian.

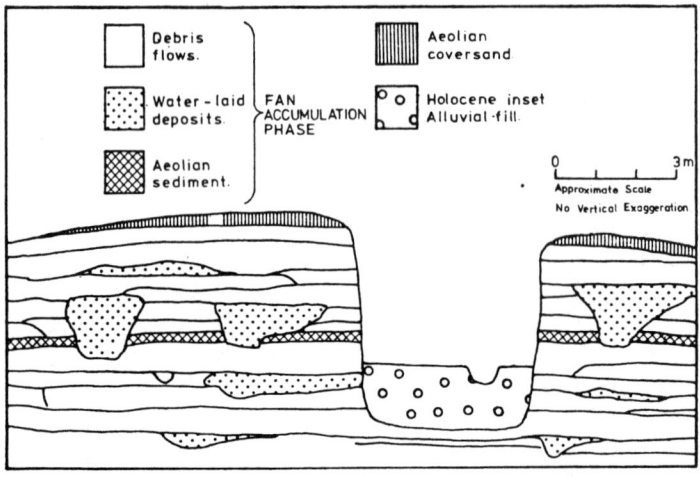

Fig. 7. Schematic transverse cross-section of a Derwent fan showing stratigraphic relationships.

A careful examination of the sediment produced two intact sponge spicules and several spicule fragments. One of the intact spicules was definitely identified as marine (N. TAIT, School of Biological Sciences, Macquarie University, pers. comm.). However BAKER (1959) has shown that sponge spicules can be blown considerable distances to be incorporated in deposits which are definitely not marine. More importantly, the deposit lies at different elevations in adjacent fans; a vertical difference of 7.3 m separates the two deposits. It is most unlikely that the deposit is marine, and therefore it is thought to be of aeolian origin. The most likely source is the Derwent River channel and floodplain.

Similar material to the yellow deposit is found within a number of fans and slope mantles in the area. These deposits are all sandy, but show varying proportions of silt and clay. Their clay fractions all suggest a provenance other than the fan catchments. These deposits are correlated because there is no evidence of more than one phase of aeolian activity during fan accumulation at any one site. If aeolian action was occurring on parts of fans during the entire period of fan accumulation, then a number of aeolian deposits should be seen in any one section through a fan. This is not the case, and so the aeolian deposits are thought to represent a period during which effective fan accumulation temporarily ceased and aeolian sedimentation was paramount. This hiatus in fan accumulation is of unknown age and duration.

From fan accumulation to fan dissection

Parts of some of the fans are veneered by aeolian coversands, as seen earlier. It is evident that the fans ceased accumulation before the source of sediment for the coversands, the Derwent River, was flooded by the marine incursion. The coversands were also dissected by trenches which incised the fans. These relationships are shown schematically in an across-fan section in Fig. 7.

Conditions during fan accumulation were completely different from conditions which followed incision. The nature of sediments within the inset fills of the trenches testifies to this difference. It is believed that the incision was directly linked to the cause of the cessation of fan accumulation. The climatic deterioration of the last glacial stimulated fan accumulation, and it is reasonable to suppose that climatic amelioration had a retarding effect on fan accumulation by allowing the spread of vegetation which hitherto had been at lower elevations. Deglaciation of the highlands was complete, at the latest, between 11,500 and 8,300 radiocarbon years B. P. (MACPHAIL & PETERSON 1975), and so the early post-glacial saw a rapid rise in temperature and precipitation with an associated rapid spread of *Eucalyptus* and *Nothofagus* forests (MACPHAIL pers. comm.).

It is argued that the fans ceased to accumulate as periglacial and nivational processes waned during climatic amelioration and revegetation of higher slopes of the fan catchments. A decrease in the net sediment discharge ratio began a phase of fan incision. RYDER, (1971 a, b) has used essentially the same reasoning to explain incised fans in the Canadian Rockies.

Some of the fans were also cliffed along their toes, and in some cases almost entire fans have been removed by a combination of toe trimming and incision. The toe trimming was partially accomplished by the postglacial rise in sea level, and the process envisaged is one of attack by waves generated in the Derwent estuary. The cliffing was probably most efficient as the sea approached its present level some 6,000 years B. P. in southeastern Australia (THOM & CHAPPELL 1975). Cliffing probably ceased as estuarine sedimentation progressively protected fans from wave attack.

While it is considered that the fan incision was primarily the result of a change in the net sediment load/water discharge ratio of the fan catchments, it is possible that some of the incisions were begun by headward retreat of gullies from cliffed toes. Incisions begun in this way remained intact because of the changed catchment conditions (cf. the Riverton incised channels). All of the fans are incised, but not all are cliffed, suggesting that the changed catchment conditions, a widespread effect, were paramount in causing an incising regime.

The sediment eroded by incision was deposited as small fans at the mouths of the incisions. Most of these small fans are now beneath sea level in the estuary, and they are still accumulating.

The inset fills

The inset fills are depicted in Fig. 7. These fills are quite different from the channel fills within the bodies of the fans. The inset fills show the usual characteristics of water-laid deposits including moderately good separation of coarse and fine particles. Much of the water-laid sedimentation within the bodies of the fans was by mixed load flow producing very poor separation of size fractions. The other major difference lies in the relatively large quantities of organic matter (charcoal, and generally disseminated organic material) in the inset fills, while the fan sediments are devoid of organic matter in any form.

An estimate of the age of the inset fills is provided by two radiocarbon dates. The first comes from charcoal from the base of a fill within the incision of Oyster Fan (Fig. 2), and it is 3840 ± 95 years B. P. (I-7931). The second comes from charcoal at a point above the base of a fill in the incision in Parramore Fan (at the end of Parramore Creek, Fig. 2), and it is 3575 ± 95 years B. P. (I-7930).

The inset fills are Late-Holocene features which are still actively aggrading. An example of what seems to be typical mode of deposition occurred on the evening of 21 March 1968 when a thunderstorm caused flooding on the surfaces of the inset fills. Particles of silt size to boulder size (up to $0.17 m^3$) were transported by stream flow (HULST 1968). Large quantities of vegetation, particularly dead trees and charcoal, accompanied the mineral sediment.

The storm had an intensity of about 90 mm/hour, as estimated from the precipitation records of Mr. RATHBONE at The Limekilns. The erosive effect of the storm was considered to have been enhanced by the disastrous bushfires of February 1967. HULST (pers. comm.) observed that the two areas which were eroded more severely than others in 1968 between Granton and New Norfolk were those which had been burnt most intensely in 1967. The bushfires presumably increased runoff from slopes, and provided many of the dead trees mentioned earlier.

The sediments deposited by the 1968 flood on the inset fills are essentially the same as those which are nearly 4,000 years old. RATHBONE (pers. comm.)

observed that floods of the type which occurred in 1968 occur about once every ten years, but the intensity of erosion varies between floods. The creeks which cross the inset fills flow about once a year, generally for short periods of a few days. It is clear from the evidence of both RATHBONE, and the sediments themselves, that debris flows no longer occur, and have not occurred for about 4,000 years. There is no doubt that the regime of the fans has changed.

The inset fills are most likely the means by which the streams maintain a profile in equilibrium with the level of the estuary. Certainly some of the incisions, in which the inset fills lie, were cut to a level below present sea level. But the dated fills lie in incisions which were cut to a level above present sea level. The former case requires that incision was almost complete well before about 9,500 years B. P. The inset fills are simply the result of Late-Holocene and contemporary slope and stream denudation, and their disposition is the result of the present position of sea-level.

History of the Derwent Fans and conclusions

The history of the Derwent Fans can be viewed in two main parts: 1, accumulation; 2 incision.

The accumulation occurred in two periods. The first consisted of debris flows and fluvial activity and was followed by deposition of yellowish aeolian sediment (labelled "aeolian sediment" within the Fan Accumulation Phase of Fig. 7). The second period of accumulation was also accomplished by debris flows and fluvial activity and was the same as the first period. This second period was followed by deposition of aeolian coversands (Fig. 7) at a number of sites.

Accumulation is attributable to a shift towards cold climatic conditions during the last glacial in Tasmania. Climatic deterioration produced periglacial activity at the highest altitudes, nivational processes at lower altitudes, and nonperiglacial colluviation at the lowest altitudes in the fan catchments. The hydrology of the catchments was greatly influenced by snow melt. Periglacial, nivational, and colluvial mass-wasting produced abundant sediment for fan accumulation within the piedmonts.

Sediment was flused from the catchments as debris flows and sediment-charged water flows. This sediment initially accumulated between ridges which extended out from both Mt Faulkner and Mt Dromedary towards the Derwent River. As sedimentation continued the valleys were filled and the ridges overtopped, so producing the fan forms.

Fan formation continued until climatic amelioration began the deglaciation of the Tasmanian highlands. The waning of periglacial and nivational conditions, and the revegetation of the fan catchments are thought to have combined to end the landscape-forming processes responsible for the fans. The end of fan deposition allowed the accumulation of aeolian coversands on the fan surfaces. These aeolian sediments are believed to have been blown from the Derwent River.

The end of the period of high sediment supply ushered in fan incision by a reduction in the sediment load/water discharge ratio. Incision is thought to have been initiated by this change of regime, but toe trimming of the fans by waves on the rising sea level in the estuary may have allowed the initiation of some incisions. The precise combination of causes is not known, but the presence of uncliffed incised fans shows that toe trimming was not the main control on incision.

Incision in many places was almost complete 9,500 years ago, and alluvial deposition within the incisions (the Holocene inset alluvial fills of Fig. 7) did not begin at one site until about 4,000 years ago. The inset fills are still aggrading and the effectiveness of sedimentation is heightened when stream flow follows a busfire. This alluviation is building small fans into the Derwent River estuary.

This investigation of historical development of a suite of alluvial fans can be used eventually, in combination with similar historical studies and inquiries into modern active fans, both to answer the two questions posed in the general introduction and to construct a generally applicable model of alluvial fan evolution.

Acknowledgments

I wish to acknowledge the assistance of Professor J. L. DAVIES throughout the period during which this work was carried out, and for his comments on a draft of the manuscript. I also wish to thank Dr. M. A. J. WILLIAMS for his comments on the manuscript. Assistance in the field in Tasmania was given by Dr JAMIE KIRKPATRICK, Mrs WENDY GRUBB and Mr ROGER KELLAWAY. The Tasmanian Public Works Department kindly provided unpublished information about the 1968 floods. Dr ERIC COLHOUN and WAYNE SIGLEO provided stimulating discussion on the field evidence and its interpretation. The work was carried out while the author held an Australian Commonwealth Post-graduate Research Scholarship in the School of Earth Sciences, Macquarie University, Sydney, Australia. Funds were also provided by Macquarie University for field expenses and radiocarbon dating.

References

ALEXANDRE, J., & P. MACAR (1960): Excursion du Jeudi 11 Juin 1959: Liège–Paraque de Fraiture–Laroche–Rochefort–Liège. – Biul. Peryglac. 9: 187–197.
ALLEN, J. R. L. (1970): Physical processes of sedimentation. – Allen and Unwin, London, 248 pp.
ANDERSSON, J. G. (1906): Solifluction, a component of subaerial denudation. – J. Geol. 14: 91–112.
BAKER, G. (1959): Opal phytoliths in some Victorian soils and 'red rain' residues. – Aust. J. Bot. 7: 64–87.
BASTIN, B., & Y. GUILLIEN (1971): Approche palynologique des grèzes litées de Sonneville et d'Echoisy (Charente). – C. R. Acad. Sci. Paris, 273 (Série D): 2063–2066.
BEATY, C. B. (1970): Age and estimated rate of accumulation of an alluvial fan, White Mountains, California, U.S.A. – Amer. J. Sci. 268: 50–77.
— (1974): Debris flows, alluvial fans, and a revitalized catastrophism. – Z. Geomorph. Suppl. 21: 39–51.
BLISSENBACH, E. (1954): Geology of alluvial fans in semi-arid regions. – Geol. Soc. Amer. Bull. 65: 175–190.
BLOOM, A. L. (1971): Glacial-eustatic and isostatic controls of sea level since the last glaciation. – The late Cenozoic glacial ages (ed. K. T. TUREKIAN), Yale Univ. Press, New Haven, 353–379.
BOUT, P. (1953): Etudes de géomorphologie dynamique en Islande. – Expéditions Polaires Françaises 1. Paris, Herman & Cie, Actualités Scientifiques et Industrielles 4: 176 pp.

Bull, W. B. (1964): Alluvial fans and near-surface subsidence in western Fresno county, California. – U.S. Geol. Survey Professional Pap. 437-A, 70 pp.
- (1972): Recognition of alluvial-fan deposits in the stratigraphic record. – Recognition of ancient sedimentary environments (eds. J. K. Rigby & W. K. Hamblin), Soc. Econ. Palaeontologists and Mineralogists, Spec. Publ. 16, 63–83.
Caine, N. (1968): The blockfields of northeastern Tasmania. – Aust. Natl. Univ., Dept. Geogr. Publ. G/6, 127 pp.
Caine, N., & J. N. Jennings (1968): Some blockstreams of the Toolong range Kosciusko state park, New South Wales. – J. Proc. Roy. Soc. N.S.W. 101: 93–103.
Chick, N. K. (1972): 1971–72 Research on Quaternary shorelines in Australia and New Zealand – summary report on the ANZAAS Quaternary shorelines committee (Tasmania). – Search 3: 412.
Church, M., & J. M. Ryder (1972): Paraglacial sedimentation: a consideration of fluvial processes conditioned by glaciation. – Geol. Soc. Amer. Bull. 83: 3059–3072.
Conybeare, C. E. B., & K. A. W. Crook (1968): Manual of sedimentary structures. – Bur. Miner. Resours. Geol. & Geophys. Aust. Bull. 102.
Cooke, R. U., & A. Warren (1973): Geomorphology in deserts. – Batsford, London, 374 pp.
Costin, A. B. (1967): Alpine ecosystems of the Australasian region. – Arctic & Alpine Environments (eds. H. E. Wright jr. & W. H. Osburn), Indiana University Press, 55–87.
Davies, J. L. (1958): The cryoplanation of Mount Wellington. – Pap. Proc. Roy. Soc. Tasmania 92: 151–154.
- (1967): Tasmanian landforms and Quaternary climates. – Landform studies from Australia and New Guinea (eds. J. N. Jennings & J. A. Mabbutt), Aust. Natl. Univ. Press, Canberra, 1–25.
- (1969): Landforms of cold climates. – Aust. Natl. Univ. Press, Canberra, 200 pp.
- (1974): Geomorphology and Quaternary environments. – Biogeography and ecology in Tasmania (ed. W. D. Williams), Junk, The Hague, 17–27.
Derbyshire, E. (1973): Periglacial phenomena in Tasmania. – Biul. Peryglac. 22: 131–148.
Dutkiewicz, L. (1967): The distribution of periglacial phenomena in N. W. Sörkapp, Spitsbergen. – Biul. Peryglac. 16: 37–83.
Eckis, R. (1928): Alluvial fans of the Cucamonga district, southern California. – J. Geol. 36: 225–247.
Embleton, C., & C. A. M. King (1968): Glacial and periglacial geomorphology. – Edward Arnold, London, 608 pp.
Fisher, R. V. (1971): Features of coarse-grained, high-concentration fluids and their deposits. – J. Sed. Pet. 41: 916–927.
Flint, R. F. (1971): Glacial and Quaternary geology. – Wiley, New York, 892 pp.
Galloway, R. W. (1970): The full-glacial climate in southwestern U.S.A. – Ann. Assoc. Amer. Geogrs. 60: 245 256.
Guillien, Y. (1973): Grèzes litées et terres grézeuses – Le Quaternaire, géodynamique stratigraphie et environement. – Travaux Français récent, 9e Congrès Intern. de l'Inqua, Christchurch, 101–104.
Hooke, R. Le B. (1967): Processes on arid-region alluvial fans. – J. Geol. 75, 438–460.
- (1972): Geomorphic evidence for late-Wisconsin and Holocene tectonic deformation, Death valley, California. – Geol. Soc. Amer. Bull. 83: 2073–2098.
Hoppe, G., & S. R. Ekman (1964): A note on the alluvial fans of Ladtjovagge. – Swedish Lapland. Geogr. Annaler 46: 338–342.
Hulst, H. van (1968): Report on damage occurring on 21st March, 1968. Lyell highway drainage – Granton – New Norfolk. – Memo (Sh 19/1-1, Sh 19/2-5) Tasmanian Public Works Department, unpublished.
Jahn, A. (1960): Some remarks on evolution of slopes on Spitsbergen. – Z. Geomorph. Suppl. 1: 49–58.
- (1961): Ilościowa analiza niektórych procesow pergyglacjalnych (Quantitative analysis of periglacial processes in Spitsbergen). – Zeszytyt Naukowe Uniw. Wroclawskiego (Nauko o Ziemi) z: 54 pp.
Johnson, A. M. (1970): Physical processes in geology. – Freeman Cooper, San Francisco, 577 pp.
Johnson, A. M., & P. H. Rahn (1970): Mobilization of debris flows. – Z. Geomorph. Suppl. 9: 168–186.

LOVEDAY, J. (1955): Reconnaissance soils map of Tasmania. – Sheet 82-Hobart. C.S.I.R.O. Div. of soils, Report 13/55.
LUSTIG, L. K. (1965): Clastic sedimentation in Deep Springs valley, California. – U.S. Geol. Surv. Professional Pap. 352-F.
MACPHAIL, M. K. (1975): Late Pleistocene Environments in Tasmania. – Search 6: 295–300.
MACPHAIL, M. K., & J. A. PETERSON (1975): New deglaciation dates from Tasmania. – Search 6: 127–130.
MALAURIE, J., & Y. GUILLIEN (1953): Le modelé cryo-nival des versants meubles de Skansen (Disko, Greenland). Interprétation générale des grèzes litées. – Soc. Géol. France Bull. Série 6: 703–721.
MILLIMAN, J. D., & K. O. EMERY (1968): Sea levels during the past 35,000 years. – Science 162: 1121–1123.
NICOLLS, K. D. (1958): Aeolian deposits in river valleys in Tasmania. – Aust. J. Sci. 21: 50–51.
PISSART, A. (1967): Les modalites de l'écoulement de l'eau sur l'ile Prince Patrick. – Biul. Peryglac. 16: 217–224.
PRICE, W. E. jr. (1974): Simulation of alluvial fan deposition by a random walk model. – Water Resources Res. 10: 263–174.
RODINE, J. D. (1974): Analysis of the mobilization of debris flows. – Final report to U.S. Army Res. Office, Durham, North Carolina. Grant No. DA-ARO-D-31-124-71-G158. Stanford University, 226 pp.
RYDER, J. M. (1971, a): The stratigraphy and morphology of paraglacial alluvial fans in south-central British Columbia. – Can. J. Earth Sci. 8: 279–298.
– (1971, b): Some aspects of the morphometry of paraglacial alluvial fans in south-central British Columbia. – Can. J. Earth Sci. 8: 1252–1264.
SCHICK, A. P. (1974): Formation and obliteration of desert stream terraces – a conceptual analysis. – Z. Geomorph. Suppl. 21: 88–105.
SOONS, J. M. (1962): A survey of periglacial features in New Zealand – Land and Livelihood. – Geographical Essays in honour of GEORGE JOBBERN (ed. M. MCCASKILL), Geogr. Soc. N. Z., Christchurch, 74–87.
SOUCHEZ, R. (1964): Sur la gélivation des calcaires et la genèse des grèzes litées. – C. R. Acad. Sci. Paris 258: 3741–3743.
SPECHT, R. L. (1972): The vegetation of South Australia. – Government Printer, Adelaide, 328 pp.
SPRY, A. H., & M. R. BANKS (eds.) (1962): Geology of Tasmania. – J. Geol. Soc. Aust. 9: 107–362.
STACE, H. C. T., et al. (1968): A handbook of Australian soils. – Rellim, South Australia, 429 pp.
THOM, B. G., & J. CHAPPELL (1975): Holocene sea levels relative to Australia. – Search 6: 90–93.
TRICART, J. (1970): Geomorphology of cold environments. – Trans. E. WATSON Macmillan, 320 pp.
VINOGRADOV, YU. B. (1969): Some aspects of the formation of mudflows and methods of computing them. – Soviet Hydrol. 5: 480–500.
WASHBURN, A. L. (1973): Periglacial processes and environments. – St. Martin's Press, New York, 320 pp.
WASSON, R. J. (1974): Intersection point deposition on alluvial fans: an Australian example. – Geogr. Annaler 56: 83–92.
– (1975): Evolution of alluvial fans in two areas of south-eastern Australia. – Unpublished PhD thesis, Macquarie University.
WATSON, E. (1970): The coastal periglacial slope deposits of the Cotentin peninsula. – Trans. Inst. Brit. Geogrs. 49: 125–144.
WILLIAMS, G. E. (1970): Piedmont sedimentation and late Quaternary chronology in the Biskra region of the northern Sahara. – Z. Geomorph. Suppl. 10: 40–63.
– (1973): Late Quaternary piedmont sedimentation, soil formation and paleoclimates in arid South Australia. – Z. Geomorph. 17: 102–125.

ERRATA

Page 161, line 37 should read: "sediment is removed from the fan, then the fan is described as incised. . . ."
Page 161, line 40 should read: "said to be entrenched.". . .

12

Copyright © 1979 by John Wiley & Sons Limited
Reprinted from Earth Surf. Process. 4:147-166 (1979)

GEOMETRY OF ALLUVIAL FANS: EFFECT OF DISCHARGE AND SEDIMENT SIZE

ROGER LeB. HOOKE AND WILLIAM L. ROHRER

Department of Geology and Geophysics, University of Minnesota, Minneapolis, Minnesota 55455, U.S.A.

Received 22 May 1977
Revised 5 April 1978

SUMMARY

The slope of an alluvial fan increases with increasing debris size and sediment concentration in the flow, and decreases with increasing discharge. Laboratory studies suggest that the discharge which controls this slope, or dominant discharge, is that which is equalled or exceeded one quarter to one third of the time that flow occurs on the fan. In contrast, the dominant discharge in perennial alluvial rivers is equalled or exceeded only about 5 per cent of the time that flow occurs in the river. The dominant discharge on fans increases with increasing debris size, reflecting the importance of threshold stress.

The slope of some natural and most laboratory alluvial fans is steepest on the flanks and gentlest along the axis. Consideration of the momentum of water debouching onto a fan at its apex suggests that the difference in slope between axis and flank should be greatest on steep fans composed of relatively non-cohesive materials because on such fans higher discharges tend to flow down the axis, whereas lower discharges can be turned to course down the flanks. On fans with gentle slopes or composed of more cohesive material the higher discharges can also be turned toward the flanks, so on such fans the difference in slope between the axis and flank is less pronounced. Field and laboratory observations support this interpretation.

Because deposition at any one time on an alluvial fan is localized, some areas aggrade while others remain at a fixed elevation. This process is treated as a Markov process with the probability of diversion from an area of active deposition into an adjacent lower area increasing as the height of the active area above the mean or 'ideal' surface increases. Analysis of data from laboratory and natural fans suggests that the amplitude of such surface irregularities is greater on fans composed of coarser material. The data on natural fans also suggest an increase in amplitude of the irregularities with increasing fan area.

KEY WORDS   Alluvial fans   Dominant discharge   Markov model

INTRODUCTION

In this paper, a continuation of some previous studies (Hooke (1967, 1968b, 1972), Hooke and Rohrer (1977)), we treat three aspects of the shape of alluvial fans. First, laboratory data showing the effect of discharge and sediment size on fan slope are presented, and the influence of sediment size on the magnitude of the dominant discharge is considered. Then the variation in slope with azimuth is discussed. Finally, a Markov-process model for deposition on fans is developed and used to interpret irregularities in fan surfaces.

Observations on alluvial fans built in the laboratory provided the stimulus for these investigations and figure importantly in the paper. A description of the laboratory studies is thus necessary.

LABORATORY APPARATUS AND PROCEDURE

Sediment was placed in a sloping V-shaped channel debouching into a 1.5×2.7 m box. Water from a constant-head tank was then run through the channel, entraining the debris and depositing it as a fan in the box (Figure 1). The channel entered the box near the centre of one of the 2·7 m sides so that fans with a maximum radius of 1·35 m could be built. Discharge was controlled with the use of a ball valve and was

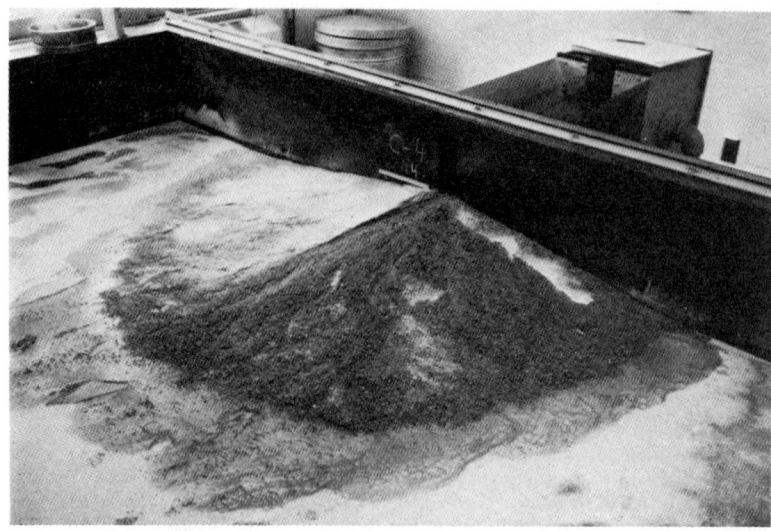

Figure 1. Photograph of laboratory fan C-4 after 44th episode. Talcum powder was sprinkled on the fan prior to the episode and remains on areas not affected by flow during the episode. Scale is 20 cm long

measured with precision-bore flowrators. As is the case on natural fans, deposition was initiated by the break in slope at the toe of the fan (Figure 1), and infiltration was negligible after the first few seconds of flow.

Each fan was built during a series of discrete depositional episodes. Prior to each episode the sand in the inlet channel was graded to a level about 6 mm higher than the apex of the fan. The slope of the sand surface in the channel was about 0·10. Water was then run through the inlet channel for a period of five minutes. After the flow was stopped, the elevation of the fan surface was measured to the nearest 0·1 mm at 22 points with the use of a point gauge. By placing a 5 cm square piece of sheet metal on the fan surface under the point gauge and pressing it down gently, measurements could be made quickly with a reproducibility of about ±0·3 mm. Measurements were made at the apex and along seven radii at distances of 25, 50, and 100 cm from the apex. The seven radii were the axis, and ±27°, ±54° and ±81° from the axis. These elevation data were used to calculate slopes at various positions on the fans. Finally, the volume of sand eroded from the inlet channel was determined by measuring the volume of sand required to regrade the channel to its pre-episode profile.

Seventeen fans were built during the studies discussed herein. These are divided into three series, designated C, E, and F (Table I). All fans in a single series were built with the same sand.

Twelve of the fans were built with constant discharges. That is, the discharge was not changed during an episode or between episodes during construction of the fan. The average number of episodes used to

Table I. Data on laboratory fans

Series	Number of fans in series	Sand size	
		Geometric mean diameter	Geometric standard deviation
C	8	0·54	2·3
E	5	0·17	1·5
F	4	1·3	3·6

build these fans was 36 and the minimum was 23. The other five fans, C-1, C-6, C-7, E-5, and F-4, were built with a distribution of discharges designed to simulate natural floods. Each episode began with a different peak discharge, and the discharge was decreased twice during the episode. The first decrease was one minute after the beginning of the episode and the second was two minutes later. For fans C-1, E-5, and F-4, the two decreases in discharge were equal in magnitude, and together they reduced the flow to either one-third of the peak discharge or to 44 cm^3/s, whichever was highest. 44 cm^3/s was lowest discharge measurable with the flowrator in use at that time. The peak discharge to be used for any given episode was selected at random from a series of peak discharges approximating a log-normal distribution, and was in no way related to the discharges in the preceding or following episodes. When the individual one-minute discharges are plotted as a cumulative frequency curve, they also form a log-normal distribution (Figure 2) but with a lower mean and standard deviation than the distribution of peak discharges. Discharges used in construction of these fans ranged from 44 to 435 cm^3/s. Fans C-6 and C-7 were constructed in a similar way but with different frequency distributions of one-minute discharges (Figure 2). Addition of a new flowrator permitted measurement of discharges as low as 6 cm^3/s; the maximum discharge used in these two runs was 568 cm^3/s. Between 75 and 103 episodes were used in building each of these five fans.

The alluvial fans built during these studies are not considered to be scale models of existing fans, and no attempt was made to determine appropriate scale factors. Instead the *similarity of process* approach (Hooke (1968a)) was used in which gross scaling relations are satisfied by ensuring that the sediment can be readily transported by the discharges used, and that the resulting fans are similar to natural fans in gross morphology (Figure 1). Observations of processes on the laboratory fans and of characteristics of

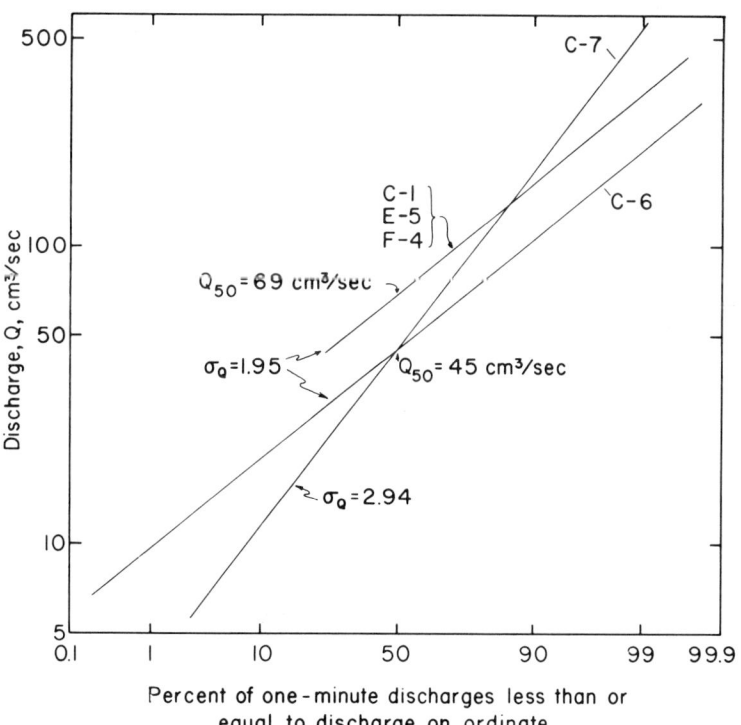

Figure 2. Distribution of discharges used to build fans C-1, C-6, C-7, E-5, and F-4. Q_{50} is discharge equalled or exceeded by 50 per cent of the discharges used. σ_Q is geometric standard deviation which is defined as $Q_{84.1}/Q_{50}$ or $Q_{50}/Q_{15.9}$

fan morphology resulting from these processes then provide a basis for interpretation of these same characteristics when observed on natural fans.

DEPENDENCE OF FAN SLOPE ON DISCHARGE AND SEDIMENT SIZE

That high discharges can transport a given sediment load on a lower slope has been known at least since G. K. Gilbert's (1914, Table 4) classic laboratory study. Later Mackin (1948) observed that slope was the dependent variable in natural systems, and emphasized the importance of discharge in determining the slope of alluvial rivers. Hooke (1968b, p. 622) used the same principle to explain why alluvial fans with larger source areas generally have lower slopes than fans built of similar material from smaller source areas; the larger source areas have proportionally higher storm discharges.

The importance of sediment size in controlling slopes of alluvial streams has also been known for a long time (Gilbert (1914)), Rubey (1938), Mackin (1948), and its effect on the slopes of natural alluvial fans have been discussed previously (Hooke (1968b, p. 625)).

The laboratory data presented in Figure 3 clearly show the influence of both discharge and sediment size on fan slope. Both effects are non-linear. The crossing of the curves for Series C and Series E at about 175 cm^3/s is anomalous and may indicate either that an equilibrium slope was not attained during construction of the fan built with the highest discharge in Series E, or that boundary effects influenced the slope in some way.

Dominant discharge

Wolman and Miller (1960) were among the first to discuss the importance of the frequency distribution of events in geomorphic studies. They found that the morphology of such diverse features as river channels, sand dunes, and beaches appeared to be controlled by events that had a recurrence interval of less than a year or two. In other words these features were not products of catastrophic events, but were instead adjusted to forces (discharges, wave heights, wind speeds etc.) which could be expected to be equalled or exceeded every year or two. Such forces are considered to be of moderate magnitude.

In any attempt to quantify geomorphic phenomena it is useful to know the magnitude of the forces to

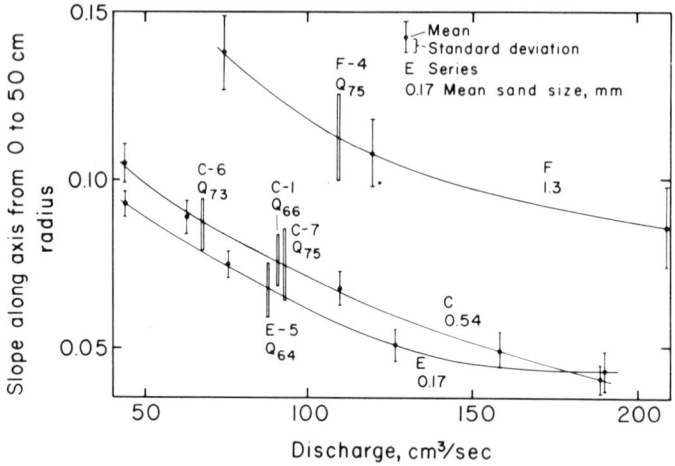

Figure 3. Relation between slope and discharge for three different sediment sizes. Slopes were calculated from elevation measurements made at the end of each episode. The average slope at any position and standard deviation of that slope were then determined. Open bars representing fans C-1, C-6, C-7, E-5, and F-4 were not used to draw curves but were later placed on the curves and the dominant discharge read from the abscissa (see text). Q_{75} is the discharge equalled or exceeded by 25 per cent of the discharges used, etc.

which any given feature is adjusted. River channels and floodplains have received more attention than other geomorphic features in this regard. Wolman and Leopold (1957), for example, showed that rivers typically overflow their banks every year or two. The relatively frequent occurrence of such bankfull flows and the uniformity of their recurrence intervals on streams in diverse geographic regions supports the intuitively reasonable assumption that river channels are somehow adjusted to the bankfull discharge. Leopold and Maddock (1953), however, used the mean annual discharge as a basis for comparing river cross sections of different size. The mean annual discharge is equalled or exceeded approximately 25 per cent of the time during the year on a large number of rivers (Leopold and Maddock (1953, p. 9)) so a comparison of several river sections on the basis of the mean annual discharge at each section would also be a comparison at approximately the same frequency of discharge. Bankfull discharge is typically 5 to 10 times the mean annual discharge, and the bankfull discharge is equalled or exceeded every year or two, or probably less than 5 per cent of the time, so comparisons on the basis of mean annual discharge are quite different from those on the basis of bankfull discharge.

Carlston (1965) found that meander wavelength was proportional to the 0·46 power of the mean annual discharge and to the 0·62 power of the bankfull discharge. The standard error was lower for the correlation with mean annual discharge (11·8 per cent) than for the correlation with bankfull discharge (25 per cent) which might be taken as support for the assumption that at least this aspect of river geometry is adjusted to mean annual discharge. However Ackers and Charlton (1970) have studied the geometry of laboratory meanders built by varying discharges, and their findings support the conclusion that meander wavelength is determined by bankfull discharge.

In the only non-fluvial study of which we are aware, Pearce (1976) found that storms of moderate intensity, 5–15 mm/h, and with durations of 1–6 h were most effective in moving sediment on bare hill slopes near Sudbury, Ontario. Such storms are equalled or exceeded about once a year.

The ambiguity in the above figures for rivers emphasizes not only the need for more systematic studies of the effects of events of different frequency, but also the need for a more rigorous definition of the magnitude of the process to which a given feature is adjusted. In the latter regard, Inglis (1949) was apparently the first to use the term *dominant discharge*, which Blench (1951, Sec. 6.16) later defined as being the 'steady discharge that would produce the same result as the actual varying discharge'. Ackers and Charlton (1970) used this definition in their laboratory studies, and Hooke (1968b, p. 625) adapted it to studies of alluvial fans by defining the dominant discharge as 'that discharge which, if it alone occurred, would produce a fan having the same slope as a fan built with a distribution of discharges'.

In Hooke's 1968 paper some of the data in Figure 3 were used to determine the magnitude and frequency of this dominant discharge on fan C-1. The open bar for fan C-1 is placed on the curve for series C fans (Figure 3) at a position corresponding to the measured mean slope of this fan, and the dominant discharge is read from the abscissa. The dominant discharge thus determined for fan C-1 is 90 cm^3/s, and this discharge was equalled or exceeded 34 per cent (Q_{66}) of the total time during which flow occurred during construction of the fan. A similar procedure is used to determine the dominant discharge for fans E-5 and F-4, both of which were built with the same distribution of discharges as used for fan C-1. The dominant discharges are 87 cm^3/s and 109^3/s, and are equalled or exceeded by 36 per cent and 25 per cent of the flows on these fans, respectively. Of interest is the fact that the magnitude of the dominant discharge increases with grain size, or in other words, with the threshold stress necessary for movement. Wolman and Miller (1960) anticipated this when they noted that catastrophic events become more important as the threshold stress increases.

The dominant discharges for fans C-6 and C-7 can also be read from Figure 3. Fan C-6 was built with a distribution of discharges having a lower mean than the distribution used for fan C-1 (Figure 2), and its dominant discharge, 67 cm^3/s, is equalled or exceeded by only 27 per cent of the flows, as compared with 34 per cent for fan C-1. The discharges used to build fan C-6, being lower, did not exceed the threshold stress by as much as on fan C-1, so this is another example of the dependence of dominant discharge on threshold stress.

Fan C-7 was built with a distribution of discharges that had the same mean as the distribution used for fan C-6, but a higher standard deviation (Figure 2). The dominant discharge on fan C-7, 93 cm^3/s, was equalled or exceeded by 25 per cent of the flows. Although this is a substantially larger discharge than the dominant discharge on fan C-6, the frequencies of these two dominant discharges are nearly identical. It thus appears from these limited data that the frequency of the dominant discharge is a function of threshold stress and mean stress but may not be a function of the standard deviation of the distribution of stresses. The term stress is used loosely here, of course, and not in a rigorous physical sense, as the actual stress may not be proportional to the first power of the discharge.

In their analysis, Wolman and Miller (1960) emphasized the amount of sediment transported by events of different magnitude. Unfortunately we did not measure the sediment discharge carried by each one-minute discharge. However, analysis of the measurements of sediment load moved during each episode on fan C-1 suggests that, as a first approximation, the sediment discharge during any given minute was proportional to the water discharge which acted during that minute. Using this relation we found that 50 per cent of the sediment was moved by discharges that were less than or equal to 98 cm^3/s and 90 per cent of the sediment was moved by discharges that were less than or equal to 243 cm^3/s. These two discharges were equalled or exceeded by 30 per cent and 3 per cent of the discharges used, respectively. These figures are a little difficult to interpret because a discharge which occurs early in an episode will move more sediment than the same discharge occurring late in an episode after the most easily moved sediment in the inlet channel has been transported down to the fan and some armouring has developed. However, the same is true of floods on natural fans. It is perhaps significant that there is fairly close agreement between the dominant discharge as defined earlier for fan C-1 (90 cm^3/s) and the discharge below which half of the sediment was moved on this fan.

We have not yet found a way to determine the magnitude, or more importantly the frequency, of the dominant discharge on natural alluvial fans. However, some data are available for laboratory and natural rivers. Ackers and Charlton (1970) found in their laboratory streams that the dominant discharge, defined in terms of meander wavelength, was equalled or exceeded 9–23 per cent of the time. In a scale model of a natural meandering stream they found that the dominant discharge was equalled or exceeded approximately 10 per cent of the time that flows exceeded the threshold stress. In terms of sediment transport, Wolman and Miller (1960, Table 1) found that 50 per cent of the sediment carried by the Rio Puerco at Rio Puerco, N.M., and by Brandywine Creek at Wilmington, Del., was moved by flows that were equalled or exceeded 5 and 3 per cent of the time, respectively. The implication of all these figures is that the laboratory fans are adjusted to somewhat more frequent discharges than the laboratory or natural rivers. (In the studies of alluvial fans, however, periods of zero flow are not included in the frequency distribution of discharges. Thus the recurrence interval of the dominant discharge on fans may be a few decades, whereas on rivers it is usually 1 to 2 years.) To what extent these differences are due to the differences in the geomorphic features involved, differences in scale or threshold stress, or differences in definition of dominant discharge, we are not presently prepared to say. However the general thesis of Wolman and Miller's work—that geomorphic features are adjusted to stresses of moderate magnitude and not to catastrophic stresses—is borne out by all of our observations.

DEPENDENCE OF FAN SLOPE ON AZIMUTH

The flanks of alluvial fans are commonly steeper than the axis. Herein we define the axis as the direction a flow would take down the fan if it were not deflected to the left or right once it reached the fan apex. The position of the axis on a fan can be estimated by examining the orientation of the lowermost few tens or hundreds of metres of the main channel debouching onto the fan from the source area. The azimuth of any other direction radiating from the apex can then be measured as an angular distance from the axis. The variation in slope with azimuth on variable-discharge laboratory fans is shown in Figure 4a. Figure 4b presents results of measurements on selected natural fans.

Observations on the laboratory fans suggest that this variation in slope is a result of a variation in average discharge with azimuth. Higher discharges tended to straighten channels on the laboratory fans, and thus their effect was concentrated near the axis. During lower-discharge flows the main channel wandered away from the axis, resulting in curved channels. Still lower discharges occurred in distributaries at higher azimuths as a result overflow of part of the flow from the main channel. The outer bank of a curved channel formed by a low discharge was commonly unable to withstand the force exerted on it by subsequent higher discharges and thus eroded rapidly, straightening the channel.

The relation between the discharge and the radial forces on the boundary of a curved channel can be studied with the use of the momentum equation of fluid mechanics. For a rectangular channel, using cylindrical co-ordinates, this equation takes the form:

$$r = \frac{\rho Q V}{k R d} \tag{1}$$

where ρ is the density of water; Q is the discharge, V is the velocity, d is the depth; r is the minimum radius of curvature of the outside bank which the channel can maintain with the prevailing discharge and bank materials; R is the critical *shear* stress at which erosion will occur on the banks, and is determined by the nature of the bank materials; and k is a dimensionless coefficient of proportionality between the tangential shear stress on the bank on one hand, and the radial shear stress on the bed plus the excess normal stress (due to superelevation) on the outside bank on the other. Measurements made in connection with a study of meandering (Hooke (1975)) suggest that k might be of the order 0·05.

From this equation we see that the minimum radius of curvature of the channel increases as Q and V increase, and decreases as the resistance of bank materials to erosion and the depth of flow increase. The variation with depth of flow might not be expected; it arises from the fact that, in the simplified model used, deeper flows distribute the force over a larger bank area and thus decrease the force per unit area,

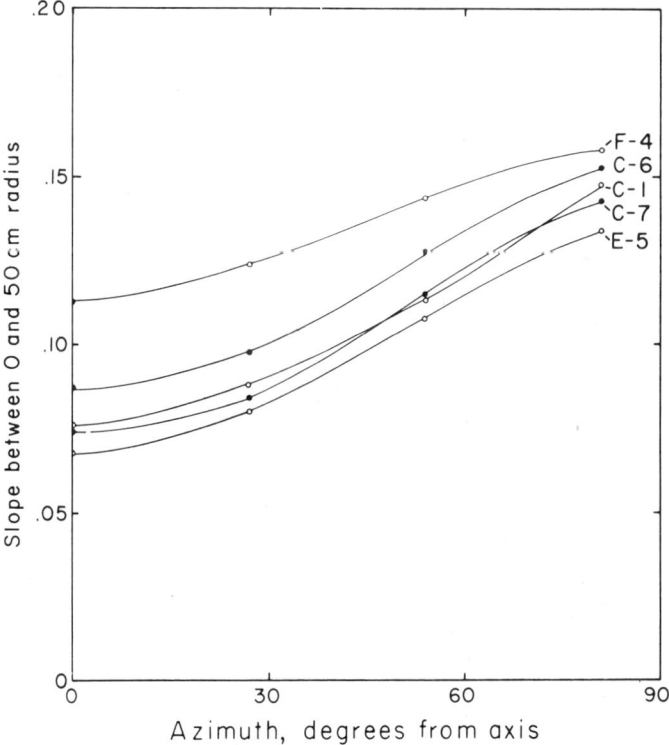

Figure 4(a). Relation between slope and azimuth for laboratory fans built with variable discharges

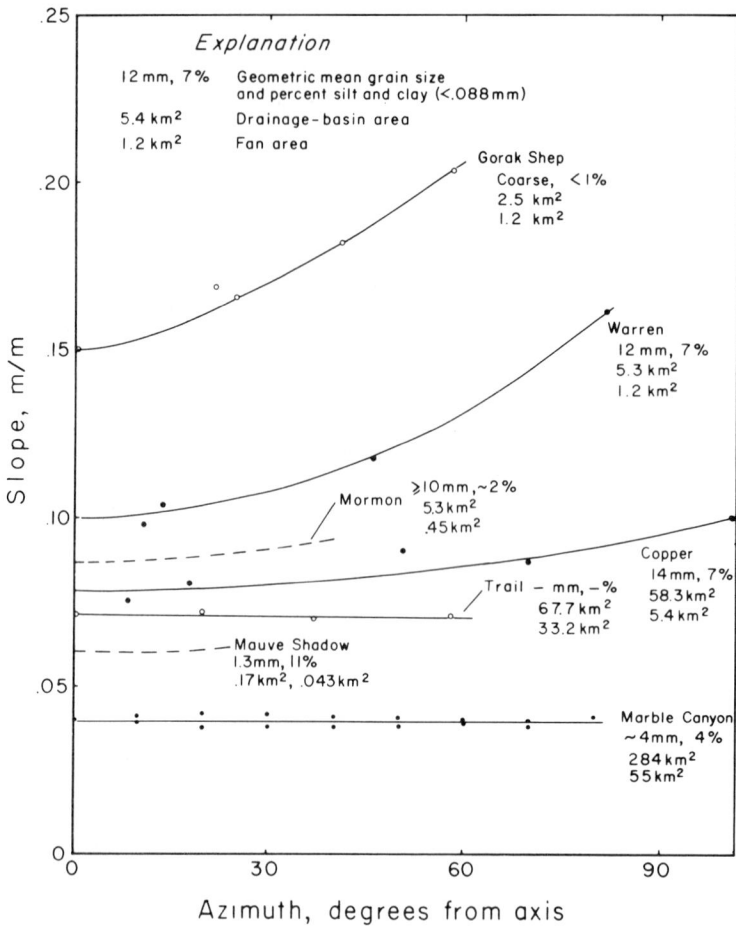

Figure 4(b). Relation between slope and azimuth on natural fans. On segmented fans (Bull (1964)) slopes were measured on the upper segment. Slopes of Marble and Trail fans were measured from 15-min. topographic maps. Slopes on Gorak Shep fan (Hooke (1967)) were measured using photogrammetric techniques. Slopes on other fans were surveyed in the field using plane table methods (Hooke (1972)). All fans are in Death Valley, California, except Gorak Shep fan which is in Eureka Valley, California.

Measurements on several other fans in California also revealed an increase in slope with azimuth, but the increase in curvature of the lines with increasing axial slope was less pronounced. These data are not plotted because (1) lines cross and diagram becomes confusing, (2) grain size data are not available, and (3) slope measurements are less accurate. Grain size measurements on Marble fan are from Conomos (1961). Other measurements were done by T. Raster (unpublished) and Hooke on representative samples of material from the main channel near the fan apex. Original samples weighed a few tens of kilograms and were coarse-sieved in the field. Part of the <12 mm fraction was returned to the laboratory for detailed analysis

or stress. This dependence may not be observed in natural systems, at least in humid regions, because the cohesiveness of bank materials commonly decreases with depth; low banks tend to be more resistant owing to the stabilizing effect of vegetation, whereas higher banks typically become undercut due to the less cohesive character of deeper material. Furthermore, M. Foley (written communication (November 1977)) has suggested that k may vary inversely with depth, thus eliminating or even reversing the sense of this dependence. However, the variation in radius of curvature with Q and R is nicely illustrated by studies which show that the radius of curvature of river bends increases with discharge (Leopold and Wolman (1960) showed that $\lambda \propto r$ where λ = wavelength, and Carlston (1965) showed that $\lambda \propto Q^{0.5}$

therefore $r \propto Q^{0.5}$), and decreases with increasing silt–clay content in the banks (Schumm (1967) found that $\lambda \propto 1/M^{0.74}$, where M = silt–clay content, so with Leopold and Wolman's (1960) result above, $r \propto 1/M^{0.74}$).

Quantitative verification of the applicability of equation (1) to the present problem of variation in fan slope with azimuth is complicated by the fact that V, k, and R are unknown. Furthermore, the channels are not rectangular and the depth of flow against the banks is very small or negligible. In addition, the path which the channel took to reach any given position on the fan, and in particular the minimum radius of curvature along that path, is also unknown. Finally the model breaks down at the highest azimuths in any case because, as mentioned, much of the flow reaching these parts of the fan results from overflow from the main channel rather than by curving of the channel.

Qualitatively, however, the relation has the desired effect; on the laboratory fans r increases with Q, so higher discharges are less readily diverted to the flanks. Thus higher discharges build the axis, and the slope along the axis is lower than on the flanks.

Some natural fans also display this variation in slope with azimuth as shown in Figure 4b. An interesting feature of this figure in the difference in curvature of the lines among the various fans. According to the hypothesis presented above, the variation in slope with azimuth should be greatest (curvature largest) for fans on which there is a large variation in effective discharge with azimuth. In terms of equation (1), this will be the case if r is large for large floods. If r is always small on a given fan, then any discharge reaching that fan could act on the flanks of the fan, and the preference of the larger discharges for the axial position would not be as pronounced.

From equation (1) it is clear that r can be large if Q or V is large or if R is small. High discharges (Q) should be associated with larger drainages but the curvature of lines in Figure 4b is least for fans from large drainages so this effect is apparently not dominant. This lack of dependence on drainage area may be a result of the fact that, as on laboratory fans, most flows reaching the flanks of natural fans result from flow in distributary channels or from overflow of the main channel near the apex (Hooke (1967)); the main channel itself rarely deviates widely from the axis (Bull (1964, p. 114)). Thus the appropriate Q to use in equation (1) would be determined by the ability of a curved bank of a distributary channel just below a bifurcation to divert part of the flow from the main channel without eroding—that is by R. High resistance of banks to erosion (R) might be associated with fine-grained sediment containing appreciable amounts of silt and clay. Unfortunately sediment size analyses are not available for all fans in Figure 4b. However there does appear to be an increase in geometric mean grain-size and a decrease in silt–clay content with increasing curvature of the lines. Thus R may be smaller on the fans exhibiting the most marked variation in slope with azimuth. Higher velocities are also expected on the steeper fans, and this may be partly responsible for the increase in curvature of lines in Figure 4b with increasing mean slope.

'LOW AREAS' ON FANS

Alluvial fans do not have smooth conical surfaces. Deposition occurs on one part of a fan for some period of time and builds up that area above the average fan surface. Eventually flow is diverted from this high area into an adjacent lower area, and the process is repeated. For convenience, we use the terms *high* area and *low* area to describe the elevation of the fan surface at a particular point relative to the elevation of an 'ideal' smooth surface at that point.

The processes by which diversion occurs on an alluvial fan are complex. Plugging of channels by debris flows (Beaty (1963)) is one relatively catastrophic process, but herein we are interested primarily in process involving water flows alone. Aggradation of a channel until overbank flow occurs is another factor contributing to diversion. Because most sediment in flows on alluvial fans is transported near the bed, the overbank flows are relatively sediment free, and thus can rapidly erode a new channel. As the new channel deepens it captures more of the main flow. If the slope of this new channel is slightly greater than that of the old, capture is more likely. Lateral erosion of the main channel, perhaps accompanied by bar building, is still another factor contributing to diversion.

Because of the complexity of this process, and because we are here interested in broad patterns of deposition over long time periods rather than in details of a single diversion event, stochastic methods are applicable. Fairly complicated stochastic models of deposition on fans can be visualized; Price (1976), for example, uses a random walk model to simulate deposition on fans and includes in his computer program several other stochastic factors such as tectonic uplift, weathering, and storm magnitude which enter into the growth and development of natural fans. Herein, however, we focus only on the probability of a diversion from a high area. Price handled this problem by assuming that the probability of flow in a certain direction was proportional to the slope in that direction. We treat the process as a Markov process; the state of the system is given by the height of the high area or depth of the low area. Laboratory data are used to determine the probabilities of transition from any given state to any other state and histograms are prepared showing the frequency with which any given state is occupied (Figure 5). These histograms are later compared with others obtained from detailed surveys on natural fans.

Laboratory data

Analysis of the data from laboratory fans begins with an estimate of the area of the fan at the end of each episode. The elevation measurements at 22 points on the fan are used for this estimate. Dividing the volume of sediment deposited during the episode by the area at the end of the episode gives the average thickness of material deposited during the episode, t_n, where n is the number of the episode. Next a particular episode, $\tilde{n}$, near the middle of the run is selected as a reference episode. The elevation of the fan surface at the end of each episode is then normalized to the reference elevation thus

$$\varepsilon_{i,n} = E_{i,n} + \sum_{j=n+1}^{\tilde{n}} t_j \qquad n < \tilde{n}$$
$$\varepsilon_{i,n} = E_{i,n} - \sum_{j=\tilde{n}+1}^{n} t_j \qquad n > \tilde{n}$$
(2)

where $E_{i,n}$ is the measured elevation at position i after episode n, and $\varepsilon_{i,n}$ is the corresponding normalized elevation. The normalized elevations at any given location, i, are then averaged to obtain an 'ideal' surface. This ideal surface is made symmetric about the axis of the fan by averaging elevations at similar locations on opposite flanks.

Deviations of the normalized surface from the ideal surface can now be studied. To approach this problem from the point of view of a Markov process, the continuum of deviations must be divided into states. To do this the deviations are tallied in 1 mm intervals. The interval from -0.5 mm to $+0.5$ mm (upper limit included) is designated state 0. The interval from -1.5 to -0.5 mm is state -1 and so forth. The transitions between states are then tabulated and a tally matrix is constructed showing the number of times any given change in state occurred. Finally, a probability matrix is obtained by dividing the number of transitions from state m to state n by the total number of transitions out of state m. Table II is such a probability matrix for laboratory fan C-1, and Figure 5 contains histograms showing the frequency of occurrence of any given state for fans C-1, E-5 and F-4.

The histograms in Figure 5 have two interesting features. Firstly, there is a pronounced broadening of the histograms with increase in grain size. This implies that highs get higher and lows get deeper on fans built of coarser sand, and probably reflects the fact that higher velocities are required to transport the coarser material. Thus on fans built of coarser material lateral slopes into lows must be higher before diversion can occur. The second interesting feature is the persistent asymmetry of the histograms. This is probably a result of channel cutting; high areas can develop only through deposition, but lows can be a result of either non-deposition or erosion. Thus low areas can be expected to be more pronounced.

The null hypothesis that the transition tally matrix might have resulted from a series of independent events (Krumbein and Dacey (1969, p. 83)) was tested with the use of a contingency table (Potter and Blakely (1968, p. 160)), Krumbein and Graybill (1965, p. 186)). This hypothesis is rejected at the 99.9 per cent level. Also tested was the hypothesis that the lengths of stay in the same state approximate a geometric distribution, as they should if the process is a first-order Markov process (Krumbein and Dacey

Table II. Transition probability matrix for fan C-1

State in	\-15	\-14	\-13	\-12	\-11	\-10	\-9	\-8	\-7	\-6	\-5	\-4	\-3	\-2	\-1
\-15			1·00												
\-14	0·20	0·40					0·20				0·20				
\-13		0·50		0·25											0·25
\-12			0·33	0·33	0·11	0·11									0·11
\-11			0·07	0·29	0·29	0·07	0·14	0·07		0·07					
\-10					0·56	0·11		0·11	0·11		0·11				
\-9						0·36	0·29	0·21				0·07			
\-8				0·05	0·05	0·30	0·15	0·05	0·10	0·05	0·05	0·10	0·05		
\-7							0·28	0·25	0·14	0·04	0·25	0·04			
\-6							0·02	0·36	0·18	0·24	0·07	0·04	0·09		
\-5					0·03			0·02	0·35	0·35	0·11	0·05	0·03	0·02	
\-4								0·02		0·03	0·32	0·30	0·12	0·11	0·05
\-3										0·01	0·06	0·29	0·42	0·12	0·02
\-2								0·02				0·05	0·36	0·28	0·11
\-1							0·03	0·01				0·08	0·25	0·35	
0												0·02	0·05	0·32	
1											0·01		0·02		0·04
2											0·01				
3													0·03	0·04	0·35
4														0·01	0·01
5															
6											0·02				

State in	0	1	2	3	4	5	6	7	8	9	10	11	12	13
\-9		0·07												
\-8		0·05												
\-7														
\-6														
\-5	0·02			0·02	0·02									
\-4		0·02		0·02	0·02		0·02							
\-3	0·02	0·01	0·01	0·01	0·02									
\-2	0·08	0·06	0·02	0·02	0·02									
\-1	0·13	0·08	0·01	0·01	0·03	0·01								
0	0·33	0·11	0·06	0·10		0·01								
1	0·29	0·35	0·13	0·09	0·02		0·02	0·01	0·01		0·01			
2	0·10	0·31	0·35	0·11	0·06	0·03	0·03							
3	0·37	0·14	0·06	0·01	0·02									
4	0·01	0·02	0·07	0·33	0·34	0·09	0·07	0·01	0·01	0·02				
5		0·02		0·14	0·34	0·31	0·12	0·03	0·03		0·02			
6		0·06		0·04	0·11	0·35	0·32	0·07		0·02	0·02			
7						0·09	0·50	0·22	0·16	0·03				
8		0·05			0·05		0·09	0·46	0·32	0·05				
9			0·07					0·36	0·29	0·21			0·07	
10									0·11	0·56	0·33			
11											0·50	0·50		
12											0·33	0·17	0·33	0·17

(1965)). Lengths of stays in each state were tabulated and compared with the expected length of stay using a chi-square test (Krumbein and Dacey (1965, p. 172)). The probability of drawing a sample with higher chi-square (poorer fit) from a population having a geometric distribution was determined for each state. The median probability was 0·65 which indicates that the lengths of stays may be approximately geometrically distributed.

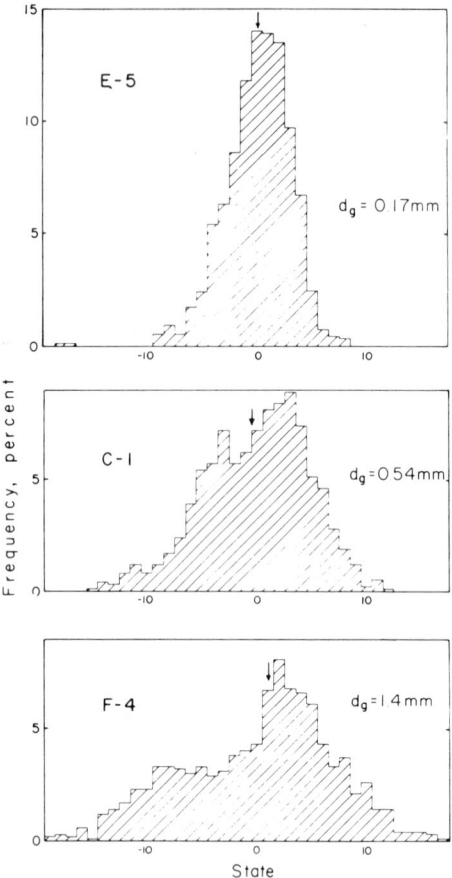

Figure 5. Histograms showing frequency of occurrence of various states on laboratory fans C-1, E-5, and F-4. Median values shown by arrow

The possibility of memory exceeding one step was also investigated by using a qualitative test described by Potter and Blakely (1968, p. 160). The test involves counting the number of times the system is in state i, N_i; the number of transitions from state i to state j, N_{ij}; the number from state k to state i, N_{ki}; and the number from state k to state i and thence to state j, N_{kij}. If

$$\frac{N_{ij}}{N_i} \approx \frac{N_{kij}}{N_{ki}} \tag{3}$$

the system has a memory of only one step, but if these ratios are not approximately equal the memory is longer than one step. The ratios are not 'approximately equal' (Figure 6a) but unfortunately there is no rigorous test by which we can estimate the level of significance of the difference (Potter and Blakely (1968)); more sophisticated tests would have to be used for such a study. However, it is noteworthy that the most common k_{ij} transitions are ones involving a progressive decrease in state. For example, transitions from state 3 to state 2 and thence to state 1 occurred twenty times on fan C-1 (Figure 6a), but transitions from state 4 directly to state 2 and thence to state 1 occurred only four times (data not shown). The twenty $3 \rightarrow 2 \rightarrow 1$ transitions were examined further and it was found that seventeen of these were parts of longer continuous sequences, some of which involved up to 17 steps.

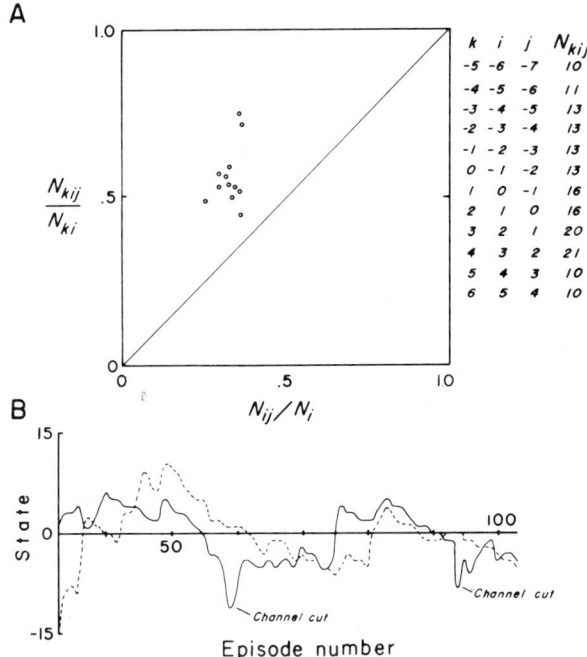

Figure 6(a). N_{kij}/N_{ki} plotted against N_{ij}/N_i for all instances in which $N_{kij} > 10$ on fan C-1. The kij transitions involved are given at right (b) change in state through time for two points on the same flank of fan C-1. Solid line is for point 50 cm from fan apex and 27° from axis. Dashed line is also for point 50 cm from fan apex, but 54° from axis

That this Markov process apparently has more than one-step memory is not unexpected. If flow does not reach one area of the fan for a long time, other parts of the fan will be built higher. Thus the area of non-deposition will get progressively lower relative to the average surface. The progressive decrease in state reflects this tendency. Eventually flow is diverted into the resulting low and an abrupt increase in state results. Figure 6b shows the change in state through time for two positions on fan C-1; gradual decreases in state followed by abrupt increases can be seen in the figure, as can episodes of channel cutting in the position near the axis. The similarity in the pattern for these two positions 23 cm apart, is consistent with field data, discussed below, which indicate that the low areas are often quite broad.

To compare the tally matrices for the various variable-discharge fans in greater detail a χ^2 parameter was calculated for each pair of matrices using the equation

$$\chi^2 = \sum \frac{(a_{ij}/n_a - b_{ij}/n_b)^2}{(a_{ij} + b_{ij}/n_a n_b)(1 - a_{ij} + b_{ij}/n_a n_b)} \tag{4}$$

where a_{ij} and b_{ij} are the values in the i, j cell of matrices a and b respectively and n_a and n_b are the total number of transitions recorded in these matrices (Dr. P. Tyragöd, oral communication (December 1969)). Values of this parameter for the various fans are shown in Table III. Unfortunately a statistical test of the significance of the differences between these matrices is not possible because an appreciable but indeterminate number of cells in each matrix are expected to be zero (Table II), and thus do not contribute to the number of degrees of freedom. However the values in Table III do show interesting trends with the series C fans being rather similar to each other and with fan F-4, the fan with the coarsest sediment, showing the greatest difference. Study of the tally matrices indicates that this difference is a result of the greater amplitude of surface irregularities on fans built of coarser material, as mentioned previously (Figure 5).

Table III. Comparison of tally matrices for various fans

	C-1	C-6	C-7	E-5	F-4
C-1	—	314	367	400	434
C-6	—	—	313	356	564
C-7	—	—	—	321	675
E-5	—	—	—	—	630

Of interest, particularly in interpreting data on natural fans, is the magnitude and frequency of changes in state at different positions on the fan surface. To study these parameters we calculated two activity indices, an activity frequency, a_f, and an activity magnitude, a_m, defined as follows

$$a_f = \frac{\sum_{i=1}^{n-1} d_{ij}}{n} \quad \begin{array}{l} d_{ij} = 1 \text{ for } S_{ij} < S_{i+1,j} \\ d_{ij} = 0 \text{ for } S_{ij} \geq S_{i+1,j} \end{array}$$

$$a_m = \frac{\sum_{i=1}^{n-1} |S_{ij} - S_{i+1,j}|}{n}$$

where S_{ij} is the state at position j after episode i, and n is the total number of episodes for which data are available at position j. a_f is a measure of the frequency with which deposition (increases in state) occurs at a given location, and a_m is a measure of the magnitude of the changes.

Activity indices for fan C-1 are presented in Table IV. It appears that the frequency of changes in state, a_f, decreases down-fan and also decreases with increasing angular distance from the axis on the upper half of the fan. An increase in state occurs once in every two episodes near the axis at a radius of 25 cm, reflecting the active erosion and subsequent backfilling of channels which occurs at this position. As before (Hooke (1967, p. 456), Tolman (1909, p. 157)) these channels are viewed as temporary conduits through which sediment is transported to loci of deposition in 'low' areas farther down fan. As these lows aggrade, the channel backfills until diversion into another low occurs, and a new temporary channel is cut. Near the toe of the fan, increases in state occur, on the average, once every five or six episodes ($a_f \approx 0.2$).

The magnitudes of the changes in state, a_m, are also highest in the mid-fan region near the axis, again reflecting channel cutting and backfilling. They too generally decrease both down-fan and with increasing angular distance from the axis, but reversals in these trends occur on the ±54° and ±81° azimuths. The increase in a_m between 25 and 50 cm from the apex on these azimuths reflects, we suspect, periodic erosion at the 50 cm positions by sediment-free water overflowing from the main channel. Such water often moved as sheet flow over the 25 cm position, and only collected into channels farther down-fan.

Both a_f and a_m are lower at the apex than at the 25-cm position on the axis because a channel is always present at the apex at the end of an episode. Deposition at the apex occurred early in the episode, if at all, when sediment concentrations in the flow were highest.

Table IV. Activity indices for fan C-1

Azimuth (degrees)	Axis				±27°			±54°			±81°		
Distance from apex (cm)	0	25	50	100	25	50	100	25	50	100	25	50	100
a_f	0.36	0.52	0.33	0.18	0.45	0.27	0.20	0.32	0.26	0.18	0.34	0.19	0.11
a_m	1.74	2.69	1.67	1.28	1.57	1.22	1.07	1.04	1.18	0.96	0.86	1.02	0.68

Uncertainties, based on comparison of values from opposite flanks, are approximately ±0.03 in a_f and ±0.07 in a_m.

GEOMETRY OF ALLUVIAL FANS   161

We observed some tendency for a_f and a_m to decrease slightly with time as the fan area increased. This is reasonable as a given episode will deposit a thicker layer of sediment on a small fan than on a large fan. The analogue in the natural environment would be that small fans from large source areas would have higher activity indices than large fans draining small source areas.

NATURAL FANS

Applications of this Markov-process model to natural fans is difficult because we can only observe the natural fan at a particular instant in time, whereas observations spanning several thousand years would be necessary to determine the transition probabilities. However, we can study the distribution of highs and lows on natural fans. To accomplish this Hooke surveyed several circular arcs on four natural fans at predetermined radial distances from the apex (Table V), using plane-table methods. The fans are all located on the east side of Death Valley, California, and were selected for study because their present surface features suggested (apparently erroneously as will be discussed below) that they were not segmented or otherwise affected by tectonic movements. The entire surface of the fan appeared to have been recently active in each case. Detailed topographic profiles were surveyed along these arcs, with elevations determined every 0·15 to 6·8 m (average≃1 m) along the arc. An 'ideal' surface was then generated to fit the data so obtained, using a multiple linear regression algorithm. The equation of the surface models was of the form:

$$E = \sum_i \beta_i R^n \theta^m \qquad \begin{array}{l} m = 0, 2 \text{ or } 4 \\ n = 0, 1 \text{ or } 2 \end{array} \qquad (6)$$

where R is the radial distance from the fan apex in metres, θ is the angular distance from the axis, E is the elevation of the fan surface, and the β_i are the regression coefficients. By choosing m even, mirror symmetry across the axis is insured. In the final equations only the four most significant of the nine possible terms in equation (5) were used because additional higher-order terms did not contribute to significant decreases in unexplained variance. The expressions obtained are given in Table VI.

Because surveyed elevations were not randomly distributed, but were instead measured along lines of constant radius and concentrated in areas of marked slope change such as channels, the influence of these

Table V. Data on natural fans

Name	Location*	Fan area (km²)	Drainage basin area (km²)	Radius of surveyed arcs (m)	Arc radius as per cent of total fan radius	Angular range (degrees)	Mean grain size (mm)	Per cent silt–clay
Mormon Point	36°3′29″ 116°45′21″	0·45	5·30	171 366	32 69	40	>10	2
Gower Gulch	36°24′38″ 116°50′19″	0·90	6·37†	366 823	32 72	40	1·8	10
Scarlet	36°25′6″ 116°50′32″	0·018	0·061	30 61 91	24 48 72	70	~3	8
Mauve Shadow	36°25′ 116°50′30″	0·043	0·17	61 122 183 244 305	19 37 56 74 93	35	1·3	11

* From U.S.G.S. 1:62,500 scale topographic maps.
† Prediversion drainage area.

Table VI. Equations of 'ideal' surfaces on four natural fans

Mormon Point
$$E = 60.96 - 0.08668R - 0.000001839R^2 - 0.01311R\theta^2 \qquad r^2 = 0.99875$$
Gower Gulch
$$E = 61.14 - 0.04312R + 0.000007627R^2 - 0.003361R\theta^2 \qquad r^2 = 0.99777$$
Scarlet
$$E = 30.52 - 0.08235R + 0.0000835R^2 + 0.0004013R\theta^4 \qquad r^2 = 0.99569$$
Mauve Shadow
$$E = 30.52 - 0.07407R + 0.00004331R^2 - 0.004947R\theta^2 \qquad r^2 = 0.99843$$

r = multiple correlation coefficient.
R is measured in metres and θ in radians.

small irregularities had to be reduced by a weighting scheme based on measurement point density. The weighting procedure insured equal influence of data points per unit surface area, and thus emphasize our concern with the distribution of broad high and low areas, rather than with single channels.

The negative coefficients of the θ-dependent terms for three of the four models (Table VI) support the slope-azimuth relation discussed previously. Slope-azimuth curves for two of the fans are plotted as dashed lines on Figure 4b. The plotted slopes are mean slopes between the apex and the arc farthest from the apex.

The positive coefficients of the R^2 terms indicate that three of the four fans are concave upward along their axes. In all three instances the concavity is greater than can be explained by random irregularities in

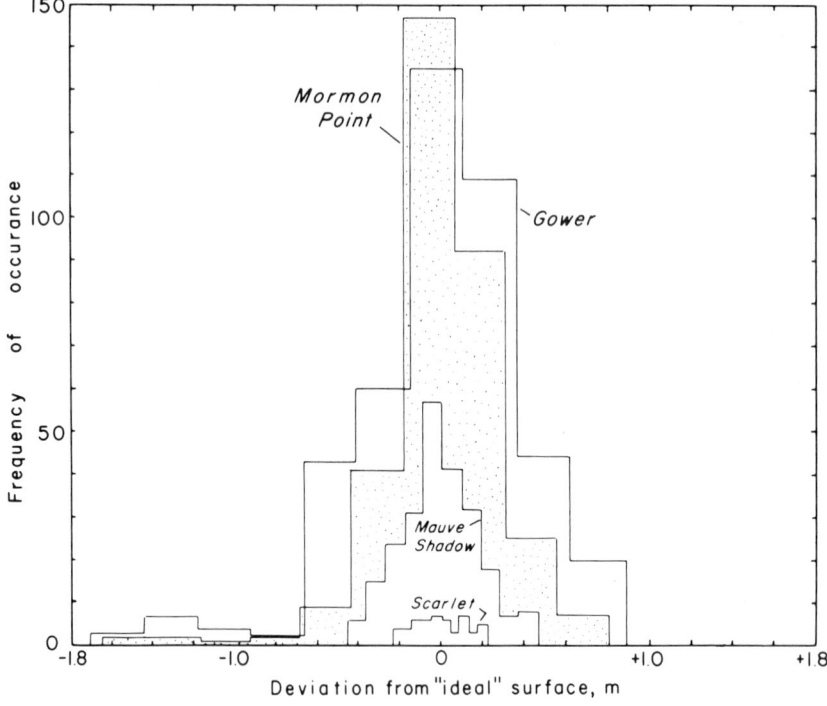

Figure 7. Histograms showing frequency of occurrence of deviations from the 'ideal' surface on four natural fans

the fan surface, although the slight convexity in the fourth fan, Mormon, can be explained in this way. One possible explanation for the concavities is that these three fans are, in fact, segmented. All three are close to one another, just south of Furnace Creek Ranch. Several kilometres farther south, fans are generally segmented with the youngest segment at the head, apparently as a result of an increase in sediment yield due to uplift of the source area (Hooke (1972)). This type of segmentation is difficult to recognize without surveyed longitudinal profiles because it does not result in fanhead incision. That recent uplift has occurred in the source areas of the three fans in question is indicated by low fault scarps near their apices (Figure 8b). Further evidence for segmentation of Mauve Shadow fan is discussed later.

Deviations of the observed fan elevations from the 'ideal' surface are plotted as histograms in Figure 7. As with the laboratory fans (Figure 5) there is an obvious asymmetry of the histograms, probably again resulting from the fact that high areas are created by aggradation whereas low areas can be formed by both non-deposition and erosion, but in contrast to the laboratory situation, the modes are in a zero or slightly negative state. The laboratory data suggest that any single point on a fan is likely to spend more time in a slightly positive state than in any other single state, whereas the field results suggest that at any one time the majority of the fan will be in a slightly negative state. This difference may reflect the fact that an average surface was used as the 'ideal' surface in the laboratory case, whereas a least-squares fitting procedure was used to obtain the 'ideal' surface for the natural fans. The elevation of the average surface would be lower because of the greater weight given to channels. Another property of the histograms is the tendency for increasing breadth with increasing fan size. Intuitively we might suppose that it would take

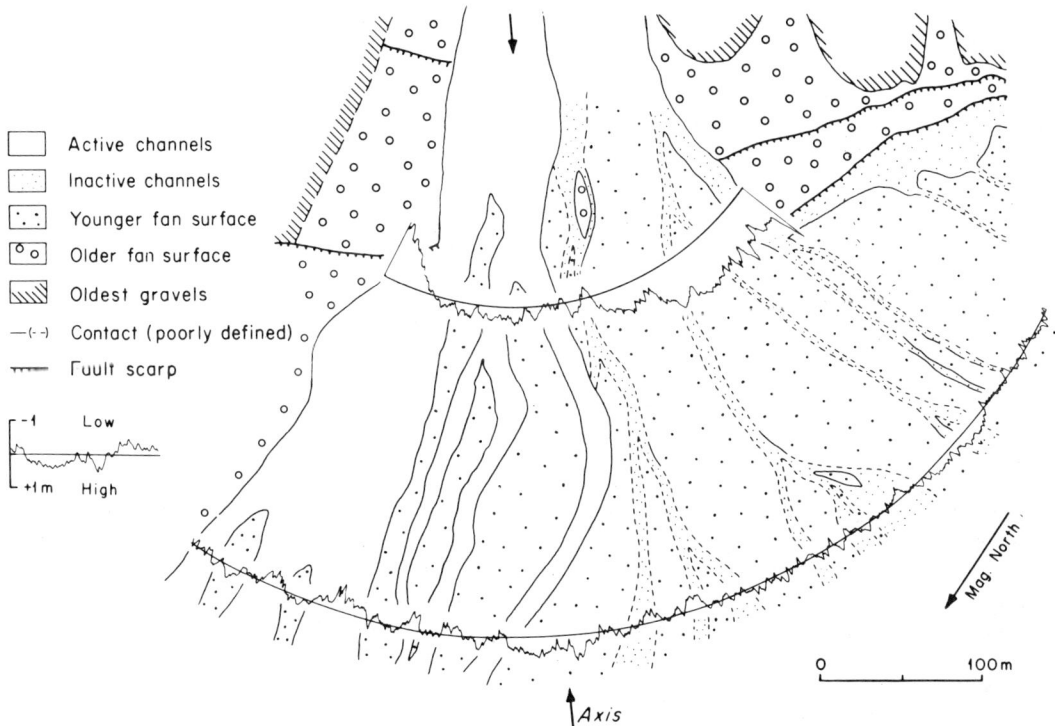

Figure 8. Gemorphic maps of two natural fans showing surveyed arcs and deviations of fan surface from 'ideal' surface along these arcs. Note that individual broad high and low areas can be traced down fan (a) Mormon fan

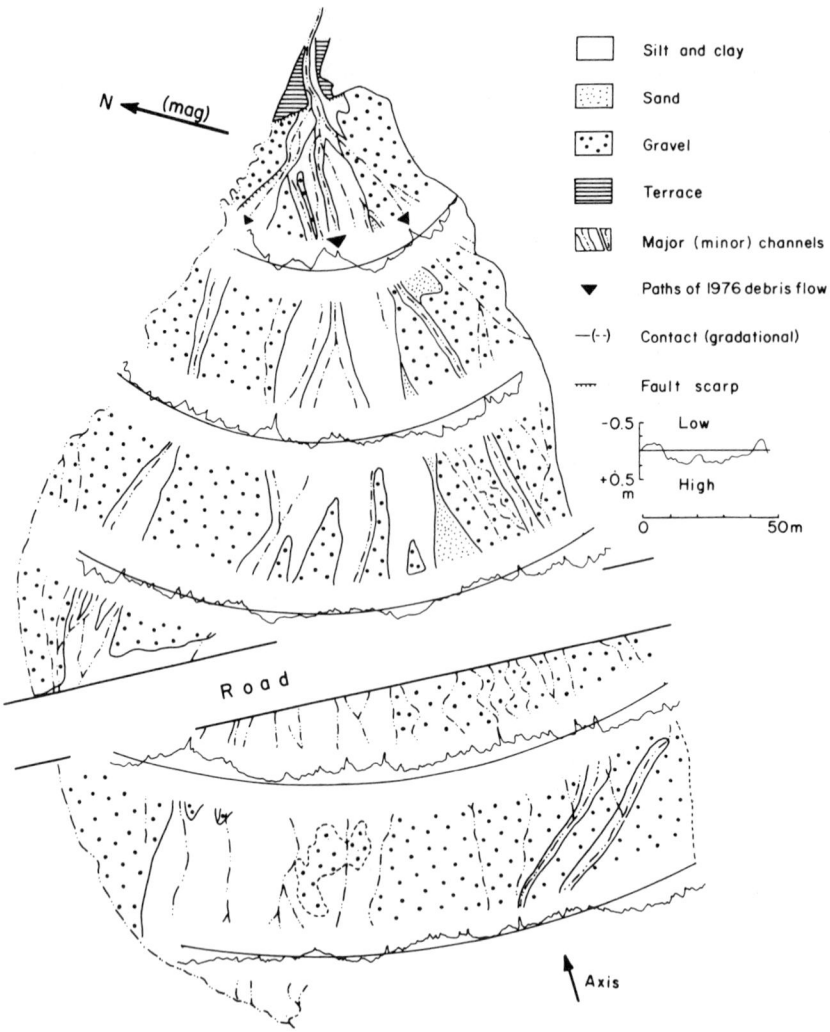

Figure 8(b). Mauve Shadow fan

longer for flows shifting back and forth on a large fan to touch all parts of the fan. Thus there would be a greater chance for topographic disparities to accrue on such a fan.

The spatial disposition of deviations from the 'ideal' surfaces can be analysed by superimposing maps of the deviations on gemorphic maps of the original fans. Such maps are shown in Figure 8 and a number of correlations between geomorphic units and highs or lows are suggested. For example on Mormon fan (Figure 8a) existing distributaries and main stem channels correspond to narrow and wider low areas respectively. It is also not surprising that active deposition is occurring on the northeastern flank, whereas slightly older inactive channels are found in the higher area to the south, suggesting a relatively recent diversion from the south flank.

Somewhat more subtle correlations are indicated on the map of Mauve Shadow fan (Figure 8b). In this case present channels locally appear to lie at greater elevations that adjacent coarser, darker and probably older deposits (see especially the arc just up-fan from the road). We might suspect that the latter areas would be prime areas for re-initiation of sedimentation in the near future; the probability of diversion into them now should be high.

The fact that some of the most recent deposition on this fan has been on high areas is consistent with our original model, according to which the most recent deposits can be found on either high areas or low areas, depending in large degree upon the length of time which has elapsed since diversion last occurred. On Mauve Shadow fan low levees separating the present channels from the older lower adjacent fan surface have prevented diversion up to now.

The surface of Mauve Shadow fan just below the road (Figure 8b) is consistently lower than the 'ideal' surface. The more numerous small channels here may reflect the influence of the road, but the broad low is believed to be a primary feature predating the road. A low in this position is precisely what we should expect to find if a segmented profile is being approximated by a smooth concave-upward curve, and is convincing evidence for segmentation in this case.

On some fans there is a marked decrease in magnitude of deviations from the 'ideal' surface in a down fan direction and greater correlation of the deviations with observed geomorphic features near the fan apex. As is the case with laboratory fans (Table IV) we would expect greater variation in surface relief near the fanhead because the entire flow must pass this position, primarily in a single channel, whereas further down fan the flow is carried in numerous smaller distributary channels. The greater and more equitably distributed drainage density in these down-fan areas is associated with a corresponding homogenization of surface relief, but the broad highs and lows remain.

Our model for deposition on fans is based primarily on observations on laboratory fans formed by fluvial processes. Debris-flow deposition is important on natural fans, and should occur in a roughly analogous manner, with sedimentation occurring preferentially in low areas. For example, a mudflow on Mauve Shadow fan in November 1976 (Figure 9), ten years after the original mapping, exploited the lower areas along the axis and north flank, and avoided the high area on the south (compare Figures 8b and 9). Debris flows build high areas in a slightly different manner, however; because they have a finite yield strength, they stop when the product of thickness and surface slope drops below a critical level for the flow in question (Hooke 1967)). Thus a high area may be formed almost instantaneously when the debris flow stops, rather than being built up gradually, perhaps over a period of years. The processes of diversion are also different for the two types of flow; diversion of a water flow from a high area is often

Figure 9. Photograph of Mauve Shadow fan taken in March 1977 and showing November 1976 mudflow deposits

accompanied by local channel cutting, but this is not the case with debris flows. For these reasons, caution should be exercised in applying the model, without further study, to natural fans on which debris flows deposit a substantial fraction of the sediment.

ACKNOWLEDGEMENTS

This study was supported by National Science Foundation Grant GA-806 and by grants from the University of Minnesota Graduate School and University of Minnesota Computer Centre. The work was done over a period of nearly a decade, and involved several students at various times. Three of these deserve especial mention for their important contributions; S. Beske-Diehl built many of the laboratory fans and J. (Koerner) Moore and T. Raster participated in both the laboratory and the field studies. We appreciate the critical reviews of W. B. Bull, M. G. Foley and S. A. Schumm.

REFERENCES

Ackers, P., and Charlton, F. G. (1970) 'Meander geometry arising from varying flows', *Journal of Hydrology*, **11**, No. 3, 230–252.
Beaty, C. B. (1963). 'Origin of alluvial fans, White Mountains, California and Nevada', *Association of American Geographers Annals*, **53**, 516–535.
Blench, T. (1951). *Hydraulics of Sediment-bearing Canals and Rivers*, Vancouver, Evans Industries Ltd.
Bull, W. B. (1964). 'Geomorphology of segmented alluvial fans in western Fresno County, California', *U.S. Geological Survey Professional Paper 352E*, 79–129.
Carlston, C. W. (1965). 'The relation of free meander geometry to stream discharge and its geomorphic implications', *American Journal of Science*, **263**, 864–885.
Conomos, T. J. (1961). 'The sedimentation of the Cottonwood-Marble Canyon alluvial fan, Death Valley, California', *unpublished Term Project*, Department of Geology, San Jose State College, Ca.
Gilbert, G. K. (1914), 'Transportation of debris by running water', *U.S. Geological Survey Professional Paper 86*, 263p.
Hooke, R. LeB. (1967). 'Processes on arid-region alluvial fans', *Journal of Geology*, **75**, 438–460.
Hooke, R. LeB. (1968a). 'Model geology: prototype and laboratory streams: discussion', *Geological Society of America Bulletin*, **79**, 391–394.
Hooke, R. LeB. (1968b). 'Steady-state relationships on arid-region alluvial fans in closed basins', *American Journal of Science*, **266**, 609–629.
Hooke, R. LeB. (1972). 'Geomorphic evidence for Late-Wisconsin and Holocene tectonic deformation, Death Valley, California', *Geological Society of America Bulletin*, **83**, 2073–2098.
Hooke R. LeB., and Rohrer, W. L. (1977). 'Relative erodibility of source area rock types, as determined from second-order variations in alluvial-fan size', *Geological Society of America Bulletin*, **88**, 1171–1182.
Hooke, R. LeB. (1975). 'Distribution of sediment and shear stress in a meander bend', *Journal of Geology*, **83**, No. 5, 543–566.
Inglis, C. G. (1949). 'The behavior and control of rivers and canals (with the aid of models), Part 1', *Central Water Power Irrigation Research Station, Poona, India, Research Publication No. 13*.
Krumbein, W. C., and Dacey, M. F. (1969). 'Markov chains and embedded Markov chains in Geology', *Mathematical Geology*, **1**, No. 1, 79–86.
Krumbein, W. C., and Graybill, F. A. (1965). *An Introduction to Statistical Models in Geology*, McGraw-Hill, Inc., 475p.
Leopold, L. B., and Maddock, T. Jr. (1953). 'The hydraulic geometry of stream channels and some physiographic implications', *U.S. Geological Survey Professional Paper 252*, 57p.
Leopold, L. B., and Wolman, M. G. (1960). 'River meanders', *Geological Society of America Bulletin*, **71**, 769–794.
Mackin, J. H. (1948). 'Concept of the graded river', *Geological Society of America Bulletin*, **59**, 463–512.
Pearce, A. J. (1976). 'Magnitude and frequency of erosion by Hortonian overland flow', *Journal of Geology*, **84**, No. 1, 65–80.
Potter, P. E., and Blakely, R. F. (1968). 'Random processes and lithologic transitions', *Journal of Geology*, **76**, No. 2, 154–170.
Price, W. E. Jr. (1976). 'A random-walk simulation model of alluvial-fan-deposition', in Merriam, D. F., (Ed.) *Random Processes in Geology*, New York, Springer-Verlag, pp. 55–62.
Rubey, W. W. (1938). 'The force required to move particles on a stream bed', *U.S. Geological Survey Professional Paper 189E*, 121–142.
Schumm, S. A. (1967). 'Meander wavelength of alluvial rivers', *Science*, **157**, 1549–1550.
Tolman, C. F. (1909). 'Erosion and deposition in the southern Arizona bolson region', *Journal of Geology*, **17**, 136–163.
Wolman, M. G., and Miller, J. P. (1960). 'Magnitude and frequency of forces in geomorphic processes', *Journal of Geology*, **68**, 54–74.
Wolman, M. G., and Leopold, L. B. (1957). 'River flood plains: some observations on their formation', *U.S. Geological Survey Professional Paper 282-C*, 87–109.

ERRATA

Page 162, Figure 7, the label on the vertical axis should read: "Frequency of occurrence."
Page 163, Figure 8, the caption should read: "Geomorphic maps. . . ."

Part III

ANCIENT ALLUVIAL FAN DEPOSITS

Editor's Comments
on Papers 13 Through 16

13 NILSEN
Old Red Sedimentation in the Buelandet-Vaerlandet Devonian District, Western Norway

14 STEEL et al.
Coarsening-upward Cycles in the Alluvium of Hornelen Basin (Devonian) Norway: Sedimentary Response to Tectonic Events

15 HEWARD
Alluvial Fan Sequence and Megasequence Models: with Examples from Westphalian D-Stephanian B Coalfields, Northern Spain

16 GLOPPEN and STEEL
The Deposits, Internal Structure and Geometry in Six Alluvial Fan-Fan Delta Bodies (Devonian-Norway)—A Study in the Significance of Bedding Sequence in Conglomerate

The interpretation of ancient stratigraphic sequences as alluvial fan deposits has developed very slowly. Before Sharp's paper on the Moncrief gravel of the Big Horn Mountains in Wyoming (Paper 1), few studies had resulted in well-defined interpretations of stratified sedimentary rocks as alluvial fan deposits. The development of the basic concepts of fan sedimentation, geometry, and stratigraphy, which began in the late 1800s, did not reach completion until the 1950s and 1960s, with the work of Blissenbach, Bull, Beaty, Denny, Bluck, and Hooke. Among the types of criteria needed to recognize ancient alluvial fan deposits were: (1) recognition and interpretation of channeled and nonchanneled depositional units, (2) understanding of debris-flow, mudflow, and streamflow processes, (3) full understanding of proximal-to-distal changes expected in maximum clast size, bedding thickness, and textural parameters such as clast roundness, and (4) techniques to determine sediment dispersal patterns, which included the understanding of the processes of formation of current-formed sedimentary structures and the development of clast fabrics. Because most of these criteria did not develop until the 1950s and 1960s, many of the earlier studies are of limited value.

However, several of the early studies of ancient sequences are of major importance. Rich (1910) described the Bishop Conglomerate of southwestern Wyoming and attempted to reconstruct both the paleoenvironment of deposition and the paleoclimate. The Bishop Conglomerate forms a widespread cap on a high, smooth-topped plateau and rests with angular unconformity on Cretaceous strata. The gravels were first thought to be of glacial origin, but Rich finally concluded that they were deposited as alluvial fans, which he termed "wash aprons." He worked out a history of fan sedimentation alternating with fan dissection by erosion.

Lawson (1913) developed his terminology for ancient alluvial fan deposits from a study of a fanglomerate near Battle Mountain in northern Nevada. This deposit, a mixture of rounded and angular clasts, rests unconformably on Paleozoic basement rocks. Lawson described lateral variations in clast size and composition of the fanglomerate and attempted a paleogeographic, paleotectonic, and paleoclimatic reconstruction for the unit. In closing, he noted the abundance of modern fan gravels in the arid southwest compared to only a few descriptions, mainly of the Triassic Newark Series, of ancient fanglomerates in the geological literature up to that time. Interestingly, he concluded that

> the failure to recognize alluvial fan formations as constituent elements of the stratigraphic column may, therefore, be explained by the supposition that the combination of bold relief and aridity was not common in the geologic past . . . the period of time extending through the Quaternary to the present is exceptional in geologic history in respect to the coexistence of these two conditions over a large portion of the continent.

Tieje (1923) studied Pennsylvanian, Permian, and Triassic redbeds exposed in the front range of Colorado that rest unconformably on Precambrian crystalline basement rocks. He concluded that the Pennsylvanian Fountain Formation, which consists of as much as 1400 m of poorly sorted, angular to subangular conglomerate and arkose, was deposited as alluvial fans on a piedmont plain along the eastern flank of ancestral Rocky Mountains. This conclusion was substantiated many years later by the work of Howard (1966), who determined that the sediment dispersal pattern for the Fountain Formation indicated a series of coalesced alluvial fan deposits.

In the 1940s, at the same time Sharp (Paper 1) was studying the Moncrief gravel in the Bighorn Mountains, increasing amounts of research on the Newark Series in the Appalachian Mountains culminated in Krynine's (1950) impressive study of Triassic sedimentation in

Editor's Comments on Papers 13 Through 16

Connecticut. He concluded that the arkosic fanglomerates that characterized these deposits had been deposited under humid and even tropical conditions. He attributed the red color of the deposits to the mature weathering in humid regions, which yields red lateritic soils. Fresh, unweathered detritus could be derived concurrently in greater amounts from the vertical erosion of deep, incised canyons that dissected rugged, uplifted source areas. Krynine thus concluded that arkosic deposits with fresh feldspars and unstable minerals could be the primary lithology in thick alluvial fan deposits of humid regions. These conclusions stimulated much discussion, because so many previous studies were of arid-region fans, and fans had been thought by many to be diagnostic of mountainous arid settings.

Concurrently with and subsequent to the many studies during the 1950s and 1960s of modern alluvial fans in arid, temperate, alpine, and Arctic regions, some of which have been reproduced herein (see Papers 2, 3, 4, 5, 6, 7, 8, and 9), workers on ancient alluvial fan deposits began to make great progress. The depositional features, geometry, stratigraphy, structure, petrography, textures, sediment dispersal patterns, and facies of alluvial fan deposits from many different areas were defined. In the western literature, studies of ancient alluvial fan deposits in the Appalachian-Caledonian chain (including the Old Red Sandstone, Catskill Formation, and New Red Sandstone) and in the western North American Cordillera predominate. I have chosen four papers for this section on ancient alluvial fan deposits that illustrate the development of major concepts regarding alluvial fan deposits. Three of the papers are on the Old Red Sandstone of western Norway, where alluvial fan sequences are remarkably well exposed, and the fourth is a general review based on detailed study of some Carboniferous basins in northern Spain.

Paper 13 presents a study of Devonian redbeds deposited along the northern margin of one of several small Old Red basins exposed in western Norway. The redbeds rest unconformably on Cambro-Silurian strata and dip homoclinally southward; they consist in ascending order of (1) a basal monomict sedimentary breccia that is poorly sorted, poorly stratified, and contains angular clasts that are poorly oriented; (2) a middle polymict conglomerate that is locally well stratified and well sorted, and contains rounded clasts that are well oriented and imbricated; and (3) an upper arkosic sandstone that is well sorted, parallel stratified, and trough cross-stratified. Using Walther's Law and compositional, textural, and paleocurrent data, Nilsen reconstructs the paleogeographic setting of the basin within the framework of humid-climate alluvial-fan sedimentation adjacent to a probably syndepositionally active basin-margin fault.

Editor's Comments on Papers 13 Through 16

Paper 14 analyzes cyclic sedimentation in the Hornelen Old Red basin of western Norway. The techniques developed in this paper represent several important advances over those utilized by Nilsen (Paper 13). In addition to studying the clast size, paleocurrents, and depositional features, Steel and his colleagues, through the use of detailed measured sections, analyze and interpret cyclicity within the flanking alluvial fan and axial alluvial floodplain deposits. The spectacular outcrops of the Hornelen basin permit major coarsening-upward cycles that are about 100 m thick to be traced across the entire basin. These cycles are deduced by Steel et al. to have been produced by major events of basin subsidence related to strike-slip faulting that caused migration of the locus of sedimentation. The recognition of major cyclic events in ancient alluvial fan deposits makes possible the closer understanding of the relations between tectonism in the source area and sedimentation in the adjoining basin. Steel worked on ancient alluvial fan deposits in Britain for his Ph.D. thesis, under the supervision of Brian Bluck at Glasgow University. After teaching at the University of Bergen for many years, he has recently taken a position with Norsk Hydro in Bergen. His coauthors were his students at the University of Bergen in the 1970s.

Paper 15 develops in greater detail the analysis of cyclic sedimentation in alluvial fan deposits. Heward examines the progress in the study of modern and ancient submarine fan deposits, which have analogous morphologic and geometric characteristics, and applies knowledge gained in the analysis of cyclicity in these deposits to alluvial fan deposits. Heward accepts the postulate that deposits of channels generally form thinning- and fining-upward cycles and that deposits of the nonchannelized or less channelized lower parts of submarine fans form thickening- and coarsening-upward cycles. He explores in considerable detail the variations in character of depositional cycles under varying conditions of uplift and subsidence, and then applies these parameters to the study of Carboniferous alluvial fans deposited in small basins in northern Spain whose depositional setting was controlled by strike-slip faulting. Heward draws many useful analogies between modern and ancient alluvial fan deposits and ancient submarine fan deposits. Heward's paper is based on a Ph.D. thesis at Oxford University; he presently works as an exploration geologist in Oman.

Paper 16 is a second, more detailed study of alluvial fan deposits in the Hornelen basin of western Norway. In this paper, debris-flow, shoreline-modified debris-flow, sheetflood, channelized streamflow, and sieve-deposited fanglomerates are carefully distinguished on the basis of fabric, bed thickness, lateral continuity, internal grading, and geometry. The prograding geometry, cyclicity, and stratigraphy of several fans are clearly outlined. Smaller, steeper fans are dominated by debris-flow deposits, and larger, more gently sloping fans are dominated by streamflow and sheetflood deposits. The cyclicity developed within the individual alluvial fan bodies is related to both autocyclic and allocyclic processes. This paper essentially represents a modern, state-of-the-art attempt to apply almost all knowledge previously gained from the study of modern and ancient alluvial fan deposits to a thorough, detailed analysis of an ancient system of alluvial fan deposits. Tor Gloppen was a student at the University of Bergen under Ron Steel and presently works for Statoil in Stavanger, Norway.

REFERENCES

Howard, J. D., 1966, Patterns of Sediment Dispersal in the Fountain Formation of Colorado, *Mountain Geologist* **3:**147–169.

Krynine, P. D., 1950, Petrology, Stratigraphy, and Origin of Triassic Sedimentary Rocks of Connecticut, *Connecticut Geol. and Nat. History Survey Bull. No. 73*, 247 p.

Lawson, A. C., 1913, The Petrographic Designation of Alluvial Fan Formations, *California Univ. Pubs. Geol. Sci.* **7:**325–334.

Rich, J. L., 1910, The Physiography of the Bishop Conglomerate, Southwestern Wyoming, *Jour. Geology* **18:**601–632.

Tieje, A. J., 1923, The Red Beds of the Front Range of Colorado: A Study in Sedimentation, *Jour. Geology* **31:**192–207.

13

Copyright © 1969 by Elsevier Scientific Publishing Company
Reprinted from *Sed. Geology* **3**:35-57 (1969)

OLD RED SEDIMENTATION IN THE BUELANDET-VAERLANDET DEVONIAN DISTRICT, WESTERN NORWAY

TOR H. NILSEN

U.S. Army Map Service, Washington, D.C. (U.S.A.)

(Received March 26, 1968)
(Resubmitted August 27, 1968)

SUMMARY

A preserved maximum thickness of 3,400 m of Devonian breccia, conglomerate, and sandstone of the Buelandet-Vaerlandet Formation crop out in the Buelandet-Vaerlandet district. The sedimentary rocks rest unconformably on Cambro–Silurian greenschist. Fragments of *Psilophyton* sp. found in the upper part of the stratigraphic sequence suggest an Early Devonian or early Middle Devonian age. The strictly continental sediments were deposited in a structurally formed half-graben by streams draining a complex crystalline source area to the north. This depositional basin was one of six separate Devonian basins located in the area of the former eugeosynclinal portion of the Caledonian geosyncline. The initial tectonism, resulting in the formation of Devonian grabens, was later followed by folding and faulting of the sedimentary basins during the Svalbardian disturbance. The tectono-sedimentologic environment of the Buelandet-Vaerlandet district is thought to have been similar to that of the Triassic Newark basins of the North American Appalachian Mountains.

The oldest member of the here-named Buelandet-Vaerlandet Formation, the Melvaer Breccia, consists of angular, chaotically arranged greenschist fragments deposited as mudflows, sheetfloods and talus accumulations on the surfaces of alluvial fans. The intermediate Vaeroy Conglomerate consists of polymict alluvial fanglomerates deposited by braided distributaries of major streams that cut deep canyons into the northerly provenance area. The youngest unit is the Sörlandet Sandstone Member, a green arkosic deposit of larger braided streams, with local gravel bars preserved. Textural, fabric and paleocurrent data indicate a facies gradation southward along the paleoslope from breccia to sandstone. Tectonic uplift of the source area contemporaneous with fan sedimentation is suggested by the thickness and coarseness of fanglomerates. Rapid deposition in upperflow regime conditions, and a temperate or humid climate are indicated.

INTRODUCTION

In the light of the plethora of recent studies on the "Old Red Continent" in the United Kingdom, Spitsbergen and Greenland, it becomes imperative that the distinctive "Old Red" sedimentary basins of Norway be restudied. The Devonian sedimentary basins of Norway contain some unusual sedimentary suites worthy of intensive investigation. The Hornelen district, for example, consisting of 25,000 m of repetitively similar sandstones deposited in a relatively small basin, is an outstanding study area (BRYHNI, 1964). The extremely thick and coarse conglomerates of the Solund district are likewise distinctive deposits (NILSEN, 1967, 1968). It is necessary that the evolution of the unusual Devonian paleogeography of Norway be understood and presented to students of the "Old Red" working in other areas. Only in this manner we can approach a systematic understanding of the "Old Red", as well as uncover those significant problems requiring further study and development.

The Buelandet (or Bulandet)-Vaerlandet district is one of six distinct areas of Devonian rocks in the Nordfjord-Sognefjord region of the western coast of Norway. Additional Devonian rocks in Norway are found near Trondheimsfjord and Röragen (Fig.1). The district is located approximately 110 km north of Bergen, and is geographically divided into three parts: (*1*) Vaerlandet (northeast), consisting of the two largest islands, Vaeroy and Melvaer, plus scattered smaller nearby islets; (*2*) Buelandet (west), consisting of numerous small islands and islets west of Olsund; and (*3*) Sörlandet (southeast), some scattered islets south of Vaeroy (Fig.2).

The islands and islets are typically low-lying and flat, being characteristic of the famous "strandflat" topography of the western coast of Norway. The "strandflat" is locally punctuated, however, by several prominent steep rounded hills that protrude above it. Vigourous marine erosion has provided excellent exposures throughout the area; only the remoteness of some of the islands, and characteristically strong seas limit study locally. As a result of protracted inclement weather, the writer did not have the opportunity of landing in Sörlandet.

KOLDERUP's (1916) classic study of the Buelandet-Vaerlandet district (one of six papers covering the Devonian districts) represents the only published detailed geological investigation of the area. The present writer spent the summers of 1964 and 1965 doing field work in the larger Solund district to the south. At the end of the second summer, at the suggestion of Prof. N-H. Kolderup of the University of Bergen, the writer made a short study of the Buelandet-Vaerlandet Devonian suite. The results of the study are presented herein. The writer's work emphasizes the paleogeography and tectono-sedimentologic history of the district, rather than details of petrography and mineralogy, which are more than adequately discussed by KOLDERUP (1916).

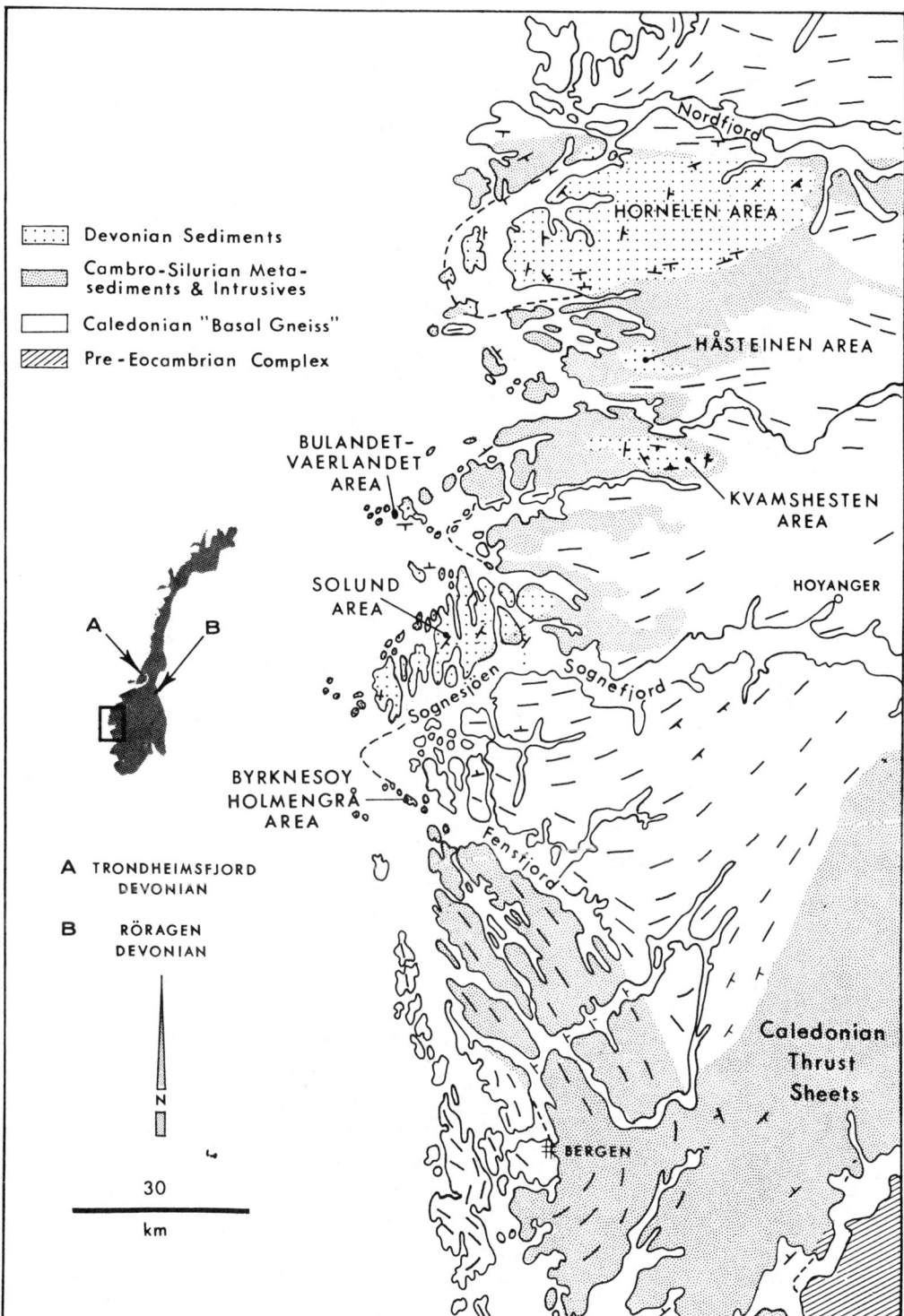

Fig.1. Geologic map of western Norway. (After Holtedahl, 1960.)

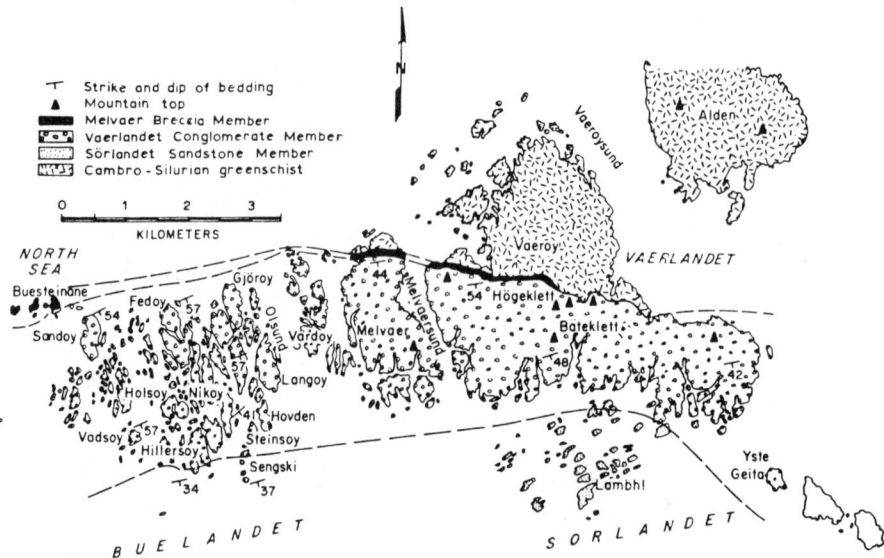

Fig.2. Geologic map of the Buelandet-Vaerlandet district.

STRUCTURAL GEOLOGY

The Devonian strata of the district strike approximately east–west and dip south-southeast, with the younger formations toward the south (Fig.2). The average dip angle is about 45°, being steeper to the north and gentler to the south. No significant folding or faulting of the homocline was noted; local, minor changes in strike and dip are thought to result from variations of primary depositional dip. The irregularity of the mapped basal contact results from: (*1*) present topographic configuration; and (*2*) the non-planar nature of the angular unconformity between Devonian sedimentary rocks and older rocks.

STRATIGRAPHY

Introduction

The Devonian rocks of Buelandet-Vaerlandet unconformably overlie a Cambro–Silurian greenschist that crops out in the northeastern part of the district. A relatively thin, basal sedimentary breccia composed of angular fragments of the underlying greenschist initiates the Devonian sequence, and grades abruptly upward into a thick polymict conglomerate unit which in turn passes upward into green pebbly arkosic sandstone. In accordance with the "Code of Stratigraphic Nomenclature for Norway" (1960), the writer proposes the name *Buelandet-Vaerlandet Formation* as the lithostratigraphic formational name for the Devonian strata of the Buelandet-Vaerlandet district, as defined above. It is further proposed that the three sub-units mentioned above be distinguished as members, i.e., (*1*) the

Melvaer Breccia Member; (2) the *Vaerlandet Conglomerate Member*; and (3) the *Sörlandet Sandstone Member*.

These names are selected from the areas where the respective members are best exposed. It is deemed desirable by the writer that the Devonian strata in each of the Norwegian districts be given formation status, for they are distinct and should be classified separately. Use of the term "Old Red Sandstone" is deemed improper except in a broad sense as an informal system name, because the true Old Red Sandstone of Great Britain belongs to a distinctly different depositional environment and bears only tenuous lithostratigraphic similarity to the "Old Red" of Norway.

The estimated thickness of the preserved stratigraphic section of the Buelander-Vaerlandet Formation is 2,450 m in Buelandet, and 3,400 m in Vaeroy-Sörlandet. It is impossible to estimate the thickness of the stratigraphic section prior to folding, uplift and erosion.

Cambro–Silurian greenschist

The underlying basement rock is a massive, dark-green to black metamorphosed volcanic rock. Little evidence of layering or foliation is present; however, to the north, on Alden, the rock is distinctly schistose (KOLDERUP, 1916). In all outcrops the greenschist is thoroughly laced by calcite-filled joints. It is presumed to be Cambro–Silurian in age by lithostratigraphic similarity to other nearby units.

Melvaer Breccia Member

This unit is best exposed and thickest at the northern tip of Melvaer, where 80–100 m of breccia are found (KOLDERUP, 1916). It thins gradually eastward and westward, being virtually absent in eastern Vaeroy. The abrupt erosional contact between the breccia and greenschist is irregular and uneven, with breccia locally filling deep crevices in the underlying greenschist. The breccia is relatively fine grained, with fragments averaging 5–10 cm in maximum length; some larger fragments of cobble and boulder size are found, however. Elongate, angular fragments predominate, with little evidence of abrasion of particles. The breccia fragments "float" in a brownish, medium- to coarse-grained, arkosic sandstone with abundant clay-sized matrix material. The dispersed framework is further characterized by a random orientation of elongate and flat breccia fragments — a chaotic, anisotropic fabric.

The breccia generally lacks stratification, and has only rare, widely spaced irregular bedding partings indicative of sedimentary layering. However, local horizons in the breccia contain partly rounded fragments within well-defined strata; long axes and flat surfaces of fragments are typically parallel with the stratification at these localities. These strata represent a modification of the characteristic depositional processes responsible for the lack of sedimentary structures in the breccia. In no locality, however, does the breccia contain fragments other than greenschist.

Fig.3. Contact between the Melvaer Breccia Member (right) and the Vaeroy Conglomerate Member (left), northern Vaeroy. The dashed line represents the contact; note sharpness of both contact and change in clast lithologies. Stratigraphic top is to the left.

KOLDERUP (1916) presents stratigraphic sections showing the upward gradation of the breccia into the Vaeroy Conglomerate Member. The sections indicate that repetitive intercalations of breccia and polymict conglomerate mark the stratigraphic boundary; Fig.3 shows a typical sharp contact, indicating the differences in appearance between the two members.

Vaeroy Conglomerate Member

The Vaeroy Conglomerate is best exposed on Vaeroy to the south of Högeklett. It characteristically is coarse (cobble- and boulder-sized clasts predominate) and very strongly lithified (Fig.4). The Vaeroy Conglomerate is polymict, and changes little throughout the area except in terms of clast size, texture and composition. It comprises about 49–68% of the preserved stratigraphic thickness of the Buelandet-Vaerlandet Formation, and attains a maximum thickness of approximately 2,260 m.

The lower portions of the Vaeroy Conglomerate contain abundant greenschist clasts that are considerably more angular than the more abundant polymict clasts. The greenschist fragments are generally restricted in position, however, to the lowest 50 m of the unit. Interstratified sandstones are found in increasing

Fig.4. Typical appearance of the Vaeroy Conglomerate Member in southern Vaeroy. Note variety of clast lithologies, particularly gneisses.

abundance upward in the sequence (to the south), where the Vaeroy Conglomerate grades into the overlying Sörlandet Sandstone. The transition takes place within a relatively short stratigraphic interval, with the boundary arbitrarily placed where sandstone becomes the dominant lithology. An exact stratigraphic boundary cannot be determined.

The matrix material of the Vaeroy Conglomerate consists of finer, more angular conglomerate and fine to coarse arkosic sand. Only a very small percentage of the matrix is silt- and clay-sized material. A crystalline mosaic of calcite cements the rock and fills most interstices. The conglomerate fabric is "closed framework", with clasts having abundant points of contact; finer conglomerate and sand matrix material is interstitial between clasts. This fabric is distinct from the "dispersed framework" of the Melvaer Breccia (Fig.3).

The dominant clast lithologies in the Vaeroy Conglomerate, found throughout its outcrop area, are (Table I) felsic gneisses (22–36%), white feldspathic quartzite (18–27%), felsic intrusives (13–27%), mafic intrusives (5–17%), garnetiferous amphibolite (6–14%), and granite pegmatite (1–9%). These date are based on random counts of 100 clasts at each of ten evenly distributed localities throughout the area of outcrop. Smaller amounts of the following clast lithologies were noted: vein quartz (0–5%), felsic extrusives (0–3%), mafic extrusives (0–7%), intermediate

TABLE I

SUMMARY OF CLAST DISTRIBUTION DATA

Clast lithology	Localities[1]										Average
	1	2	3	4	5	6	7	8	9	10	
White feldspathic quartzite	26	27	17	20	18	18	8	20	14	18	18.6
Dark-coloured quartzite	0	0	0	0	0	0	0	0	4	0	0.4
Metaconglomerate	0	0	0	0	0	0	3	1	4	0	0.8
Vein quartz	0	0	0	1	0	2	0	0	5	2	1.0
Felsic non-quartzose gneiss	9	11	15	8	4	7	4	4	8	2	7.2
Felsic quartzose gneiss	23	25	21	21	22	22	23	21	14	26	21.8
Quartz-mica schist	0	0	1	2	0	0	0	0	0	0	0.3
Amphibolite	10	8	7	8	8	12	12	14	8	6	9.3
Intermediate gneiss	0	0	0	0	2	4	0	0	0	0	0.6
Granite pegmatite	1	5	5	4	6	5	1	7	5	9	4.8
Felsic igneous intrusive	13	16	15	17	21	16	27	22	21	18	18.6
Felsic igneous extrusive	1	2	2	0	3	2	0	0	0	2	1.2
Intermediate igneous intrusive	1	1	2	2	2	2	1	0	2	0	1.3
Mafic igneous intrusive	14	5	15	17	8	10	14	9	15	11	11.8
Mafic igneous extrusive	2	0	0	0	6	0	7	2	0	6	2.3
Totals	100	100	100	100	100	100	100	100	100	100	100.0

[1] Localities are numbered consecutively from west to east across the Buelandet-Vaerlandet district (Fig.2). One hundred clasts were counted randomly at each locality.
Localities: *1* = Northern Sandoy; *2* = Vadsoy; *3* = Fedoy; *4* = Nikoy; *5* = Northern Melvaer; *6* = Southern Melvaer; *7* = Northwestern Vaeroy; *8* = Southwestern Vaeroy; *9* = Near Börekletten; *10* = Southeastern Vaeroy.

intrusives (0–2%), quartz–muscovite schist (0–2%), dark-coloured quartzite (0–4%), and metaconglomerate (0–4%).

Contoured maps of the individual clast distribution patterns indicate the following relationships: (*1*) felsic gneisses are uniformly distributed; (*2*) white feldspathic quartzite clasts decrease in abundance toward the northeast; (*3*) felsic intrusives increase in abundance toward the northeast; (*4*) mafic intrusives are irregularly distributed; (*5*) amphibolite clasts are most abundant in the central portions of the area; and (*6*) granite pegmatites are most abundant in the eastern portions of the area. Less abundant clasts have irregular distribution patterns. The source area for this material was a crystalline complex of meta-igneous, meta-sedimentary and igneous rock types with smaller amounts of essentially non-metamorphosed felsic and mafic extrusives or near-surface intrusives.

Sörlandet Sandstone Member

This unit crops out in Sörlandet and in the southernmost islands of Buelandet (Fig.2). The sandstone is arkosic, typically light green in colour, and characterized

Fig.5. Interbedded Vaeroy-type conglomerate and Sörlandet-type sandstone in the southern portion of Hillersoy, south Buelandet.

by several types of stratification. The Sörlandet Member is most commonly coarse to medium grained, but does locally contain thinly stratified to laminated siltstone and fine-grained sandstone. Scattered pebbles, with compositions similar to those of clasts found in the underlying Vaeroy Conglomerate, are common (Fig.5). The Sörlandet Sandstone is fairly well sorted, with a crystalline mosaic of calcite cement between grains. The unit is finer grained toward the south; this may reflect both increasing distance of sediment transport from the source areas, and progressively finer grain sizes found in the younger portions of the Buelandet-Vaerlandet Formation.

The individual sand grains are angular, with the dominant compositions being lithic fragments, quartz, plagioclase, orthoclase, and perthite. KOLDERUP (1916) also noted the presence of epidote, sericite, chlorite, calcite, magnetite, zircon, and titanite. No statistical determinations of variations in sandstone composition were made by the writer. The clastic sand detritus is typically fresh, showing no evidence of chemical paleoweathering.

Using the sandstone classification system of Gilbert (WILLIAMS et al., 1954), the Sörlandet Sandstone is a lithic arenite, i.e., a sandstone with low matrix content and abundant lithic fragments. The finer grained varieties contain higher percentages of quartz and feldspar, with accordingly lower percentages of lithic fragments. The origin of the green color of the rocks was not determined.

On the island of Lamholmen in Sörlandet, KOLDERUP (1916) found abundant remains of *Psilophyton* sp. and other plant fragments. NATHORST (1915), KOLDERUP (1915), and KIAER (1918) discussed correlation of the identified specimens with other areas of western Europe and Norway. The flora were thought to be from the Early Devonian or early Middle Devonian, which is distinctly younger than that attributed to the floral and fish remains of the Hornelen and Kvamshesten areas, thought to be late Middle Devonian (JARVIK, 1949). However, in the light of recent studies of Devonian flora, and *Psilophyton* in particular (HUEBER, 1967), the specimens from Lamholmen should receive additional study to clarify the biostratigraphic position of the Buelandet-Vaerlandet Formation.

SEDIMENTOLOGY

Sedimentary structures

The general lack of stratification, the dispersed framework and abundance of clay-sized matrix material of the Melvaer Breccia suggests deposition by continental sheetflood–mudflow processes.

The Vaeroy Conglomerate does not contain abundant bedding structures– it consists of large thicknesses of rather homogeneous closed framework conglomerate. Stratification is generally lacking, with bedding surfaces rare, widely spaced, irregular and discontinuous. Those bedding surfaces that are preserved in the Vaeroy Conglomerate must represent extended periods of non-deposition, perhaps intraformational diastems or paraconformities. Locally the presence of thin sandstone lenses permits the strike and dip of the unit to be determined. The Vaeroy Member is thought to represent deposition by running water in the upper-flow regime, with fine, clay-sized material winnowed out and deposited further down the paleoslope.

The thin sandstone lenses within the Vaeroy Conglomerate are less than 0.5 m in thickness, but extend laterally as much as 20 or 30 m in width and length. They wedge out in all directions on the surface of accumulation, but tend to be preferentially elongated north–south. Stratification within the lenses is typically flat-stratification or very low-inclination tabular or low-plunging solitary trough cross-stratification. The strata are thin, averaging 1–3 cm in thickness. The cross-strata have amplitudes from 3 to 25 cm with very long lengths (up to 5 or 10 m) in relation to their amplitudes. It is impossible to make paleocurrent determinations from most of these sandstone lenses because of their low inclination angles. The bases of the lenses are conformable on sandy conglomerate or pebbly sandstone, and have the upper surfaces typically truncated unconformably by conglomerate that may be very coarse.

The nature of the stratification in the sandstone lenses suggests deposition in the upper part of the lower-flow regime and in the transition zone into the upper-flow regime. The sandstone beds represent periods or areas on the depositional surface of less intense current conditions and transport of clastic debris. The elongate sandstone lenses are thought to resemble swale-fill deposits behind gravel bars in areas of braided stream deposition. Under such conditions the sand deposits would tend to become elongate downstream until covered again by gravel flooding or shifting in stream patterns.

The Sörlandet Sandstone represents a further change in depositional conditions. The zone of interstratified sandstone and conglomerate contains irregularly bedded layers of conglomerate up to 1 m in thickness, alternating with flat-stratified or trough-stratified sandstones that gradually increase in thickness toward the south. The flat-stratified sandstones are most common, and contain primary current lineation locally. The trough cross-strata have extremely low plunge angles, and are characteristically solitary in occurrence. However, as the sandstone increases in abundance upward in the section, the trough cross-strata become more festoon in nature, with low-amplitude troughs dominant. Primary current lineation is also present in trough cross-strata and is parallel with the plunge directions of the solitary troughs. The primary current lineation consists of low linear parallel ridges up to 3 mm in height, 1 m or more in length, and 1–5 cm apart.

The Sörlandet Sandstone contains primarily low-amplitude festoon trough cross-strata, with lesser amounts of tabular cross-strata and a minimum of flat-stratification. The plunge angles are very low, such that it is almost impossible to use the structures for paleocurrent determinations. Amplitudes and inclination angles for those tabular cross-strata utilized for paleocurrent determinations are summarized in Table II. The values are rather low in relation to published values from other sandstones. It must also be noted that the mean inclination angles and

TABLE II

SUMMARY OF TABULAR CROSS-STRATIFICATION PALEOCURRENT DATA

Locality[1]	N	Amp.	Inclin.	Curr.	L %	S.D.	0.95 c.i.
Vadsoy-Nikoy	6	11.0	12.8	193	86	46	50
Hillersoy	5	9.8	12.4	207	82	49	54
Middle Sengskjaer	6	5.2	19.3	76	43	76	88
Steinsoy	13	7.6	15.8	140	45	74	44
Totals	30	8.5	15.3	169	35	81	30

[1] Localities are listed above consecutively from west to east across the Buelandet-Vaerlandet district (Fig.2), with map representation in Fig.12.

Abbreviations used: N = number of measurements per locality; *Amp.* = mean amplitude of cross-strata in cm; *Inclin.* = mean inclination angle of cross-strata in degrees; *Curr.* = mean current direction as determined by vector method (CURRAY, 1956); $L\%$ = percentage value of L, or vector consistency (CURRAY, 1956); *S.D.* = standard deviation of mean current direction, calculated mathematically; *0.95 c.i.* = 95% confidence interval for mean current direction, calculated mathematically.

amplitudes for *all* tabular cross-strata is probably considerably lower, because only those cross-strata with large enough inclination angles to be measured easily are included in Table II.

Irregularly spaced solitary pebbles or accumulations of pebbles are found in the Sörlandet Sandstone. These pebble accumulations are generally elongate in a north–south direction. Most commonly they are one pebble thick, but may locally be several pebbles thick (up to 30 cm). The Sandstone Member is thought to have been stream-deposited, with the gravel lenses representing preserved gravel bars formed by stream currents segregating the coarser debris into elongate deposits parallel to current flow.

Texture

The Buelandet-Vaerlandet Formation decreases regularly in maximum clast size toward the south, or upward in the stratigraphic section (excluding the Melvaer Breccia). The contoured map of maximum clast size (Fig.6) is based on measurements of the ten largest diameters of clasts at each of nineteen randomly selected localities. The largest boulders in the Vaeroy Conglomerate are almost 2 m in longest diameter; the maximum clast size in the Sörlandet Sandstone in southern Buelandet is 30–40 cm. The standard deviations of the means of the ten largest clasts at each locality are proportional to the mean itself, i.e., the coarser conglom-

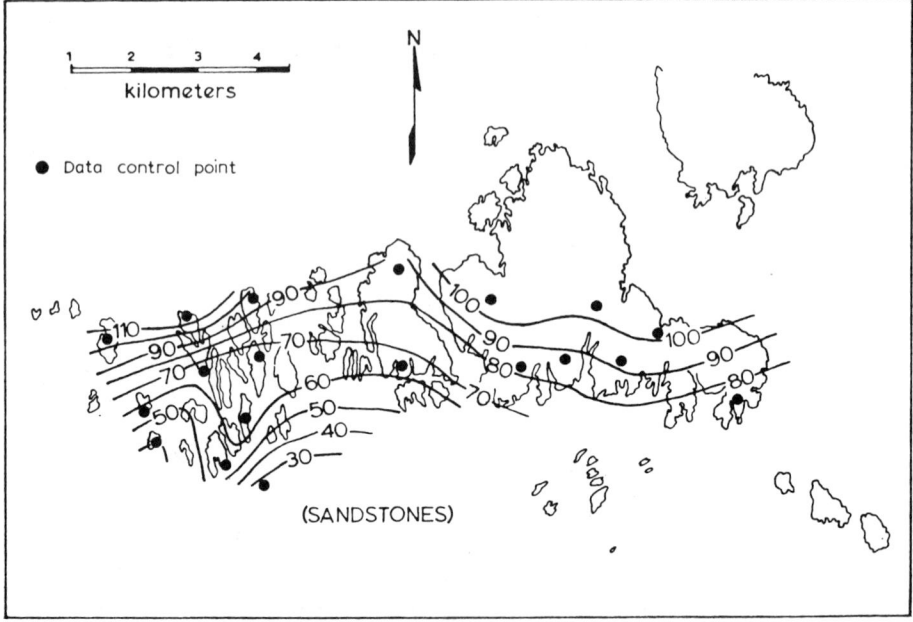

Fig.6. Maximum clast size map of the Buelandet-Vaerlandet Formation. The data represent the mean values of the ten longest diameters of boulders, cobbles or pebbles at each locality. The contour interval is 10 cm.

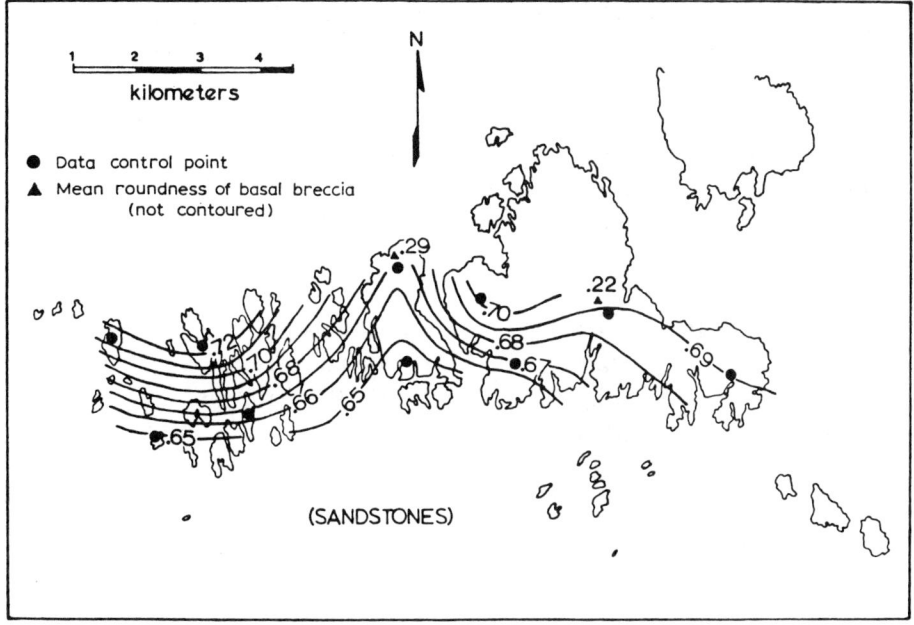

Fig.7. Clast roundness map of the Buelandet-Vaerlandet Formation. The data represent the mean roundness values of twenty-five randomly selected clasts, measured on bedding surfaces using the comparative charts of KRUMBEIN and SLOSS (1958, fig.4–9). The contour interval is 0.01 roundness unit.

erates have larger variations in size of maximum clasts than the finer conglomerates. The spread in standard deviations ranges from 2.1 to 14.9 cm. The east–west trends of the contour lines preclude sediment transport from east or west, because this direction of transport would cause the presence of at least some north–south trending contours in laterally equivalent strata.

Abrasion of clasts resulted in variations of roundness and sphericity in the Buelandet-Vaerlandet Formation. The two-dimensional visual comparison charts of KRUMBEIN and SLOSS (1958) were used to measure roundness and sphericity; 25 randomly selected clasts (larger than 5 cm in longest dimension) were measured at twelve scattered localities. The roundness contour map (Fig.7) indicates a large difference in mean roundness between the Melvaer Breccia (approximately 0.25 roundness unit) and the Vaeroy Conglomerate (approximately 0.68 roundness unit). In addition, the mean standard deviation for the Melvear Breccia mean roundness values is 1.78 roundness units, compared to 0.95 for the Vaeroy Conglomerate. The greater variation of roundness values in the monomict Melvaer Breccia reflects the mixture of clasts with little or no rounding (0.1–0.2 roundness unit) and a smaller amount of moderately rounded clasts (0.4–0.5 roundness unit). The polymict Vaeroy Conglomerate, on the other hand, contains wholly well-rounded clasts with roundness values consistently between 0.6 and 0.8 roundness unit.

The progressive decrease in roundness toward the south is difficult to evaluate in terms of sediment transport, because the strata in the exposed section from north to south are of progressively younger age. The writer believes that the roundness decrease is primarily a function of decreasing grain size upward stratigraphically, i.e., roundness is directly proportional to grain size. This relationship has been demonstrated in other areas, as well as experimentally. The fact that the contour patterns trend east–west, however, does indicate either a northerly or southerly transport direction.

Clast sphericity was measured in a manner identical to that used for clast roundness. A contoured map of mean sphericity indicates substantially lower values for the Melvear Breccia than the Vaeroy Conglomerate (Fig.8). This results from the inherited schistosity of all the Melvear Breccia clasts, the greenschist fragments being consistently platy rather than equidimensional or elongate. The mean sphericity of the Vaeroy Conglomerate increases steadily toward the south. This change is probably related to decreasing clast size toward the south. The standard deviations of the mean values of sphericity of the Breccia fragments is greater than that of the Conglomerate fragments. This relationship reflects the greater uniformity of clast shape within the well-abraded polymict Conglomerate. The Breccia fragments present varied shapes to the viewer because of the chaotic orientation of the fragments.

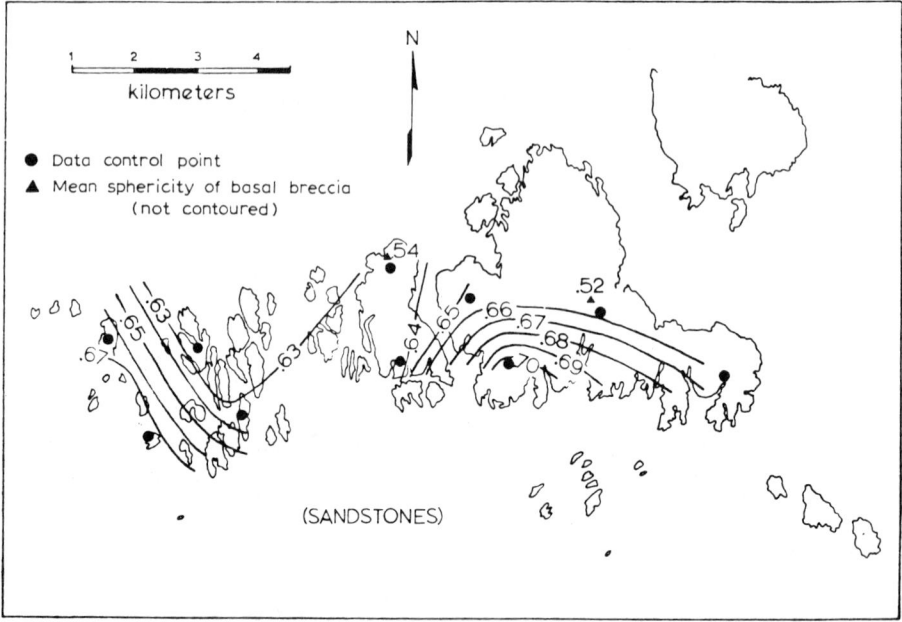

Fig.8. Clast sphericity map of the Buelandet-Vaerlandet Formation. The data represent the mean sphericity values of twenty-five randomly selected clasts, measured on bedding surfaces using the comparative charts of KRUMBEIN and SLOSS (1958, fig.4–9). The contour interval is 0.01 sphericity unit.

The data indicate an inverse relationship between mean clast roundness and mean clast sphericity. The causes of this phenomenon in the Buelandet-Vaerlandet Formation are not clear, but must be related to the abrasive processes taking place during clast transport.

Sedimentary fabric

The well-defined fabric of the Vaeroy Conglomerate was examined at several localities. Rose diagrams of the orientation of elongate clasts (elongation ratios greater than 2/1) on bedding surfaces indicate a strong preferred orientation of long axes (Fig.9). Fifty clasts were randomly selected for measurement at four localities, with plunge directions later rotated to a horizontal surface on a stereographic net. The rose diagrams have a primary north–south maximum and a secondary east–west maximum. The preferred fabric reflects strong, persistent, unidirectional current flow with a north–south orientation.

The numerals at the ends of some of the rose diagrams of Fig.9 indicate the number of elongate tapered clasts having their blunt ends facing northward or southward, respectively. The preferential orientation of blunt ends toward the north, with twenty-five clasts randomly examined at three localities, suggests current flow from north to south.

Clast imbrication was studied in detail at Hillersoy and northwestern Vaeroy. The imbrication was measured on vertical joint surfaces parallel to the north–south direction of preferred orientation of elongate clasts on bedding surfaces. Sandstone lenses in the Vaeroy Conglomerate permitted accurate determination

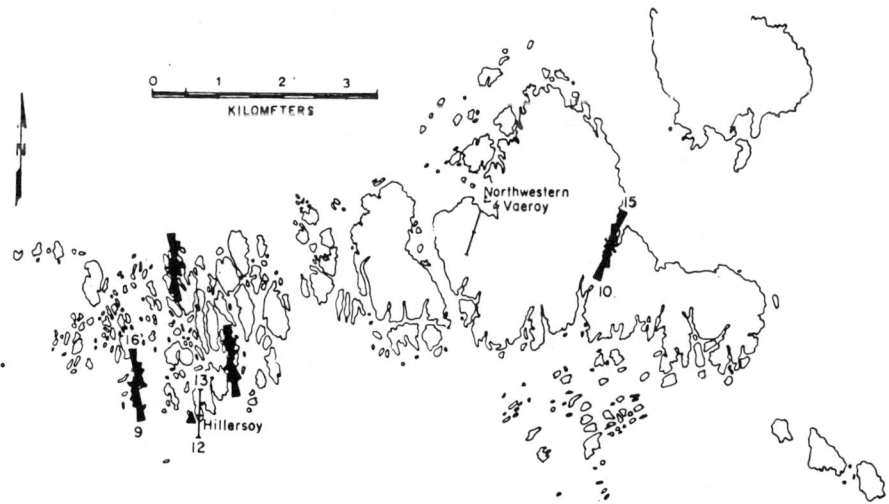

Fig.9. Map showing the orientation of the clast long axes on bedding surfaces of the Buelandet-Vaerlandet Formation. A total of fifty elongate clasts were measured at each locality for long-axis orientation; twenty-five tapered elongate clasts were measured at three localities for blunt-end orientation.

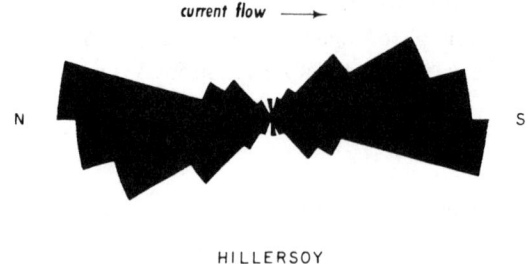

HILLERSOY

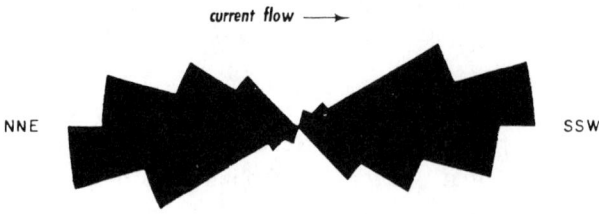

NORTHWESTERN VAEROY

Fig.10. Clast imbrication plots, Vaeroy Conglomerate Member, Hillersoy and northwestern Vaeroy. Per locality fifty randomly selected clasts were measured; the bedding surface is represented by the horizontal.

of bedding orientation at these two localities; the general lack of bedding indicators elsewhere prohibited additional two-dimensional imbrication studies. It must be emphasized that the imbrication is a statistical one, not always clearly visible in the field; in fact, one rarely sees imbricated clasts resting directly upon one another. However, unbiased measurements clearly indicate the presence of the imbrication. The strike directions of the vertical joints, along which the imbrication measurements were made, are indicated by solid lines at Hillersoy and northwestern Vaeroy in Fig.9. Rose diagram plots of the apparent dip directions of clasts presenting elongate cross-sectional areas on the joint surfaces show a statistically preferred imbrication toward the south (Fig.10). Fifty clasts were randomly selected for measurement at each locality. Thus, the imbrication data available also suggest current flow from north to south.

The darkened triangle near Hillersoy in Fig.9 shows the location of some three-dimensional determinations of conglomerate fabric. The measurements were made to check the fabric determinations from the various two-dimensional studies described above. The Vaeroy Conglomerate is locally not thoroughly lithified, permitting extraction of clasts from the matrix and cement. Fortunately, at Hillersoy, the presence of sandstones permitted determination of bedding orientation. The three-dimensional orientation with respect to bedding of 25 flat disc-shaped

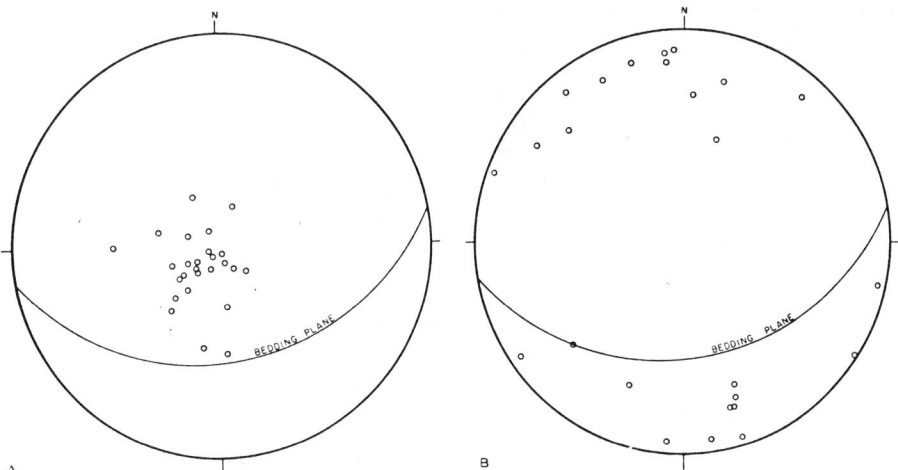

Fig.11. Stereographic projections of the three-dimensional orientations of flat and elongate clasts in the Vaeroy Conglomerate Member, Hillersoy, southern Buelandet. A. Flattened clasts (poles to maximum projection area of clasts). B. Elongate clasts (plunge directions of long axes of clasts).

All plots are on Lower Hemisphere stereonet, with the inclined surface representing the strike and dip of the bedding. The plots are based on twenty-five measurements, and indicate the statistically preferred imbrication of flattened clasts toward the south, and the statistically preferred alignment of elongate clasts in a north–south direction, plunging primarily toward the northwest.

and 25 elongate rod-shaped clasts was determined, using random sampling procedures. The stereographic projections are presented in Fig.11.

The plot of poles to the maximum projection areas of the flattened or disc-shaped clasts indicates a very strong imbrication toward the south (Fig.11A). The plot of plunge orientations of elongate or rod-shaped clasts yields a concentration of plunge directions slightly west of north, reflecting imbrication toward the south and orientation of long axes primarily in a north–south direction (Fig.11B). These data conform very closely with the two-dimensional fabric determinations, and a current flow toward the south.

All the fabric data from the Vaeroy Conglomerate suggest transport of detritus from north to south. Brief micrscopic study of sandstone fabric from oriented thin sections indicates that the sandstone bodies have basically similar fabrics to those of the conglomerates.

PALEOCURRENTS

Thirty paleocurrent measurements from tabular cross-strata were made in the Buelandet-Vaerlandet Formation. The measurements were from the Sörlandet Sandstone in southern Buelandet (Fig.12). The data are summarized in Table II. The mean paleocurrent direction is 169°, with a standard deviation of 81°. No

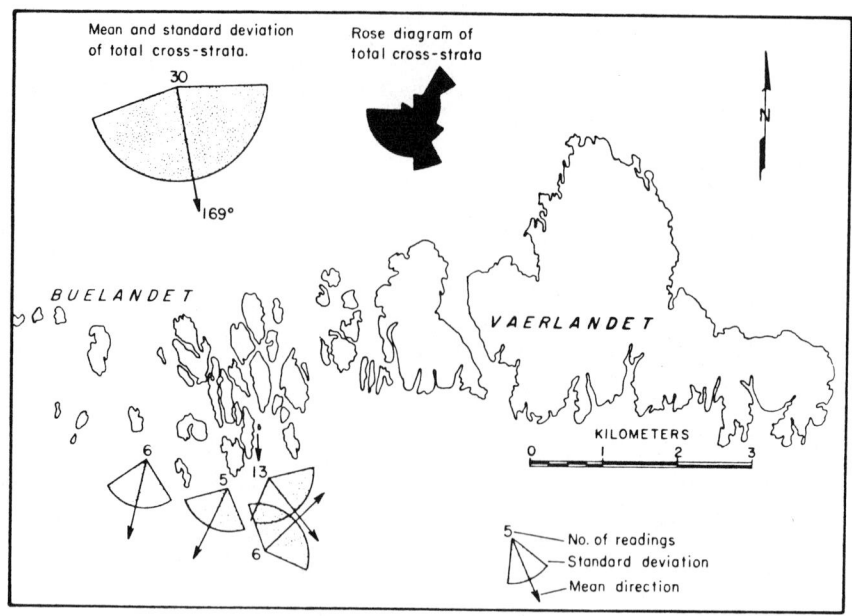

Fig.12. Paleocurrent pattern map of the Buelandet-Vaerlandet Formation. The single arrow represents a primary current lineation direction.

measurements of trough axes from trough cross-strata could be made because of the low plunge angles. The single arrow in Fig.12 represents the orientation of primary current lineation in southern Buelandet.

The data indicate considerable scatter of paleocurrent orientations. However, although the measurements are limited in number, it is clear that the current flow during deposition of the Sörlandet Sandstone in the western part of the area was from north to south. This conforms with the transport directions determined from the Vaeroy Conglomerate by studies of textural variations, sedimentary structures (e.g., elongation of sandstone lenses), and fabric orientation. The Sörlandet Sandstone, then, does not represent a significant change in the tectono-sedimentologic regime, merely greater distance from the northerly source areas and probably deposition by streams with lower gradients.

DEPOSITIONAL ENVIRONMENT AND TECTONIC RELATIONSHIPS

No evidence was found in any part of the Buelandet-Vaerlandet Formation to suggest other than continental sedimentation. Sediment dispersal patterns, stratigraphic relationships and sedimentary structures indicate deposition by streams draining a highland area to the north. The large thickness of conglomerate and breccia suggest that the provenance was being uplifted contemporaneously with sedimentation. The coarseness of the Vaeroy Conglomerate indicates high relief in the source area, with erosion and deposition both proceeding rapidly. Preserved abraded stems of *Psilophyton* sp. found in the Sörlandet Sandstone prove the existence of a primitive plant cover in at least part of the source area.

The transitional, interbedded boundaries between the members of the Buelandet-Vaerlandet Formation define some lateral facies relationships between the members. It is probable that, along a single time-stratigraphic horizon, the Melvaer Breccia passes laterally southward into the Vaeroy Conglomerate, which in turn passes laterally southward into the Sörlandet Sandstone. Fig.13 represents the writer's interpretation of the Devonian paleogeography of the area, with finer sediment deposited to the south.

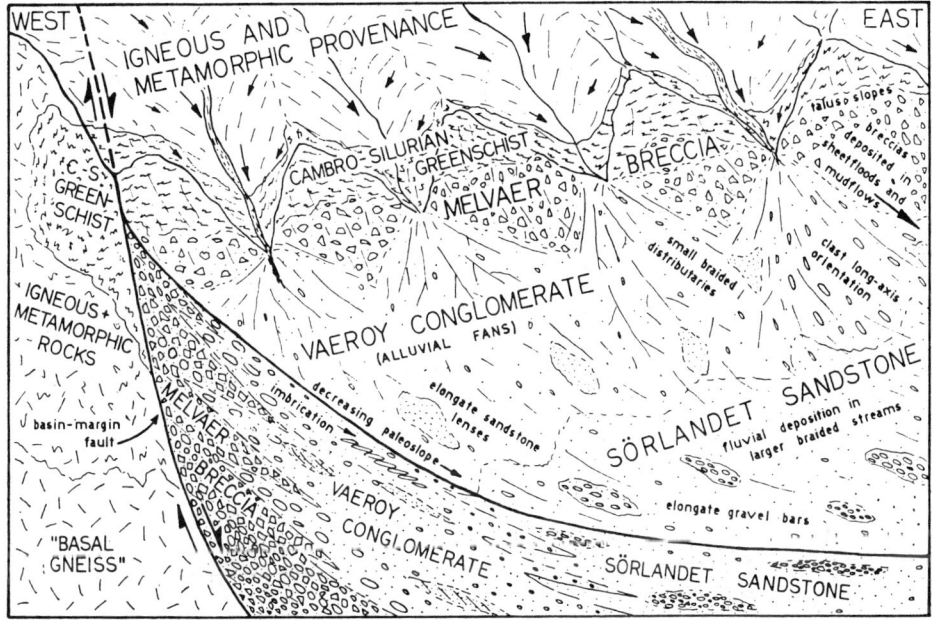

Fig.13. Devonian paleographic diagram of the Buelandet-Vaerlandet district.

The sedimentary basin is interpreted as a half-graben similar to that of the Solund district. Local crustal tension resulting in graben formation after the Caledonian orogeny of the Silurian seems to characterize the Devonian tectonics of western Norway (BRYHNI, 1964; NILSEN, 1968; D. Peacock, personal communication, 1965). The bounding normal fault of the Buelandet-Vaerlandet area trended east–west, following older structural trends (Fig.1). Faulting upraised an igneous and metamorphic complex that provided a source of abundant polymict clastic detritus. Uplift and deposition must have been relatively continuous, because the thick section of Vaeroy Conglomerate is relatively homogeneous; few recognizable significant depositional breaks or changes in sedimentation were noted.

The formation of the structural basin represented an initial phase of the

Svalbardian disturbance in western Norway, which was later followed by post-depositional folding. The Svalbardian disturbance, as originally defined by VOGT (1928), refers to post-Caledonian Devonian tectonic movements in Norway. Vogt correlated pre-Upper Devonian tectonism in Spitsbergen and Bear Island with that in Norway. He suggested that Svalbardian movements had affected the entire west coast of Norway, as well as older basement rocks in south-central Norway. HOLTEDAHL (1960, p.293) considered the deformation to be probably less extensive. The Devonian strata in the Buelandet-Vaerlandet district indicate the presence of pre-, syn- and post-depositional tectonism; each is herein considered to be part of the Devonian Svalbardian disturbance.

Erosion of deep canyons in the uplifted source area by the south-flowing streams cut into deeper-lying crystalline rocks, including the Caledonian "basal gneiss" (Fig.1). This vertical downcutting provided a continuous supply of fresh, unweathered detritus. Weathering was taking place in inter-canyon areas, however, as attested to by the presence of red-brown clayey matrix material in the Melvaer Breccia and plant fragments, which must have grown on some type of soil. Relatively little of this weathered material was incorporated in the Vaeroy Conglomerate, however.

The streams emerged from the canyons onto a broad, more gently-sloping depositional surface. The change in slope caused the streams to fan out into many distributaries, where most of the bed load was deposited and distributed over the depositional area as alluvial fans. The paleodepositional slope was concave upward, gradually decreasing toward the south. The decrease in dip angle to the south, in the present orientation of the Buelandet-Vaerlandet Formation, is probably in part a manifestation of the original concave-upward paleoslope. During floods or seasonal periods of high rainfall, the distributary streams carried abundant gravel and sands under upper-flow regime conditions down the fan surfaces. The rapid flow and high bed load yielded braided distributaries, preferred clast orientation, and southward-facing imbrication.

The steep paleoslopes and high capacities of the streams yielded rapid abrasion of the transported bed load. During transport from drainage basin to depositional site, the polymict clasts acquired rather uniform textural characteristics. During periods of diminished rainfall, when streams were less competent, sand was deposited in thin, elongate lenses on the gravel depositional interfaces. Some lenses were probably deposited behind gravel bars as swale-fill sand deposits. These sand lenses migrated downstream, becoming elongate in a north–south direction. The sheets or lenses contained either flat-strata or solitary cross-strata. Shifting depositional patterns on the surfaces of the coalescing fan complexes also produced areas where sands were deposited even during periods of maximum rainfall. These areas of sand accumulation were probably located primarily between adjacent individual fans in an environment where maximum flow conditions did not prevail.

The Melvaer Breccia accumulated as sheetflood, mudflow, and talus deposits on the steep slopes of inter-canyon areas. Periods of high rainfall washed surficial fine, angular weathered material downslope as viscous mudflows. The mudflows locally spread out over the surfaces of conglomerate accumulation as sheetfloods, resulting in fan-like apron breccia deposits. Soil products, clay, silt, organic material and coarser greenschist fragments were all included in these sudden, rapid deposits of chaotic breccia. The resultant contacts between the mudflows and stream-deposited conglomerates were sharp and clearly defined. The numerous interstratifications of breccia and conglomerate (KOLDERUP, 1916, fig. 4–7, 9) represent individual or superposed mudflows.

The local presence of irregular bedding surfaces, abraded clasts, and oriented fabric along some stratigraphic horizons within the breccia accumulations are thought to represent reworked mudflows. Water running over the surfaces of recently deposited mudflows would result in erosion, abrasion and clast orientation. The succeeding mudflow preserved the intervening diastem as an irregular bedding surface. The bréccia beds locally wedge out southward into the Vaeroy Conglomerate. The Cambro–Silurian greenschist must have been the only unit cropping out along the fault scarp, because it constitutes the only clast lithology in the Melvaer Breccia.

Southward, the Sörlandet Sandstone was deposited in larger braided stream channels that probably meandered considerably across the southward-dipping paleoslope. Gravel was segregated into elongate gravel bars that were lens shaped in cross-section. Flat stratification, primary current lineation and trough-type cross-stratification attest to relatively high flow-regime conditions in this environment. Fine clay and silt-sized material, as well as mica flakes, if supplied by the source area, were carried still further to the south by these streams.

Continued uplift and erosion of the fault scarp further northward with time resulted in a stratigraphic onlap ("transgression") of the sandstone and conglomerate northward (Fig.13). The Vaeroy Conglomerate, however, remained the thickest depositional unit, as well as most widespread. Shifting sedimentation patterns in the individual fans caused coalescence of the fans into a broad piedmont surface characterized by swells and hollows along the depositional strike. These surface irregularities produced some of the present irregularities in strike and dip of the Vaeroy Conglomerate.

The Buelandet-Vaerlandet district was later folded or tilted to its present strike and dip orientation by post-depositional tectonic movements of the Svalbardian disturbance. The original north-bounding fault, although not exposed, is probably concealed within the Cambro-Silurian greenschist, or along the greenschist–breccia contact. It may in fact be a series of smaller step faults rather than a single large normal fault. The large, south-bounding fault of the Kvamshesten district, trending parallel with Dalsfjord, may extend into the Buelandet-Vaerlandet district. Perhaps it forms the bounding fault of the half-graben where it dies out westward.

The former crystalline source area to the north presently underlies the North Sea. The Håsteinen and Ryggstenen islands, 5–10 km north of Buelandet-Vaerlandet, however, do contain a variety of crystalline rocks (N. H. Kolderup, 1926, unpublished field notes from Askvoll Herad, on file at University of Bergen, p.8–9). These scattered islets are probably remnants of the former highland provenance.

In terms of depositional environment and tectonic relationships, the Buelandet-Vaerlandet district is quite similar to the Solund district located 15 km to the south. The possibility that they are the opposite sides of a larger, east-west-trending graben, comparable in size to the Hornelen graben, is very real indeed, particularly because the Solund Conglomerate was derived from a bordering highland to its south.

There is no evidence in the Buelandet-Vaerlandet district to suggest a climate other than temperate, with abundant, albeit seasonal, rainfall. No desert varnish, hematitic deposits, glacially faceted or striated clasts were noted in the sediments. Consolidation and lithification of the Buelandet-Vaerlandet Formation yielded a rock more resistant, in general, than the surrounding basement rocks; thus, the Devonian strata at present are topographic ridge-formers.

CONCLUSIONS

Sedimentary, structural and stratigraphic analyses of the Buelandet–Vaerlandet Formation yielded the Devonian paleographic interpretation diagrammatically outlined in Fig.13. Rapid uplift and erosion resulted in deposition of a broad coalesced alluvial fan piedmont at the base of a contemporaneously uplifting source area to the north. Abundant rainfall and relatively steep paleoslopes resulted in deposition of fanglomerates with anisotropic fabrics-imbrication and orientation of long axes and blunt ends of elongate clasts. Sandstones characterized by flat-strata, solitary cross-strata, primary current lineation and segregated gravel bars were deposited by braided streams to the south of the fanglomerates.

Viscous mudflows characterized by isotropic dispersed framework packing and monomict compositions were deposited close to the source area, with material derived from the nearby weathered slopes of the source area. Modern environmental equivalents are to be found in gravel fan accumulations at the bases of steep mountain slopes in humid or temperate regions of the world. An ancient equivalent, although much larger in size, is perhaps found in the Triassic Newark basins of the Appalachian Mountains of eastern North America (KRYNINE, 1950).

ACKNOWLEDGEMENTS

The writer is indebted to Professor Robert H. Dott, Jr., his thesis advisor at the University of Wisconsin, U.S.A., and Professor N.-H. Kolderup, University of Bergen, Norway, for their assistance in his thesis study (the Solund district), with which this paper is closely allied. The study was supported by two National Science Foundation Summer Fellowships, a Penrose Bequest from the Geological Society of America, and a Sigma Xi Grant-in-Aid of Research.

REFERENCES

BRYHNI, I., 1964. Relasjonen mellom senkaledonsk tektonikk og sedimentasjon ved Hornelens og Håsteinens devon (English summary). *Norg. Geol. Undersökelse*, 223:10–25.

CURRAY, J. R., 1956. The analysis of two-dimensional orientation data. *J. Geol.*, 64:117–131.

HOLTEDAHL, O., 1960. Devonian, including Downtownian in the Hitra district, etc. In: O. HOLTEDAHL (Editor), *Geology of Norway — Norg. Geol. Undersökelse*, 208:285–297.

HUEBER, F. M., 1967. Psilophyton: the genus and the concept. *Abstr. Proc. Intern. Symp. Devonian System, Calgary, Canada*, pp.72–73.

JARVIK, E., 1949. On the Middle Devonian Crossopterygians from the Hornelen field in western Norway. *Univ. Bergen Aarbok 1948, Naturvidenskap. Raekke*, 8:48 pp.

KIAER, J., 1918. Fiskerester fra den devoniske sandsten paa Norges vestkyst (English sum.). *Bergens Museums Aarbok 1917–1918, Naturvidenskap. Raekke*, 7:17 pp.

KOLDERUP, C. F., 1915. Vestlandets devonfelter og deres plante fossiler. *Naturen*, 1915: 217–232.

KOLDERUP, C. F., 1916. Bulandets og Vaerlandets konglomerat og sandstensfelt (English sum.). *Bergen Museums Aarbok 1915–1916, Naturvidenskap. Raekke*, 3:26 pp.

KRUMBEIN, W. C. and SLOSS, L. L., 1958. *Stratigraphy and Sedimentation*. Freeman, San Francisco, Calif., 497 pp.

KRYNINE, P. D., 1950. Petrology, stratigraphy and origin of Triassic sedimentary rocks of Connecticut. *Conn., State Geol. Nat. Hist. Surv. Bull.*, 73:247 pp.

NATHORST, A. G., 1915. Zur Devonflora des westlichen Norwegens. *Bergens Museums Aarbok 1914–1915, Naturvidenskap. Raekke*, 9:12–34.

NILSEN, T. H., 1967. *The Relationship of Sedimentation to Tectonics in the Solund Devonian District of Southwestern Norway*. Thesis, Univ. Wisconsin, Madison, Wisc., 168 pp. (unpublished).

NILSEN, T. H., 1968. Old red sedimentation in the Solund district, western Norway. *Proc. Intern. Symp. Devonian System, Calgary, Canada*, 2:1101–1115.

VOGT, Th., 1928. Den norske fjellkjedes revolusjons-historie. *Norsk Geol. Tidsskr.*, 10:97–115.

WILLIAMS, H., TURNER, F. J. and GILBERT, C. M., 1954. *Petrography*. Freeman, San Francisco, Calif., 406 pp.

ERRATA

Page 53, Figure 13, the caption should read: "Devonian paleogeographic diagram. ..."

Page 56, line 22 should read: "Vaerlandet Formation yielded the Devonian paleogeographic interpretation. ..."

14

Copyright © 1977 by R. J. Steel, S. Maehle, H. Nilsen, S. L. Roe and A. Spinnagr
Reprinted from Geol. Soc. America Bull. **88**:1124–1134 (1977)

Coarsening-upward cycles in the alluvium of Hornelen Basin (Devonian) Norway: Sedimentary response to tectonic events

R. J. STEEL
S. MÆHLE*
H. NILSEN*
S. L. RØE*
Å. SPINNANGR*

Geological Institute, University of Bergen, 5000 Bergen, Norway

ABSTRACT

Hornelen Basin (Devonian) is filled with ~25 km of sediments, mostly sandstones. These sedimentary rocks are spectacularly organized into more than 150 basin-wide cycles, each on the order of 100 m thick, most of which coarsen upward. The cycles are otherwise complex, consisting of marginally derived fanglomerates and laterally equivalent, longitudinally dispersed alluvial plain sediments.

The basin-wide nature of the cycles, the fact that the coarsening upward occurred at the same time in both marginal and axial facies, and because successive alluvial fan bodies coarsen upward whether they are composed of debris flow or of stream deposits suggest that the cycles are allocyclic and that they are the basin's response to the lowering of its floor. In their marginal development, the cycles are commonly segmented, consisting of coarsening-upward subcycles of the order of 10 to 25 m thick. The geometry and internal details of these suggest that they also were tectonically generated.

It is likely that the 10 to 25-m coarsening-upward sequences, representing aggrading base-level conditions, were the basic sedimentary response to basin-floor subsidence. The 100-m cycles represent additional complexity in style of subsidence. Progressive eastward overlap of successive 100-m units suggests that at this interval the locus of subsidence abruptly shifted in a proximal direction, by ~0.25 km.

A dextral wrench fault model is proposed to account for this pattern of basin filling.

INTRODUCTION

Descriptions of sedimentary successions with inferences about tectonic control on sedimentation are legion, and this is hardly surprising in view of the fact that the major control on almost all sedimentation is tectonic. The critical question is how the control operates (Blatt and others, 1972, p. 591). A weakness in many of the published cases is that arguments are generalized and usually uninformative as to how a relationship develops and is maintained between the two sets of processes.

We suggest that some insight to the relationship may be gained by closer examination of trends of sedimentation in time in alluvium-filled basins, particularly in basins whose geologic context and general sedimentary attributes are such as to suggest, a priori, that tectonism was a dominant external control during sedimentation.

* Present addresses: (Mæhle and Spinnangr) Statoil, Lagårdsvn. 78, 4000 Stavanger, Norway. (Nilsen and Røe) Oljedirektoratet, Lagårdsvn. 80, 4000 Stavanger, Norway.

Cases in point, presently being restudied, are the three main Devonian basins of western Norway (Solund, Kvamshesten, and Hornelen Basins). All three were supposedly tectonically controlled (Nilsen, 1968; Skjerlie, 1971; Bryhni, 1964b), although in none of these studies has that conclusion been made by inductive reasoning, because no detailed facies analyses or detailed sedimentary successions have been published. In these cases, a general feature of interest is the difference in coarseness of the alluvial sediment pile in the different basins. Solund Basin is filled with 4 km of conglomerates, whereas Hornelen Basin apparently contains 25 km of sediment that is made up largely of sandstones. If the notion that tectonism was the dominant control of sedimentation is true (and varying climatic or lithologic controls can probably be ruled out because the basins are sited close together and each lies within an equally varied assemblage of source rocks), then there is clearly scope for a more detailed discussion of how differences of tectonic style or intensity have produced such a varied sedimentary response (Steel, 1976).

In the present discussion, we are concerned primarily with a particular feature of the internal organization of one of these basins, Hornelen Basin. This is a basin-wide cyclicity, usually expressed as repeated coarsening-upward sequences of the order of 100 m thick and that is developed equally well in the sandstone portion of the fill (basin-axis) as in the conglomerate portion (marginal). In making a case for the tectonic generation of this cyclicity, we emphasize its development in alluvial plain facies. The internal details and geometry of individual fan-floodplain bodies give relatively short-term evidence of the sedimentary response to individual periods of faulting or subsidence. The geometrical relationships of successive bodies (the way in which they are stacked) give longer term evidence concerning the locus and changing style of tectonism in time.

HORNELEN BASIN

Hornelen Basin is relatively small (< 2,000 km²) and is filled with Devonian alluvial sediments of two main types. Minor amounts of conglomerate occupy the marginal areas adjacent to the present northern and southern bounding faults (Fig. 1). These deposits were dispersed laterally across the margins on alluvial fans which, on the basis of paleocurrents, sediment coarseness, and pebble size changes, headed either from the present fault lines or from parallel faults not far outside the present basin boundaries (see Fig. 8). Most of the basin, however, is filled by sandstones which probably accumulated from westward-flowing rivers on a broad alluvial plain (see Fig. 8) (Bryhni, 1964b). A marginal interfingering of these deposits with the fanglomerates suggests that at certain times the entire width of the basin was flooded by the longitudinal alluvial plain system.

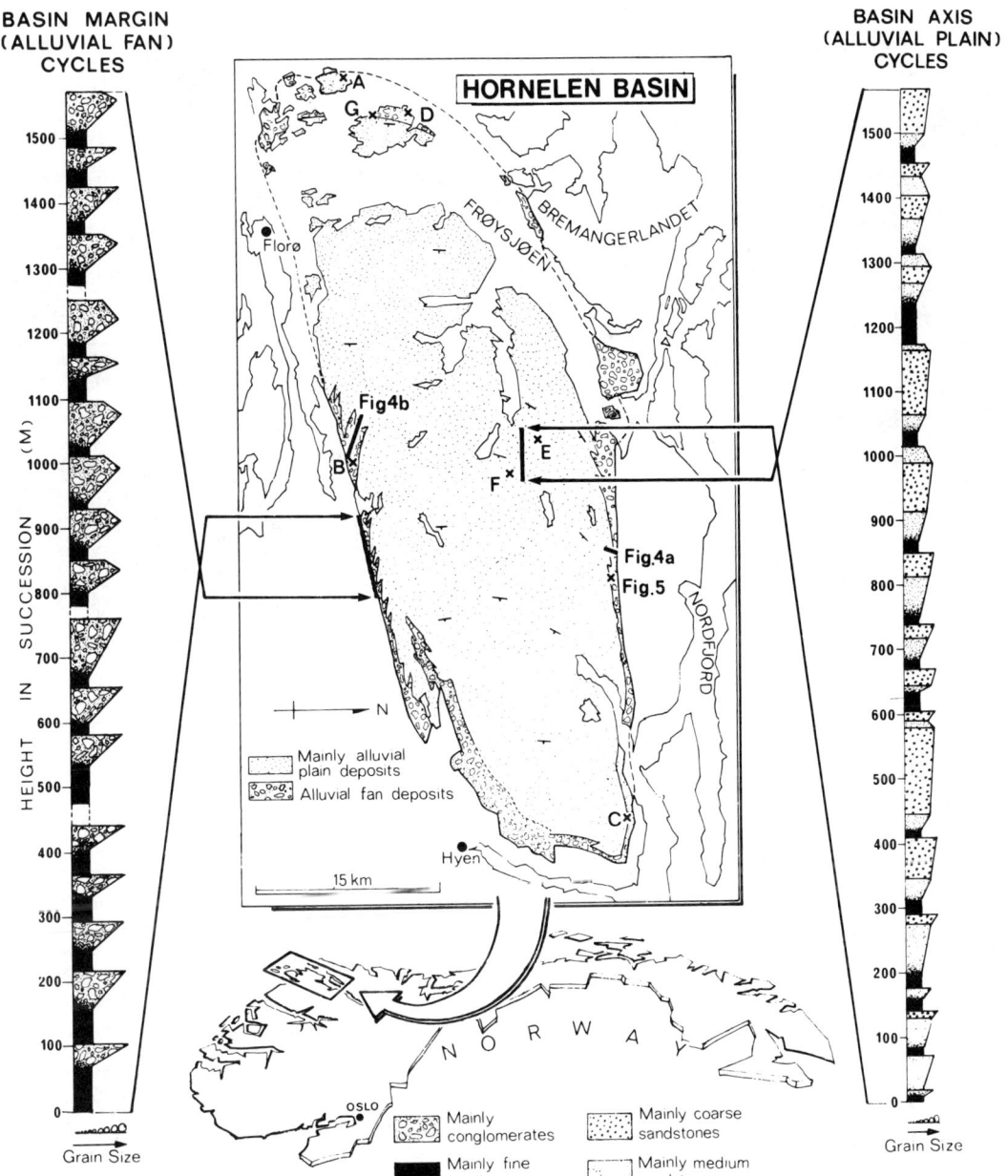

Figure 1. Simplified map of Hornelen Basin (after Kildal, 1970). Note the remarkable cyclicity permeating the 25-km-thick basin succession. Both axial and marginal deposits are dominated by coarsening-upward sequences. Lettering refers to the locations of sections shown on Figures 3 through 6. The bedding generally dips east at 20° to 40° but is subvertical on the northern margin.

Figure 2. View eastward along part of the axial region of Hornelen Basin showing the basin-wide cyclicity. Scarps are of the order of 100 m high. Note the transition from fine-grained (immediate foreground) (Fig. 6, loc. E) to coarse-grained alluvial plain facies (middle distance) (Fig. 6, loc. F). Letters refer to Figures 3 through 6.

The two most striking features of the basin succession are the apparently enormous stratigraphic thickness (~ 25 km) (Kolderup, 1927; Bryhni, 1964a) and the presence of a prominent cyclicity through much of this (Figs. 1 and 2). We are particularly interested here in the latter feature. Superb exposure across the basin provides an unusual opportunity for examining both the sediments and the cyclicity, as it is traced through the various laterally equivalent facies.

We attempt to examine the cyclothems in both their marginal (alluvial fan sediments) and axial (alluvial plain sediments) development, but we discuss mainly the marginal facies, because as these are very coarse grained, they are likely to reflect the details of marginal tectonics more clearly.

MEASUREMENT OF CYCLOTHEMS

We use the terms "cyclothem" or "cycle" in the very general sense recommended by Duff and others (1967, p. 2). Because of the critical importance of the correct delimitation of cyclothems and the correct recognition of their internal trends, we make the following points concerning field measurements:

Maximum particle size and bed thickness measurements in conglomerates were made in the manner and for the reasons outlined by Bluck (1967). The plotted values of particle size represent a mean value of the ten largest clasts (after the omission of any "outsized" clasts) from an area over several metres on either side of the point of bed-thickness measurement. Bed-by-bed maximum particle size readings together with sandstone percent averages proved to be critical in the recognition of coarsening-upward trends within fan bodies and therefore also in the delimiting of fan cyclothems. In places where the uppermost fraction of a conglomerate sequence becomes finer grained, the cyclothem boundary was placed above this portion but immediately below the overlying fine-grained sandstones and siltstones.

In the sandstone succession of the basin axis region, cyclothems are less easy to define and have probably been previously identified incorrectly as fining upward (Bryhni, 1964c), presumably the simplest interpretation of a succession seen to consist of an apparent alternation of fine- and coarse-grained "members." Detailed measurements through numerous coarse "members" have since demonstrated that they are internally dominated by a coarsening upward and that there is usually a transition up from the underlying fine "member." The latter usually overlies the coarsest portion of an underlying sequence. This position of the cyclothem boundaries was further demonstrated when they were traced laterally into the bounding surfaces of well-defined marginal fan cyclothems.

MARGINAL SEQUENCES (ALLUVIAL FAN BODIES)

The marginal deposits are likely to have accumulated mainly on alluvial fans. This environment is suggested by the coarseness of the

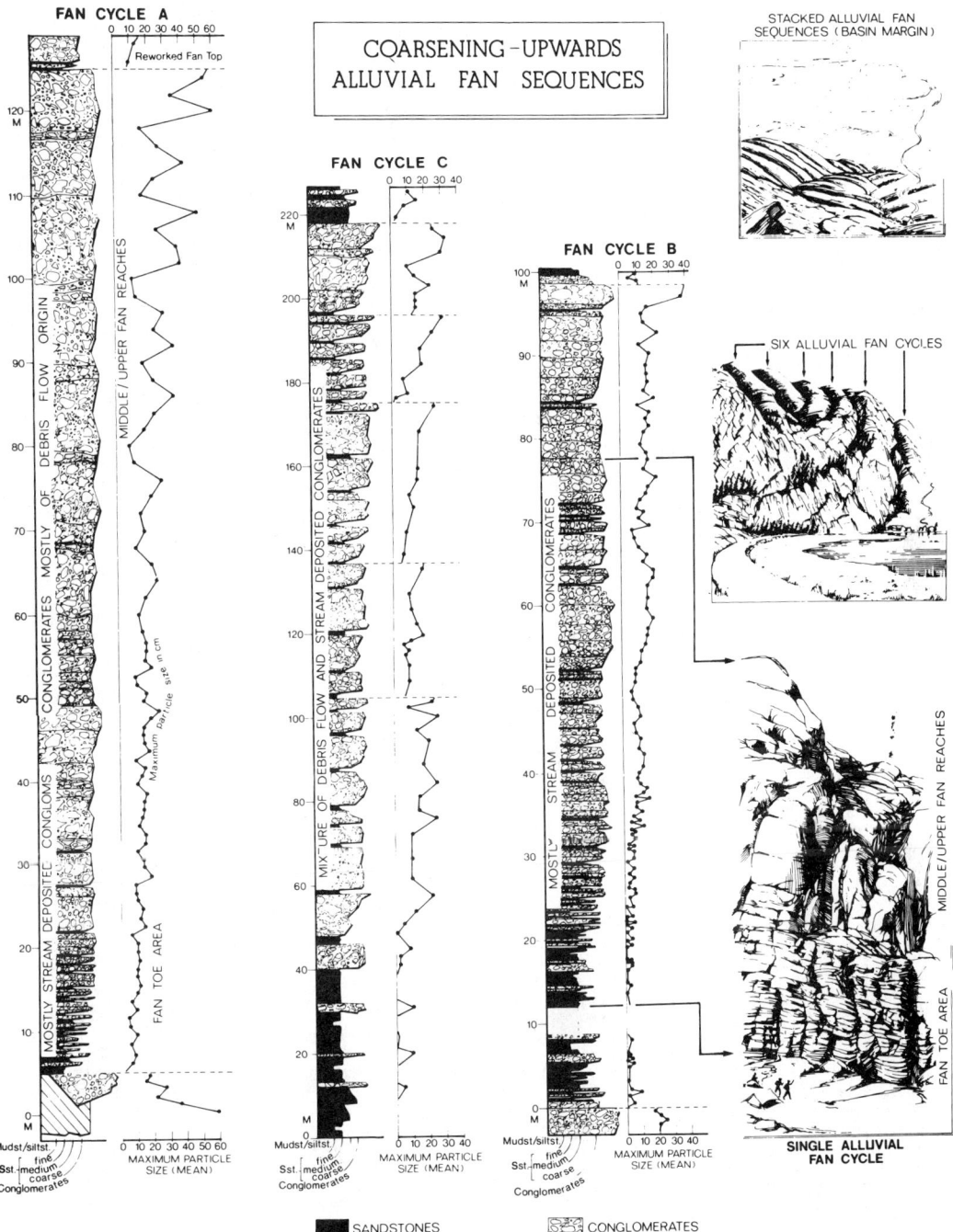

Figure 3. Some of the internal details of three alluvial fan sequences. Note the general coarsening upward as well as increasing maximum particle size upward. Profile C is clearly segmented into a number of coarsening-upward units.

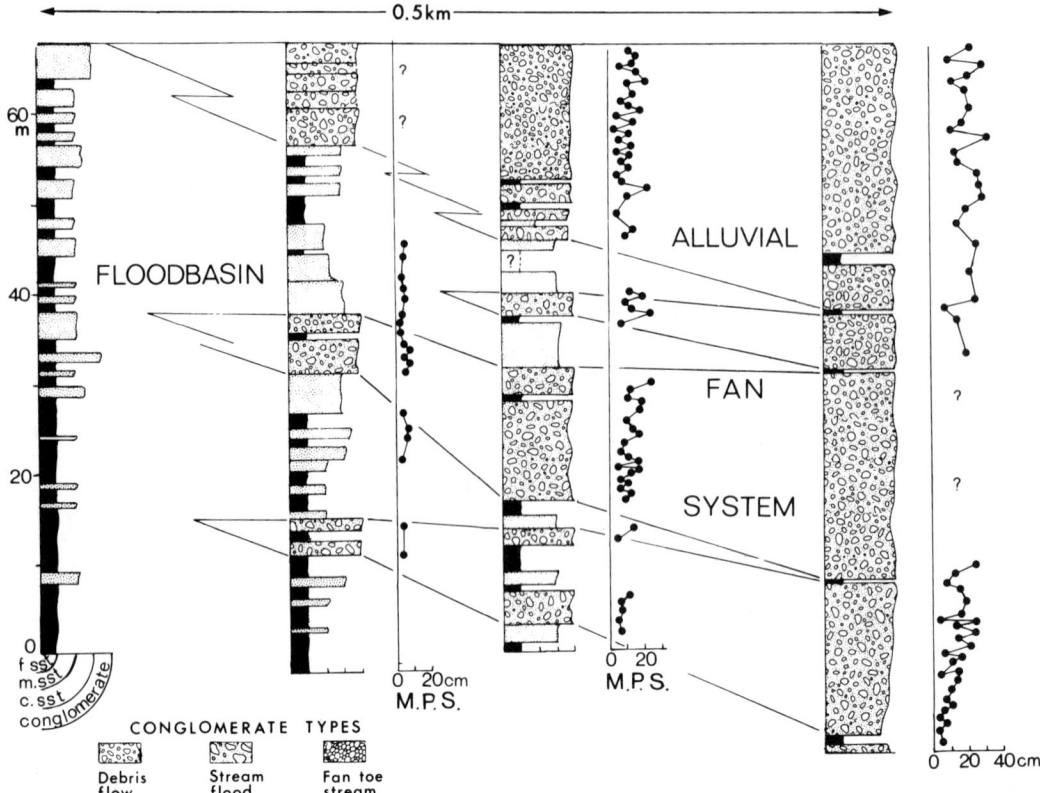

Figure 4a. Internal details of a composite, debris-flow–dominated alluvial fan wedge which is interfingering with floodbasin fines along the northern margin of Hornelen Basin. Note the progressive change in sediment coarseness as well as in maximum particle size (MPS) both distally and upward in the fan body.

sediments (Fig. 3), the wedge shape (in section) of the bodies (Fig. 4), progressive downfan decrease in maximum clast size throughout the wedge (Fig. 4), and frequent dominance of clasts from local "basement," evidence of subaerial exposure (for example, mudcracks, rainprints) and the paleocurrents which sometimes indicate a "fan-like" dispersal out from the basin margins. Some cyclothems in the marginal areas also include fine-grained sediments derived from the basin-axis floodplain system. This material sometimes wedges out before the margin is reached, but, when present, occupies the lowest part of the cyclothem (Figs. 3, 4).

The alluvial-fan deposits have been studied at three levels of organization. At the most general level, there are successions of the order of kilometres thick, which consist of stacked alluvial fan bodies, mostly of the same type. For example, successions of fan bodies of predominantly debris-flow deposits (Fig. 3, loc. A; Fig. 4a), of predominantly stream deposits (Fig. 3, loc. B; Fig. 4b), or of mixed types (Fig. 3, loc. C) have been identified along different marginal segments of the basin. Secondly, the general organization within individual fans is of interest. Along the southern margin of the basin, fan bodies (~5-km radius and 50 to 80 m thick in proximal reach) are highly asymmetric, dominated by coarsening and thickening-upward trends (Figs. 3, 4b), while those bodies along the northern margin are frequently of smaller radius (~1 km), thicker (100 to 200 m), and are less asymmetric (Fig. 4a). In the latter instances, however, a general coarsening upward can still be seen. In addition, individual fan sequences commonly consist of a number of subunits (10 to 25 m thick), and each of these also may be dominated by coarsening- and thickening-upward trends (Fig. 3, loc. C; Fig. 5). Finally, the sediments themselves, at bed or sedimentation-unit level, have been studied, and various types of debris-flow, mudflow, and streamflow deposits have been recognized.

Details of the various sediment facies at bed level are beyond the scope of the present discussions (see Maehle, 1975; Nilsen, 1975; Spinnangr, 1975) but will be treated in detail elsewhere. The ensuing discussion is concerned mainly with the second level of organization, because the individual, coarsening-upward fan body is believed to be the unit of major importance in the basin succession and is of tectonic significance.

Some 55 alluvial fan bodies along the margins of the basin have been examined. Of these, 90 percent show the following features:

(1) A general coarsening-upward, indicated by decreasing sandstone (discrete bed) percentage (Figs. 3, 4). (2) Coarsening upward of the conglomerates themselves as indicated by increasing

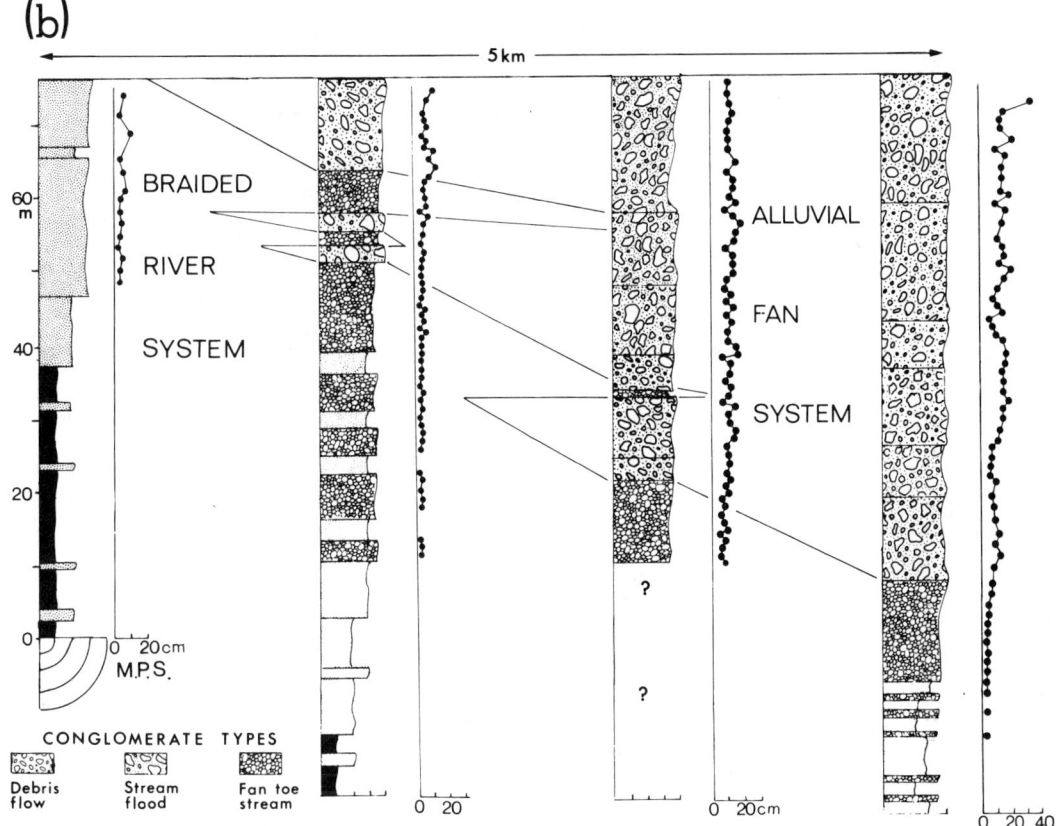

Figure 4b. Internal details of a stream-dominated alluvial fan body on the southern margin of Hornelen Basin. Note the contrasts with the fan of Figure 4a in facies, fan radius, overall cross-sectional geometry, and downfan rate of change of MPS, despite the similar coarsening upward and fining outward. Legend as in Figure 4a.

mean maximum clast size (Figs. 3, 4). (3) A general thickening-upward of conglomerate beds as implied from a positive bed thickness–maximum particle size relationship. (4) A sharp and apparently planar upper boundary to the fan body (Fig. 4). The coarsest fanglomerates are abruptly overlain by the finest alluvial plain sediments. (5) Lower boundary is gradational because there is a gradual coarsening upward from alluvial plain sandstones to the finest (distal) fanglomerates. This has resulted from a lateral interfingering of fan and alluvial plain sediments and from the gradual enlargement of fan radius through time so that the base of the fan body is now diachronous (Fig. 4).

The coarsening upward is not restricted to a particular sector of any fan body, and so the responsible mechanism is not a local one. The general grain-size and bed-thickness trends therefore suggest that successive dispersions on the fan surfaces were of increased competence and capacity. Moreover, because successive dispersions contained greater volumes of debris, individual fans had a characteristic history of radial enlargement. Because of fan-area –source-area relationships (Bull, 1964; Hooke, 1968), there is also implied an increasing drainage area with time.

The question arises as to the cause of the increased volume of debris shed onto the fans with time. In a Devonian piedmont situation, this may be attributed to time changes in relief, climate, or bedrock geology in the drainage area, probably in that order of importance (Blatt and others, 1972). Bedrock controls were probably unimportant because there are no significant vertical changes in pebble type in individual fan sequences. Of relief and climatic controls, the former was probably more important for the following reasons:

(1) Coarse-grained, conglomeratic fan sequences, each of the order of 100 m, were repeatedly built and stacked above each other. (2) The marginal fan cycles are notably asymmetric (Fig. 1); that is, they coarsen upward. As discussed below, this asymmetry is more easily explained by a lateral (tectonic) shifting of the locus of subsidence and sedimentation than by a climatic hypothesis. (3) The coarsening upward in the fan sequences appears to be independent of the dominant type of deposit from which the fans were built. Sequences dominated by debris-flow deposits, stream deposits, or mixtures of this show similar time trends (Fig. 3). Climatic change probably tends to alter the dominant depositional process during fan building (Lustig, 1965). (4) Hornelen Basin presently has prominent fault boundaries (Figs. 1 and 8), and subsidence was certainly a major control on the accumulation of 25 km of sediment within this relatively small area.

273

It is suggested that, in the areas adjacent to the basin margins, the effect of a relative subsidence of the basin floor was the creation of relief, increased drainage area in the new upland region and unstable slopes. It is further suggested that the main tectonic event is marked by the sharp planar top of any fan body, while the overlying coarsening-upward sequence records the gradual export of stored debris and the gradual building of the next fan outward in an attempt to re-establish an equilibrium profile across the basin margin. The coarsening upward itself simply records the tendency for relatively distal deposits to be overlaid by more proximal deposits through time, as the radius of the fan enlarges. Because the coarsening upward is a feature of the entire fan body, growth appears to have taken place fairly uniformly over the fan surface.

As regards the tectonic mechanism, it is of interest to speculate about the frequency or intensity of fault movement needed to produce a single coarsening-upward sequence. In fan sequences many times thicker than those here, it has been suggested that coarsening upward may have resulted from faulting with a history of progressively greater movements (Steel and Wilson, 1975). In the present case, with sequences of the order of 100 m, we suggest that several short-lived but intense movements along the same fault line may have been sufficient to begin and maintain continuous fan growth. It is possible that there was a lag between each tectonic event and the sedimentary response so that a uniform series of fault movements caused an accumulation and progressive storage of gravel, with a resultant gradual enlargement of fan size. If we assume that it is unrealistic to postulate a single fault scarp of the order of 50 to 100 m high, then it is tempting to correlate the subunits in many of the fan sequences with discrete episodes of fault movement. The segmented nature of many of the fan sequences is clear (Fig. 3, loc. C; Fig. 5), and in some fans, the subcycles show not only a coarsening upward but also a thickening upward of component beds (Fig. 5). Of course, these subcycles may have resulted from an autocyclic mechanism on the fan, such as from the natural intermittent lateral shifting of the main dispersal system. There are, however, two attributes of the subcycles which favor an origin from external events. First, they appear to have a sheet-like geometry which suggests that they represent a unit of accretion over most of the fan surface. In addition, in some areas of the basin, the subcyclicity appears to penetrate the alluvial plain facies beyond the fan toes. Secondly, in some segmented profiles, particularly along the northern margin, the base of each subcycle consists of fine-grained sandstones and siltstones derived from the basin-axis dispersal system and not from the fans themselves (Fig. 5). This feature again emphasizes that the subcyclicity is not simply an internal feature of the fan prism but penetrates from the marginal into the basin-axis facies.

The Hornelen type of fan sequence contrasts strongly with other described Devonian (Bluck, 1967) and Triassic (Steel, 1974) basin-margin fan sequences that are characterized by a fining upward. It is likely that such sequences, particularly if they are relatively thin, could have been generated either against basin margins where the rate of subsidence was minimal (and sourceward scarp retreat was important) or across basin margins which were actively expanding by a sourceward shifting of the locus of faulting and subsidence (Steel and Wilson, 1975; Steel and others, 1975).

AXIAL SEQUENCES (ALLUVIAL-PLAIN SAND BODIES)

Approximately 90 percent (by volume) of the sediments in Hornelen Basin are sandstones which accumulated from a westward-flowing fluvial system (Figs. 1, 8b). They form a variety of facies, the sum of which may be loosely referred to as an alluvial-plain association. Detailed study of these sediments is yet at

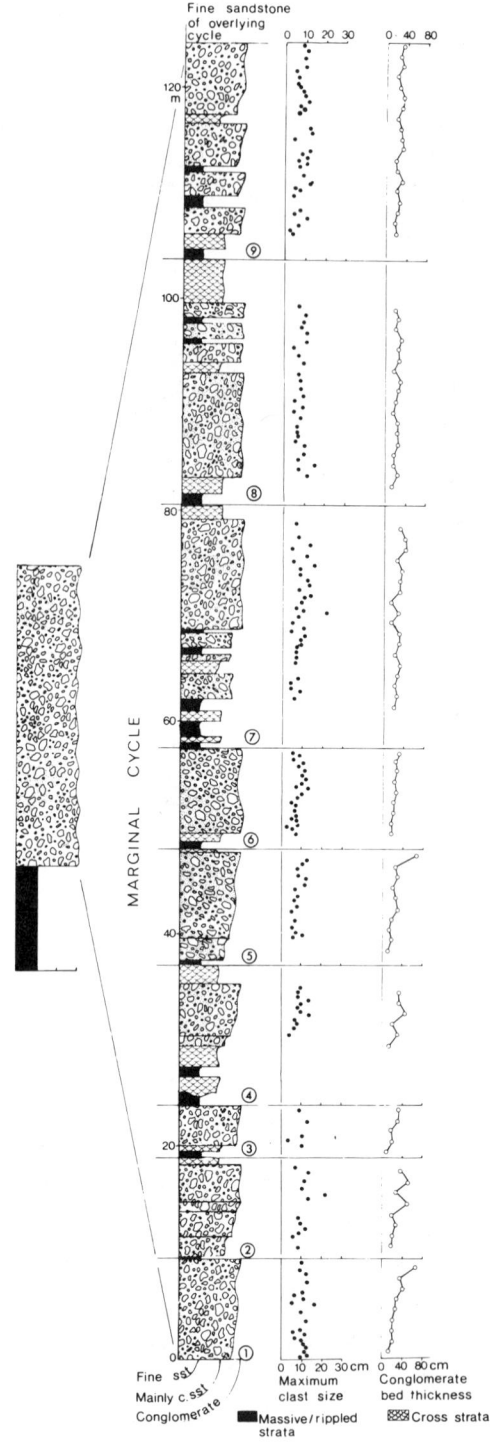

Figure 5. Details of the coarsening upward and thickening upward within subunits in the fanglomerate portion of a cyclothem on the northern margin. The fine sandstone at the base of the subunits is of floodbasin origin and not derived from the fans.

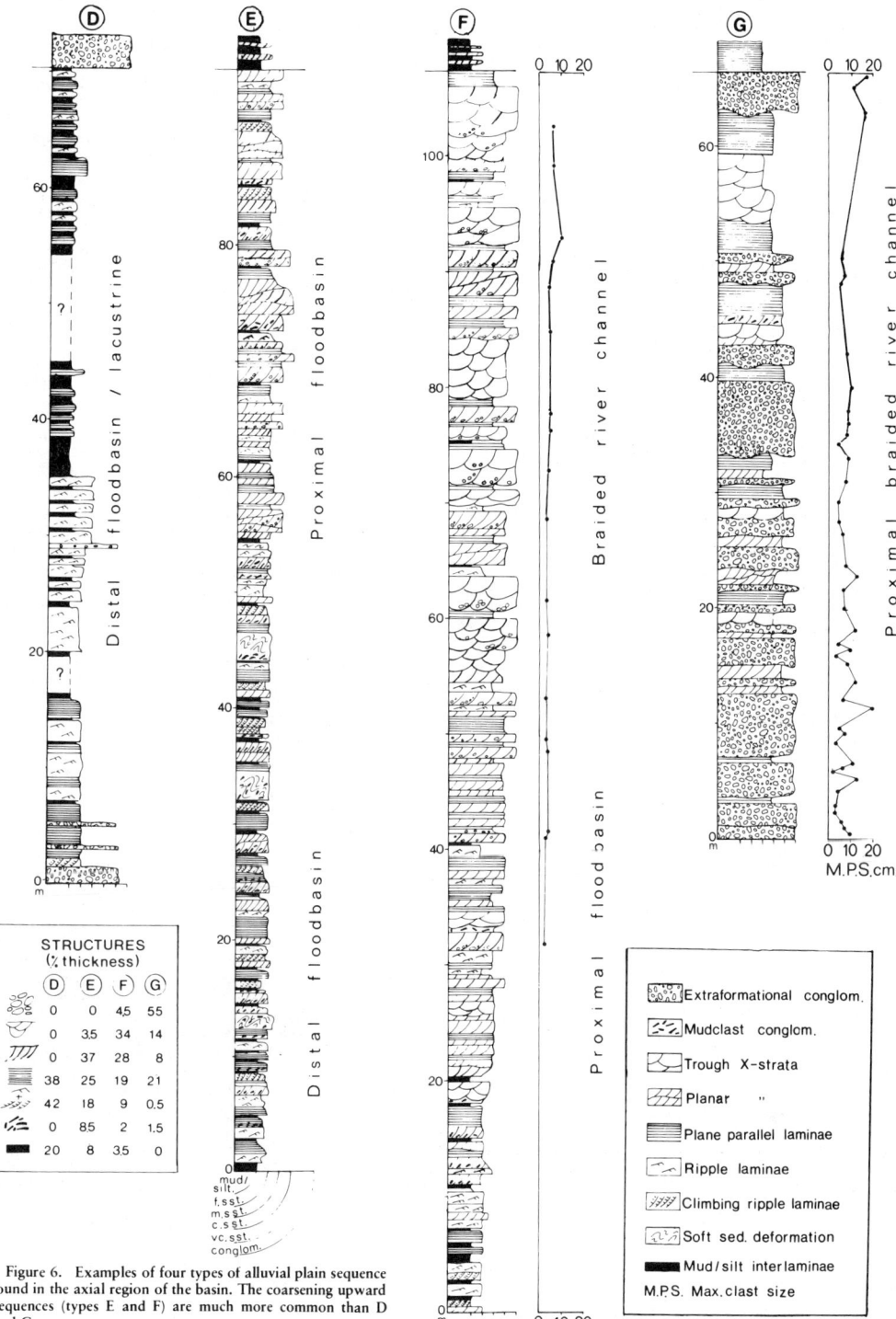

Figure 6. Examples of four types of alluvial plain sequence found in the axial region of the basin. The coarsening upward sequences (types E and F) are much more common than D and G.

a preliminary stage, but a number of facies of progressively more distal aspect (progressively finer grained) have been recognized. It is assumed that sequences with abundant polymict conglomerates are most proximal while those with greater amounts of ripple lamination and mudclast-conglomerate – mudstone strata are more distal (Fig. 6). If these criteria are correct, it is of additional interest to note that both plane parallel lamination and planar cross-stratification show some increase in amount distally while trough cross-strata decrease (see summary of structures, Fig. 6; see also Smith, 1970).

Like the alluvial fan deposits, these facies sequences are vertically repeated many times in different parts of the basin (Fig. 1). Cycle thickness varies from 70 to 140 m. Most of the sequences exhibit a coarsening upward which reflects transition from relatively distal to relatively proximal parts of the alluvial-plain system (Fig. 6, profiles E and F). On the other hand, the few outcropping examples of the most proximal (Fig. 6, profile G) and the most distal sequences (Fig. 6, profile D), which are usually relatively thin, appear to show no similar vertical organization. Cycles containing these end-member types represent less than 5 percent of the total in the basin.

It has not been possible to examine a single cycle for more than several kilometres longitudinally in the basin, partly because of inaccessibility and because the succession has an eastward tectonic dip parallel to the main axial paleoflow and toward the proximal reaches. Therefore no single cycle has a demonstrable sourceward facies change such as that shown by the profiles D through G in Figure 6. However, a great number of cyclothems can be examined vertically; these do show a range of variation typified by these sequences and suggest that any one cyclothem has the geometry and facies differentiation approximating that shown in Figure 7a. In addition, individual cyclothems can be examined laterally and demonstrable facies changes from braided river channel deposits into floodbasin deposits are consistent with the proposed model (Fig. 7b). Examination of aerial photographs of the basin suggest that axial cyclothems may have a minimum east-west extent of 10 to 12 km.

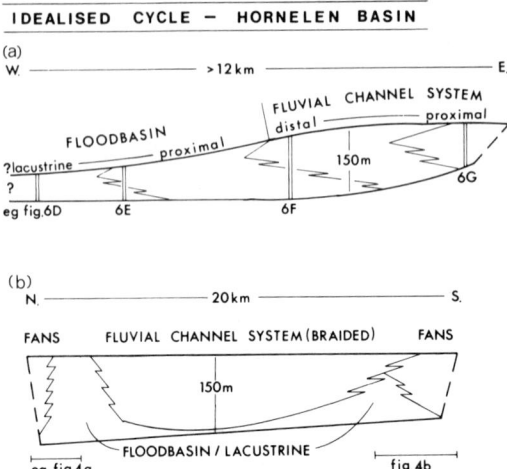

Figure 7. Suggested longitudinal and lateral geometry and facies variation in the Hornelen Basin cycles. Section b can be observed; section a is hypothetical.

The general coarsening upward of most of the alluvial plain cyclothems is also interpreted in terms of a general (westward) progradation of the basin-axis system, so that within a single cyclothem, the deposits of, for example, proximal channel reaches gradually encroached on distal reaches. The apparent arrangement of floodbasin-lacustrine deposits both laterally and distally equivalent to the deposits of the main channel system suggests a closed-basin situation.

Any individual, axially transported alluvial-plain sequence can be traced laterally into a marginal alluvial fan sequence. The axial cyclicity is, therefore, not independent of that developed marginally and, for reasons already discussed, we suggest that cyclothems in their entire width (basin-wide) are the product of discrete episodes of basin subsidence. The basin infill consists of more than 150 cyclothems, implying a similar number of major episodes of alluvial-plain progradation and probably the same number of major episodes of basin subsidence.

The alluvial plain cyclothems also often consist of a number of coarsening-upward subcycles, of the same order of size as those in the fan sequences (Fig. 6, profile F). Their development here, however, is less easily argued as attributable to allocyclic mechanism, although in certain well-exposed areas, they can be "walked-out" into subunits on the fans.

FAULTING, SUBSIDENCE, AND SEDIMENTATION

In the previous sections, the individual cyclothem has been examined, and the development of facies variation within it has been briefly discussed. The evidence suggests that the development of a typical 100-m-thick cyclothem represented aggrading base-level conditions in the basin in response to a lowering of its floor. Further, internal details of cyclothems suggest that the amplitude of single subsidence events was commonly of the order of 10 to 25 m, as reflected by individual cyclothems being composed of six to ten subunits.

The question then arises as to the tectonic significance of the boundaries between cyclothems as compared to those between subcyclothems. We suggest, because of the spatial relationship of adjacent subcyclothems (in fan sequences), that the latter boundaries separate intervals during which units of sediment were vertically stacked — that is, the group of subunits within a single cyclothem originated from approximately the same focus. On the other hand, the cyclothem boundaries appear to represent intervals during which there was an important eastward shift in the locus of both marginal and axial dispersal systems.

Bryhni (1964a) suspected that the cyclothems overlap each other eastward in Hornelen Basin; he noted that a series of progressively younger alluvial fan bodies on the southern margin appeared to be displaced successively eastward. Further examination of this and other similar areas suggest to us that the amount of overlap of successive cyclothems is commonly less than 0.25 km. In addition to this evidence, a periodic, rapid eastward shift of the main locus of subsidence could account for the remarkable planar tops of cyclothems and for their strong asymmetry. A discontinuous eastward migration of the marginal fans and of the proximal reaches of the axial fluvial system would result in a more exaggerated grain-size asymmetry in the cycles than would be the case if they were stacked vertically. During the eastward shifting of the locus of sedimentation, it might be expected that the underlying sequence should be locally tilted eastward. In only one area has a slight unconformity, with this sense, been recorded at the base of a cyclothem.

The validity of the above model for Hornelen Basin development may also be argued indirectly, from the enormous stratigraphic thickness of the basin succession. The lack of a significantly higher

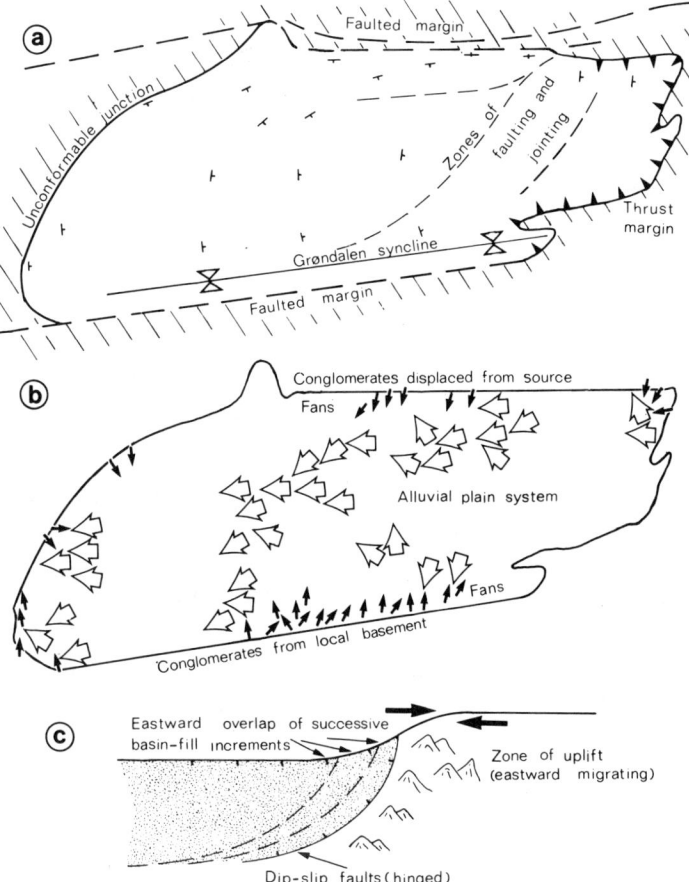

Figure 8. a. Some of the main structural features of Hornelen Basin; b. paleocurrent systems (solid arrows are marginal fanglomerate dispersal; open arrows are axial sandstone system dispersal); c. a speculative model of basin development along the southern flank of a right-slip wrench system.

degree of metamorphism at the base of the pile than at the top (Bryhni, 1964b) suggests that the basin was not 25 km deep. Assuming a simple repetition of the type of cyclothem shown in Figure 7, present estimates of the minimum east-west extent of individual cyclothems (12 km) and of the amount of eastward overlap between adjacent cyclothems (0.25 km), together with an average cycle thickness of 120 m, suggest that the basement may not have been deeper than 6 km below the surface of the basin.

The argument may be taken a stage further. A basin in which the overall rate of vertical subsidence was less than the overall rate of lateral expansion could be described as one developing under the control of a wrench fault system. Examples of basins developing in response to wrench fault systems have been figured and discussed by Crowell (1974), mainly in terms of pullapart basins formed at releasing double bends or sag basins formed adjacent to rising uplands on restraining double bends. In the case of Hornelen Basin, the observed eastward overlap of successive basin-fill increments does not in itself necessarily imply that basin development was controlled by strike-slip faulting, but the following features (Fig. 8) are consistent with such a notion, and they suggest that the basin may have formed on the southern flank of a right-slip (oblique) fault system approximately along the present line of Nordfjord:

1. Because of the predominantly strike-slip movement and dip-slip movement along the northern and southern margins, respectively, of the model basin (Fig. 8c), it is predicted that a reasonable correspondance exists between conglomeratic pebble composition and adjacent "basement" lithology across the southern margin, but that anomalies exist across the northern margin. These predictions are partially confirmed where easily identified augen-gneiss pebbles in conglomerates along the western part of the northern margin conglomeratic belt now lie adjacent to schists, with the parent augen-gneiss displaced eastward (right-slip) by more than 5 to 10 km. In younger conglomerates along the same margin, gabbroic and granodioritic clasts can be less easily matched with basement and may imply considerably great right-slip movement. On the other hand, along the central part of the southern margin of the

basin, a change from quartzite to gneiss as dominant pebble type appears to correspond fairly well with a similar change in adjacent "basement" lithology (Fig. 8b).

2. The faulted southern margin of the model basin is less regular and linear than the northern margin because the development of the southern margin is dependent on the coalescence of a number of hinged, curved southwest-northeast dip-slip faults (Fig. 8c). The latter define the southern boundary of the area of sag caused by the strike-slip movement of basement blocks against each other along a restrictive double bend (see Fig. 4 in Crowell, 1974). The southeastern end of Hornelen Basin, as defined now by the fringing fanglomerates, appears to have resulted from at least two, curved, subparallel fault systems (Fig. 1). In addition, there are prominent, curved northeast-southwest zones of jointing presently cutting the basin, possibly reflecting reactivated basement faulting (Fig. 8a).

3. The facts that the bulk of Hornelen Basin sediments are sandsized and that these filled the basin longitudinally from its east end (Fig. 8b) are reasonably consistent with the strike-slip fault model. In contrast, Solund Basin, of similar size and subparallel to Hornelen Basin but lying some 60 km to the south, is filled with 5 km of coarse conglomerates which were dispersed laterally into the basin. This grain size and mode of filling is more typical of a rift basin, dominated by dip-slip tectonics (Steel, 1976).

CONCLUSIONS

The basic response to vertical movement of Hornelen Basin floor is deduced to have been the creation of marginal relief, gradually increasing sediment discharge into the basin, and the progradation of sedimentary bodies in which a characteristic coarsening-upward sequence was generated.

The major bodies produced by such a period of subsidence are composite, being constructed of marginally dispersed fanglomerates and longitudinally dispersed fluvial sandstones, but they are basin-wide and commonly of the order of 100 m thick. These major bodies are composed, in turn, of thinner tabular units, also coarsening upward, of the order of 10 to 25 m, which are believed to reflect more closely the maximum amplitude of individual phases of subsidence.

The main significance of the 100-m unit is that this vertical interval represents the period after which the locus of subsidence suffered a rapid lateral migration, commonly of the order of 0.25 km. This periodic eastward migration caused an overlap of the basement and the generation of a stratigraphic thickness of some 25 km in the basin.

A possible model for the development of this basin sequence with its spectacular cyclicity is one involving discontinuous strike-slip movement along a major right-slip wrench system.

ACKNOWLEDGMENTS

We thank our colleagues at the Geological Institute (Department A) for stimulating discussions about Hornelen Basin. Professors Crowell (California), Bluck (Glasgow), Sturt and Kvale (Bergen), and Tesseyre (Wroclaw) read the manuscript and suggested improvements. We are also grateful to Professor A. Whiteman, who initiated the West Norway Devonian Project, and to Norges Teknisk-Naturvitenskapelige Forskningsråd, who largely financed it.

REFERENCES CITED

Blatt, H., Middleton, G., and Murray, R., 1972, Origin of sedimentary rocks: Englewood Cliffs, New Jersey, Prentice Hall, 634 p.
Bluck, B. J., 1967, Deposition of some Upper Old Red Sandstone in the Clyde area: A study in the significance of bedding: Scottish Jour. Geology, v. 3, p. 139–167.
Bryhni, I., 1964a, Migrating basins on the Old Red Continent: Nature, v. 202, p. 284–285.
—— 1964b, Relasjonen mellom senkaledonsk tektonikk og sedimentasjon ved Hornelens og Håsteinens devon: Norges Geol. Undersøkelse, v. 223, p. 10–25.
—— 1964c, Sediment structures in the Hornelen series: Norsk Geol. Tiddskr., v. 44, p. 486–488.
Bull, W. B., 1964, Alluvial fans and near surface subsidence in West Fresno, California: U.S. Geol. Survey Prof. Paper 437A, 71 p.
—— 1968, Alluvial fans: Jour. Geol. Education, v. 16, p. 101–106.
Crowell, J. C., 1974, Origin of late Cenozoic basins in southern California, in Dickinson, W. R., ed., Tectonics and sedimentation: Soc. Econ. Paleontologists and Mineralogists Spec. Pub. 22, p. 190–204.
Duff, P.McL.D., Hallam, A., and Walton, E. K., 1967, Cyclic sedimentation: Amsterdam, Elsevier, 280 p.
Hooke, R.L.B., 1968, Steady-state relationships on arid region alluvial fans in closed basins: Am. Jour. Sci., v. 266, p. 609–629.
Kildal, E. S., 1970, Geologisk kart over Norge, berggrunn kart, 1:250,000: Norsk Utgave, Norges Geol. Undersøkelse.
Kolderup, C. F., 1927, Hornelens Devonfelt: Bergens Mus. Arb. 1926, Naturvidensk. Raekke nr. 6.
Lustig, L. K., 1965, Clastic sedimentation in Deep Springs Valley, California: U.S. Geol. Survey, Prof. Paper 35F, 131 p.
Maehle, S., 1975, Devonian conglomerate-sandstone facies relationships and their palaeogeographic importance along the margin of Hornelen Basin, between Storevann and Grøndalen, Sunnfjord [Cand. Real thesis]: Bergen, Norway, Univ. Bergen, 163 p.
Nilsen, H. R., 1975, Sedimentological studies along the central part of the southern margin (Haukå-Storevatn) of Hornelen Devonian basin, western Norway [Cand. Real thesis]: Bergen, Norway, Univ. Bergen, 223 p.
Nilsen, T. H., 1968, The relationship of sedimentation to tectonics in the Solund area of southwestern Norway: Norges Geol. Undersøkelse, v. 259, 108 p.
Spinnangr, Å., 1975, Some sedimentary and stratigraphic studies of the Devonian strata across the western part of Hornelen Basin, Western Norway [Cand. Real thesis]; Bergen, Norway, Univ. Bergen, 247 p.
Skjerlie, F. J., 1971, Sedimentasjon og tektonisk utvikling i Kvamshestens devonfelt, Vest-Norge: Norges Geol. Undersøkelse, v. 270, p. 77–108.
Smith, N. D., 1970, The braided stream depositional environment: Comparison of the Platte River with some Silurian clastic rocks, north-central Appalachians: Geol. Soc. America Bull., v. 81, p. 2993–3014.
Steel, R. J., 1974, New Red Sandstone floodplain and piedmont sedimentation in the Hebridean province, Scotland: Jour. Sed. Petrology, v. 44, p. 336–357.
—— 1976, Devonian basins of western Norway — Sedimentary response to tectonism and varying tectonic context: Tectonophysics, v. 36, p. 207–224.
Steel, R. J., and Wilson, A. C., 1975, Sedimentation and tectonism (?Permo-Triassic) on the margin of the North Minch Basin: Geol. Soc. London Jour., v. 131, p. 183–202.
Steel, R. J., Nicholson, R. N., and Kalander, L. K., 1975, Triassic sedimentation and palaeogeography in central Skye: Scottish Jour. Geology, v. 11, p. 1–13.

MANUSCRIPT RECEIVED BY THE SOCIETY DECEMBER 8, 1975
REVISED MANUSCRIPT RECEIVED JULY 10, 1976
MANUSCRIPT ACCEPTED SEPTEMBER 2, 1976

15

Copyright © 1978 by the Canadian Society of Petroleum Geologists
Reprinted from pages 669-702 of *Fluvial Sedimentology,* A. D. Miall, ed., Canadian Society of Petroleum Geologists Memoir No. 5.

ALLUVIAL FAN SEQUENCE AND MEGASEQUENCE MODELS:
with examples from Westphalian D — Stephanian B coalfields, Northern Spain

ALAN P. HEWARD[1]

ABSTRACT

Sequences, megasequences and basin-fill sequences of progressively changing character appear common to many alluvial fan accumulations. Their study can provide evidence of depositional processes (sequences), short term fan behaviour (sequences), longer term fan behaviour (megasequences) and the depositional basin setting (basin-fill sequences). The occurrence and identification of these sequences appear dependent on down-fan trends in sediment character and depositional process, the localisation, switching and migration of the region of active fan sedimentation, a regularity of depositional event magnitude, and an absence of reworking by subsequent floods.

"The prime requisite of fan formation is the setting of highland and lowland side by side", Denny (1967). Where alluvial fans are dependent solely on initial fault scarp or erosional topography, geographically limited, relatively thin accumulations of decreasing grain size result. Fan sediments of greater geographical extent and thickness, repeated vertical stacking of sequences and megasequences, and the occurrence of alluvial fan basin-fill sequences result from continued movement along fault lines.

INTRODUCTION

"An alluvial fan is a body of detrital sediments built up by a mountain stream at the base of a mountain front" and commonly having the shape of a segment of a cone (Blissenbach, 1954, p. 176). Sedimentation is characterised by an accumulation of debris within a drainage basin and its sporadic transference to the mountain front. Beaty (1970), for example, estimated that one 'typical' debris flow every 350 years was sufficient to explain the growth of a White Mountain fan over the past 700,000 years. Deposition results from deceleration due to the increase in flow width and decrease in flow depth as floods emerge from the confines of the feeder canyon or fanhead channel (Bull, 1964a, 1968; Denny, 1965).

This paper attempts to add to the understanding of ancient alluvial fan deposits through the discussion of vertical sequences of related beds. Such sequences of progressively changing character are of fundamental importance for, as Walker (1970) emphasised in a discussion of turbidite sequences "a trend of any type forces the geologist to think in terms of long period control of . . . sedimentation". In this analysis previous descriptions are utilised[2] of both ancient alluvial fan and submarine fan vertical sequences and of modern and ancient alluvial fan deposits, and an attempt is made to marry the development of vertical sequences with models of alluvial fan behaviour. Some of the concepts derived are then applied to examples of alluvial fan vertical sequences from Westphalian D — Stephanian B coalfields, northern Spain.[3]

---

[1]Department of Geology and Mineralogy, Parks Road, Oxford, OX1 3PR, England; Present address: Department of Geological Sciences, University of Durham, South Road, Durham, DH1 3LE, England.

[2]My usage of submarine fan sequence data stems from their more extensive documentation, and from a belief that in important respects these point source accumulations of gravity induced sediment flows are similar.

[3]Since this analysis was completed relevant new data has been published by Bull (1977) and Wasson (1977a, b) on alluvial fan sediments, and by Martini and Sagri (1977) on submarine fan sequences.

Table 1 : Ancient Alluvial Fan Sequences

Author and Author's Terminology	Scale	Style	Suggested Causes of Sequences.
BLUCK (1967) Upper Old Red Sandstone, Clyde area, western Scotland.			
Sequences	eg. 140 m	Fining (and thinning) upward, mudflow - braided stream deposits.	Recession of source area, or gradual reduction in slope (relief) of source area.
WESSEL (1969; not seen: data from Klein, 1975) Upper Triassic, north-central Massachusetts, U.S.A.			
Sequences.	3-6 m	Coarsening upward single event. Coarsening upward multiple event.	Fan progradation.
WILLIAMS (1969) Torridonian (late Precambrian), northwest Scotland.			
		Upward fining.	Continuous retreat of source area.
		(Thickening and coarsening).	Distal-proximal fan relationship (his fig.14).
MIALL (1970) Devonian, Prince of Wales Island, Arctic Canada.			
	eg. > 40 m	Upward coarsening associated with an 'inverted stratigraphy' in clast types.	Rejuvenation and downcutting in source area.
DEEGAN (1973) Lower Carboniferous, Kirkcudbrightshire, Scotland.			
Megacycles or large-scale cycles.	10's - 100's m	Fining upward.	Basin subsidence and fan aggradation. Progressive decrease in depositional gradient causing finer detritus to be supplied.
		Coarsening upward.	Relatively gradual basin subsidence.
SCHLUGER (1973) Devonian, Perry Formation, New Brunswick, Canada, and Maine, U.S.A.			
Sequences.	10's - 100's m	Upward fining, conglomerate-sandstone-mudstone.	Gradual reduction of highland and lateral shift, alluvial fan-marginal lacustrine sedimentation.
STEEL (1974) New Red Sandstone, Hebridean Province, Scotland.			
Sequences.	10's m	Upward fining, upward increase in % of sandstone beds, upward decrease in extra-formational v. intra-formational clasts, mudflow-streamflood-braided stream.	Initial fault movement followed by lowering of relief and recession of source area.
		Vertically stacked upward fining etc. sequences as above.	Tectonic movements causing drainage basins to be periodically rejuvenated.
BRYHNI and SKJERLIE (1975) Middle Devonian, Old Red Sandstone, Kvamshesten district, western Norway.			
Rhythms or sequences.	10-40 m	Upward coarsening, siltstone-sandstone.	Formed at foot of fans, each rhythm indicates deposition in increasingly higher flow regimes, caused by progressive downcurrent spreading of braid bar sediments resulting from tectonic or climatic changes in source region.
STEEL and WILSON (1975) ?Permo-Triassic, Stornaway Formation, Lewis, northwest Scotland.			
Sequences.	100's m	Fining upward, mudflow - streamflood - braided stream.	Basin margin faulting of gradually decreasing intensity accompanied by sourceward migration of locus of sedimentation, probably implying a lowering of relief within drainage basin.
		Coarsening upward, braided stream - streamflood - mudflow.	Increasing rate of uplift in source area, probably resulting in fan progradation.
STEEL (1976) Devonian, western Norway			
Sequences, cycles, or cyclothems.	50 - > 200 m	Coarsening upward and decrease in amount of interbedded sandstone.	Major period of fan-building in response to fault movement, probably basin subsidence.
Units.	10 m	Coarsening upward, sometimes capped by slight fining upward.	Prograding fan-building lobes, may be in response to fault movement in fan head region.
STEEL ET AL. (1977) Devonian, Bornelen Basin, western Norway.			
Cycles, cyclothems or sequences.	50-80 m	Coarsening and thickening upward, decreasing sandstone percentage, strongly asymmetric.	Fan progradation and increasingly greater volumes of debris, marked asymmetry due to abrupt lateral tectonic shifting of the locus of subsidence and sedimentation.
	100 - 200 m	Coarsening and thickening upward, decreasing sandstone percentage, less asymmetric.	Fan progradation and increasingly greater volumes of debris, lesser asymmetry.
Subcycles, subunits, subcyclothems or sequences.	10-25 m	Coarsening and thickening upward.	Progradation of main dispersal system in response to discrete episodes of fault movement, or an autocyclic mechanism.

Previous Descriptions of Alluvial Fan and Submarine Fan Sequences

Tables 1 and 2 list many of the published descriptions of alluvial fan and submarine fan vertical sequences, *defined on progressive changes in grain size, bed thickness, interpreted depositional processes and clast types*. As can be seen, a confusing terminology exists of cycles, rhythms, sequences, megasequences, megacycles, large scale cycles, cyclothems etc. Throughout this paper the term sequence is used in a general sense, and includes: sequences in a specific sense, m-10's m thick, consisting of a single bed or a series of related beds; megasequences, 10's - 100's m thick consisting of arrangements of related beds and sequences; and basin-fill sequences, 100's - 1000's m thick, consisting of arrangements of sequences and megasequences.

Two types of *alluvial fan sequences and megasequences* have been described, fining upward and coarsening upward (Table 1). Fining upward sequences and megasequences, often accompanied by increasingly distal depositional processes, have generally been considered to result from the reduction of source area relief and/or scarp retreat. Steel and Wilson (1975) suggested that very thick fining upward megasequences might result from basin margin faulting of gradually decreasing intensity. Coarsening upward sequences and megasequences, paralleled by increasing bed thicknesses and by increasingly proximal processes, have been attributed to the progradation of fan building lobes or of the fan itself. Very thick coarsening upward megasequences were interpreted by Steel and Wilson (1975) to result from faulting with a history of progressively greater movements.

Considerably greater attention has been focussed on *submarine fan sequences and megasequences* (Table 2). Early workers (e.g. Kimura, 1966; Sestini, 1970; Sagri, 1972) proposed numerous factors which might control their development, including variations in the rate of subsidence, rhythmic tectonic movements, changes in sediment supply through variations in depositional or erosional rates in source areas, bottom topography, changes in slope angles, combination of turbidity currents from different sources, changes in the spatial arrangement of successive turbidites, and varying distances of the point of deposition from the source of the turbidity current. More recently, however, the specific ideas of Mutti, Ricci-Lucchi and Walker have received fairly general acceptance (Table 2). Fining and thinning upward turbidite and deep water conglomerate sequences have been considered to result from gradual channel abandonment (e.g. Mutti and Ricci-Lucchi, 1972; Mutti, 1974, 1977; Ricci-Lucchi, 1975), although Walker (1975a, 1977) noted that a gradual diminution of sediment supply from the source was equally likely. Van Vliet (in press) emphasised the incompatability of fining and thinning upward sequences occurring in gradually abandoned channels, and progradational coarsening and thickening upward depositional lobe sequences forming below channel mouths. He followed Ghibaudo and Mutti (1973) and Mutti (1974) in suggesting that fining and thinning upward sequences might form by lateral accretion on very low angle point bars. Walker (1975b), in contrast, presented evidence that gradual channel filling may have occurred subsequent to abandonment, a conclusion reached also by Nelson *et al.* (1977).

Coarsening and thickening upward sequences have been considered to result from the progradation of fan depositional lobes, their asymmetry, or symmetry reflecting the abrupt or gradual nature of fan lobe abandonment (Table 2). Coarsening upward megasequences have likewise been attributed to phases of fan progradation. Mutti (1977) and van Vliet (in press) also described refined versions of the basic coarsening and thickening upward sequence considered to be indicative of specific locations on submarine fans (e.g. the thickening upward mouth bar cycles of Mutti, or the abruptly based middle fan megacycles, the gradationally based outer fan megacycles and the diluted fan fringe megacycles of van Vliet).

With the exception of Kimura (1966), Bluck (1967), Sestini (1970), Sagri (1972), Steel (1974), Walker (1975a, 1977) and Steel *et al.* (1977) who also consider gradual changes in

Table 2 : Ancient Submarine Fan Sequences
(sand grade turbidite deposits except where indicated)

Author and Author's Terminology	Scale	Style	Suggested Cause of Sequences
KIMURA (1966) Permo-Triassic, Sambosan Group, and Jurassic-Cretaceous, Shimanto Group, Japan			
Major cycles.	100's m	'Increasing type', coarsening and thickening upward. 'Decreasing type', thinning and fining upward.	Regressions allowing sand into depositional basins. Transgressions excluding sand from depositional basins.
Minor cycles.	10's-100's m	'Increasing type', coarsening and thickening upward, 'decreasing type', thinning and fining upward. Tendency for 'increasing type' to occur during regressions and 'decreasing type' to accompany transgressions.	Series of suggestions: 1. Angle of slope may change intensity of turbidity currents. 2. Change in rate of sedimentation in turbidity current source area. When rate is high, turbidite may transport more sediment. 3. A stronger trigger may produce stronger turbidity currents. 4. Distance of point of deposition from source of turbidity current.
SESTINI (1970) Upper Cretaceous and Tertiary, Italy (northern Apennines), Greece and Turkey, Palaeozoic of Great Britain			
Megarhythms or rhythms.	20-80 m	Periodic changes in stratofacies.	1. Changes in sediment supply. 2. Changes in tectonic activity. 3. Interbedding of turbidites from differing sources.
Cycles or sub-cycles.	m-10's m	Thinning or thickening upward	1. Varying sediment supply. 2. Spatial arrangement of successive turbidites such that progressively thinner or thicker beds accumulate.
MUTTI and RICCI-LUCCHI (1972) Tertiary, northern Apennines, Italy			
Large scale fan sequence (their fig. 14).	100's m		Progradation of fan system.
Megasequence.	10's m	Positive; grain size and bed thickness decrease upwards, channellised.	Gradual abandonment of fan channel.
		Negative; grain size and bed thickness increase upwards, minor channels near top.	Prograding depositional lobes.
SAGRI (1972) Upper Cretaceous and Tertiary, northern Apennines, Italy			
Rhythms or megarhythms.	30-60 m	Thick turbidites at base, upper parts argillaceous with thin beds of fine grained sediments.	Various factors: 1. Rhythmic alternation of tectonic movements and tectonic quiescence. 2. Combination of turbidity currents from different sources. 3. Reduction in clastic inflow due to the increasing stability of continental shelf. 4. Variation in bottom topography.
WALKER and MUTTI (1973) Turbidite facies and facies associations in general			
Large scale fan sequence (their fig.12).			Progradation of fan system over basin plain.
Slope channel, inner fan channel, and middle fan channel sequences.		Thinning and fining upward.	Gradual abandonment of fan distributary channel.
Middle fan sequence.		Thickening and coarsening upward.	Prograding and aggrading fan depositional lobes.
MUTTI (1974) Tertiary, Apennines, Italy, and Island of Rhodes			
Major fan cycles.	100's m	Thickening and (or) coarsening upward.	Fan progradation.
Middle fan cycles.	10's m	Thinning and (or) fining upward, channellised.	Gradual channel abandonment, vertical (or possibly lateral) accretion.
Outer fan cycles.	10's m	Thickening and coarsening upward, minor channels near top.	Prograding fan lobes.

Table 2: continued

RICCI-LUCCHI (1975) Tertiary, Apennines, Italy

1st order cycles, bounded vertically by slope deposits.	3000 m	Marnoso-arenacea, thickening and coarsening upward.	Fan progradation, basin plain-outer fan - middle and inner fan.
	3000 m	Laga, thinning and fining upward.	Fan recession, inner and middle fan - outer fan and minor basin plain deposits.
2nd order cycles.	2-70 m	Asymmetric - positive, thinning and fining upward, mostly channellised.	Gradual channel abandonment.
		Asymmetric - negative, thickening and coarsening upward, mostly non-channelised, minor channels near top.	Progradation and rapid abandonment of fan lobes.
		Symmetric, thickening and coarsening, and thinning and fining. Composite.	Progradation and gradual abandonment of fan lobes.

WALKER (1975a) Upper Cretaceous, Wheeler Gorge, California, U.S.A.

Sequences.	25-45 m	Fining and thinning upward, conglomeratic at base.	Questions Italian workers gradually abandoned channel hypothesis and suggests a gradual diminution of supply in hinterland, giving rise to progressively finer grained and smaller flows, is equally likely.

WALKER (1975b) Miocene, Capistrano Formation, San Clemente, California, U.S.A.

Sequences.	m-10's m	Channellised fining and thinning upward.	Gradual channel abandonment.
		Channellised thickening and coarsening upward.	Progradation of turbidite lobe down a channel.

(Walker's examples occur in channels where mudstone/siltstone drapes provide evidence that channels were cut and abandoned prior to infill)

MUTTI (1977) Eocene, Hecho Group, south-central Pyrenees, Spain

Channel-fill sequences.		Fining and thinning upward.	Gradual channel abandonment.

(channels cut prior to infilling and are characteristically straight rather than meandering)

Channel-mouth bar cycles.		Thickening upward cycle of distinctive thin bedded sediments.	Downcurrent progradation of mouth bar.
Outer fan cycles.		Thickening upward symmetric.	Lobe and lobe fringe progradation and sudden shifting of related feeder system.
		Thickening upward asymmetric.	Lobe and lobe fringe progradation and gradual shifting of related feeder system.

NELSON, MUTTI and RICCI-LUCCHI (1977) Upper Cretaceous, Wheeler Gorge, California, U.S.A.

Upper and middle fan sequences.		Fining upward.	Following channel abandonment, result from infill of channel by overbank flows from adjacent channel.

WALKER (1977) Upper Mesozoic, southwestern Oregon, U.S.A.

Large scale sequences.	50-100 m	Thinning and fining upward.	
Inner fan? sequences.	4-85 m	Thinning and fining upward, conglomeratic at base.	Progressive channel abandonment or long-term changes in the source area leading to smaller and finer grained flows.

RUPKE (in press) Upper Carboniferous, Cantabrian Mountains, northern Spain

1st order basin fill sequences.	1000's m		
2nd order sequences.	100's m	Thickening and coarsening upward facies triplet, conglomeratic at top.	Complete cycle of fan lobe progradation and avulsion.
3rd order sequences.	m-10's m	Thickening upward.	Progradation of outer fan lobes.
		Symmetrical thickening and thinning upward.	Progradation and lateral shifting of middle fan depositional lobes.
		Thinning upward.	Construction of inner fan valley levees.

VAN VLIET (in press) Lower Tertiary, Guipuzcoa, northern Spain

Inner fan megacycles (not present in this area).		Channellised fining and thinning upward.	Channel lateral accretion.
Middle fan megacycles.	20-70 m	Abruptly based, coarsening and thickening upward, channelling near top.	Avulsion into interlobe depression followed by progradation of depositional lobe.
Outer fan megacycles.	10-50 m	Gradationally based, coarsening and thickening upward.	Progradation of depositional lobe, when strongly asymmetrical indicating rapid abandonment.
Fan fringe megacycles.		Gradationally based, coarsening and thickening upward, diluted with thin distal turbidites particularly near base.	Progradation of depositional lobe, dilution by thin turbidites from adjacent lobe(s) until depositional relief causes their exclusion.

the magnitude of causative or depositional events, most authors have related alluvial fan or submarine fan vertical sequences to the accumulation of increasingly proximal or distal deposits through the progradation, lateral migration, gradual abandonment or recession of the location of active sedimentation. Comparing the background of the interpretations for alluvial fan and submarine fan vertical sequences, alluvial fan interpretations draw on geomorphological 'cycle of erosion' concepts, whilst the majority of those of submarine fans arise from limited modern morphological data (e.g. Normark, 1970) and from analogies with deltaic mouth bars, and fluvial and distributary channels. Whilst the sequences are essentially similar in character and similar criteria are used in their definition, the resulting interpretations show considerable variance, particularly for fining and thinning upward sequences.

CHARACTERISTICS OF ALLUVIAL FANS IMPORTANT TO VERTICAL SEQUENCE DEVELOPMENT

Down-fan and down-flow trends in grain size, bed thickness, sorting, clast shape, number of channels, and processes (Table 3, Fig. 1), abundantly recorded from modern alluvial fans and from ancient fan deposits, are pre-requisites of vertical sequences, in that gradual or abrupt progradation or recession of all or part of the fan, leads to the super-position of deposits of gradually or abruptly changing character. More rapid down-fan trends may, perhaps, preferentially favour the development of sequences on smaller fans (Fig. 1b).

Table 3 : Down-Fan and Down-Flow Trends

Modern Alluvial Fans	Ancient Alluvial Fan Deposits
i. DECREASE IN GRAIN SIZE DOWN-FAN OR DOWN-FLOW (Fig.1) Eckis (1928), Chawner (1935), Sharp and Nobles (1953), Bull (1963,1968,1972), Beaty (1963), Bluck (1964), Ruhe (1964), Denny (1965,1967), Lustig (1965), Hunt and Mabey (1966), Scott and Gravlee (1968), Williams (1970), McPherson and Hirst (1972), Meckel (1975), Tanner (1976).	Bluck (1965,1967), Laming (1966), Meckel (1967), Nilsen (1968, 1969), Williams (1969), Miall (1970), Wilson (1970), McGowen and Groat (1971), Groat (1972), Schluger (1973), Steel (1974, 1976), Steel and Wilson (1975), Steel et al. (1977).
ii. DECREASE IN BED THICKNESS DOWN-FAN OR DOWN-FLOW (Fig.1c) Sharp and Nobles (1953), Beaty (1963,1974), Bull (1964a,b,1972).	Bluck (1965,1967), Williams (1969), Steel (1974).
iii. INCREASE IN SORTING DOWN-FAN OR DOWN-FLOW Blissenbach (1954), Scott and Gravlee (1968), Bull (1972), McPherson and Hirst (1972).	
iv. CHANGES IN CLAST SHAPE DOWN-FAN OR DOWN-FLOW (Trends ambiguous and variable due to original clast shapes, lithologies, distance of transport, weathering and abrasion on abandoned fan areas, and selective transport of different shaped clasts) Blissenbach (1954), Bluck (1964), Lustig (1965), Scott and Gravlee (1968), McPherson and Hirst (1972).	Bluck (1965), Laming (1966), Nilsen (1968,1969), Miall (1970), Schluger (1973). N.B. Meckel (1967), and McGowen and Groat (1971) noted no trends in clast shape.
v. DOWN-FAN DISTRIBUTION OF CHANNELS a. Decrease in number and depth apex to toe: Beaty (1963), Bull (1964a,b,1972). b. More abundant near mountain front - in mid-fan downstream from washes heading in pavement - lower part of fan: Denny (1965). c. More abundant in down-fan regions where infiltrated water emerges: Hooke (1967).	
vi. DOWN-FAN CHANGES IN TRANSPORTING AND DEPOSITIONAL PROCESSES Beaty (1963,1974), Denny (1965,1967), Hooke (1967).	Bluck (1967), Steel (1974), Steel and Wilson (1975).

284

a. Decrease in Grain Size Down-fan : lithological variations on Bucaramanga Fan, Colombia (after Tanner, 1976)

	near fan apex	near fan toe	decline in grain size
Gneiss	1 m	10-20 cm	30 cm/km
Schist	1 m	North part of fan	20 cm/km
		South part of fan	10 cm/km
Vein Quartz	20-45 cm	10-13 cm	8 cm/km

b. Decrease in Grain Size Down-fan : variation due to fan size (after Denny, 1965)

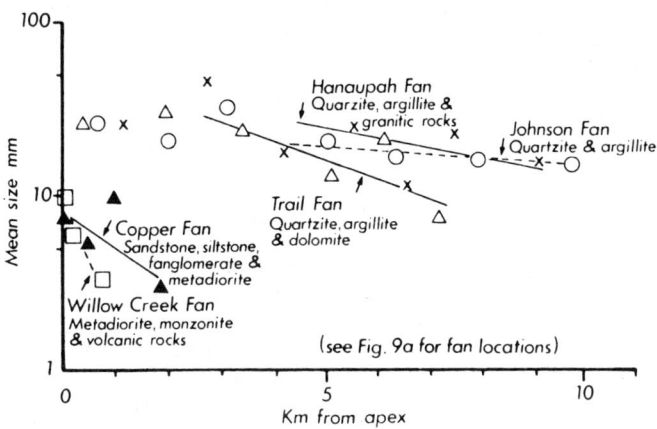

c. Decrease in Grain Size & Bed Thickness Down-flow (after Sharp and Nobles, 1953)

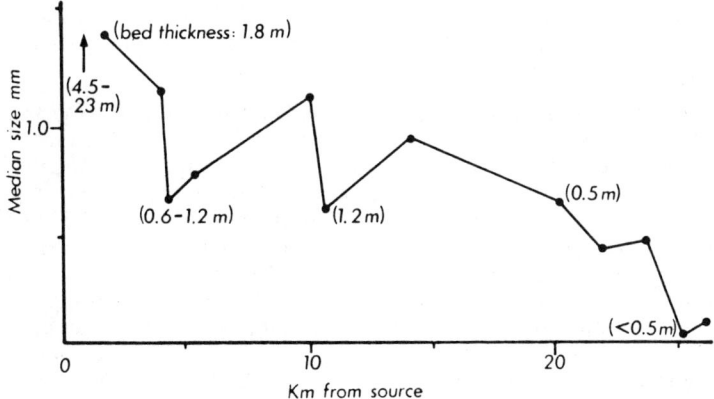

Fig. 1. Examples of down-fan decrease in grain size, and down-fan decrease in grain size and bed thickness from recent alluvial fan deposits.

The permanence of the locus of deposition on a fan surface, important to the development and magnitude of fan sequences, depends on the presence of entrenched channels. The latter typify two regions of a fan (Eckis, 1928), the fan apex where the fanhead trench controls the site of active fan sedimentation, and the fan toe where entrenched channels determine the location of secondary fans (Fig. 7). Over a considerable period of time the cutting, infilling, migration and abandonment of channels results in fairly even, overall, fan sedimentation (Beaty, 1963, 1970, 1974; Denny, 1965, 1967; Lustig, 1965; Hooke, 1967, 1968; Bull, 1968, 1972). In the short and medium term, however, the presence or absence of a fanhead trench is one of the most important features controlling the site of sedimentation (e.g. Buwalda, 1951; Bull, 1968). Numerous factors have been considered as causes of *fanhead entrenchment*. These factors, the estimated duration of their influence, and their possibly characteristic depositional products are summarised in Table 4, and discussed in relation to vertical sequences in a later section.

At any time much of the fan surface is inactive, being the site of weathering, pedogenesis and erosion (Bluck, 1964; Denny, 1965, 1967; Lustig, 1965; Hooke, 1967, 1968, 1972). The resultant non-depositional horizons may occur within, or form the boundaries between, fan sequences.

THE INDIVIDUALITY OF THE ALLUVIAL FAN FLOOD EVENT — FACT OR FICTION

Many students of modern alluvial fans have noted the variability of fan flood events, in terms of amount of precipitation required to initiate an event, the character of the flood discharge and the volume of sediment deposited. Thus Beaty (1963, p. 526) remarked that "debris flowage is undoubtedly a highly irregular process, dependent upon the chance concentration of a cell of heavy rain in a drainage basin the truck canyon of which is floored with unconsolidated materials". Likewise Bull (1964a, p.A4) commented that "floods are controlled by the areal distribution, intensity, and duration of rainfall, and by vegetable cover, lithology, and slopes of the drainage basin. The resulting flow may range from clear water to viscous mud".

Observations, such as these, perhaps prompted Bull (1972, p. 66) to state that "each bed of a fan represents a single depositional event that has resulted from one of a wide spectrum of precipitation and erosion events within the source area. The runoff that is supplied to the main stream channel leading to the fan may be the result of rainfall over the entire basin, or snowmelt runoff from all or part of the basin. Thus, differences in runoff characteristics, source and amount of sediment load, mode of transport, and other factors vary greatly and are reflected in the individual beds preserved in a fan".

Despite the apparent individual characteristics of modern fan flood events, vertical sequences of progressively changing grain size, bed thickness and depositional process characterise many ancient alluvial fan deposits (and submarine fan deposits; Tables 1 and 2; Figs. 12-14). Such sequences are generally interpreted to represent the gradual progradation, avulsion, abandonment, or lateral migration of depositional elements; rather than fractionation of sediment within a depositional element, or progressive changes in the sediment supplied. If the former are the case, the occurrence and identification of sequences require a regularity of depositional event magnitude, for with greater variation sequences would be obscured by the variable size, grain size and properties of the deposits. This possible discrepancy between modern observations and ancient fan sequences can be reconciled in that modern observations refer to flood events of very variable depositional importance and cover only a short time span. In contrast the volumetrically important event of a fan sequence probably occur once/100's or 1000's of years. However, the regularity in magnitude (volume of deposits), within certain limits, of these larger depositional events requires further explanation.

Curry (1966), Statham (1976) and Renwick (1977), when describing modern debris flow events, noted that the exceptional rainfall which preceded flows was no more exceptional than precipitation in previous years which had not resulted in debris flowage. Beaty (1963), Johnson and Rahn (1970) and Scott (1971) documented the accumulation of debris within alluvial fan feeder canyons by landsliding, slumping and rain and rill wash, and Statham (1976) and Beaty (1963, p. 526) noted that "a buildup of fresh debris on their floors must take place before they can again yield large amounts of rubble". These observations when combined with Schumm's (1973) concepts of geomorphic thresholds, can explain why exceptional precipitation frequently does not initiate large-scale fan depositional events (e.g. debris flows) and also why major depositional events for a single fan system are likely to be of similar magnitude.

Consider Fig. 2a (based on Schumm, 1973) where increasing sediment instability is plotted against time. As time progresses colluvial debris accumulates within the feeder canyon, the whole mass becoming increasingly unstable (line 1). At the threshold of instability (line 2) failure will occur and liberate the accumulated volume of sediment (X).

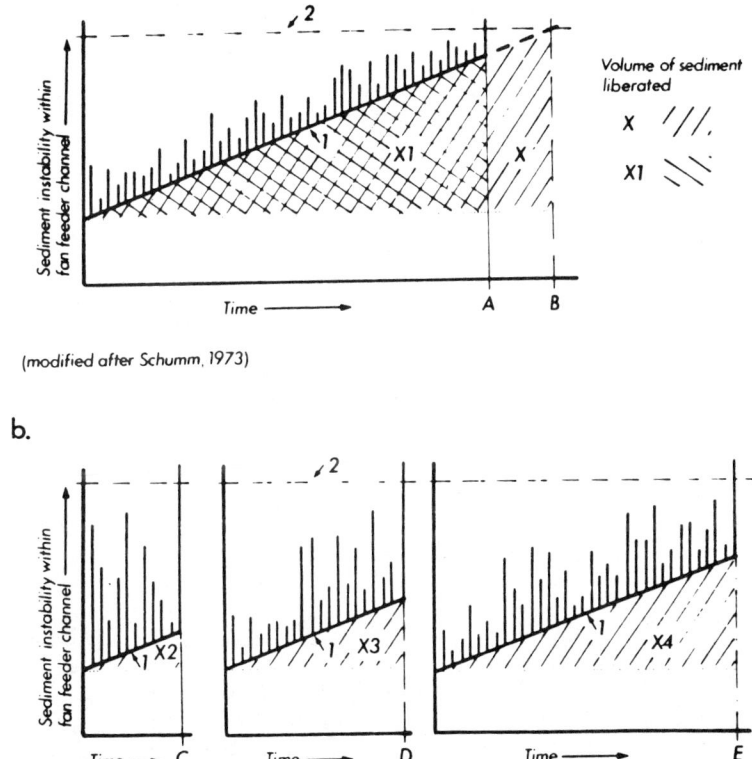

Fig. 2. (a) Hypothetical relationship between instability of the sediment accumulating within an alluvial fan feeder canyon, and time. Superimposed on line 1 are vertical lines representing extra instability induced by major precipitational events. When the ascending line of sediment instability (line 1), intersects line 2, the instability threshold, failure will occur (time B). However, failure occurs at time A as the result of a precipitational event. Diagonally hatched areas X and X1 represent the volume of accumulated sediments liberated at the time of failure. (b) As above, except extra instability induced by precipitational events is more variable and extreme, relative to the instability threshold (line 2). Hence, the considerable variability in volume of accumulated sediments liberated (X2, X3, X4) at the times of failure (C, D, E).

287

Superimposed on line 1 are vertical lines representing the extra instability induced by exceptional precipitation. At time A the threshold of instability is exceeded and flow occurs (failure would anyway have occurred at time B). The amount of extra instability induced by flood events in Fig. 2a is minor relative to the level of the instability threshold (line 2), with the result that when failure occurs similar volumes of accumulated sediment are liberated (e.g. X or X1). When the system lies below the instability threshold exceptional precipitation cannot induce flowage, so explaining the observations of Curry (1966) and Renwick (1977). Scott (1971), when describing the 1969 debris flows near Glendora, California, commented that the most striking aspect of these events was the degree to which channels in all parts of the watershed had been scoured to bedrock. He noted that six months after the storms, debris was again accumulating by gravitational processes.

In Fig. 2b the extra instability of each flood event is larger relative to the instability threshold (line 2) and very variable volumes of sediment could be liberated (e.g. X2, X3, X4). Sequences of progressively changing character representing fan segment progradation, migration, abandonment etc., however, imply an inherent regularity in event magnitude. That such sequences commonly occur in alluvial fan and submarine fan deposits suggests that drainage basin settings of the type illustrated by Fig. 2a may be

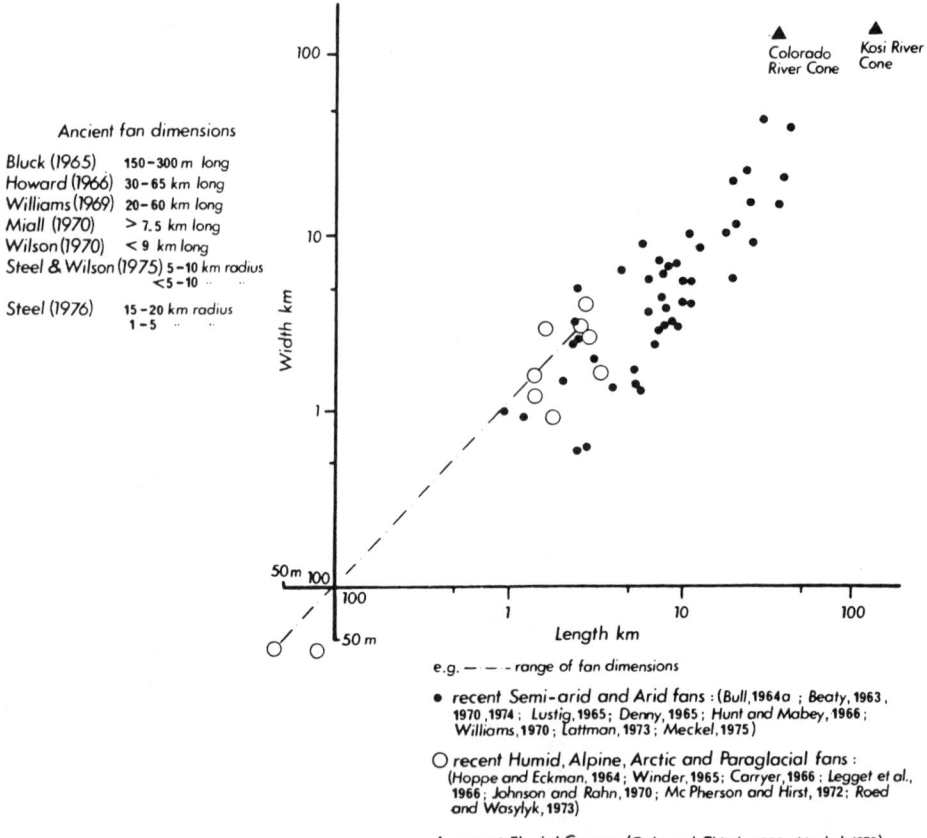

Fig. 3. Dimensions of recent and ancient alluvial fans. (Note log scales; n = 58). Anstey (1965) provides numerous additional dimensions of recent semi-arid and arid fans from the western United States and West Pakistan.

more normal in nature. For flood deposits which have suffered no subsequent reworking, the concepts of geomorphic thresholds and regularity in depositional event magnitude can help explain the maximum particle size/bed thickness relations (flood competence) described from ancient alluvial fan (Bluck, 1967; Steel, 1974) and submarine fan sediments (Walker, 1977).

TECTONIC SETTINGS OF ALLUVIAL FANS

"The prime requisite of fan formation is the setting of highland and lowland side by side", Denny (1967, p. 83). Alluvial fans normally extend from mountain fronts into valleys or bordering lowland areas for only 5-20 km (Fig. 3). Most modern fans are associated with fault scarps (Davis, 1925; Eckis, 1928; Sharp and Nobles, 1953; Blissenbach, 1954; Hunt and Mabey, 1966; Rahn, 1967; Beaty, 1970; Scott, 1971, 1973; Hooke, 1972; Wasson, 1974; Meckel, 1975; Tanner, 1976). Faults provide the initial relief required for fan formation and continued movement can lead to further accumulation and preservation of fan sediments.

There are three contrasting basin margin/alluvial fan settings (Fig. 4) which govern the thickness and extent of alluvial fan successions, and also the character and arrangement of internal vertical sequences (discussed later).

Adjacent to a *relatively permanent basin margin fault* alluvial fan sediments are stacked up along a prominent fault or fault zone (Fig. 4a). This type of basin margin/alluvial fan setting commonly occurs on the downthrown side of major strike-slip faults and may persist through several geological epochs. Crowell (1973, 1974) described the alluvial fan Violin Breccia from this type of setting in Ridge Basin, California. The Violin Breccia is 10,000 m thick, 1500 m along strike, extends only a km or so into Ridge Basin, and is now displaced laterally by strike-slip movement 28 km from its distinctive source area. Crowell (1974) suggested that the strewing out of coarse debris across a strike-slip fault, with a dip-slip component, may occur commonly in such basins. Tanner (1976), similarly documented the Bucaramanga fault, a major strike-slip fault in Colombia. Six conglomeratic alluvial fan formations, Permo-Triassic to recent in age, and 1000's m thick occur on the downthrown side of this fault. Steel (1976), considered that Devonian alluvial fan deposits from the Hornelen Basin, Norway, accumulated adjacent to a strike-slip fault. He noted that the fans commonly extended 1 km from the basin margin and that internal vertical sequences were comparatively thick.

Limited back-faulting of the basin margin (Fig. 4b) results in linear, moderate thickness, alluvial fan accumulations along prominent basin margin faults, particularly those bounding grabens or half grabens (Sharp, 1948; Howard, 1966; Belt, 1968; Miall, 1970; Groat, 1972; Schluger, 1973; Hubert *et al.*, 1976). Continued uplift and dissection of the source area is commonly indicated by an 'inverted stratigraphy' of conglomerate clast types (Sharp, 1948; Miall, 1970; Wilson, 1970). Back-faulting results in limited directional younging of alluvial fan formations. Sharp (1948) and Miall (1970) have respectively described Eocene (48 km long, 5 km wide and 1200 m thick) and Devonian (93 km long, 16 km wide and 300 m thick) alluvial fan accumulations of this type.

Repeated back-faulting of the basin margin (Fig. 4c) can result in geographically extensive coarse grained alluvial fan successions which young in the direction of back-faulting. Although the total stratigraphic thickness of such accumulations may be very considerable, the thickness deposited and preserved on any single downthrown block is notably less. Steel and Wilson (1975) first described Permo-Triassic alluvial fan deposits in terms of this setting and commented that under such conditions of progressive back-faulting, simple 'inverted stratigraphies' of conglomerate clast types were less likely to occur. More recently Steel (1976) and Steel *et al.* (1977) have interpreted the Devonian 5 km thick conglomerate fill of the Solund Basin, and the majority of the 25 km thick sandstone fill of the Hornelen Basin, Norway, in terms of this model.

a. Relatively Permanent Basin Margin Fault or Fault Zone
(after Crowell, 1973)

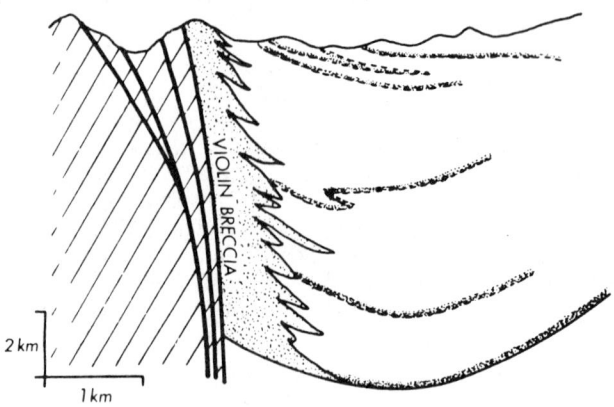

b. Limited Back-faulting of Basin Margin
(after Belt, 1968)

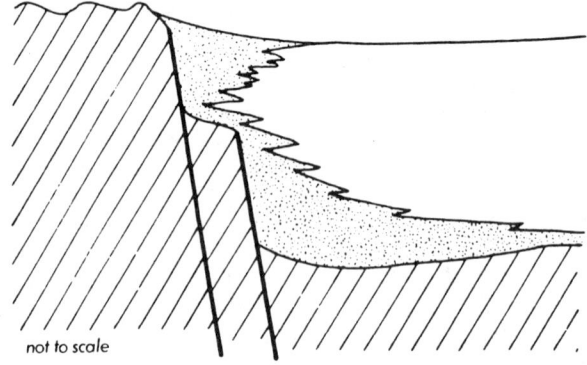

c. Repeated Back-faulting of Basin Margin
(after Steel and Wilson, 1975)

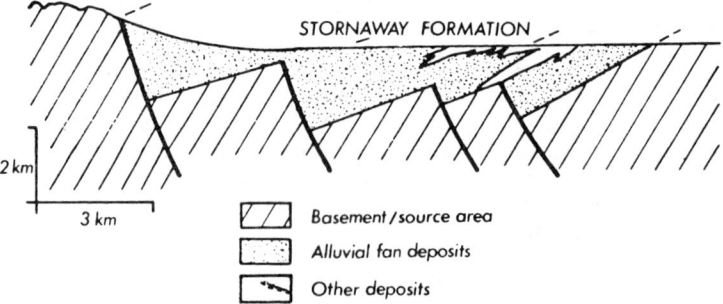

Fig. 4. Alluvial fan / basin margin settings.

a. Response to Initial Topography

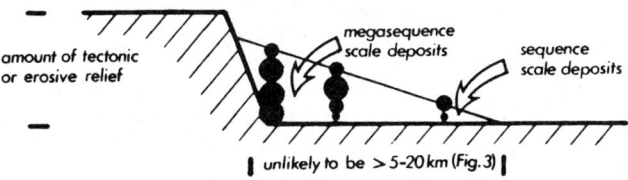

b. Short-Moderate Duration Fanhead Entrenchment: resulting from intrinsic or climatic factors (Table 4)

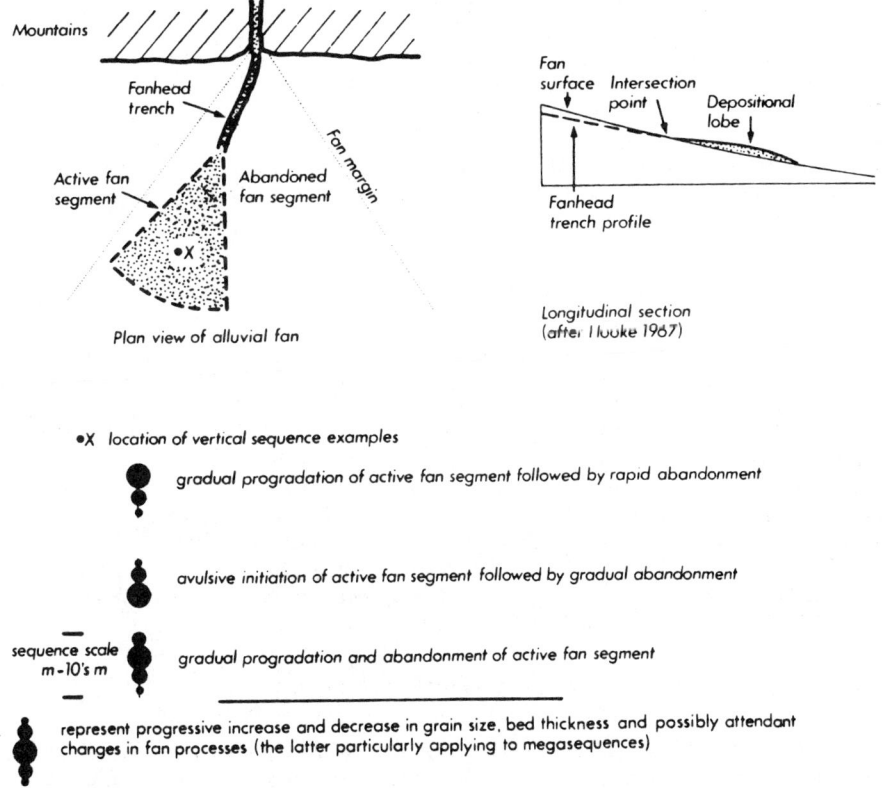

Fig. 5. Hypothetical alluvial fan behavioural models. (a) Response to initial topography. (b) Short-moderate duration fanhead entrenchment.

HYPOTHETICAL ALLUVIAL FAN BEHAVIOURAL MODELS

This section attempts to marry some of the interpretations of ancient fan sequences (Tables 1 and 2) with characteristics of modern alluvial fans and their possible behavioural patterns. In such a comparison one is faced with the problem of determining the relative importance of tectonic, climatic and sedimentologic events. Schumm (1976) predicted that five scales of fining upward sequence would be associated with erosional evolution (cycles in his description); 1st order sequences reflecting the initiating tectonic uplift, 2nd order sequences (occurring within the 1st order sequences etc.) resulting from isostatic rebound or climatic change, 3rd order sequences representing the influence of intrinsic thresholds (Schumm, 1973, 1976), and the 4th and 5th order sequences reflecting complex response, seasonal and flood effects. The models below follow the relative order of magnitude of Schumm's (1976) and Bull's (1968, p. 102-103) events.

Response to initial topography (Fig. 5a)

From a starting point of tectonic or erosional topography, alluvial fans may form where streams leave highland and enter lowland regions. As fans build upward and outward, older distal fan deposits are progressively overlain by younger mid-fan and proximal fan sediments. The products of original down-fan trends (Table 3, Fig. 1) become arranged in megasequences and sequences of increased proximality (coarsening, thickening and increasingly proximal processes upward). In the absence of uplift, as source area relief is reduced, lesser amounts of finer grained debris become available, resulting in a partially symmetrical megasequence (Fig. 5a). Close to the mountains a relatively constant, coarser grained megasequence may eventually fine upward. The ultimate thickness of the fan accumulation is the extent of the original topography, and fan deposits formed under such tectonically non-reactivated settings have been reported 10's - 100's and occasionally 1000's m in thickness (e.g. Allen, 1965; Selley, 1965; Laming, 1966). Megasequences near the mountain front are equivalent to sequences near the fan toe. An 'inverted stratigraphy' of conglomerate clast types may occur as successive rock types are eroded in the source area (e.g. Laming, 1966).

Secondary fans resulting from localised drainage of abandoned areas of primary fan

Secondary fans occurring at the toes of primary accumulations can be of diverse character and origin (Fig. 7). In their simplest form they result from drainage established on abandoned areas of the primary fan and debris is entirely reworked from the primary fan (Fig. 7a.2; Blissenbach, 1954; Ruhe, 1964; Denny, 1967; Meckel, 1975). Such accumulations of weathered and reworked fan material, which may include multi-cycle conglomerate clasts (Tanner, 1976), are better sorted and finer grained than the primary fan (e.g. Ruhe, 1964, primary fan 20 mm median clast diameter, secondary fan 1.5 mm median clast size at comparable distance from source). Secondary fans of this type are unlikely to be of great volumetric significance and may be typified by asymmetrical — partially symmetrical coarsening upward sequences[1], the nature of the sequence termination reflecting the character of abandonment.

Short-moderate duration fanhead entrenchment (Fig. 5b)
— resulting from intrinsic or climatic factors

For varying periods of time deposition on alluvial fans may be localised through the entrenchment of fan channels, particularly the fan feeder channel. Fanhead entrenchment

---

[1]Throughout this section, the terms coarsening or fining upward also imply parallel trends in bed thickness and depositional processes accompanying the increased proximality or distality (Table 3, Fig. 1).

Table 4 : Fanhead Entrenchment

	Factors Causing Entrenchment	Estimated Duration of Influence *1	Depositional Results *1
	INTRINSIC FACTORS (those inherent within the system)		
i.	Variable nature of storm events (Denny, 1967), extensive scour and fill can occur during a storm (Sharp and Nobles, 1953; Beaty, 1970,1974; Scott, 1971,1973), or at the declining floodwater stage (Blackwelder, 1928; Beaty, 1963, 1974).	Short-lived	Accumulation of small volume of similar grade sediment below the intersection point (intersection point is where fanhead trench merges with the alluvial fan surface, Hooke, 1967).
ii.	Natural result of an alternation of debris flow and stream flow processes (Bluck, 1964; Hooke, 1967).	Short-lived	Accumulation of small volume of similar grade sediment below the intersection point.
iii.	When the locus of deposition shifts to a topographically low area that has not received sediment for some time (Hooke, 1967).	Short-lived	Accumulation of small volume of similar grade sediment below the intersection point.
iv	Build up of slope of fan apex until it exceeds the stability threshold. Trenching then flushes sediment down-fan and reduces slope near apex (Schumm, 1973; Weaver and Schumm, 1974).	Short-lived	Accumulation of small volume of similar grade sediment below the intersection point.
v.	Capture of fan feeder channel by an adjacent channel or a minor channel heading in abandoned area of fan (Denny, 1965,1967; Goreau and Burke, 1966; Hooke, 1967,1968; Troxel 1974).	Short - prolonged	Accumulation of small-considerable volume of similar or coarser grade sediment, perhaps transported by a differing process. Channel capture may be indicated by abrupt differences in clast type (Hunt and Mabey, 1966).
	EXTRINSIC FACTORS (those external to the system)		
vi (CLIMATE CHANGES)	An increase in the volume of storm floodwater (Eckis, 1928), due to a temporary increase in precipitation (Bull, 1964b,c).	Short - moderate duration	Accumulation of small-moderate volume of similar or coarser grade sediment, perhaps transported by a differing process.
vii (CLIMATE CHANGES)	A decrease in the volume of storm floodwater due to a climatic change towards increased aridity (Antevs, 1952; Lustig, 1965; Williams, 1970).	Short - moderate duration	Accumulation of small-moderate volume of similar or differing grade sediment, perhaps transported by a differing process. Deposition may be confined to channels (Lustig 1965).
viii (MAN)	As a consequence of overgrazing and human activities (eg. Antevs, 1952; Schumm and Hadley, 1957).	Prolonged	Fortunately, irrelevant to geological past.
ix (DECREASING DEBRIS SUPPLY)	Resulting from a decrease in debris supply (paraglacial fans in particular, Carryer, 1966; Ryder, 1971), perhaps due to a lowering of relief or to the exposure of more resistant rock types in the drainage basin (Eckis, 1928).	Prolonged	The decreasing amounts of debris supplied to fan and reworking of weathered former fan deposits will probably result in the accumulation of small-moderate volume of increasingly fine grained debris.
x (DECREASING DEBRIS SUPPLY)	As the product of downcutting during the cycle of erosion. As highland areas are reduced in altitude the slopes of stream channels are also reduced causing entrenchment of the fanhead (Eckis, 1928; Carryer,1966).	Prolonged	The decreasing amounts of debris supplied to fan and reworking of weathered former fan deposits will probably result in the accumulation of small-moderate volume of increasingly fine grained debris.
xi (LOWERING OF BASE LEVEL)	Resulting from lowering of base level of the fan or of the depositional basin, by the truncation and dissection of a fan by a major river and its tributaries (Drew, 1873; Eckis, 1928; Blissenbach 1954; Ryder, 1971), or by a change in basin character from internal to external drainage (Denny, 1967; Groat, 1972).	Prolonged	Moderate-large accumulation of debris derived from source area and reworked fan. Extent of eventual preservation questionable.
xii (TECTONIC CHANGES)	Following continued tectonic relative uplift of source area (Bull, 1964b; Hunt and Mabey, 1966; Denny, 1967; Williams, 1970), or tilting (Hooke, 1972).	Prolonged	Large accumulation of debris (probably relatively coarse grained) transferred from actively dissected drainage basin onto the fan. Gradual changes in clast type may reflect progressive dissection of source area (eg. Miall, 1970).

*1 predictions of author generally based on previous writer's comments.

results in an active fan segment, whilst adjacent temporarily abandoned fan surfaces are subject to weathering, pedogenesis and erosion (Fig. 5b).

Various factors which can lead to short-moderate duration fanhead entrenchment are listed in Table 4. These are mainly intrinsic factors, the variable nature of storm events, the alternation of differing transportational processes, and the development of depositional topography, the products of which are probably indistinguishable. However, entrenchment resulting from the capture of the fan feeder channel may be indicated by changes in clast type and clast size within the accumulated sediments (Hunt and Mabey, 1966). Such entrenchment may be prolonged and result in complete abandonment of the primary fan (Goreau and Burke, 1966). Entrenchment following climatic changes may be apparent through differing transporting processes being operative. Whilst adjacent fans will probably have differing geomorphic thresholds and hence response periods following climatic changes, such changes should occur regionally (Hooke, 1967; Schumm, 1973, 1976).

Short-moderate duration fanhead entrenchment probably results in the accumulation of small-moderate volumes of sediment below the intersection point (Hooke, 1967; Fig. 5b). Such accumulations have proximal-distal characteristics as in Table 3, and internal vertical sequences will reflect the gradual or abrupt initiation and termination of sedimentation (Fig. 5b). Such m-10's m thick alluvial fan accumulations appear comparable to the suprafan lobes of some modern submarine fans (e.g. Normark, 1970), whose progradation, migration, and gradual or abrupt abandonment are considered to result in submarine fan vertical sequences of similar scale and style (Table 2; Mutti and Ricci-Lucchi, 1972; Walker and Mutti, 1973; Mutti, 1974, 1977; Ricci-Lucchi, 1975; Rupke, 1977, van Vliet, in press).

Secondary fans resulting from stream capture

The feeder streams of secondary fans which head in and rework abandoned areas of the fan may eventually capture the feeder channel of the primary fan and receive sediment directly from the source area. Secondary fans of this type may be identifiable due to the abrupt change from finer grained reworked fan sediments to coarser debris derived directly from the source area, perhaps by differing processes (Fig. 7a.3).

Prolonged fanhead entrenchment (Fig. 6a)
— resulting from decreasing sediment supply or lowering of base level (Table 4).

Prolonged fanhead entrenchment results in the active fan segment occurring at the fan toe in the form of a secondary fan. Secondary fans due to entrenchment following a decrease in sediment supply are predominantly of debris reworked from primary accumulations (Carryer, 1966). Those resulting from a lowering of base level may initially be of reworked debris, but with their gradient advantage they are likely, eventually, to capture the supply from the source area (Fig. 7b). Asymmetric coarsening upward sequences, or symmetrical coarsening and fining upward sequences may typify such accumulations due to gradual progradation and abandonment. However, on those fans characterised by a decline in amount and grade of source area debris, fining upward sequences may predominate. Fan accumulations of these types are unlikely to be of great volumetric importance and in the case of those initiated by lowering of base level, the eventual preservation of such deposits is questionable (Fig. 7b; Groat, 1972).

Scarp retreat and lowering of relief (Fig. 6b)

Scarp retreat and the lowering of relief results in the reworking of the proximal parts of successive fan accumulations as a broad alluvial fan apron develops. At the site of the original progradational fan accumulation a remnant coarsening upward sequence may occur. However, fining upward sequences predominate as localities become more distant

a. *Prolonged Fanhead Entrenchment* : resulting from decreasing sediment supply, the latter perhaps due to the advanced state in the cycle of erosion; or due to lowering of base level (Table 4)

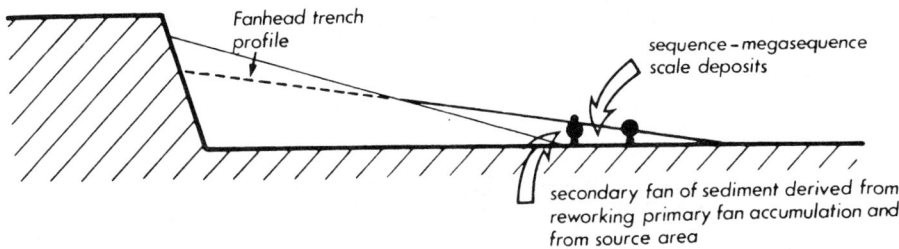

b. *Scarp Retreat and Lowering of Relief* :

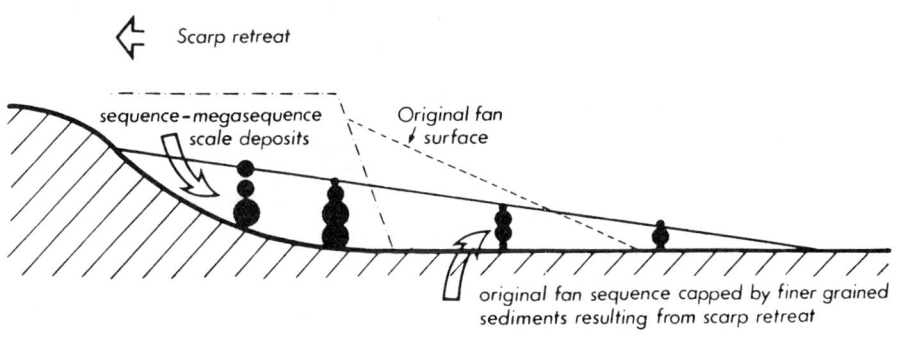

e.g. ● represent progressive increase and decrease in grain size, bed thickness and possibly attendant changes in fan processes (the latter particularly applying to megasequences)

Fig. 6. Hypothetical alluvial fan behavioural models (continued). (a) Prolonged fanhead entrenchment. (b) Scarp retreat and lowering of relief.

from a source area of reduced relief (Fig. 6b; Bluck, 1967; Williams, 1969; Deegan, 1973; Steel, 1974; Steel and Wilson, 1975). Deposits 10's m (e.g. Steel, 1974, Fig. 14) to 100's m thick may result (Bluck, 1967, Fig. 13). Changes in conglomerate clast types may occur as differing lithologies are eroded.

Response to tectonic uplift (Fig. 8)

Recent alluvial fans respond to tectonic uplift in two ways, (1) where relative uplift exceeds streams dissection, active fan segments occur immediately adjacent to the fan apex (Fig. 8a), (2) where the rate of stream dissection exceeds relative uplift, prolonged entrenchment occurs (Table 4) and the active fan segment occurs at the fan toe (Fig 8b). In each case progradation of the active fan segment results in an asymmetrical coarsening upward sequence or megasequence which may be terminated by a further phase of uplift. Progressive offlap of fan segments as in Fig. 8a, or an eventual decline in the grade of supplied debris prior to the next phase of uplift may result in sequences and

295

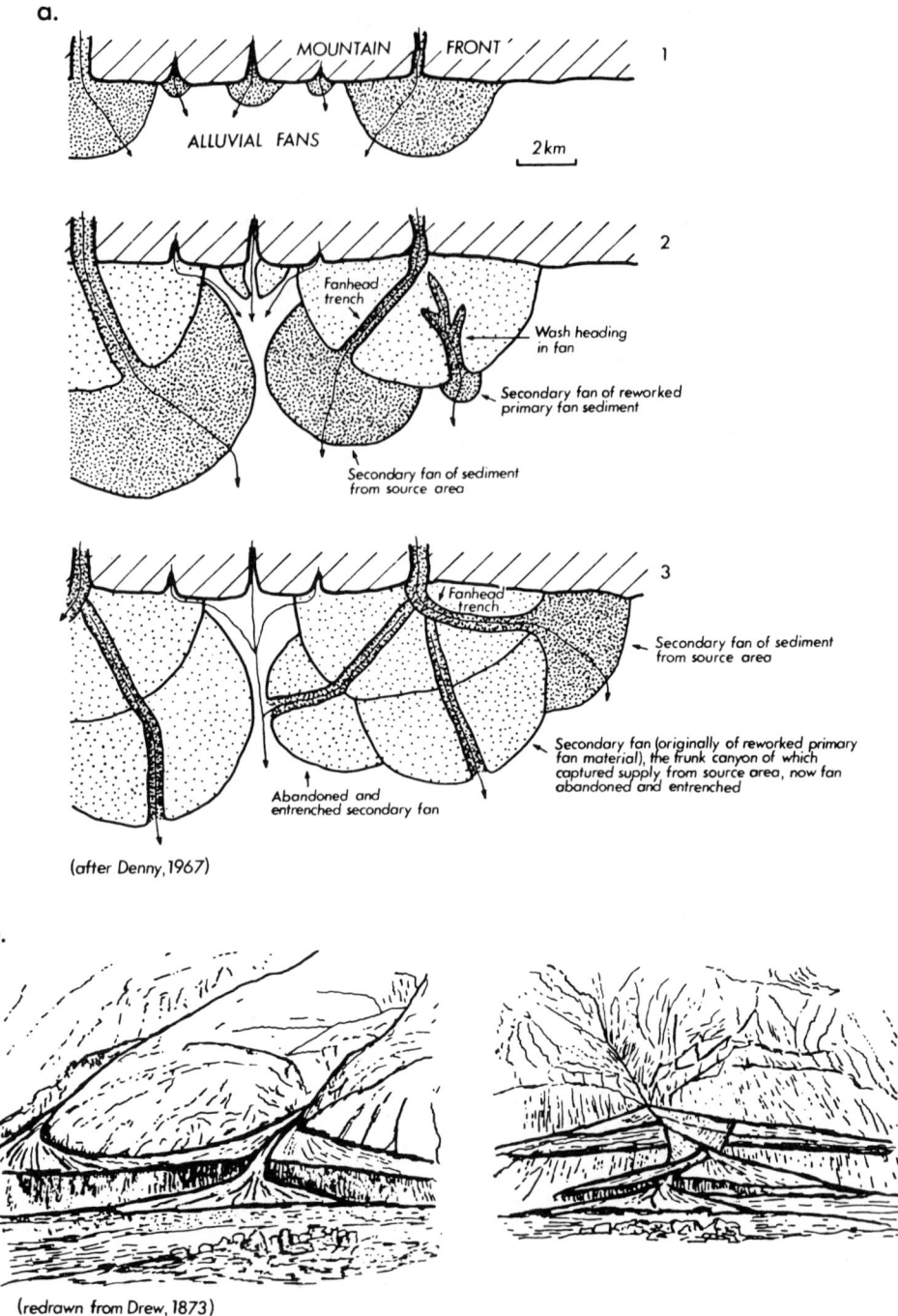

Fig. 7. Secondary fans. (a) Development of alluvial fans, after Denny (1967), illustrating secondary fans of diverse origins. (b) Secondary and triple fans, Upper-Indus Basin, resulting from the lowering of base level by the erosion of fan toes by degrading trunk river.

a. <u>Response to Tectonic Uplift</u> : relative uplift exceeds rate of stream dissection

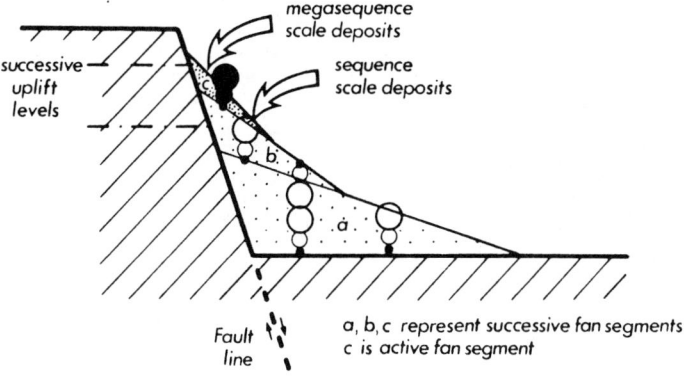

b. <u>Response to Tectonic Uplift</u> : prolonged entrenchment (Table 4), as stream dissection exceeds rate of relative uplift

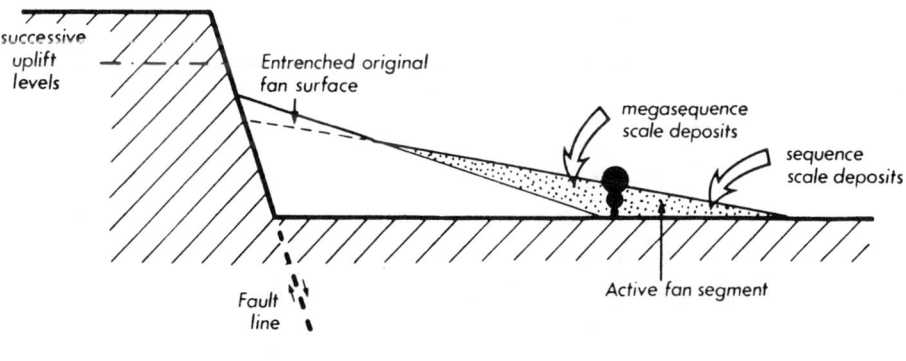

e.g. ● represent progressive increase and decrease in grain size, bed thickness and possibly attendant changes in fan processes (the latter particularly applying to megasequences)

Fig. 8. Hypothetical alluvial fan behavioural models (continued). (a) Response to uplift, where relative uplift exceeds rate of stream dissection, (b) Response to uplift, where stream dissection exceeds rate of relative uplift.

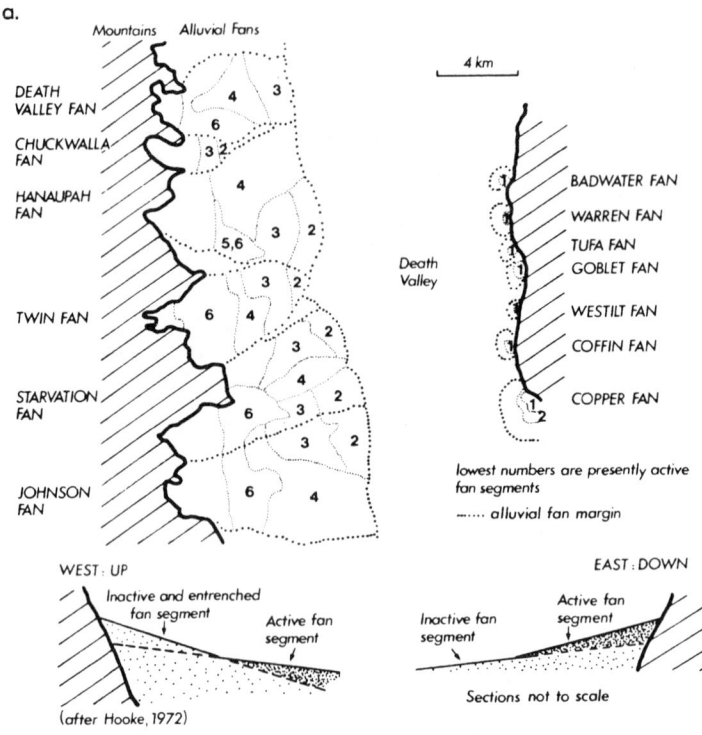
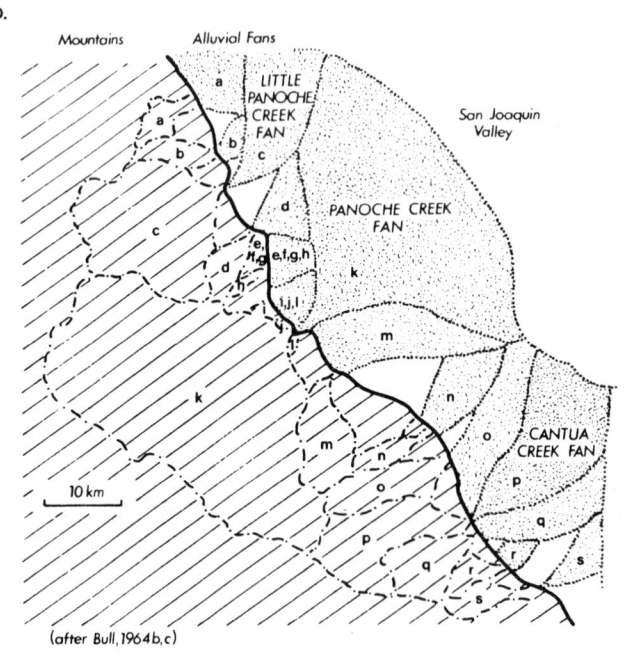

Fig. 9. Examples of the responses of alluvial fans to relative uplift (see text for details).

> **Table 5 : Summary of Alluvial Fan Behavioural Models**
>
> Response to Initial Topography (Fig.5a):
>
> Asymmetrical - partially symmetrical coarsening[*1] upward megasequence near fan apex - sequence near fan toe.
>
> Immediately adjacent to mountain front relatively constant coarser grained megasequence may eventually fine upward.
>
> Secondary fans resulting from streams reworking abandoned areas of primary fan.
>
> Short-Moderate Duration Fanhead Entrenchment (Fig.5b):
>
> resulting from - variable nature of storm events, alternation of differing transportational processes, development of fan topography, capture of fan feeder channel, or climatic changes.
>
> Asymmetrical or symmetrical coarsening or fining upward sequence reflecting gradual or abrupt initiation or termination of sediment supply.
>
> Secondary fans resulting from the capture of fan feeder channel and deriving sediment directly from the source area.
>
> Prolonged Fanhead Entrenchment (Fig.6a):
>
> resulting from - decrease in sediment supply, or lowering of base level.
>
> Asymmetrical coarsening upward sequence, or symmetrical coarsening and fining upward sequence. Fining upward sequences may predominate or fans characterised by a decline in amount and grade of source area debris.
>
> Scarp Retreat and Lowering of Relief (Fig.6b):
>
> Asymmetrical fining upward sequence - megasequence.
>
> Response to Tectonic Uplift (Fig.8):
>
> resulting in - deposition immediately adjacent to fan apex, or prolonged entrenchment and deposition concentrated at fan toe.
>
> Asymmetrical - partially symmetrical coarsening upward megasequence (proximal locations) - sequence (distal locations).
>
> [*1] In addition to grain size changes, parallel trends in bed thickness and depositional processes are also implied, accompanying the increased proximality or distality (Table 3, Fig.1).

megasequences of a more symmetrical character. Large accumulations of debris, perhaps including 'inverted stratigraphies' of conglomerate clast types, probably result from active drainage basin dissection following repeated uplift.

These responses of recent alluvial fans to uplift are derived from the excellent descriptions of Bull (1964b) and Hooke (1972). Bull (1964b) described fans in Fresno County, California, fed predominantly by ephemeral streams (Fig. 9b). Little Panoche Creek fan, Panoche Creek fan and Cantua Creek fan have drainage basins extending further into the mountains and are fed by intermittent streams. With the exception of Little Panoche Creek fan and Wildcat Canyon fan (b and Fig. 9b), all the other fans have responded to relative uplift of the mountains by developing active fan segments, with progressivley steeper slopes, adjacent to the fan apex. On Little Panoche Creek fan and Wildcat Canyon fan stream dissection has exceeded the rate of uplift such that the fans are strongly entrenched and the active fan segment occurs near the fan toe.

Similarly, Hooke (1972) documented fans on the east and west sides of Death Valley, California (Fig. 9a). Progressive tilting of Death Valley (west side up, east side down) causes fans on the west side to be large, with active fan segments developed at the fan toes. In contrast fans on the east side are smaller and have active segments adjacent to the apex.

Table 5 summarises the fan behavioural models described in this section.

Vertical Arrangement of Alluvial Fan Sequences and Megasequences

If alluvial fans are dependent solely on initial erosive or tectonic topography only relatively thin successions of decreasing grain size accumulate (e.g. Allen, 1965; Selley, 1965; Laming, 1966; Bluck, 1967; Williams, 1969). Greater thicknesses of fan sediments and more complex vertical arrangements of fan sequences result from continued uplift along fault lines (e.g. Belt, 1968; Nilsen, 1969; Steel and Wilson, 1975; Steel, 1976; Steel *et al.*, 1977). "The presence of a thick sequence of fanglomerate in the geologic column indicates that deformation and sedimentation were concurrent processes", Denny (1965, p. 58). The minor faulting and slight angular unconformities recorded commonly from alluvial fan successions provide evidence of such deformation (Eckis, 1928; Sharp, 1948; Hunt and Mabey, 1966; Bluck, 1967; Miall, 1970; McGowen and Groat, 1971; Bryhni and Skjerlie, 1975; Steel and Wilson, 1975, Riba, 1976; Tanner, 1976; Steel *et al.*, 1977).

Except for the suggestion by Steel (1974, 1976) and Steel *et al.*, (1977), that stacked arrangements of sequences result from periodic drainage basin rejuvenation caused by tectonic movements, there has been little discussion of the vertical arrangement of fan sequences. Several distinct types of arrangement can be envisaged reflecting the basin margin/alluvial fan setting and hence type of depositional basin (Fig. 4). In Fig. 4a, whilst the effects of repeated fault movement may be evident in sediments occurring at some distance out into the depositional basin, fan vertical sequences or megasequences will persistently be stacked adjacent to the fault line. Strike-slip faulting may result in some lateral displacement of successive fan accumulations. In Fig. 4b, limited back-faulting of the basin margin may result in a series of sequences stacked up against a single fault line until the succeeding fault takes over as the basin margin. At this time the sediments occurring in a proximal position adjacent to the first fault may lie at a very distal locality relative to the second. Abrupt changes from proximal to distal fan deposits, or from fan to laterally equivalent environments (e.g. fluvial, deltaic, lacustrine etc.) may indicate this type of setting. Figure 4c, repeated back-faulting of basin margin, illustrates a more extreme version of the previous case in which abrupt changes may be more common. The geographical extent of alluvial fan sediments (cf. Fig. 3), complexity of successive fan palaeocurrents (e.g. Steel and Wilson, 1975) and progressive basin marginward displacement of fan accumulations (Steel *et al.*, 1977) may be further indicators of this latter setting.

Examples of Alluvial Fan Sequences, Megasequences and Basin-Fill Sequences from Westphalian D — Stephanian B Coalfields, Northern Spain

The Upper Carboniferous succession in the Cantabrian Mountains, northern Spain, excedes 15,000 m in thickness and becomes increasingly continental upward. Alluvial fan and lacustrine sediments predominate during the Stephanian. Sequences, megasequences and basin-fill sequences have been recorded from the La Magdalena, Matallana (Ciñera-Matallana) and Sabero coalfields (Heward, in press) and the Tejerina syncline (work in progress; Fig. 10).

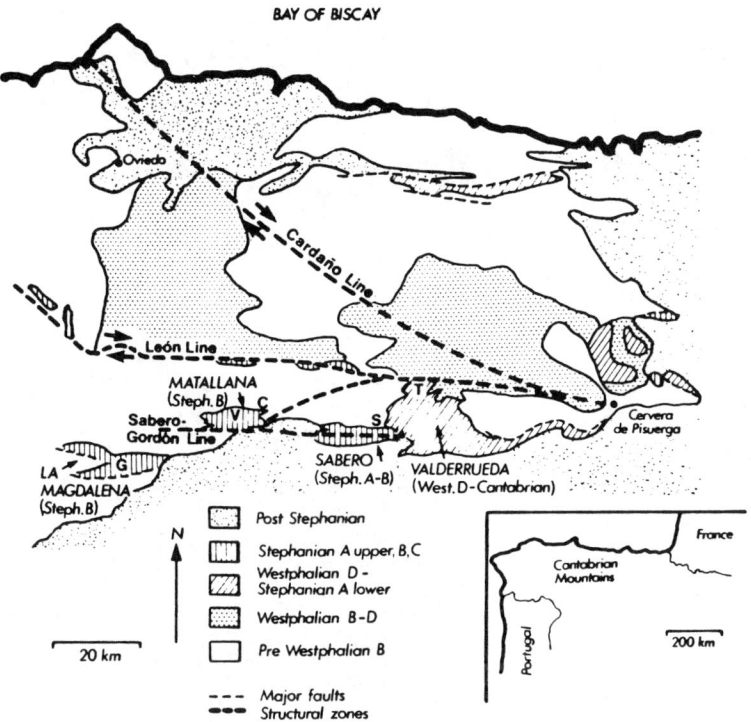

Fig. 10. Geological map of the Cantabrian Mountains, northern Spain. Stratigraphic ages (based on floras and faunas) for the coalfields from Wagner (1970) and Knight (1975). C = Correcillas, G = Garaño, S = Saelices, T = Tejerina, V = Vegacervera, refer to locations mentioned in text or figure captions.

Sequences

Three types of sequence-scale accumulations occur, consisting of single beds or series of related beds. They are m - 10's m thick, have progressive grain size trends and other characteristics, and are separated by non-depositional or slow depositional horizons, such as coals, prominent rootlet beds, or lacustrine shales.

Single beds attaining sequence scale and having internal grain size and morphological trends are generally conglomeratic. In the examples illustrated (Fig. 11), the well rounded conglomerates are poorly sorted and unstratified. They are composed of relatively homogeneous clast supported conglomerate but contain occasional cross-bedded sandstone lenses. These conglomerates are inversely graded or inversely coarse tail graded (Fig. 11) and have an imbricate fabric in which long axes parallel the flow direction determined from adjacent cross-bedding. The sharp non-erosive bases, lateral extent, absence of stratification, lack of sorting, presence of inverse grading, and nature of the fabric, are all in accord with the transportation and deposition of these conglomerates by a debris flow mechanism (Johnson, 1970; Fisher, 1971; Middleton and Hampton, 1973; Walker, 1975c; Enos, 1977). *These single event vertical sequences provide evidence of processes operative on these Stephanian alluvial fans.*

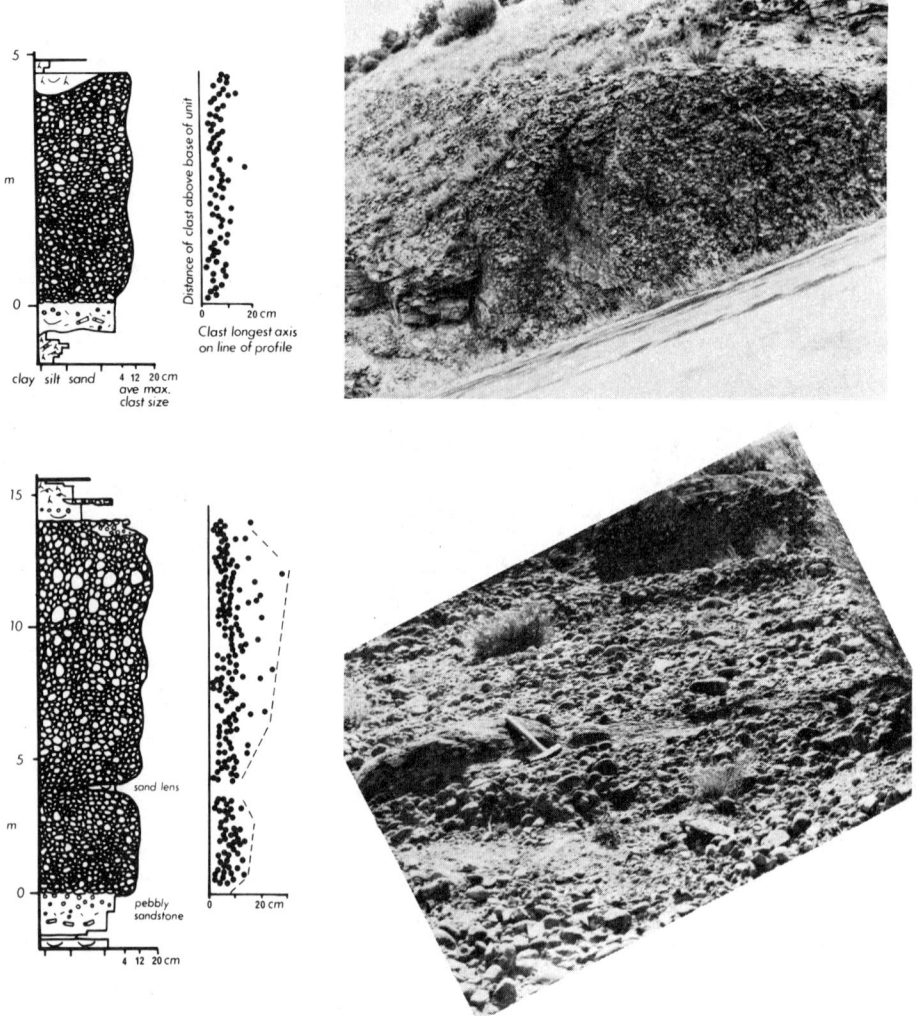

Fig. 11. Examples of poorly sorted, inversely graded, clast supported unstratified conglomerate sequences from the La Magdalena coalfield. The average maximum clast size in these conglomerate sequences was measured from the long axes of 25 largest clasts, 1-5 m either side of the section line, at intervals normally not greater than 1 m. Both conglomerate sequences young to top right; hammer handle 30 cm long. La Magdalena — Mora de Luna road, north of Garaño. (Legend on Fig. 12).

15 - 65 m thick sequences of conglomerate beds comprise the second type of vertical sequence recognised (Fig. 13a). Laterally extensive conglomerate beds are crudely organised into sequences in which maximum clast size, overall clast size and bed thickness decline upward, and matrix support becomes more prominent upwards (fining and thinning upward sequences). Conglomerate beds are poorly sorted, lack stratification and again have characteristics of debris flow deposits. The sequences may represent gradually abandoned, relatively proximal, depositional lobes (following short-moderate duration fanhead entrenchment, Table 5). Their character, scale and stacked nature appear to argue against origins resulting from decreasing sediment supply, lowering of

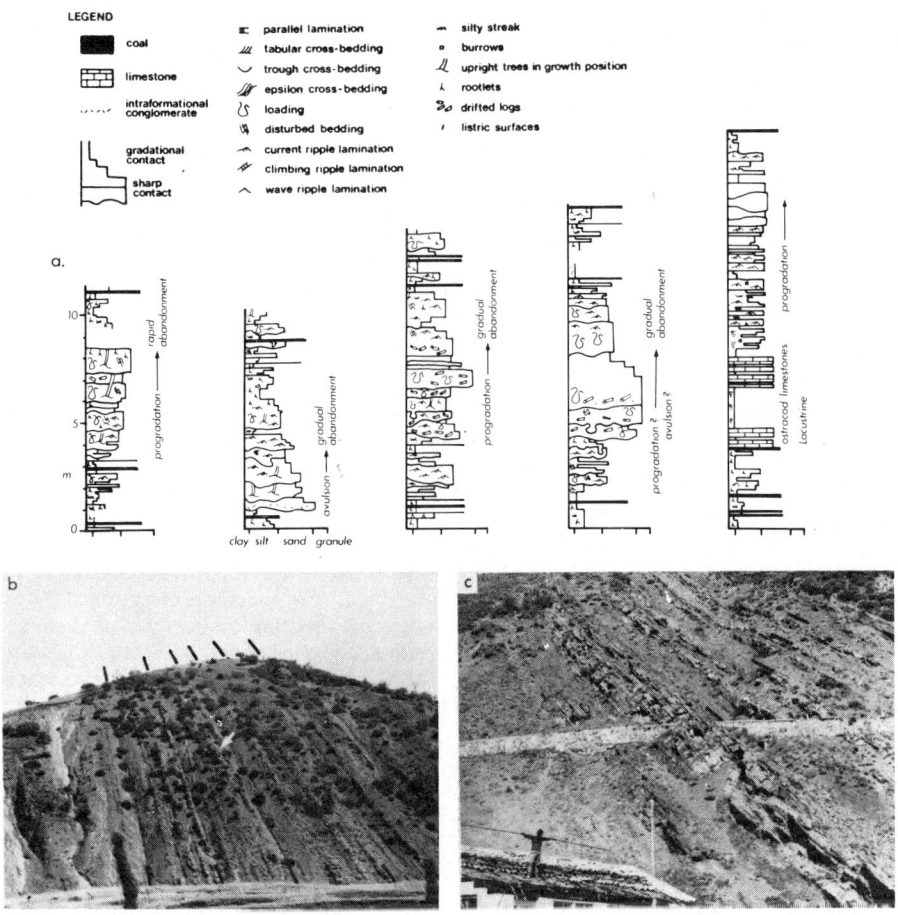

Fig. 12. Examples of distal fan sheetflood sandstone coarsening and thickening, and fining and thinning upward sequences. (a) With the exception of the last sequence from near La Magdalena, La Magdalena — Mora de Luna road. Last sequence from Vegacervera, Matallana coalfield. (b) Asymmetrical coarsening and thickening upward sheetflood sandstone sequences, except sequence containing channel (arrowed, 1.5 m deep) which is partially symmetrical. Succession youngs to top right. North of Saelices, Sabero coalfield. (c) Above prominent lacustrine coarsening upward sequence (10 m thick), series of asymmetrical lacustrine / distal fan coarsening and thickening upward sequences. Succession youngs to top right; aqueduct traverses centre of photograph. Vegacervera, Matallana coalfield.

base level, or from scarp retreat. These conglomeratic alluvial fan fining and thinning upward sequences appear closely comparable to submarine fan sequences of similar style, generally considered to result from gradual channel abandonment (Table 2). The conglomerate beds of this example are traceable for 100's m - km and show no evidence of internal channelling, even though they occur within a palaeovalley eroded in pre-Stephanian sediments (2 km wide and 300 m deep). The crude organisation of the described alluvial fan sequences (Fig. 13a) and their irregular thicknesses may result from a drainage basin setting of the type illustrated by Fig. 2b (rather than Fig. 2a). Alternatively, it could be argued that the sequences are too poorly defined to really warrant distinction. Closer analysis is clearly required.

The third type of sequence observed are 1.5 - 15 m thick and consist of a series of laterally extensive sheet sandstones separated by siltstones, mudstones, rootlet beds and coals (Fig. 12). The 10 cm - 1.5 m thick sheet sandstones are sharply based, occasionally solemarked (grooves) and individual beds fine upward. Palaeocurrents, from current ripple cross-lamination within and grooves on the bases of these sandstones, parallel the general direction of sediment dispersal. Well preserved plants occur within the sandstones and upright (in growth position) and tilted fossil tree trunks abound (Fig. 12a). These features, in combination with abundant syn-sedimentary loading and water escape structures, indicate rapid accumulation. The sandstone beds are arranged in sequences separated by coals, lacustrine shales, or prominent rootlet horizons. Sequences of beds may be asymmetrical, and coarsen and thicken, or fine and thin upward; or may be symmetrical, and consist of the superimposition of the two (Fig. 12). Some of the coarsening and thickening upward sequences may directly overlie lacustrine shales (Fig. 12a, c).

The rapidly accumulated sheet sandstones are interpreted as sheetflood deposits (Heward, in press). Their fine grain size and interbedding with lacustrine deposits, suggest a distal fan position. Such distal fan sediments could be derived directly from the source area or might be reworked from proximal fan areas (Denny, 1965; Rahn, 1967). However, the organisation of these sheetflood sandstones into sequences suggest that they are distal representatives of depositional lobes undergoing progradation or abandonment, rather than a succession of variable storm events reworking older proximal fan deposits. The sequences probably result from short-moderate duration fanhead entrenchment (Table 5) and closer observation may allow distinction of possible causes of such entrenchment (e.g. of sediment type, depositional process, character and thickness of sequence compared with surrounding sequences etc.). The fine grain size of these sequences (compared with associated deposits), similar sequence thickness and repetitive stacked nature (Fig. 12b, c) would appear to preclude origins from more prolonged fan head entrenchment resulting from decreasing sediment supply, lowering of a base level, or a response to tectonic uplift (Table 5). As has been noted with submarine fan sequences of similar type, minor channels (as in Fig. 12b) occur near the top of some coarsening and thickening upward sequences (Table II; Mutti and Ricci-Lucchi, 1972; Mutti, 1974, 1977; Ricci-Lucchi, 1975; van Vliet, in press).

The above multiple event vertical sequences appear to provide evidence of short term fan behaviour; more careful study of their character and sediment type may allow elucidation of its causes. In other fan deposits, multiple event sequences may also reflect fractionation of sediment within a fan environment and hence, indicate depositional processes (e.g. lateral migration of a meandering fan channel, gradual abandonment of a fan channel, levee progradation etc.).

Megasequences

Four types of alluvial fan megasequences can be recognised within the described coalfield deposits. In each case they are asymmetrical and are characterised by progressive changes in grain size, bed thickness, interpreted depositional processes and/or depositional environments (Figs. 13, 14). *Megasequences appear to provide some insight into longer term alluvial fan behaviour.*

Figure 13a illustrates the first type, conglomeratic megasequences, 110 - 120 m thick in which maximum clast size, overall clast size and bed thickness decrease upwards, and matrix supported conglomerates become increasingly abundant, compared with the predominant clast supported conglomerates. These megasequences probably were deposited at progressively more distal locations, relative to the source area, caused by scarp retreat and/or a lowering of relief (Table 5). Their thickness and stacked nature argue against gradual fan segment abandonment following short-moderate duration fanhead entrenchment, or from prolonged entrenchment due to decreasing sediment supply or lowering of base level.

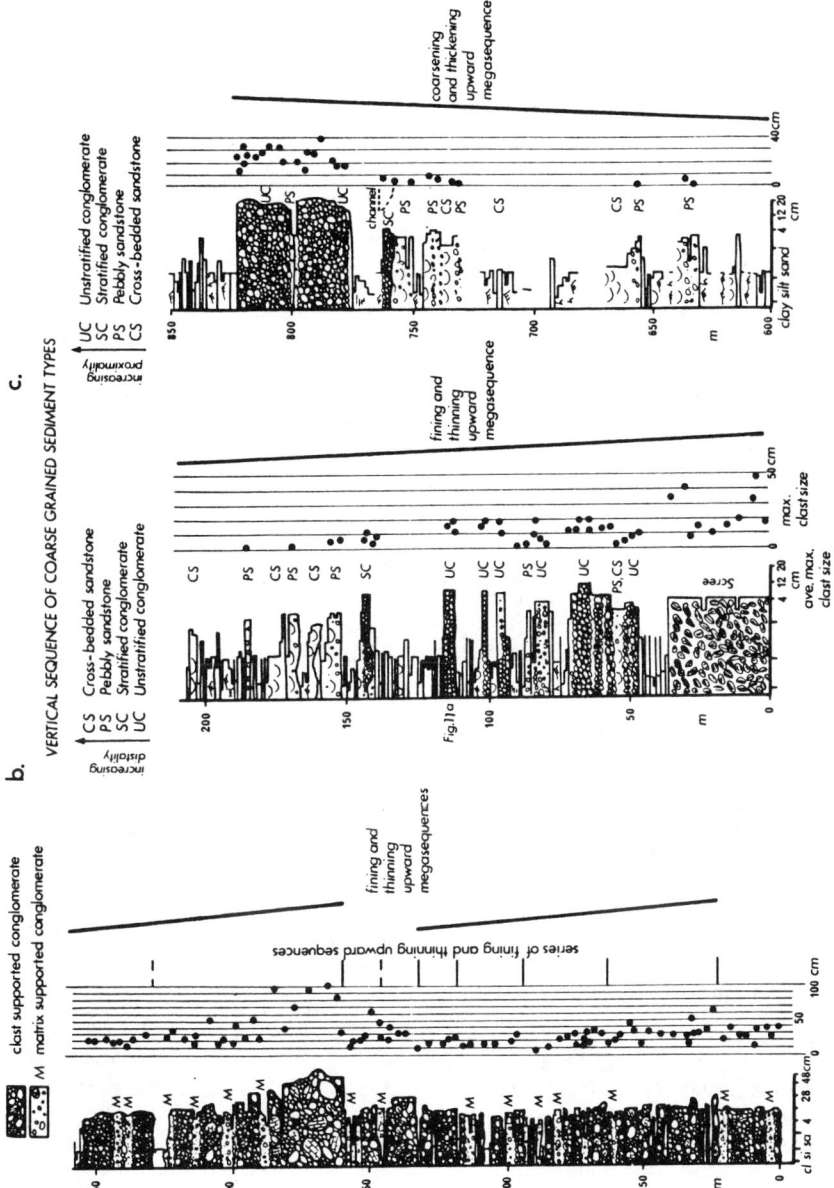

Fig. 13. Examples of alluvial fan megasequences. (a) Fining and thinning upward megasequences from the Correcillas conglomeratic valley-fill at the base of the Matallana coalfield succession. (b and c) Fining and thinning, and coarsening and thickening upward megasequences from the La Magdalena coalfield succession (Fig. 14c). The average maximum clast size in these conglomeratic megasequences was measured from the long axes of the 25 largest clasts, 1-5 m either side of the section line, at intervals normally not greater than 1 m. Maximum clast size represents the largest long axis encountered whilst measuring the above.

A fining upward megasequence of different character occurs near the base of the La Magdalena coalfield succession north of Garaño (Figs. 10, 13b). In this megasequence, fining and decreasing bed thickness are associated with a trend from scree deposits — predominantly unstratified conglomerates — stratified conglomerates — pebbly and cross-bedded sandstones. These deposits are interpreted as resulting from progressively down-flow processes (Heward, in press), and hence indicate increasing distality. Mudstones and siltstones with abundant well preserved plant remains and occasionally thin coal seams occur interbedded throughout. This fining and thinning upward megasequence is interpreted in a similar fashion to the previous description.

Immediately overlying this fining upward megasequence are two coarsening upward megasequences, 210 and 220 m thick (Fig. 14c). Figure 13c illustrates the second of these, where an increase in maximum clast size, general conglomerate clast size and bed thickness is paralleled by a transition from pebbly and cross-bedded sandstones, through stratified conglomerates, and into unstratified conglomerates (increasing proximality). These coarsening and thickening upward megasequences are considered to reflect progradation in response to tectonic uplift, deposition being concentrated by prolonged entrenchment at the fan toe (Table 5). Their scale, consistency of the very maximum clast size (40-50 cm), absence of multi-cycle clasts, stacked nature of the megasequences and their position within the coalfield succession would appear to suggest that a secondary fan origin resulting from reworking the primary fan accumulation, or an origin resulting from short-moderate duration fanhead entrenchment, was unlikely.

The fourth type of megasequence, coarsens upward over 300 - 650 m, from thick, laterally extensive lacustrine shales into coal bearing distal fan sheetflood sandstones and occasional fine grained conglomerates (Fig. 14c; Knight, 1975, and pers. comm.). Four megasequences of this type, from the Sabero coalfield record fan progradation following rapid basin deepening and probably result from relative tectonic uplift (Table 5). Their scale and stacked nature suggest they are not the distal fan products of short-moderate duration, or prolonged fanhead entrenchment.

Basin-fill sequences

The depositional model proposed for these Stephanian coalfields is illustrated in Fig. 14a (Heward, in press). Slight modifications to this model are required to explain the Westphalian D — Cantabrian, Tejerina succession, where quartzitic conglomerates interfinger with limestone rich conglomerates, stratified conglomerates are more abundant, and distal fan sediments interfinger with marine shales. In the following discussion of basin-fill sequences, composite vertical successions are presented for Matallana and Sabero, whilst those of Tejerina and La Magdalena represent continuous vertical sequences (Fig. 14). In none of the coalfields has detailed analysis of single fan morphologies been attempted, to date.

Three general aspects of the basin-fill sequences of Fig. 14 appear worthy of note. Firstly, each succession is 1500 - 2500 m in thickness and consists entirely of alluvial fan deposits, with thin intercalations of lacustrine or marine shales. The accumulation and preservation of such thicknesses of fan sediment indicate repeated fault movements. Overall, these basin-fill sequences fine upward, a feature common to thick alluvial fan accumulations as the depositional basin becomes distant from the source area, or as source area relief becomes subdued (Howard, 1966; Nilsen, 1969; McGowen and Groat, 1971). Secondly, each succession has a marked horizon where conglomerates and coarse grained proximal fan deposits are abruptly overlain by very fine grained distal fan, lacustrine or marine shales. This abrupt change occurs at varying levels within each succession and may reflect back-faulting (Fig. 4b, 4c), thus, deposits initially close to a fault line suddenly lie distant to the succeeding fault line. Thirdly, each succession comprises a series of stacked sequences and megasequences (Fig. 14). The stacking of alluvial fan sequences and megasequences, whatever their internal characteristics,

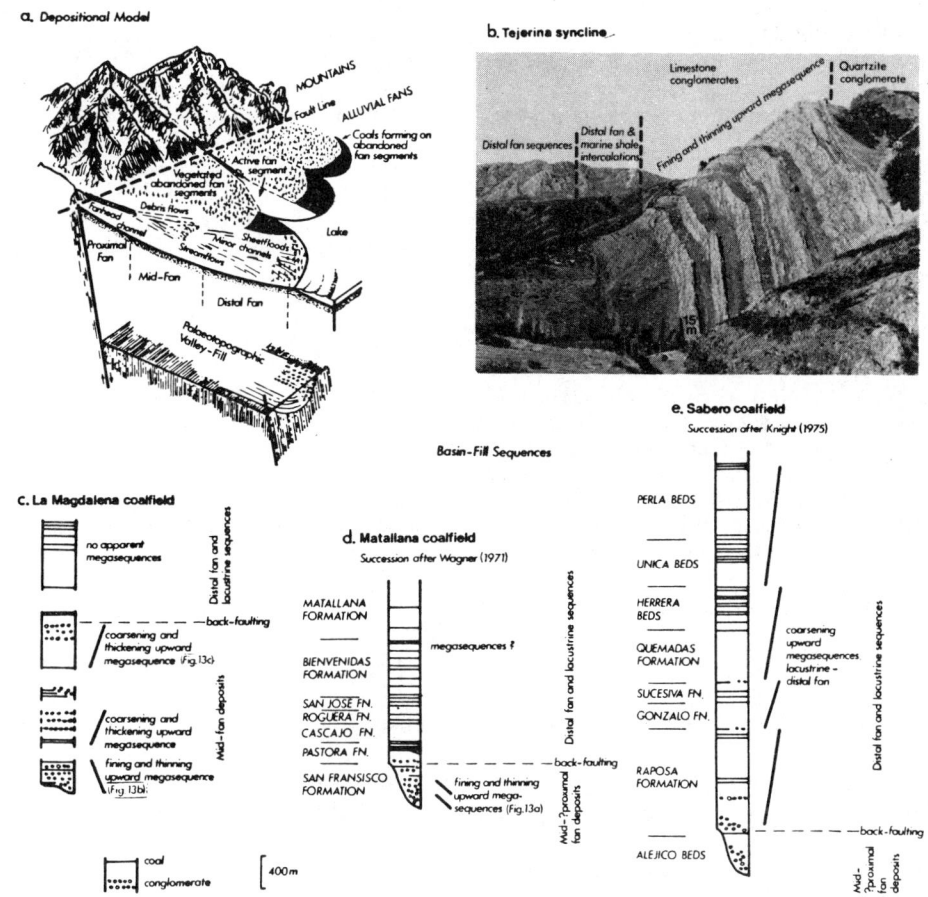

Fig. 14. Depositional model and basin-fill sequences (see text for details). Photograph (b) illustrates part of the Tejerina syncline alluvial fan succession and youngs to bottom left.

probably indicates repeated basin margin faulting and source area rejuvenation (Steel, 1974, 1976; Steel *et. al.*, 1977).

CONCLUSIONS

The functions of vertical sequence or facies models were summarised by Walker (1975d). If properly applied they can be powerful tools in the development of sedimentological understanding. However, all too commonly we are content to re-document an existing model for strata of a different age, in a different geographical area, and the comparative power of the model and even the processes and controls upon which the model depends are forgotten. The intention of this analysis has been the collation of some of the processes and controls which might affect alluvial fan vertical sequences, and to illustrate the necessity for detailed analysis of individual fan sequences rather than the establishment of a 'general or local sequence model' through the distillation of many.

Sequences, megasequences and basin-fill sequences of progressively changing character appear common to many alluvial fan accumulations (Table 1). Their study can provide evidence of depositional processess (sequences), short term fan behaviour (sequences), longer term fan behaviour (megasequences) and the depositional basin setting (basin-fill sequences). The occurrence and identification of these sequences appear dependent on downfan trends in sediment character and depositional process, the localisation, switching and migration of the region of active fan sedimentation, a regularity of depositional event magnitude, and an absence of reworking by subsequent floods. At the present day, alluvial fans occur most commonly under semi-arid climatic conditions (Blissenbach, 1954; Bull, 1972). In the geological past one can only speculate on the climatic setting most favourable to alluvial fan formation (Schumm, 1968). However, under conditions of increased rainfall and more permanent streams, flood debris may be subject to more intense stream reworking and primary sequences of fan flood events be modified or obscured.

Acknowledgments

This study represents an extension of ideas presented in the author's D. Phil. thesis. Harold Reading supervised this thesis and Shell International Petroleum Company Ltd. provided a post-graduate grant. The paper was written whilst the author was employed at Koninklijke/Shell Exploratie en Produktie Laboratorium, Rijswijk, The Netherlands. Howard Johnson, John Knight, Mike Leeder, Bruce Levell, Harold Reading and Ron Steel kindly criticised the manuscript. Some of the ideas expressed were alien to certain of the above and the author is solely responsible for misconceptions.

References

Allen, J. R. L., 1965, The sedimentation and palaeogeography of the Old Red Sandstone of Anglesey, North Wales: Yorks. Geol. Soc. Proc., v. 35, p. 139-185.

Anstey, R. L., 1965, Physical characteristics of alluvial fans: U.S. Army Natic Laboratories, Tech. Rept. ES-20, 109p.

Antevs, E., 1952, Arroyo — cutting and filling: J. Geol., v. 60, p. 375-385.

Beaty, C. B., 1963, Origin of alluvial fans, White Mountains, California and Nevada: Ann. Assoc. Am. Geogr., v. 53, p. 516-535.

———, 1970, Age and estimated rate of accumulation of an alluvial fan, White Mountains, California: Am. J. Sci., v. 268, p. 50-77.

———, 1974, Debris flows, alluvial fans and a revitalised catastrophism: Z. Geomorph. Suppl. Bd. 21, p. 39-51.

Belt, E. S., 1968, Carboniferous continental sedimentation, Atlantic Provinces, Canada: *in* G. deV. Klein, *ed.*, Late Paleozoic and Mesozoic Continental Sedimentation, Northeastern North America; Geol. Soc. Am. Spec. Paper 106, p. 127-176.

Blackwelder, E., 1928, Mudflow as a geological agent in semi-arid mountains: Geol. Soc. Am. Bull., v. 39, 465-484.

Blissenbach, E., 1954, Geology of alluvial fans in semi-arid regions: Geol. Soc. Am. Bull., v. 65, p. 175-190.

Bluck, B. J., 1964, Sedimentation of an alluvial fan in southern Nevada: J. Sediment Petrol. v. 34, p. 395-400.

———, 1965, The sedimentary history of some Triassic conglomerates in the Vale of Glamorgan, South Wales: Sedimentology, v. 4, p. 225-245.

———, 1967, Deposition of some Upper Old Red Sandstone conglomerates in the Clyde area: A study of the significance of bedding: Scott. J. Geol., v. 3, p. 139-167.

Bryhni, I., and Skjerlie, F. J., 1975, Syndepositional tectonism in the Kvamshesten district (Old Red Sandstone), western Norway: Geol. Mag., v. 112, p. 593-600.

Bull, W. B., 1963, Alluvial fan deposits in western Fresno County, California: J. Geol., v. 71, p. 243-251.

———, 1964a, Alluvial fans and near surface subsidence in western Fresno County, California: U.S. Geol. Survey Prof. Paper 437-A, 70p.

———, 1964b, Geomorphology of segmented alluvial fans in western Fresno County, California: U.S. Geol. Survey Prof. Paper 352-E, p. 89-129.

———, 1964c, History and causes of channel trenching in western Fresno County, California: Am. J. Sci., v. 262, p. 249-258.

———, 1968, Alluvial Fans: J. Geol. Ed., v. 16, p. 101-106.

———, 1972, Recognition of alluvial fan deposits in the stratigraphic record: *in* J. K. Rigby and W. K. Hamblin, *eds.* Recognition of Ancient Sedimentary Environments; Soc. Econ. Palaeont. Mineral. Spec. Pub. 16, p. 63-83.

———, 1977, The alluvial fan environment: Progress in Phys. Geogr., v. 1, p. 222-270.

Buwalda, J. P., 1951, Transport of coarse material on alluvial fans: Geol. Soc. Am. Bull., v. 62, p. 1497.

Carryer, S. J., 1966, A note on the formation of alluvial fans: New Zealand Jour. Geol. Geophys., v. 9, p. 91-94.

Chawner, W. D., 1935, Alluvial fan flooding, the Montrose, California flood of 1934: Geogr. Rev., v. 25, p. 77-88.

Crowell, J. C., 1973, Ridge Basin southern California, Sedimentary Facies Change in Tertiary Rocks California Transverse and Southern Coast Ranges: Soc. Econ. Palaeont. Mineral. field trip guide, p. 1-7.

———, 1974, Sedimentation along the San Andreas Fault, California: *in* R. H. Dott, Jr., and R. H. Shaver, *eds.*, Modern and Ancient Geosynclinal Sedimentation; Soc. Econ. Palaeont. Mineral. Spec. Pub. 19, p. 292-303.

Curry, R. R., 1966, Observations of alpine mudflows in the Tenmile Range, central Colorado: Geol. Soc. Am. Bull., v. 77, p. 771-776.

Davis, W. M., 1925, The basin range problem: Nat. Acad. Sci. Proc., v. 11, p. 387-392.

Deegan, C. E., 1973, Tectonic control of sedimentation at the margin of a Carboniferous depositional basin in Kirkudbrightshire: Scott. J. Geol., v. 9, p. 1-28.

Denny, C. S., 1965, Alluvial fans in the Death Valley region, California and Nevada: U. S. Geol. Survey Prof. Paper 466, 62p.

———, 1967, Fans and pediments: Am. J. Sci., v. 265, p. 81-105.

Drew, 1873, Alluvial and lacustrine deposits and glacial records of the Upper Indus basin: Quart. J. Geol. Soc. London, v. 29, p. 441-471.

Eckis, R., 1928, Alluvial fans in the Cucamonga district, southern California: J. Geol., v. 36, p. 111-141.

Enos, P., 1977, Flow regimes in debris flow: Sedimentology, v. 24, p. 133-142.

Fisher, R. V., 1971, Features of coarse-grained, high-concentration fluids and their deposits: J. Sediment. Petrol., v. 41, p. 916-927.

Ghibaudo, G., and Mutti, E., 1973, Facies ed interpretazione paleoambientale delle Arenarie di Ranzano nei dintorni di Specchio (Val Pessola, Apennino Parmense): Mem. Soc. Geol. Italia, v. 12, p. 251-265.

Gole, C. V., and Chitale, S. V., 1966, Inland delta building activity of the Kosi River: J. Hydraulics Div., Am. Soc. Civil Eng., v. 92 (HY2), p. 111-126.

Goreau, T., and Burke, K., 1966, Pleistocene and Holocene geology of the island shelf near Kingston, Jamaica: Marine Geology, v. 4, p. 207-225.

Groat, C. G., 1972, Presidio Bolson, Trans-Pecos Texas and adjacent Mexico: Geology of a desert basin aquifer system: Bur. Econ. Geol. Univ. Texas, Rept. Invest., No. 76, 46p.

Heward, A. P., in press, Alluvial fan and lacustrine sediments from the Stephanian A and B (La Magdalena, Ciñera-Matallana and Sabero) coalfields, northern Spain: Sedimentology.

Hooke, R. Le B., 1967, Processes on arid-region alluvial fans: J. Geol., v. 75, p. 438-460.

———, 1968, Steady-state relationships on arid-region alluvial fans in closed basins: Am. J. Sci., v. 266, p. 609-629.

———, 1972, Geomorphic evidence for Late Wisconsin and Holocene tectonic deformation, Death Valley, California: Geol. Soc. Am. Bull., v. 83, p. 2073-2098.

Hoppe, G., and Ekman, S. R., 1964, A note on the alluvial fans of Ladtjovagge, Swedish Lapland: Geogr. Ann., v. 46, p. 338-342.

Howard, J. D., 1966, Patterns of sediment dispersal in the Fountain Formation of Colorado: Mountain Geologist, v. 3, p. 147-153.

Hubert, J. F., Reed, A. A., and Carey, P. J., 1976, Paleogeography of the East Berlin Formation, Newark Group, Connecticut Valley: Am. J. Sci., v. 276, p. 1183-1207.

Hunt, C. B., and Mabey, D. R., 1966, Stratigraphy and structure, Death Valley, California: U. S. Geol. Survey Prof. Paper 494-A, 162p.

Johnson, A. M., 1970, Physical processes in geology: Freeman, Cooper and Co., San Fransisco, 577p.

———, and Rahn, P. H., 1970, Mobilisation of debris flows: in New Contributions to Slope Evolution; Z. Geomorph. Suppl. Bd. 9, p. 168-186.

Kimura, T., 1966, Thickness distribution of sandstone beds and cyclic sedimentations in the turbidite sequences at two localities in Japan: Bull. Earthquake Res. Inst. Tokyo, v. 44, p. 561-607.

Klein, G. de V., 1975, Sandstone depositional models for exploration for fossil fuels: Continuing Education Publishing Company, Champaign, Illinois, 109p.

Knight, J. A., 1975, The systematics and stratigraphic aspects of the Stephanian flora of the Sabero coalfield, Part 1: The stratigraphy and general geology of the Sabero coalfield: Unpub. Ph.D. Thesis, Univ. Sheffield.

Laming, D. J. C., 1966, Imbrication, palaeocurrents and other sedimentary features in the Lower New Red Sandstone, Devonshire, England: J. Sediment. Petrol., v. 36, p. 940-959.

Lattman, L. H., 1973, Calcium carbonate cementation of alluvial fans in southern Nevada: Geol. Soc. Am. Bull., v. 84, p. 3013-3028.

Legget, R. F., Brown, R. J. E., and Johson, G. H., 1966, Alluvial fan formation near Aklavik, North West Territories, Canada: Geol. Soc. Am. Bull., v. 77, p. 15-30.

Lustig, L. K., 1965, Clastic sedimentation in Deep Springs Valley, California: U.S. Geol. Survey Prof. Paper 352-F, p. 131-192.

Martini, I. P., and Sagri, M., 1977, Sedimentary fillings of ancient deep-sea channels: Two examples from the Northern Apennines (Italy): J. Sediment. Petrol., v. 47, p. 1542-1553.

McGowen, J. H., and Groat, C. G., 1971, Van Horn Sandstone, west Texas: An alluvial fan model for mineral exploration: Bur. Econ. Geol. Univ. Texas, Rept. Invest. No. 72, 57p.

McPherson, H. J., and Hirst, F., 1972, Sediment changes on two alluvial fans in the Canadian Rocky Mountains: in H. O. Slaymaker and H. J. McPherson, eds., Mountain Geomorphology; Geomorphological Processes in the Canadian Cordillera; British Columbia, Geographical Series, 14, p. 161-175.

Meckel, L. D., 1967, Origin of Pottsville conglomerates (Pennsylvanian) in the Central Appalachians: Geol. Soc. Am. Bull., v. 78, p. 223-258.

———, 1975, Holocene sand bodies in the Colorado Delta area, northern Gulf of California: in M. L. Broussard, ed., Deltas Models for Exploration; Houston Geol. Soc., p. 239-265.

Miall, A. D., 1970, Devonian alluvial fans, Prince of Wales Island, Arctic Canada: J. Sediment. Petrol., v. 40, p. 556-571.

Middleton, G. V., and Hampton, M. A., 1973, Sediment gravity flows: Mechanics of flows and deposition: in G. V. Middleton, and A. H. Bouma, co-chairmen, Turbidites and Deep-Water Sedimentation; Soc. Econ. Paleont. Mineral. Short Course 1, Anaheim, p. 1-38.

Mutti, E., 1974, Examples of ancient deep-sea fan deposits from circum-Mediterranean geosynclines: in R. H. Dott, Jr., and R. H. Shaver, eds., Modern and Ancient Geosynclinal Sedimentation; Soc. Econ. Palaeont. Mineral. Spec. Pub. 19, p. 92-105.

———, 1977, Distinctive thin-bedded turbidite facies and related depositional environments in the Eocene Hecho Group (south-central Pyrenees, Spain): Sedimentology, v. 24, p. 107-131.

———, and Ricci-Lucchi, F., 1972, Le torbiditi dell'Appennino settentrionale: Introduzione all'analisi di facies: Soc. Geol. Italiana Mem., v. 11. p. 161-199.

Nelson, H., Mutti, E., and Ricci-Lucchi, F., 1977, Discussion: Upper Cretaceous resedimented conglomerates at Wheeler Gorge, California: Description and field guide: J. Sediment. Petrol., v. 47, p. 926-934.

Nilsen, T. H., 1968, The relationship of sedimentation to tectonics in the Solund district of southwestern Norway: Universitetsforlaget, Oslo Norges Geologiske Undersokelse No. 359, 108p.

———, 1969, Old Red sedimentation in the Buelandet-Vaerlandet Devonian district, western Norway: Sediment. Geology., v. 3, p. 35-57.

Normark, W. R., 1970, Growth patterns of deep-sea fans: Am. Assoc. Petrol. Geol. Bull., v. 54, p. 2170-2195.

Rahn, P. H., 1967, Sheetfloods, streamfloods and the formation of pediments: Ann. Assoc. Am. Geogr., v. 57, p. 593-604.

Renwick, W. H., 1977, Erosion caused by intense rainfall in a small catchment in New York State: Geology, v. 5, p. 361-364.

Riba, O., 1976, Syntectonic unconformities of the Alto Cardener, Spanish Pyrenees: A genetic interpretation: Sediment. Geol., v. 15, p. 213-233.

Ricci-Lucchi, F., 1975, Depositional cycles in two turbidite formations of northern Appenines (Italy): J. Sediment. Petrol., v. 45, p. 3-43.

Roed, M. A., and Wasylyk, D. G., 1973, Age of inactive alluvial fans — Bow River Valley, Alberta: Can. J. Earth Sci., v. 10, 1834-1840.

Ruhe, R. V., 1964, Landscape morphology and alluvial deposits in southern New Mexico: Ann. Assoc. Am. Geogr., v. 54, p. 147-159.

Rupke, N. A., 1977, Growth of an ancient deep-sea fan: J. Geol., v. 85, p. 725-744.

Ryder, J. M., 1971, The stratigraphy and morphology of paraglacial alluvial fans in south-central British Columbia: Can. J. Earth Sci., v. 8, p. 279-298.

Sagri, M., 1972, Rhythmic sedimentation in the turbidite sequences of the Northern Apennines (Italy): 24th Intl. Geol. Congr. Proc. Sect. 6, p. 82-88.

Schluger, P. R., 1973, Stratigraphy and sedimentary environments of the Devonian Perry Formation, New Brunswick, Canada, and Maine, U. S. A.: Geol. Soc. Am. Bull., v. 84, p. 2533-2548.

Schumm, S. A., 1968, Speculations concerning palaeohydrologic controls of terrestrial sedimentation: Geol. Soc. Am. Bull., v. 79, p. 1573-1588.

———, 1973, Geomorphic thresholds and complex response of drainage systems: in M. Morisawa, ed., Fluvial Geomorphology; Pub. in Geomorphology, SUNY, Binghampton, N.Y., p. 299-310.

———, 1976, Episodic erosion: A modification of the geomorphic cycle: in W. N. Melhorn, and R. C. Flemal, eds. Theories of Landform Development; Pub. in Geomorphology, SUNY, Binghampton, N. Y., p. 69-85.

———, and Hadley, R. F., 1957, Arroyos and the semi-arid cycle of erosion: Am. J. Sci., v. 255, p. 161-174.

Scott, K. M., 1971, Origin and sedimentology of 1969 debris flows near Glendora, California: U. S. Geol. Survey Prof. Paper 750-C, p. C242-C247.

———, 1973, Scour and fill in Tujunga Wash — A fanhead valley in urban California: U. S. Geol. Survey Prof. Paper 732-B, 29p.

———, and Gravlee, G. C., Jr., 1968, Flood surge of the Rubicon River, California — hydrology, hydraulics and boulder transport: U. S. Geol. Survey Prof. Paper 422-M, 40p.

Selley, R. C., 1965, Diagnostic characteristics of fluviatile sediments of the Torridonian Formation (Precambrian) of northwest Scotland: J. Sediment. Petrol., v. 35, p. 366-380.

Sestini, G., 1970, Vertical variations in flysch and turbidite sequences: a review: J. Earth Sci., Leeds, v. 8, p. 15-30.

Sharp, R. P., 1948, Early Tertiary Fanglomerate, Big Horn Mountains, Wyoming: J. Geol., v. 56, p. 1-15.

———, and Nobles, L. H., 1953, Mudflow of 1941 at Wrightwood, southern California: Geol. Soc. Am. Bull., v. 64, 547-560.

Statham, I., 1976, Debris flows on vegetated screes in the Black Mountain, Carmarthenshire: Earth Surf. Proc., v. 1, p. 173-180.

Steel, R. J., 1974, New Red Sandstone floodplain and piedmont sedimentation in the Hebridean province, Scotland: J. Sediment. Petrol., v. 44, p. 336-357.

———, 1976, Devonian basins of western Norway — sedimentary response to tectonism and to varying tectonic context: Tectonophysics, v. 36, p. 207-224.

———, and Wilson, A. C., 1975, Sedimentation and tectonism (?Permo-Triassic) on the margin of the North Minch Basin, Lewis: J. Geol. Soc. London, v. 131, p. 183-202.

———, Maehle, S., Nilsen, H., Roe, S. L., Spinnangr, A., 1977, Coarsening-upward cycles in the alluvium of Hornelen Basin (Devonian) Norway: Sedimentary response to tectonic events: Geol. Soc. Am. Bull., v. 88, p. 1124-1134.

Tanner, W. F., 1976, Tectonically significant pebble types: sheared, pocked, and second-cycle examples: Sediment. Geol., v. 16, p. 69-83.

Troxel, B. W., 1974, Man-made diversion of Furnace Creek Wash, Zabriske Point, Death Valley, California: California Geol., Oct. 1974, p. 219-223.

Vliet, A. van, in press, The early Tertiary deep-water fans of Guipuzcoa: in D. J. Stanley and G. Kelling, eds., Submarine Canyon and Fan Sedimentation, Dowden, Hutchison and Ross.

Wagner, R. H., 1970, An outline of the Carboniferous stratigraphy of northwest Spain: Congr. Coll. Univ. Liège, v. 55, p. 429-463.

———, 1971, The stratigraphy and structure of the Ciñera-Matallana coalfield (prov. León, N.W. Spain): Trabajos de Geologia Fac. Ci. Univ. Oviedo, v. 4, p. 385-429.

Walker, R. G., 1970, Review of the geometry and facies organisation of turbidites and turbidite bearing basins: in J. Lajoie, ed., Flysch Sedimentology in North America: Geol. Assoc. Canada Spec. Paper 7, p. 219-251.

———, 1975a, Upper Cretaceous resedimented conglomerates at Wheeler Gorge, California: Description and field guide: J. Sediment. Petrol., v. 45, p. 105-112.

———, 1975b, Nested submarine-fan channels in the Capistrano Formation, San Clemente, California: Geol. Soc. Am. Bull., v. 86, p. 915-924.

———, 1975c, Conglomerate: Sedimentary structures and facies models: in Depositional Environments as Interpreted from Primary Sedimentary Structures and Stratification Sequences; Soc. Econ. Palaeont. Mineral. Short Course 2, p. 133-161.

———, 1975d, From sedimentary structures to facies models: example from fluvial environments: in Depositional Environments as Interpreted from Primary Sedimentary Structures and Stratification Sequences; Soc. Econ. Palaeont. Mineral. Short Course 2, p. 63-79.

———, 1977, Deposition of upper Mesozoic resedimented conglomerates and associated turbidites in southwestern Oregon: Geol. Soc. Am. Bull., v. 88, p. 273-285.

———, and Mutti, E., 1973, Turbidite facies and facies associations, in G. V. Middleton, and A. H Bouma, co-chairmen, Turbidites and Deep Water Sedimentation: Soc. Econ. Palaeont Mineral. Short Course 1, Anaheim, p. 110-157.

Wasson, R. J., 1974, Intersection point deposition on alluvial fans: An Australian example: Geoogr. Ann., v. 56, p. 83-92.

———, 1977a, Catchment processes and the evolution of alluvial fans in the lower Derwent Valley, Tasmania: Z. Geomorph., Bd. 21, p. 147-168.

———, 1977b, Late-glacial alluvial fan sedimentation in the Lower Derwent Valley, Tasmania: Sedimentology, v. 24, p. 781-799.

Weaver, W., and Schumm, S. A., 1974, Fan-head trenching: An example of a geomorphic threshold: Geol. Soc. Am., Abs. with Program, v. 6, p. 481.

Williams, G. E., 1969, Characteristics and origin of a Precambrian pediment: J. Geol., v. 77, p. 183-207.

———, 1970, Piedmont sedimentation and late Quaternary chronology in the Biskra region of the northern Sahara: Z. Geomorph., v. 10, p. 40-60.

Wilson, M. D., 1970, Upper Cretaceous-Palaeocene synorogenic conglomerates of south-western Montana: Am. Assoc. Petrol. Geol. Bull., v. 54, p. 1843-1867.

Winder, C. G., 1965, Alluvial cone construction by alpine mudflow in a humid temperate region: Can. J. Earth Sci., v. 2, p. 270-277.

ERRATUM

Page 676, line 28 should read: "vegetation cover, lithology, ..."

THE DEPOSITS, INTERNAL STRUCTURE AND GEOMETRY IN SIX ALLUVIAL FAN–FAN DELTA BODIES (DEVONIAN-NORWAY)–A STUDY IN THE SIGNIFICANCE OF BEDDING SEQUENCE IN CONGLOMERATES

TOR G. GLOPPEN AND RON J. STEEL

Statoil, Stavanger, Norway; and Geological Institute, University of Bergen, Norway

INTRODUCTION

Hornelen Basin (Middle Devonian) in western Norway, is notable for its great depth (6-8 km) despite its relative small size (25 km × 70 km) and for the unusual stratigraphic thickness (25 km) of its alluvial infill. It has also been the most intensively studied of the eight or nine small Devonian basins in Norway. Recent work in the Hornelen Basin includes studies of general aspects of the basin and the cyclic organisation of its infill (Steel, 1976; Steel et al, 1977; Bryhni, 1978), analyses of the fluvial sandstones and of the associated floodbasin and lacustrine deposits which occupy the greatest volume of the basins (Steel and Aasheim, 1978), a detailed study of a single, mass-flow dominated marginal fan delta (Larsen and Steel, 1978) and an analysis of the features which suggest that the basin originated and evolved by strike-slip faulting (Steel and Gloppen, 1980).

The theme of the relatively small, marginal alluvial fans and fan deltas (Fig. 1) is taken up here. These are important because their degree of exposure often allows them to be mapped in detail and because their internal structure and sequence contains a useful key to the understanding of tectonic movements along the basin margins. Hornelen fans allow examination both of fan body geometry and of the internal distribution of various fanglomerate types as well as opportunity to study the interaction of the fans with the adjacent, impinging alluvial and lacustrine facies.

The various types of mass flow and waterlaid fan deposits are first described, then their distribution within six selected fan bodies is documented. The use of conglomerates in basin analysis, in reading details of the basin's tectonic history is also discussed. If a fan contains evidence of its lowermost reaches of having been under standing water, for example in the case of interfingering with lacustrine deposits or in having a distal facies of subaqueous mass flows, we refer to it has a fan delta (Holmes and Holmes, 1978). Although Rust (1979) has advised against this terminology, we hold to it here because there is considerable evidence in Hornelen Basin that the subaqueous facies of fans can vary considerably from their subaerial counterparts; not least because of an interaction with fine-grained, basinal sediment, often causing local mixing and resedimentation, even where the associated lakes were clearly ephemeral (Larsen and Steel, 1978).

313

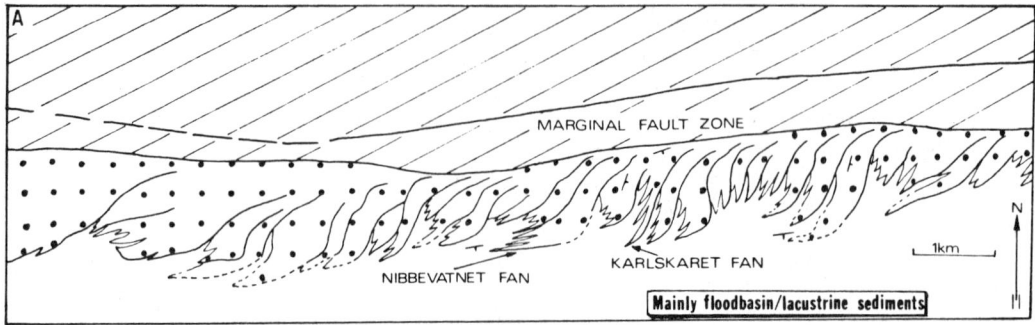
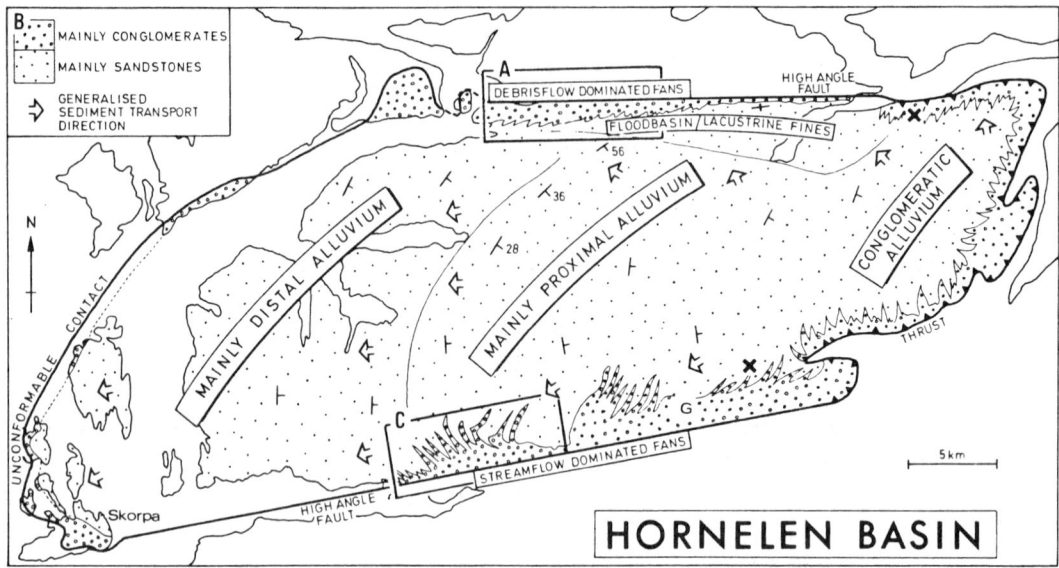
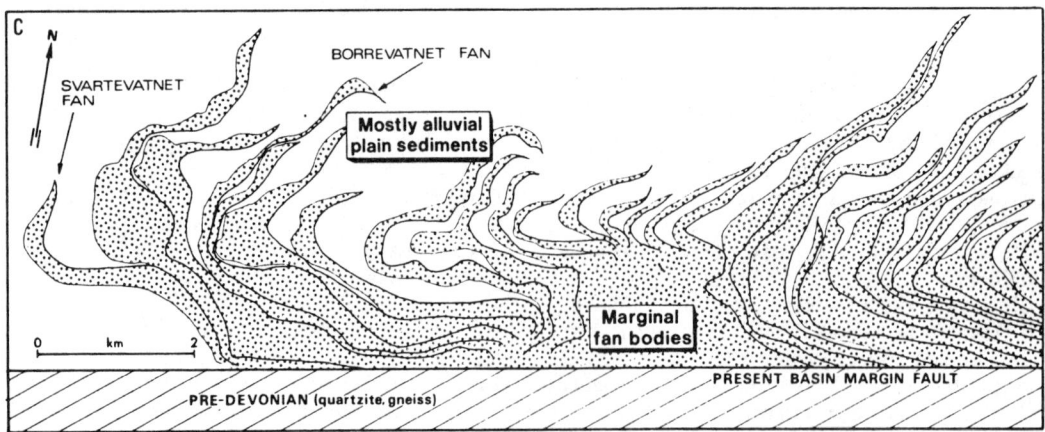

Fig. 1.— The main lithofacies in Hornelen Basin, with detailed maps of some of the fans along northern (A) and southern (C) margins. X denotes Hjortestegvatnet (north) and Lassenipa Fans (south). Position of Grøndalen is marked by G.

FAN DEPOSITS IN HORNELEN BASIN

Debris Flow Deposits

This group of conglomerate beds are either clast- or matrix-supported, poorly sorted, lack any pervasive stratification (though the beds themselves are commonly well-defined) and show a significant relationship between maximum particle size (MPS) and bed thickness (BTh). They are interpreted as debris flow deposits despite the common lack of any large amounts of clay or silt-sized matrix. Consistent with this interpretation is the high MPS/BTh ratio, the common presence of large clasts projecting above the top of the bed (Fig. 2) and the lack of any significant erosion between beds. The latter two features strongly suggest a high matrix strength and laminar flow, at least during the final stages of deposition. This group of beds is further divided into two, dependent on texture, fabric and grading:

Subaerial (high viscosity) flows are coarse and usually clast-supported, have an MPS/BTh ratio of 3 or less (Fig. 3), have either an unordered fabric or a tendency for elongate clasts to lie subparallel to the bed boundaries (Fig. 4c) (though a number of examples of vertical clastic orientation also occur, Fig. 4a), and are largely ungraded except for the lowermost few cm of the bed which is occasionally inversely graded (Fig. 4b). The lack of imbrication and grading in the beds is taken as an indication of lack of clast-to-clast movement during flow. The grain size (MPS) and bed thickness characteristics of these beds are summarised in Figure 3, which also shows clearly that the values of these parameters decrease significantly in a direction away from the marginal faults (downfan).

An important variant of this type of flow occurs where the coarse conglomerate bed is irregularly and abruptly overlain (draped) by a capping of fine, massive conglomerate or laminated granule sandstone (Fig. 4c). This capping is thinner than the underlying bed. It is also generally better sorted, and consists largely of grain sizes scarcely represented in the uppermost part of the underlying bed. Because of the latter, the capping is interpreted as the product of waning-stage flow and winnowing of the debris top by waterflow such as has been described by Jahns (1947) and Johnson (1970).

Subaqueous (low viscosity) flows have resulted from relatively low sediment concentration debris flow, are generally finer grained, have a higher percentage of matrix (Fig. 5a), and have a lower MPS/BTh ratio (though there is still a significant MPS/BTh correlation; Fig. 6) than their high viscosity counterparts. In addition, they are often imbricated (Fig. 5b, c), are commonly inversely or inverse-to-normally graded throughout the bed (Fig. 5b), and usually have a capping of massive or rippled, fine sandstone which grades up from the matrix of the underlying conglomerate (Fig. 5b). Various types of fine-grained, graded flows in this category have already been described in some detail (Larsen and Steel, 1978) from the toe region of Karlskaret Fan. Larsen and Steel (1978) emphasized the close association of this type of bed with fine-grained floodbasin and lacustrine deposits, especially along the northern margin of Hornelen Basin. The well-developed grading and imbrication indicates clast-to-clast movement throughout the bed during flow, despite the common matrix-supported texture. This together with lack of reworked tops, the lower maximum particle size for a given bed thickness (compared with the subaerial flows) and the intricate interfingering with lacustrine deposits is critical in suggesting relatively low viscosity, *subaqueous* flows. The sandy interval into which the debris flow grades, probably originated from a more dilute, turbidity current which travelled above and followed the higher viscosity flow, and is a feature also typical of many submarine resedimented conglomerates (Nemec et al, 1980). In some cases this rippled interval appears to encompass conglomeratic fingers of this facies (Fig. 7) particularly common on the small fans along Hornelen Basin's northern margin, and are much less common on the mixed fluvial-debris flow fans along the southern margin. Furthermore, they are more or less confined to the distal reaches of the fans (Fig. 14) and therefore occur as a well-marked, facies belt of diamictite-like character between the alluvial fanglomerates proper and the impinging fine deposits of the basin's axial system.

Shoreline-modified Debris Flows

Closely associated with both subaqueous debris flow deposits and the fine-grained basinal sediments there occurs a patchily developed conglomerate facies which has been interpreted as shoreline gravels. These conglomerates are better sorted, sometimes texturally bimodal, and the clasts are

FIG. 2.—Typical debris flow with large clast projecting above top of bed. Overlying flow has inversely graded base.

SVARTEVATNET FAN

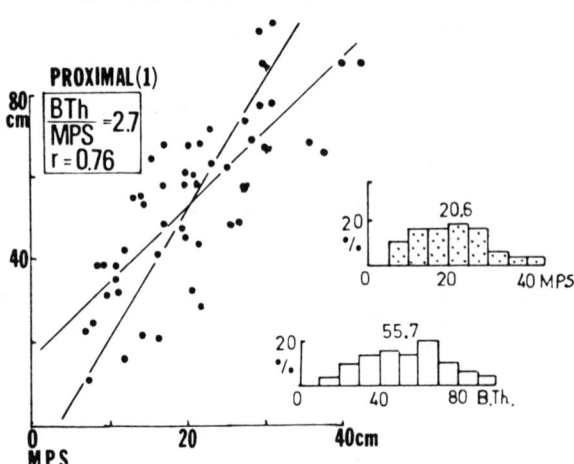
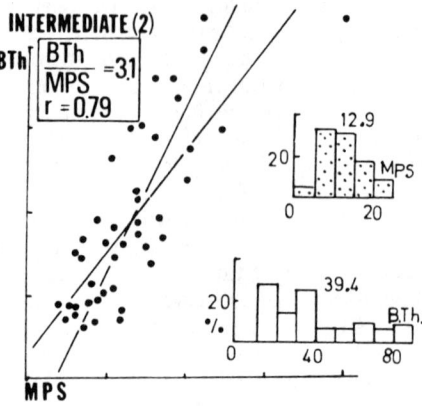

HJORTESTEGVATNET FAN

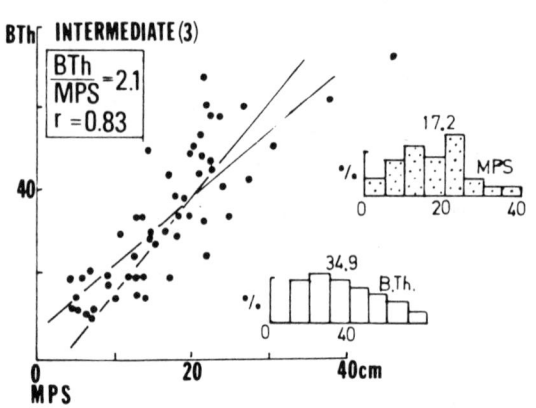
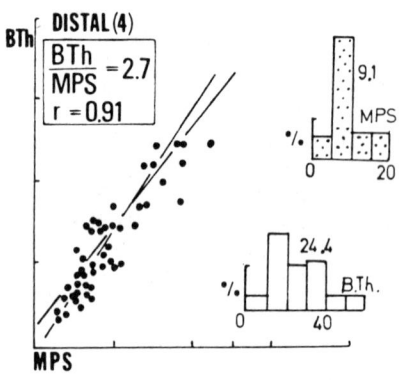

NIBBEVATNET FAN

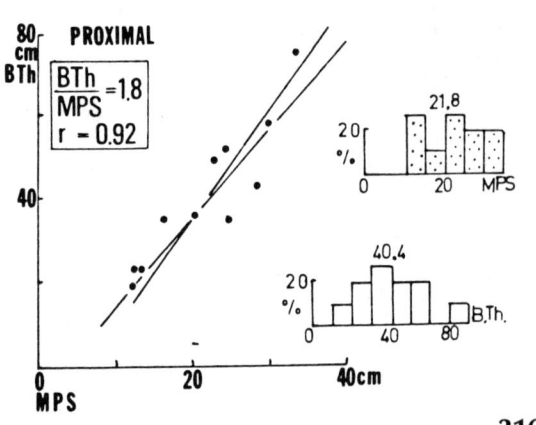
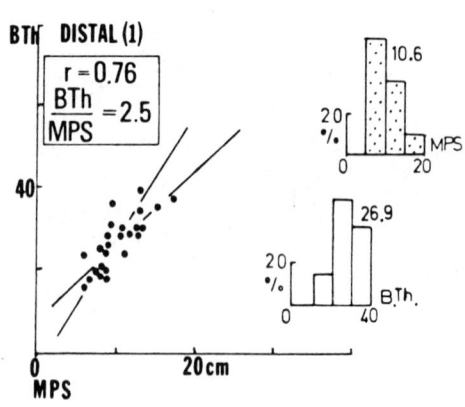

Fig. 4. — Typical *subaerial* debris flows, showing vertical clast orientation (a), poor sorting and inverse grading at base of bed (b), and fine conglomerate/sandstone cappings to debris flow beds (c).

Fig. 3. — Bed thickness (BTh) and maximum particle size (MPS) characteristics of the high viscosity (subaerial) debris flows from three fans. Note the various downfan trends and the relationship between BTh and MPS.

better rounded than in the associated fanglomerates and often show a distinct 'fitted' pebble fabric (Fig. 8). They tend to be overlain and pass laterally into low-angle, laminated, granule sandstone (Fig. 8a) and are interpreted as the product of relatively low energy wave action along the lacustrine shoreline on the fans. The extensive interfingering of lacustrine and fan deposits along the northern margin implies repeated drowning of the fans during periods of instability and tilting of the basin floor (Steel and Aasheim, 1978). Examination of many such subaqueous facies transitions give rise to a facies

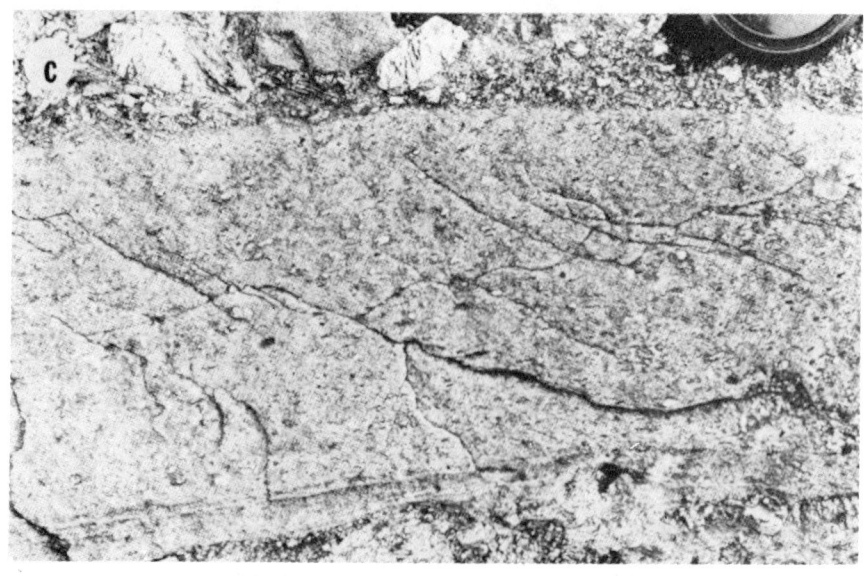

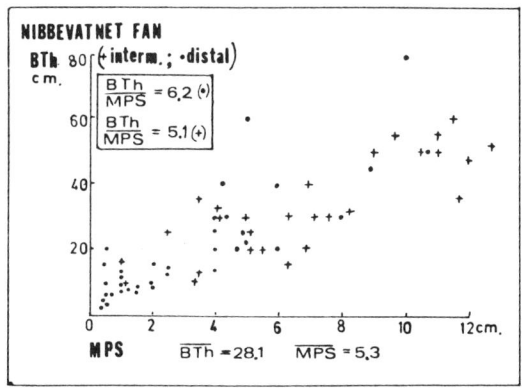

FIG. 6.—Thickness and size characteristics of *subaqueous* debris flows. Compare with Figure 3 but note differences in scale.

sequence which has been interpreted in terms of a landward migration of the lake shoreline on the fan surface, as shown in Figure 8d. The patchy development of this facies is probably due to (a) interference by debris flow action (sinking or tilting of the basin floor may have caused the fan process response almost as quickly as a rise in lake level) or (b) lack of persistent or significant wave activity.

Distribution of Debris Flows on Fans

Studies on recent fans where both streamflow and debris flow processes are present have shown that the proportion of debris flow deposits decreases downfan (Hooke, 1967). This has been recorded also from ancient alluvial fan sequences (e.g. Steel and Wilson, 1975) and is confirmed in the present study (Figs. 12 and 13). Considering only debris flows, it is also reasonable to assume that proximal flows are usually more fluid (and presumably, therefore, better graded and imbricated) than distal ones, due to persistent loss of water to the subsurface with time (Rust, 1979). In many fans of the present study we find quite the opposite trend, though this of course, is due to those fans being of small radius and having their lower reaches below lake level periodically. This general feature is consistent with our interpretation of many of the distal debris flows being subaqueous.

The distribution of the various types of deposits within the Hornelen fan bodies are further discussed below.

Waterlaid Deposits

This group of conglomerates and pebbly sandstones is significantly better stratified, has better developed clast imbrication and rounding (Fig. 9), is often better sorted and is more closely interbedded with laminated sandstones (Fig. 9b,c) than the debris flow deposits. Two main types are distinguished, on the basis of the scale (set thickness) and types of the dominant stratification. We find no difficulty in distinguishing these two types of waterlaid deposit and follow the terminology for recent fans by Bull (1972). There are, of course, many instances of transition where no clear cut distinction can be made. In addition a third, rare type (sieve deposits) is described.

Stream channel or streamflood deposits are conglomerates or pebbly sandstones in horizontally (Fig. 9b) or cross-stratified (Fig. 9c) sets varying from 15 to 100 cm thick. The latter type of set tends to be isolate and of planar wedge character, with common reactivation surfaces. The lower set boundary is usually markedly erosional. These deposits are interpreted as the infillings of the main stream channels on the fans. In the mixed fluvial-debris flow fans in Hornelen Basin the stream chan-

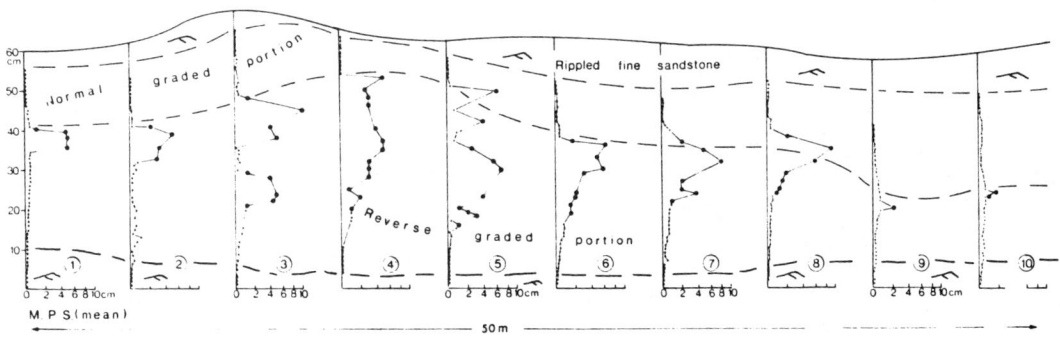

FIG. 7.—The typical lensoid form of a coarse-grained, subaqueous debris flow. Note that the flow is enveloped by ripple laminated, fine sandstone. Sampling transverse to flow.

FIG. 5.—Typical *subaqueous* debris flow beds showing matrix-supported texture (a), inverse grading through most of the bed (b), and imbrication of granules in a fine-grained flow (c).

319

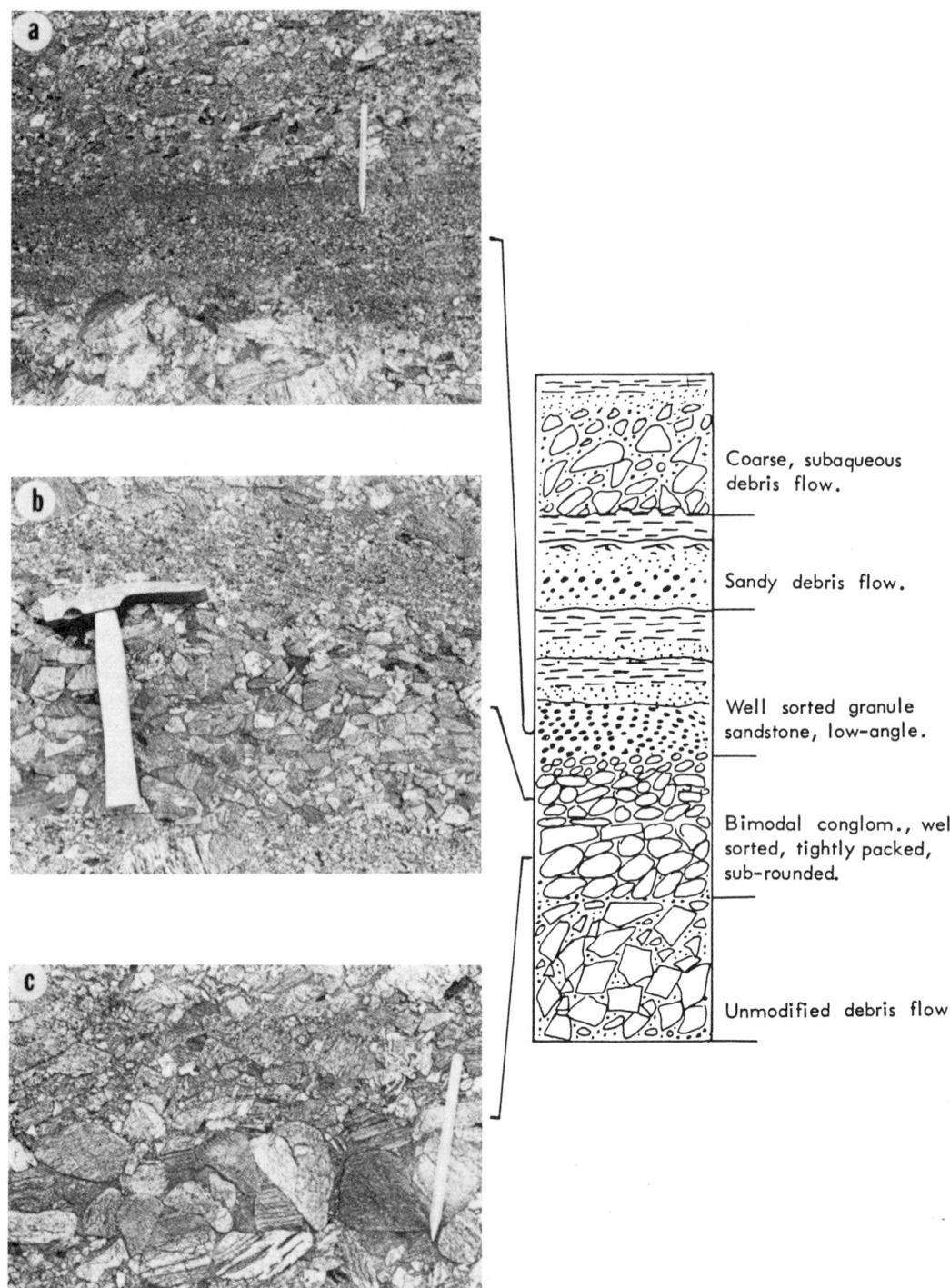

ALLUVIAL FAN–FAN DELTA BODIES

FIG. 9.—Stream deposited fanglomerates, showing relatively good sorting and rounding (a), association with laminated sandstones (b) and planar cross-stratification (c).

nel deposits tend to occur on the middle and lower fan reaches (Fig. 12).

Sheetflood deposits are dominated by sand, granule and very small pebble grain-sizes and are characterized by an alternation of coarse and fine-grained sheets of sediment (Fig. 10a). The all-pervasive, low-angle or flat lamination, together with the thinness of sets exaggerates the sheetlike appearance of the facies. Actually set boundaries are erosive and most sets extend less than 20 m laterally. A notable feature is the occasional occurrence of isolate, outsize clasts (Fig. 10b). This facies is interpreted as having originated from a network of very shallow, braided channels and low, longitudinal bars, such as commonly form and spread out from the end of the main stream channels on fans

---

FIG. 8.—Features of low energy shoreline deposits on the fans. Note the tightly fitted clast fabric (c), good sorting (b) and laminated granule sandstone (a). The vertical sequence is that developed by lacustrine transgression and is mirrored by similar *lateral* facies variations.

FIG. 10.—Sheetflood deposits with typical thin interbedding and flat lamination (a), and isolate, outsize pebbles (b).

(Bull 1972). The outsize clasts presumably result from the abrupt drop in competence associated with the increased flow width at the mouths of the main channels. This facies occurs in some of the larger debris flow-dominated alluvial fans, as well as in the mixed fans. It is apparently absent from fans with significant quantities of subaqueous debris flows. It is always restricted to the lower and outer reaches of fan lobes and bodies.

Sieve deposits, according to Hooke (1967) and Bull (1972), developed in the form of small lobes on fans which receive little sand or mud from their source areas. In recent fans such lobes are hummocky and consist of well-sorted debris, often relatively angular because of preferential occurrence on fans supplied by, for example, jointed quartzite. In Hornelen Basin we have looked for such deposits along the southern margin of the basin, where quartzitic gneisses dominate as source rocks. Fan bodies from Grøndalen and Skorpa (Fig. 1) contain some sequences which we speculatively interpret as sieve deposits and remobilised sieve deposits. These are thick (<1 m), massive units of pebble and small pebble conglomerates. They are often well-sorted (Fig. 11a) sometimes have an 'open' framework but commonly have been later in-

FIG. 11.—Possible sieve deposits. Note sorting and angularity of quartzite pebbles (a) and lensoid, coarse deposit with markedly bimodal texture (b).

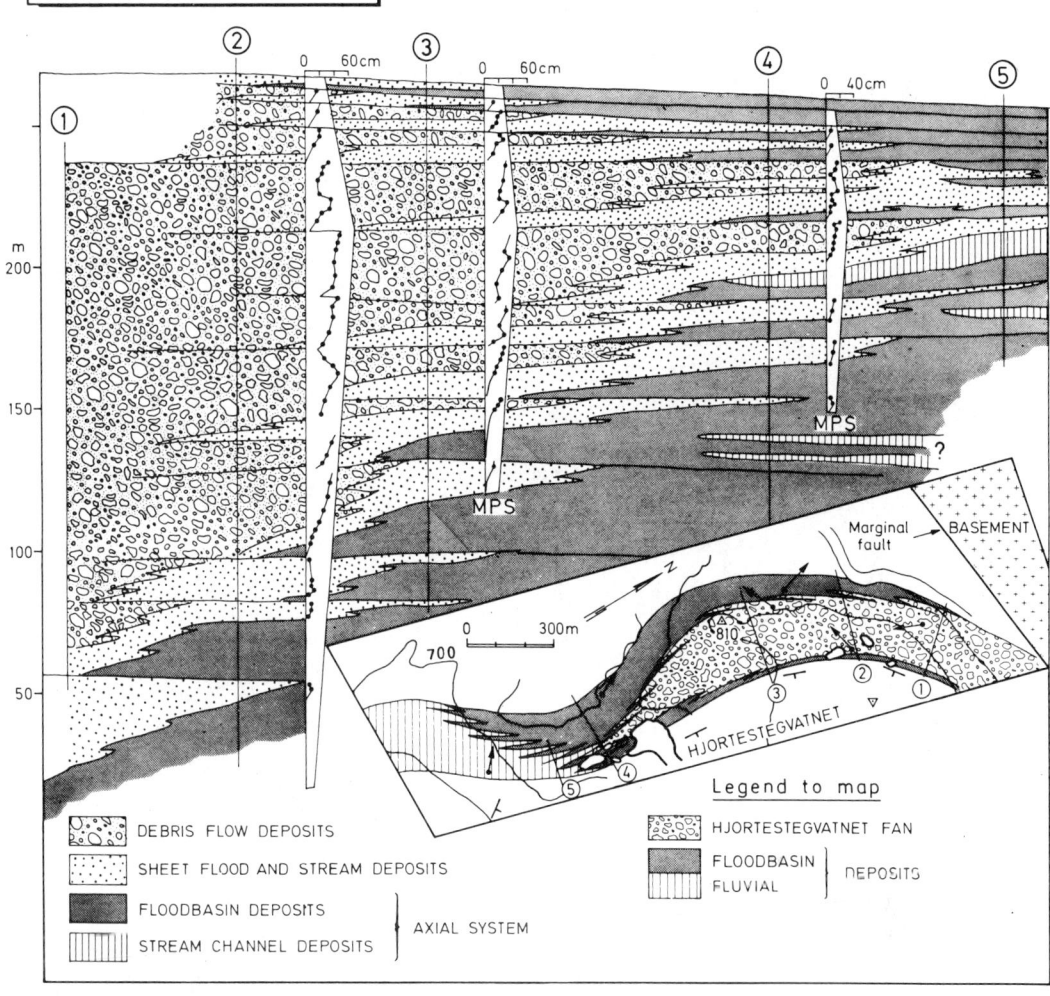

FIG. 12.—Internal details of Hjortestegvatnet Fan. Profiles 1-5 are located in the inset map.

filtrated by fine, red siltstone, creating a markedly bimodal deposit (Fig. 11b). The units are clearly lensoid and sometimes banked against larger boulders or blocks. In some instances these deposits have been remobilised *after* a late stage infiltration by lacustrine silts, because they now form inversely graded beds or have suffered movement to the extent that the laminated, silty matrix is now deformed.

DEPOSITS OF THE ADJACENT ENVIRONMENTS

The small marginal fans of Hornelen Basin represent only a small proportion of the basin succession. The dominating sedimentary system was the axial, westwards and northwestwards prograding, sandy fan delta complex, whose various alluvial and lacustrine facies have been described in detail by Steel and Aasheim (1978). Of particular relevance to the marginal fans are the fine-grained lacustrine and floodbasin facies which periodically draped large parts of the northern margin and therefore mixed and became remobilised to varying degrees with the fan-derived materials, as documented by Larsen and Steel (1978). Only in the middle and upper levels of some of the fan bodies, representing times when the axial fluvial system had achieved considerable westwards progradation, is there limited interfingering with coarser grained braided alluvium of the axial system (e.g. Figs. 12, 13a).

The interfingering relationship between the fine-grained deposits and the fanglomerates closely reflects the rate and style of subsidence of the basin floor. The latter controls the rate and direction of aggradation of the marginal fan wedges relative to the axial alluvium. This varied through time as well

a SVARTEVATNET FAN
m
100
80
60
40
20
0
0 40 80cm
0 40cm
0 40cm
MPS
UNDERLYING FAN
500 m
N
Ulldalsnipa
726
760
SVARTE-VATNET
MARGINAL FAULT
BASEMENT
SVARTEVATNET FAN
MAINLY FLOODBASIN DEPOSITS
b LASSENIPA FAN
0 20cm
0 20cm
0 20 40cm
m
80
60
40
20
0
MPS
MPS
TOP OF UNDERLYING FAN
LASSENIPA FAN
MAINLY FLOODBASIN DEPOSITS
N
500 m
LASSENIPA
1024
946
MARGINAL FAULT
BASEMENT
MAINLY DEBRIS FLOW DEPOSITS
MAINLY SHEETFLOOD AND STREAM DEPOSITS
FAN DEPOSITS
FLOODBASIN
STREAM CHANNEL
AXIAL ALLUVIUM

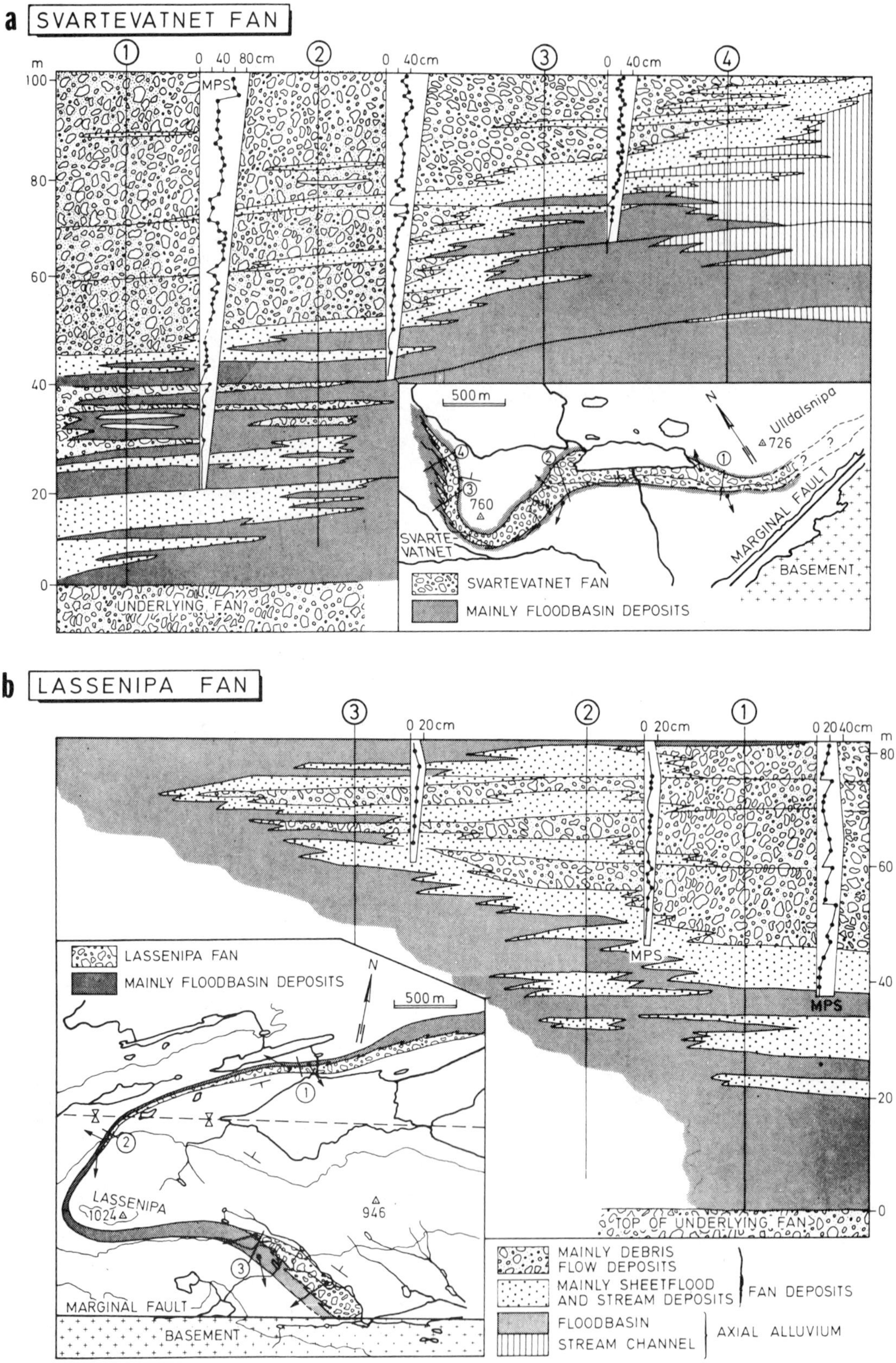

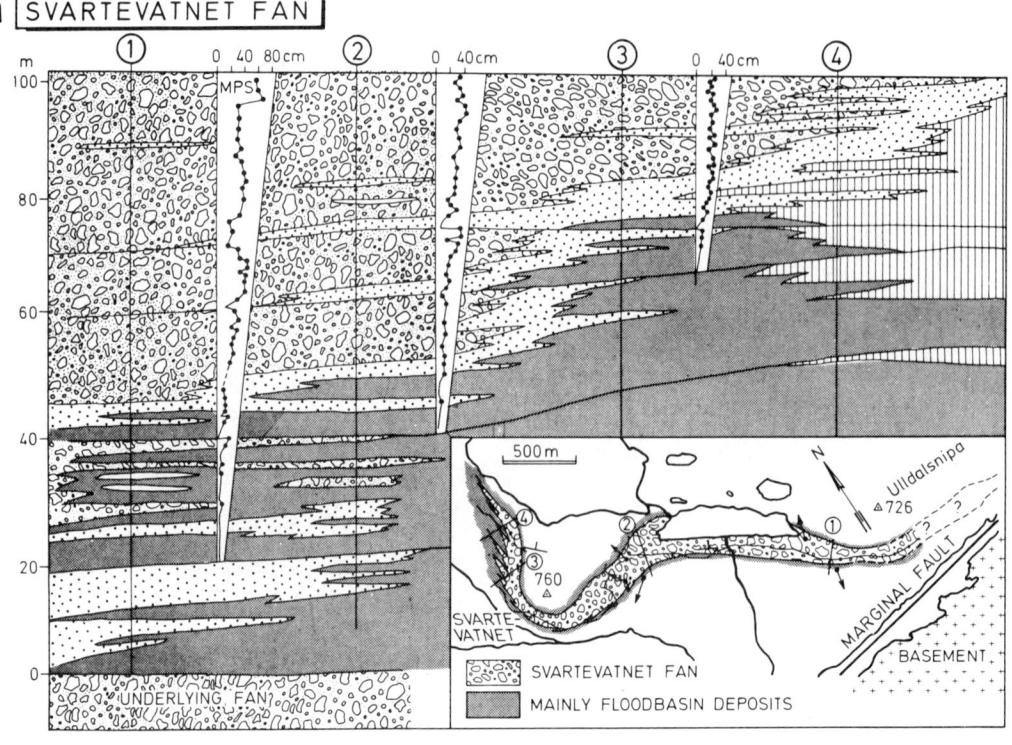

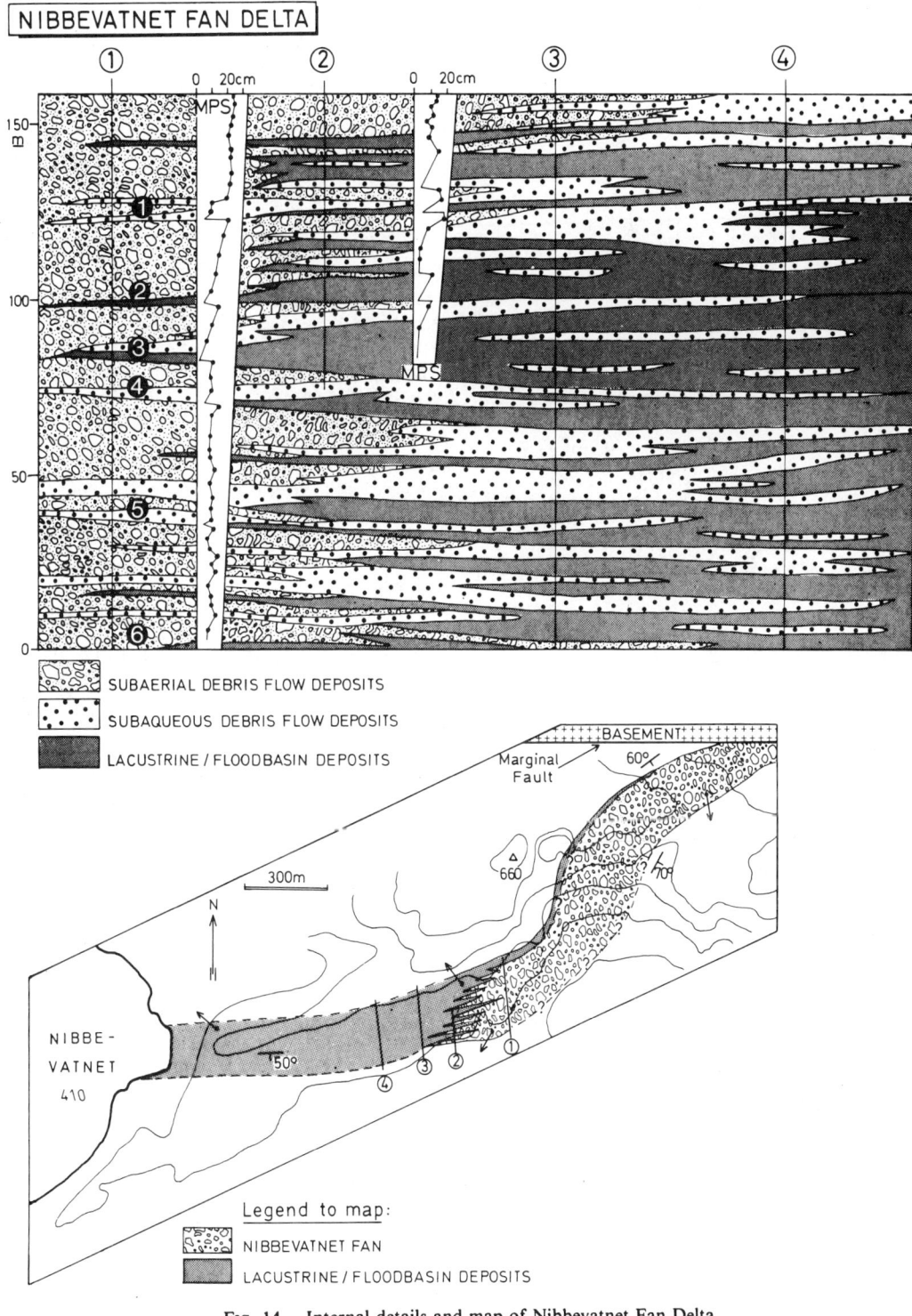

FIG. 14.—Internal details and map of Nibbevatnet Fan Delta.

FIG. 13.—Internal details and maps of Svartevatnet (a) and Lassenipa Fans (b).

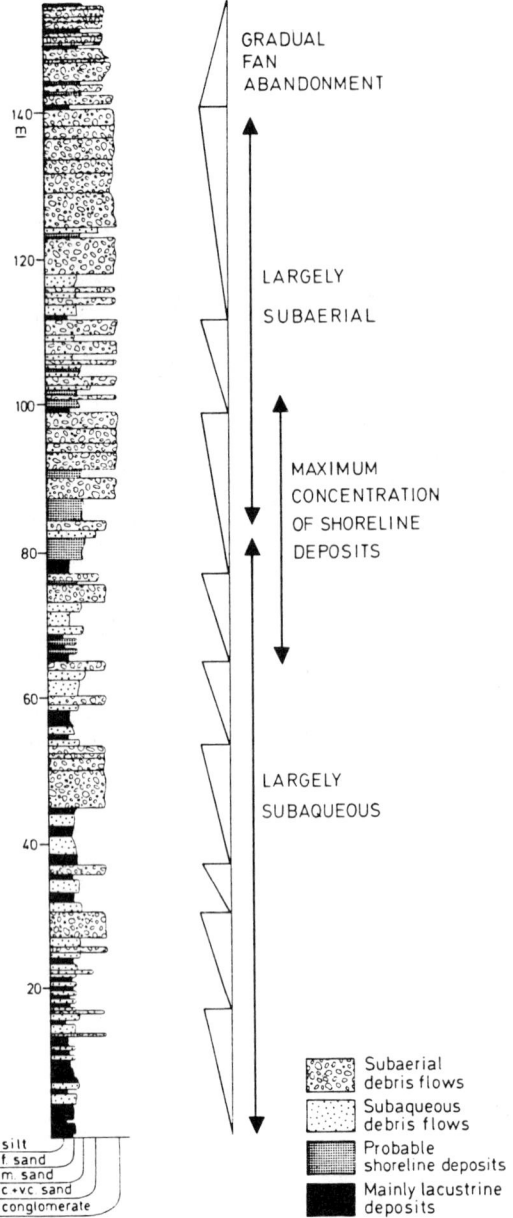

FIG. 15. — Vertical log through an intermediate section of Karlskaret Fan Delta (modified from Figure 2 in Larsen and Steel, 1977).

as from margin to margin, as discussed below in connection with the description of the geometry of various types of fan bodies.

FAN BODIES:
INTERNAL AND EXTERNAL CHARACTERISTICS

The excellent exposure of Hornelen Basin has allowed a detailed mapping of the marginal fans (e.g. Fig. 1). This resulted in both a mapping of deposits within the fans themselves and an examination of the feather-edge of the fans. The latter aspect, rather neglected in the literature, emphasised for us the importance of the dynamic interaction of adjacent environments and the generation of mixed facies along their interface. Of particular interest was the discovery that thin layers of siltstone and fine sandstone which interbedded with the conglomerates, even relatively high on the fans, were not fan-derived but originated from the axial system.

The Six Fans

Maps and sections/profiles of the internal facies, structure and geometry of six selected fans are shown in Figures 12-16. Hjortestegvatnet, Nibbevatnet and Karlskaret Fans are representative for the northern margin of the basin, and Svartevatnet, Borrevatnet and Lassenipa Fans for the southern margin. The following general features and trends within these fan bodies are noted:

a) Where both mass flow and fluvial deposits are represented on a fan the former are volumetrically more important proximally and are increasingly replaced by fluvial deposits distally (Figs. 12 and 13).

b) Maximum particle size (MPS, mean of 10 largest clasts, locally per bed) as well as bed thickness, decreases downfan (Figs. 3, 12 and 13a), though this happens more evenly in fans with a significant fluvial component than in those little influenced by streams. Estimates of MPS downfan decrease for debris flows vary from about 14 cm/km on the small fans to 5 cm/km on the larger fans. Figures for bed thickness vary from about 17 cm/km to 11 cm/km respectively (Table 1). Note that the gradients are not true radial ones.

c) As a corollary of (b) and of the clear relationship of MPS to BTh, and also demonstrable in the field, all fan bodies are wedge-shaped in two dimensions and thicken towards the marginal fault zones. Thinning takes place from the base of the fan body upwards, by means of interfingering (on a surprisingly small scale) with sediments of the adjacent system. The rate of thinning was tectonically controlled, and dependent on the relative rates of aggradation on fan and adjacent environment, as discussed below.

d) The dynamic interaction between fan and adjacent system blurs tendencies to increased sorting and textural maturity downfan and, in cases of fan deltas (Nibbevatnet and Karlskaret Fans; Figs. 14, 15) actually causes a marked textural inversion downfan.

e) The stacking of fan bodies along the basin margins resulted in major interfingering of fanglomerate with axial sediment and a cyclicity on a 100-200 m scale (Figs. 1, 16). In addition, *within* individual fan bodies, smaller scale (5-25

FIG. 16.—Borrevatnet Fan showing upward coarsening sequence (a) and remarkably flat top to fan body (b). (c) shows stacking of fan bodies at Grøndalen.

m), upward-coarsening cycles are developed (Figs. 12, 15, 17). Both types of cyclothem are discussed below.

In Hjortestegvatnet and Lassenipa Fans (Figs. 12 and 13b) debris flows dominate the proximal areas while the distal areas were dominated by waterlaid, largely sheetflood deposits.

Nibbevatnet and Karlskaret Fans (Figs. 14 and 15) are entirely dominated by mass flow deposits, also distally, where subaqueous mass flows are abundant.

Fluvial events were rare in these fans and can only be recognised in the form of reworked tops to debris flows. In Svartevatnet and Borrevatnet Fans (Figs. 13a and 16), however, fluvial events extensively influenced fan development. Debris flows are still common proximally but these were extensively reworked as shown by winnowed conglomerate units and much interbedded, laminated sandstone. The middle and lower reaches of these fans were dominated by fluvial processes, resulting in thick sequences of both sheetflood and channel-fill deposits.

Cyclicity Within Fan Bodies

The interfingering of floodbasin/lacustrine and alluvial sediments with the fan deposits has resulted in cyclicity in the fan pile in both large and small

Table 1.—Summary of Maximum Particle Size and Bed Thickness Characteristics for Subaerial Debris Flow Deposits on 4 of the Studied Fans. The Position of the Control Points on Each Fan Is Noted in Figures 12-14

	Hjortestegvatnet		Nibbevatnet		Svartevatnet		Lassenipa	
	Interm.	Dist.	Prox.	Dist.	Prox.	Interm.	Prox.	Dist.
$\overline{MPS}$	17.2	9.1	21.8	10.6	20.6	12.9	16.3	9.1
$\overline{BTh}$	34.9	24.4	40.4	26.9	55.7	39.4	38.4	21.4
$\overline{BTh} : \overline{MPS}$	2.0	2.7	1.8	2.5	2.7	3.1	2.3	2.3
Downfan $\overline{MPS}$-decrease	11.6 cm/km		14 cm/km		5.1 cm/km		?	
Downfan $\overline{BTh}$-decrease	15.1 cm/km		16.8 cm/km		10.8 cm/km		?	

scales. Two types of small-scale cycle have been observed within fan bodies (Fig. 17):
a) Cycles which are 5–25 m thick, shown an upwards coarsening and thickening of beds and have an abrupt top against the overlying fine sediments. These cycles are present in the lower 75 percent of all fans and dominate completely most fan bodies. In addition, when stacked vertically they show an overall upwards coarsening and thickening.
b) Cycles which are 3–10 m thick, show an upwards coarsening but also have a fining-upward capping. These, when stacked, show an overall upwards fining and thinning.

Representative examples of these cycles and their overall pattern within two fan bodies are shown in Figures 15 and 17. In view of the sedimentation context here, a small fault-controlled basin which accumulated 25 km of alluvial deposits, there are two most likely explanations for the small-scale cyclicity. Either they represent a direct response (rapid fan retreat followed by floodbasin aggradation then by gradual fan progradation) to periods of base level lowering against the marginal faults or they are due partly to this and partly to lateral fan lobe shifting. If the latter was important it clearly happened within a context of basin floor subsidence and may easily have been triggered by the tectonic movements. Further support to the notion of direct or indirect tectonic control of cyclicity comes from the lateral persistence of cycle boundaries out of the fan facies and into the axially transported alluvium, as documented from some areas (Steel and Gloppen, 1980). Those cycles with an upward-fining cap presumably represent a more gradual abandonment or retreat of fan lobes.

Fan Geometry

All of the fans are more or less wedge-shaped in vertical sections, predictably so in view of the relatively fixed position of the marginal fault zones through time. Some of them, for example Svartevatnet and Borrevatnet Fans (Figs. 13a, 16b) are terminated upwards by a sharp, planar, probably isochronous surface; others by an upwards fining portion (e.g. Hjortestegvatnet and Lassenipa Fans) which grades them into the fine sediment at the base of the overlying fan body (Figs. 12, 13b, 17). The tops of the fan bodies are the boundaries between the major fan cycles and represent either abrupt or gradual fan abandonment, interpreted below, in either case, as the result of *lateral* migration of the basin depocentre.

All of the fans are characterised by having a diachronously rising base, simply reflecting the general progradation of the fan onto the adjacent aggrading floodbasin (Fig. 18). The *rate* of rising of the fan base varies considerably, however, both between fans and within individual fans, dependent on the rate of sedimentation on the fan relative to that on adjacent floodbasin. A rapidly rising base (e.g. Nibbevatnet Fan, Fig. 14) indicates little difference in sedimentation rates on fan and floodbasin, while a more slowly rising interface (e.g. upper part of Svartevatnet Fan, Fig. 13a) indicates relatively high sedimentation rates of the fan. In most of the fans, and particularly those along the northern margin, the fine-grained sediments of the axial system penetrate deeply into the fan body, forming the base of each small-scale cycle. As argued above, these repeated, minor periods of fan abandonment (and therefore maximum floodbasin aggradation) reflect the periodic, vertical component of basin floor movement.

Attempts were made to quantify some of these differences in fan geometry in terms of the rate of proximal-distal migration of the fan-floodbasin interface along a fan radius, using a model given by Hooke (1968). The model assumes a closed basin and that there was uniform deposition over the fan surface within the time interval considered. The latter assumption is probably not valid here, so the figures obtained are used only as a measure of comparison *between* the various Hornelen Basin

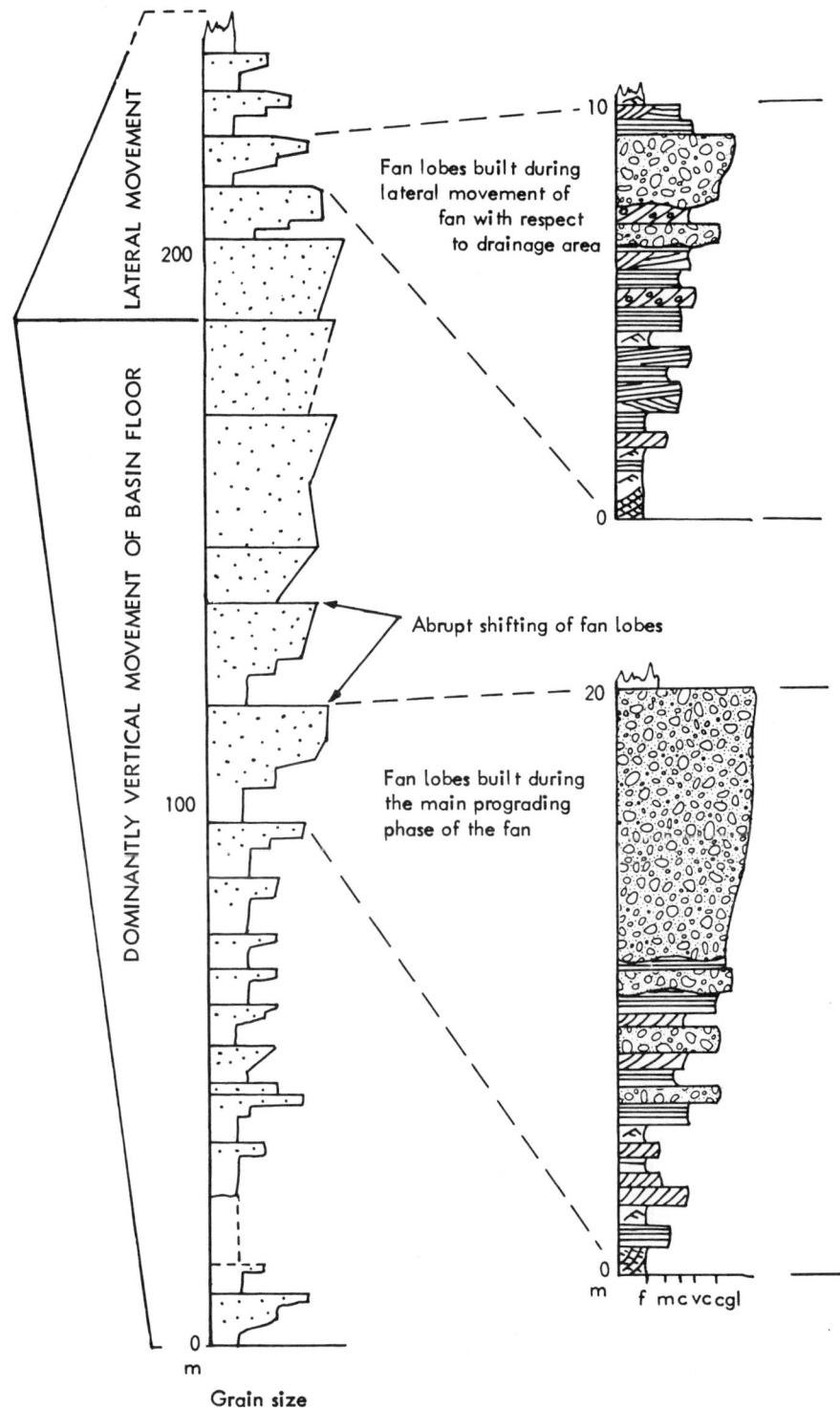

FIG. 17.—A vertical profile through Hjortestegvatnet Fan showing details of two types of component upward coarsening sequence and the overall upwards coarsening-fining of the fan body (Steel and Gloppen 1980).

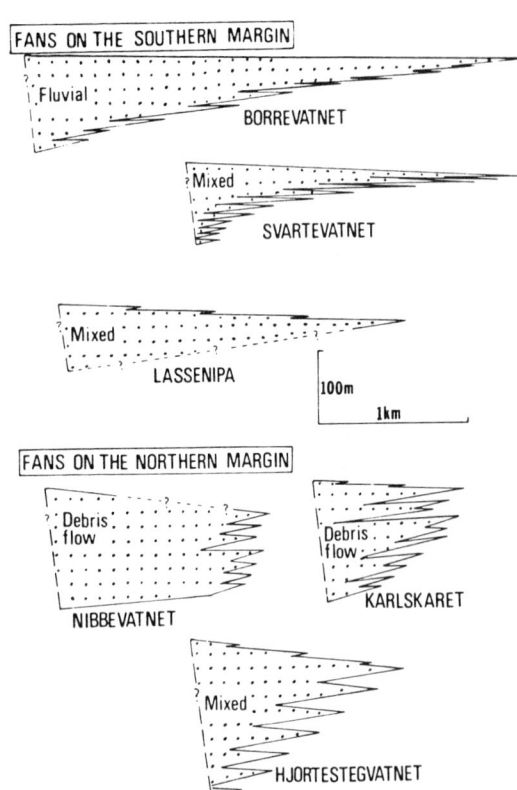

FIG. 18. — Schematic illustration of the varied geometry of the six studied fans. The fans with a fluvial component have a larger radius and a lower angle interface with the adjacent deposits.

fans. The sedimentation rate on the fans are calculated to vary from 8 percent higher (near vertical interface, eg. Nibbevatnet Fan) to 61 percent higher (Svartevatnet Fan) than sedimentation rates in adjacent environments (Fig. 18).

Lake Levels on the Fans

There is evidence both from the fanglomerates themselves (shoreline gravels, subaqueous mass flows) and from some of the interfingering fine-grained sediments (wave-generated ripple lamination, cm.thin sand/silt density flows) for the existence of ephemeral lakes which sometimes drowned the lower and middle reaches of some of the fans, hence the fan-delta designation. Lacustrine conditions were particularly common along the northern margin of the basin (Nibbevatnet and Karlskaret Fans) but also occasionally in the south (Lassenipa Fan).

It has occasionally been possible to estimate the depth of such lakes by identifying lake shoreline deposits, measuring or estimating their radial distance from the fan toe and estimating fan slope.

One case, on Nibbevatnet Fan, is illustrated in Figure 19A, where water depth estimate varies between 7 m and 70 m (minimum) depending on the fan slope estimate and on whether the shoreline deposits lie on an isochronous fan surface or are within the retreating uppermost portion of the fan body. In general, good exposure and mappable fan bodies are necessary before such estimates can be made, though in the case of Nibbevatnet Fan ambiguity arose because of poor exposures at this level in the critical intermediate region.

Characteristically on such fan deltas there is abundant evidence of greatly fluctuating lake levels. Shoreline or subaqueous debris flows on the middle reaches of a fan can be overlain by fluvial deposits extending well down the lower fan reaches. As discussed in more detail by Larsen and Steel (1978) the association of abundant soft sediment deformation with evidence of incomplete mixing between floodbasin and fan deposits and with re-sedimented (subaqueous) flows suggests very unstable conditions on the fan surfaces, probably

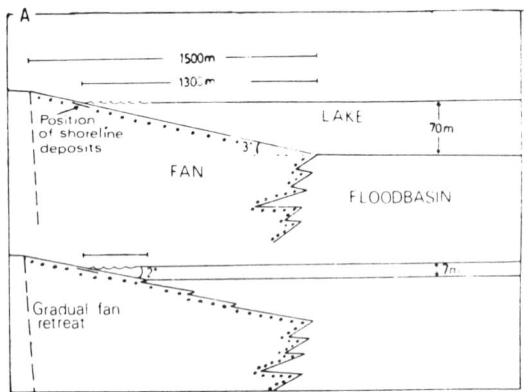

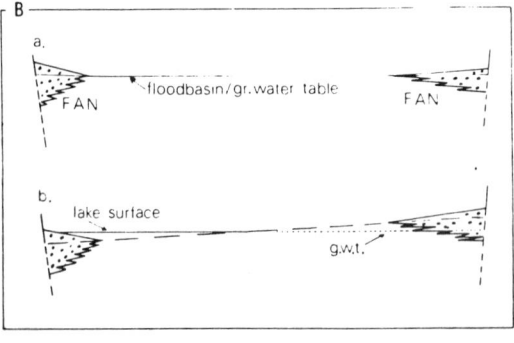

FIG. 19. — A. Estimations of lake depth using position of identified shoreline gravels relative to fan toe and fan slope.
B. Illustration of how lakes could have been created and dispersed by tilting of the basin floor along the northern margin.

margin, would, as soon as the groundwater table has adjusted, create lakes up to 50 m deep, more than enough to drown large portions of the fans along the northern margin. In the permeable sands of the basin centre such a water table readjustment would probably be extremely rapid (days or weeks). Along the northern margin, on the other hand, the alternation of permeable fan sediment with impermeable lake and floodbasin deposits would cause periodic build-up of pore water pressure leading to instability and the abundance of deformation structures now found on and around the fan toes (Hesjedal, 1979). Independent evidence of increased gradients of soft sediment deformation towards the northern margin of the basin further supports this notion (Hesjedal, 1979).

GENERAL TECTONIC IMPLICATIONS FROM HORNELEN BASINS CONGLOMERATE SEQUENCES

Basin margin conglomerates have been used to draw tectonic conclusions at a number of levels (Fig. 20):

a) on a scale of fan sequences; the basic tectonic response
b) on a basinwide scale, the contrast between northern and southern margins
c) on a scale of basin development

Individual Fan Sequences

Within the discussion of cyclic sedimentation above it was emphasised that the 5–25 m cyclicity in the basin alluvium probably reflects the background of persistent lowering of the basin floor. Because of the drop in base level caused by such movement or flurries of movement there was a rapid eastwards retreat of the axial system, as well as retreat of the small marginal fans, and the consequent "drowning" of most of the basinal area by the fine-grained floodbasin and lacustrine sediments. Because of the marginal relief now created, however, the longer term response to these same movements was a progradation of *all* the alluvial systems, producing the characteristic upward coarsening signatures. It is probably a moot point to attempt to distinguish sequences of autocyclic and allocyclic nature within the context of sedimentation envisaged here.

Because of their larger scale the 100–200 m cycles are more striking and can often be seen to be basinwide in nature. It has been argued, however, that sequences on this scale probably do not have any additional significance as regards the vertical movement of the basin floor, because they are too thick compared to known and realistic amounts of fault movement or even accumulated fault throw after flurries of movement. It is therefore likely, particularly in view of independent evidence of mappable eastwards onlap of successive units on this scale

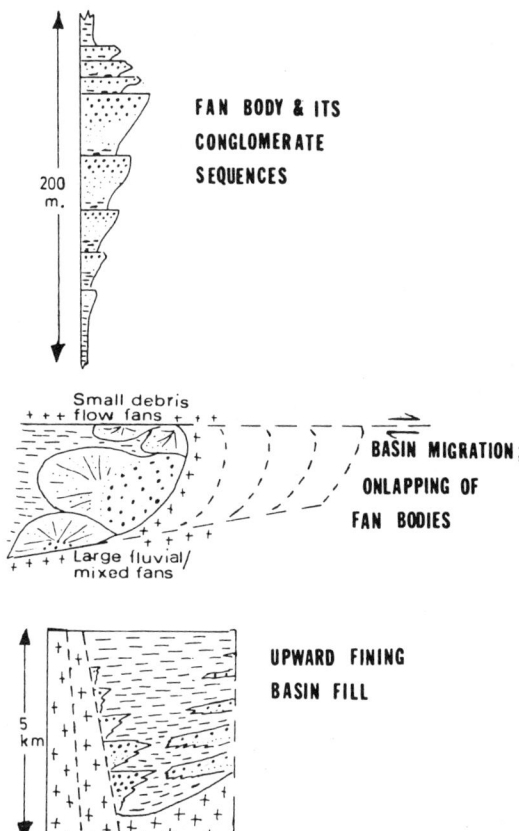

FIG. 20.—Summary of how conglomerates have aided the study of synsedimentary tectonics in Hornelen Basin. Small-scale upward coarsening sequences record the downwarping of the basin floor; fan contrasts between northern and southern margins reflect a northward component of tilt; and the overall upward fining of the basin succession together with the eastwards onlap and large, upward coarsening cycles result from the lateral migration of the basin with time.

due to rapid inundation and retreat of lake water.

The most likely causes of these rapid lake level fluctuations are large floods, longer term climatic variations or tectonic tilting of the basin floor. The first two are less likely on account of the enormous volumes of water necessary for even modest lake level rises and because of the absence of any increased amounts of fluvial deposits at equivalent levels proximally, or of other grain size changes which should have accompanied changed weathering conditions in the drainage areas. Tectonic tilting of the basin floor, on the other hand, is extremely likely in the context of this deep, narrow basin and a periodic northwards component of tilt has been independently proven, as discussed below. Figure 19B illustrates how even minor tilting of the basin floor (e.g. 0.3 %), with hinging along the southern

(Steel and Gloppen, 1980 Fig. 10), that these intervals mark times of significant lateral movement (strike-slip component) of the basin floor. The envisaged scheme is illustrated in Figure 17 where the lateral movement of the depocentre may be either gradual or abrupt, but in either case superimposed on a more continued background of vertical movement.

Contrasts Between Northern and Southern Margin Fans

Careful examination of the distribution of the fan type and size and fan body thickness can provide useful information about basin geometry and its tectonic development.

Although all of the various types of deposits can be found in fans of both the northern and southern margins there is a clear tendency for the fans of the former margin to be dominated by debris flows. Waterlaid deposits, on the other hand, are much more common along the southern margin, particularly west of Grøndalen. In addition, the mapping shows that the fans in the north normally have a smaller radius (often less than half the size) than those in the south. The latter is not merely apparent, but is a true radius difference as confirmed by considerably steeper slopes on the northern fans. This is also consistent with observations on recent fans, where increase in the amounts of mass flow is generally accompanied by steeper slopes and smaller fan areas (Bull 1968).

The tectonic implication of these differences in fan size and facies on opposite margins is that there was a northward component in basin floor palaeoslope in addition to the westward one. It is argued that the large fluvial fans developed on long, low northward slopes while the small, debris flow fans fringed the much steeper slopes in the north, a situation superficially analogous to the present day distribution of fans in Death Valley (Hooke 1972). Consistent with this hypothesis for Hornelen Basin is the fact that very many of the large-scale cycles which cross the basin can be seen to thicken significantly northwards. This implies more rapid rate of subsidence along the northern as compared to the southern edge of the basin floor.

Basin Development with Time

On a basinwide scale the conglomerates and conglomeratic bodies of Hornelen Basin are an important key to understanding basin evolution. The various arguments for an eastwards migration of the basin through time and for the likely strike-slip faulting mechanism have been detailed by Steel and Gloppen (1980). Probably most important is the evidence of mappable eastwards onlap of successive fanglomerate bodies and the striking organisation of the basin infilling into repeated upward coarsening cyclothems, each some 100-200 m thick (Fig. 20). As regards the overall behaviour of the marginal conglomerates in this type of basin it should be noted that (a) they occupy narrow, stationary but extensive belts parallel to the boundary faults (in contrast to rift basins which often expand marginally with time, creating a much wider conglomerate drape) and (b) the overall vertical trend is one of upwards fining, due to the axial migration of the basin with time (Fig. 20).

ACKNOWLEDGMENTS

This work is partly the result of a Cand. Real. dissertation (TGG) and partly the conclusion of conglomerate studies in Hornelen Basin since 1973. We thank Signe-Line Røe and Vidar Larsen for access to their unpublished data on Nibbevatnet and Karlskaret Fans respectively. The work has been financed by the University of Bergen and Norges Tekniske Naturvitenskapelige Forskningsråd and has been completed while one of us (RJS) was in receipt of NATO Grant 1724 for fanglomerate studies. We thank our colleagues in Bergen and Stavanger for continued lively discussion on Norway's Devonian conglomerates and Ellen Irgens and Jane Ellingsen for drafting.

REFERENCES

BRYHNI, I., 1978, Flood deposits in the Hornelen Basin, west Norway (Old Red Sandstone): Norsk Geol. Tidsskr., v. 58, p. 273-300.
BULL, W. B., 1968, Alluvial fans: Jour. Geol. Education, v. 6, p. 101-106.
────── 1972, Recognition of alluvial fan deposits in the stratigraphic record. *In*: J. K. Rugby and W. K. Hamblin, eds., Recognition of ancient sedimentary environments: Soc. Econ. Paleontologists and Mineralogists, Spec. Publ. No. 16, p. 63-84.
────── 1977, The alluvial fan environment: Progr. Phys. Geog., v. 1, p. 220-270.
JAHNS, R. H., 1947, Geological features of the Connecticut Valley, Massachusetts, as related to recent floods: U. S. Geol. Survey Water Supply Paper 6, 158 p.
JOHNSON, A. M., 1970, Physical processes in geology: Freeman, Cooper and Co., San Francisco, 577 p.
HESJEDAL, A., 1979, Studies of the structures and trigger mechanisms for soft sediment deformation in Hornelen Basin (Devonian), western Norway. Cand. real. Thesis, University of Bergen, 214 p.
HOLMES, A. AND HOLMES, D., 1978, Principles of physical geology: Nelson, Middlesex, England, 358 p.
HOOKE, R. L., 1967, Processes on arid region alluvial fans: Jour. Geol., v. 75, p. 438-465.

Hooke, R. L., 1968, Steady-state relationships on arid region alluvial fans in closed basins: Am. Jour. Sci., v. 266, p. 609-629.

―― 1972, Geomorphic evidence for Late-Wisconsin and Holocene deformation, Death Valley, California: Geol. Soc. Am. Bull., v. 83, p. 2073-2098.

Larsen, V. and Steel, R. J., 1978, The sedimentary history of a debris flow-dominated alluvial fan—a study of textural inversion: Sedimentology, v. 25, p. 37-59.

Nemec, W., Porebski, S. and Steel, R. J., *In press*, Texture and structure of resedimented conglomerates—examples from Ksiaz Formation (Famennian-Tournaisian), southwestern Poland: Sedimentology.

Rust, B. R., 1979, Facies models 2: coarse alluvial deposits. *In*: R. G. Walker, ed., Facies Models: Geoscience Canada, Reprint Series 1, p. 9-21.

Steel, R. J., 1976, Devonian basins of western Norway, sedimentary response to tectonism and varying tectonic context: Tectonophysics, v. 36, p. 207-224.

―― and Wilson, A. C., 1975, Sedimentation and tectonism (?Permo-Triassic) on the margin of the North Minch Basin: Jour. Geol. Soc. London, v. 131, p. 183-202.

―― Mæhle, S., Nilsen, H. R., Røe, S. L. and Spinnangr, Å., 1977, Coarsening upward cycles in the alluvium of Hornelen Basin (Devonian, Norway)—sedimentary response to tectonic events: Geol. Soc. Am. Bull. v. 88, p. 1124-1134.

―― and Aasheim, S., 1978, Alluvial sand deposition in a rapidly subsiding basin (Devonian, Norway). *In*: A. D. Miall, ed., Fluvial Sedimentology: Can. Soc. Petrol. Geol., Memoir 5, p. 385-412.

Steel, R. J., and Gloppen, T. G., 1980, Late Caledonian (Devonian) basin formation, western Norway—signs of strike-slip tectonics during infilling. *In*: H. Reading and P. F. Ballance, eds., Sedimentation in oblique-slip mobile zones. Spec. Publ. Int. Assoc. Sediment., No. 4, p. 79-103.

ERRATUM

Page 51, line 52 in the righthand column (the heading) should read: "Shoreline modified debris flows."

Part IV

COMPARATIVE STUDIES OF MODERN AND ANCIENT ALLUVIAL FAN DEPOSITS

Editor's Comments
on Paper 17

17   BULL
    Recognition of Alluvial-Fan Deposits in the Stratigraphic Record

    A number of workers have made comparative studies of modern and ancient alluvial fan deposits. Most students of modern fans have examined the exposures of unconsolidated sediments in arroyos and channels in order to understand more about ongoing deposition in active channels and on active depositional surfaces. Many workers on modern fans have utilized borehole and geophysical information to better understand the three-dimensional geometry and growth history of modern fans. As a result, almost all studies of modern alluvial fan deposits involve a certain amount of stratigraphic and sedimentologic work that is very similar in nature to that conducted by workers on ancient alluvial fan deposits. Many workers on modern fans have also worked on older deposits, and their studies of the modern enabled them to make better observations of the sedimentary rocks and to ask more significant questions about the ancient deposits.
    Similarly, many workers who have studied ancient alluvial fans have had at least some experience studying modern fans and observing the depositional, erosive, and weathering processes on modern fans. It should be clear that the more observations of different sizes of alluvial fans from different tectonic and climatic settings a geologist makes, the better equipped he will be to make paleogeographic, paleotectonic, and paleoclimatic observations of ancient stratigraphic sequences. Probably every student of an ancient alluvial fan deposit has wished that he could find a well-studied equivalent modern fan with which to make comparisons. Unfortunately, very few studies in that much detail have been made; however, as we learn more about paleoclimates and paleotectonic settings, more direct comparisons of modern and ancient alluvial fan deposits will be possible.

Editor's Comments on Paper 17

Because of the many early studies of well-exposed modern alluvial fans and pediments in the southwestern United States and adjoining Mexico, where the faulted Basin and Range province has yielded extensive alluvial fan deposition, opinions and general conclusions about alluvial fans by geomorphologists, sedimentologists, and stratigraphers have very heavily emphasized the arid-climate setting. Before 1960, few modern alluvial fans from temperate, alpine, humid, or Arctic settings had been studied. As a result, in almost all studies of ancient alluvial fan deposits prior to the work of Krynine (1950), interpretations were based on arid climates. We now know considerably more about alluvial fans from other climatic regimes, as has been demonstrated by some of the papers reproduced in this volume. However, there remain lingering problems for the student of ancient fan deposits regarding paleoclimatic setting—i.e., what are the reliable criteria to be used in such deposits to determine the climate? How does one go about determining the paleoclimate of Precambrian deposits for which the paleolatitude and continental setting may be virtually unknown? What is the paleoclimatic significance of the red color of so many alluvial fan deposits? Are debris-flow deposits really more abundant in arid-region settings or can they be equally or more abundant in other climatic settings that experience seasonal rainfall? These and many other major questions still leave the interpretation of paleoclimates as a major problem to students of ancient alluvial fan deposits. Many of these questions will perhaps always remain difficult to answer for workers in some ancient rock sequences.

Blissenbach (1954), Bull (Paper 17, 1977), and Nilsen (1982) have attempted systematic descriptions and comparisons of modern and ancient alluvial fan deposits based on extensive reviews of the literature. Shorter comparative papers can be found as chapters in some sedimentology textbooks (Reineck and Singh, 1975; Friedman and Sanders, 1978; Collinson, 1978; Blatt, Middleton, and Murray, 1980; Leeder, 1982). In addition, comparative studies have been fundamental parts of many papers, such as Paper 15 by Heward in this volume.

Paper 17 has been selected for reproduction herein because it is a very concise, descriptive, and useful summary of modern and ancient alluvial fan deposits. Bull's paper published in 1977 covers much of the same ground in more detail and in much greater length. Nilsen's (1982) paper is also quite long and incorporates a large number of color photographs to show examples of modern and ancient alluvial fan deposits. These papers are recommended to the interested reader, but were much too long to be reproduced herein, and did not lend themselves to extensive excerpting.

In Paper 17, Bull examines the morphology, stratigraphy, depositional processes, and grain-size distributions and sorting of fan deposits. He also focuses on proximal-to-distal changes in stratigraphy, maximum clast size, bedding thickness, clast roundness, sedimentary structures, geometry, and ground-water properties. The lateral and vertical interfingering of fan deposits with adjacent deposits is also discussed. Bull compares and contrasts alluvial fan deposits with other types of alluvial deposits, and discusses several ancient alluvial fan deposits in some detail, chiefly the Precambrian Torridonian Sandstone of northern Scotland and the Miocene and Pliocene Ridge Basin Group of southern California. Bull concludes his very useful contribution with a listing of diagnostic criteria that can be used to recognize alluvial fan deposits.

As a conclusion to this volume, I would like to expand upon Bull's list of criteria with a list published ten years later (Nilsen, 1982). It is clear that alluvial fan deposits are recognized primarily by their physical rather than biological or chemical characteristics. Fossils generally are extremely rare, and chemical sediments such as evaporites and bedded carbonates typically are not present. The most important physical characteristics include the types of facies present, textural and compositional immaturity, and coarseness of the deposits. The major physical criteria for the recognition of alluvial fan deposits are listed below; although each criterion is not necessarily unique to fans, collectively they can be used for the recognition of fan deposits.

1. The deposits are located relatively close to their source area.
2. Deposition is dominantly by unidirectional, high-energy fluid flow.
3. The deposits are typically very poorly sorted and may contain a great range in grain size.
4. Clasts are poorly rounded, reflecting the short distance of transport.
5. The deposits are compositionally immature and have a great range in composition, depending upon the types of rocks present in the source area.
6. The deposits are characterized by major changes in lateral and vertical facies, particularly in the downfan direction.
7. The deposits are characterized by rapid downfan decrease in both average and maximum clast size.
8. The deposits are generally oxidized because they have never been placed in reducing conditions; characteristic colors are red, brown, yellow, or orange.
9. The deposits generally contain very small amounts of organic matter because of the oxidizing conditions of sedimentation.
10. The deposits generally contain no fossils except for scattered vertebrate bones and plant fragments.

11. The deposits generally contain a limited suite of sedimentary structures, most commonly medium- to large-scale cross-strata and planar stratification.
12. The depositional bodies have a lenticular or wedge-shaped geometry and typically form clastic wedges.
13. The deposits may be characterized by a radial sediment dispersal system.

REFERENCES

Blatt, H., G. Middleton, and R. Murray, 1980, *Origin of Sedimentary Rocks,* 2nd ed. Prentice-Hall, Inc., Englewood Cliffs, N.J., 782p.

Blissenbach, E., 1954, Geology of Alluvial Fans in Semiarid Regions, *Geol. Soc. America Bull.* **65:**175-190.

Bull, W. B., 1977, The Alluvial Fan Environment, *Prog. Physical Geog.* **1:**222-270.

Collinson, J. D., 1978, Alluvial Sediments, in *Sedimentary Environments and Facies,* H. G. Reading, ed., Blackwell Scientific Publications, London, pp. 15-60.

Friedman, G. M., and J. E. Sanders, 1978, *Principles of Sedimentology,* John Wiley and Sons, New York, 792p.

Krynine, P. D., 1950, Petrology, Stratigraphy, and Origin of Triassic Sedimentary Rocks of Connecticut, *Connecticut Geol. and Nat. History Survey Bull. No. 73,* 247p.

Leeder, M. R., 1982, *Sedimentology, Process and Product,* George Allen and Unwin, Ltd., London, 344p.

Nilsen, T. H., 1982, Alluvial Fan Deposits, in *Sandstone Depositional Environments,* P. A. Scholle and D. Spearing, eds., Am. Assoc. Petroleum Geologists Mem. 31, p. 49-86.

Reineck, H. E., and I. B. Singh, 1975, *Depositional Sedimentary Environments,* Springer-Verlag, Berlin, 439p.

Copyright © 1972 by the Society of Economic Paleontologists and Mineralogists
Reprinted from pages 63–83 of *Recognition of Ancient Sedimentary Environments,*
J. K. Rigby and W. K. Hamblin, eds., Soc. Econ. Paleontologists and Mineralogists
Spec. Pub. 16, 340p., by permission of the publisher

RECOGNITION OF ALLUVIAL-FAN DEPOSITS IN THE STRATIGRAPHIC RECORD

WILLIAM B. BULL
University of Arizona

INTRODUCTION

Alluvial fans are distinctive terrestrial stratigraphic units. A fan is a deposit whose surface forms a segment of a cone that radiates downslope from the point where the stream leaves the mountains. Although the surface is not preserved in the stratigraphic record, distinctive suites of alluvial-fan deposits are preserved in many parts of the world.

Alluvial fans are important economically. The deposits form the principal ground-water reservoir in many areas, and the recharge of many groundwater basins is through the alluvial fans that fringe the basin. The surfaces of many fans are highly desirable for agricultural, urban, and industrial uses.

Modern alluvial fans have the following definitive characteristics that aid in interpretation of alluvial-fan deposits preserved in the stratigraphic record. Each fan is derived from a source area with a drainage network that transports the erosional products of the source area to the fan in a single trunk stream. The result of fluvial deposition by the stream is a cone-shaped deposit. The plan view of the deposit commonly is fan-shaped, and the contours bow downslope from the fan apex. Overall radial profiles are commonly concave, and cross-fan profiles are convex. Bedrock knobs, such as are commonly associated with pediments, rarely protrude through the thick alluvium. If present, they are most common near the fanhead.

A vertical aerial view of an alluvial fan having the above characteristics is shown in Figure 1. The Copper Canyon fan forms a 180° cone because of the absence of large fans that would restrict lateral expansion of the fan. Generally, adjacent fans restrict the lateral extent of the individual deposits. Thus, most fan deposits occur as a series of coalescing alluvial cones that form a piedmont slope that is sometimes called a bajada. Small coalescing fans are shown in the upper part of Figure 1.

Alluvial fans are most widespread in the drier parts of the world, but also occur in humid regions such as Japan (Murata, 1966); the Himalayan Mountains (Drew, 1873); Canada (Winder, 1965); and in the Arctic regions of Scandanavia (Hoppe and Ekman, 1964) and Canada (Legget, and others, 1966).

The effect of source-area characteristics on fan size has lead to the development of equations that relate the two. For example variations of the coefficient in the equation

$$A_f = cA_d^n$$

(where A_f and A_d are the fan and drainage-basin areas) reflects the erodibility of the source rocks and the tectonic environment affecting the erosional-depositional system (Bull, 1962a, 1968,

Fig. 1.—Vertical view of the Copper Canyon fan, Death Valley, California. Photograph by Howard Chapman, U.S. Geological Survey.

p. 104–105; Denny, 1965, p. 38; Melton, 1965; Hawley and Wilson, 1965; Hooke, 1968, p. 609–621).

Purpose and Scope

This article discusses the characteristics of alluvial-fan deposits that may be useful in the recognition of alluvial-fan deposits in the geologic record.

Characteristics that can be determined in small outcrops or core samples will be described first, to provide criteria for the identification of five types of fluvial processes that construct fans. Then the stratigraphy of alluvial fans will be discussed to provide the three-dimensional framework needed to complete the identification of the alluvial-fan depositional environment.

ACKNOWLEDGMENTS

The author is indebted to Roger LeB. Hooke of the University of Minnesota and W. K. Hamblin of Brigham Young University for critical review of the manuscript.

ALLUVIAL-FAN DEPOSITS

Although some depositional environments can be identified by a combination of physical, chemical, and biological data, alluvial fans are identified mainly by a distinctive suite of physical properties. A few general chemical and biological characteristics will be noted at this point, then the remainder of the article will be devoted to the physical properties of fan deposits.

Two types of chemical data pertain to alluvial fans—the degree of oxidation and the salts that have been deposited in, or have accumulated in, alluvial-fan deposits. Most fan deposits are oxidized. Therefore, lack of reduced deposits is a typical characteristic of fans that have accumulated in the arid and semiarid parts of the world. Fine-grained oxidized fan deposits that contain visible organic matter have been described by Miller, Green, and Davis (1971). Meade (1967, p. 6) notes that the oxidized color of fan deposits can be retained even after burial to more than 1,500 feet below the water table.

Alluvial-fan deposits commonly contain salts such as gypsum and calcite that have been deposited with the sediments or have accumulated as a result of weathering of the surficial materials. Because a fan is the product of a single stream, marked variations of soluble salts may occur in adjacent fans if the lithologies of the source areas are markedly different. Chemical characteristics can be used to distinguish between fans from different source areas. For example, Bull, (1964a, p. 61) lists mean gypsum contents of 2.2 and 0.03 percent for adjacent fans. Burial below the groundwater table permits circulation of groundwater to change the chemical character of the contained salts by solution and deposition. Maximum salt contents commonly occur near the downslope edge of fans, because this may be an area of ground-water discharge or of flow of saline water from adjacent playa deposits. The concentric banding near the downslope edge of the fan shown in Figure 1 is the result of larger salt and moisture contents and more intense chemical weathering, than the upslope parts of the fan.

The solution of calcium carbonate in the upper part and deposition in the lower part of the soil profile is characteristic of fan deposits that are not receiving additional increments of sediment and have insufficient water infiltrating through the soil to remove the authigenic carbonate. Detailed descriptions of carbonate accumulations in desert soils, such as can occur on alluvial fans, are given by Gile, Peterson, and Grossman (1966). Alternating periods of deposition and soil-profile formation are characteristic of most alluvial fans as the depositional area shifts from one part of the fan to another during the process of constructing the cone-shaped deposit. If sufficient time is available for weathering to occur between periods of deposition, a series of soil profiles will result, each profile representing a time of no deposition. An example of multiple buried soil profiles is shown in Figure 2. Three zones of carbonate accumulation are exposed in the 35-foot bank where the main stream channel has trenched the fanhead. The fan deposits consist entirely of water-laid sediments derived from a carbonate-rock source area.

Fossils not only are rare in alluvial-fan deposits but are of little use in differentiating the fan depositional environment from other continental depositional environments such as flood plains. Plant fragments are rare in gravelly alluvial-fan deposits. Fragments of organic matter have not been noted in many fan deposits, but

FIG. 2.—Caliche layers in the Wheeler Wash fan, south side of the Spring Mountains, Nevada.

the presence of organic material should not be used as an indicator of the lack of an alluvial-fan environment. Legget and others (1966, p. 20) describe alluvial-fan deposits in northern Canada containing decomposed and undecomposed organic matter in silts having a reduced color. Meade (1967, p. 7) describes core-hole samples from alluvial fans in the western San Joaquin Valley, California, that contain disseminated small fragments of organic material in sand, silt, and clay that has an oxidized color. The present climate of the area consists of hot, dry summers and a mean annual rainfall of 6–8 inches that occurs during the winter months.

It is the physical characteristics that provide the surest means of identifying alluvial-fan deposits in the stratigraphic record. Some modes of deposition on alluvial fans are common in other depositional environments. Stream-channel deposits are found in several depositional environments, but sheetflood and sieve deposits are most likely to be preserved if deposited on an alluvial fan. Debris flows also can occur in several different environments. They are found on hillslopes (an erosional environment) and in river valleys. However, debris flows generally are stored only temporarily in river valleys, because they are associated with terrains that produce rapid runoff and hence are likely to be eroded by subsequent water flows. Little erosion occurs on an actively aggrading alluvial fan; thus, the fan environment is ideal for preservation of debris flows in the stratigraphic record.

The tendency of most modes of fan deposition to occur as sheets results in a diagnostic geometry of bedding that helps identify modes of deposition—such as mudflows—as occuring on alluvial fans rather than in other depositional environments. This important aspect is discussed in the sections on bedding and stratigraphic geometry.

Alluvial fans have greatly differing lithologies. Some fans consist of organic silt (Legget and others, 1966, p. 21), others consist largely of pebble- to boulder-size material without fines (Hooke, 1967, p. 456). More than one mode of deposition occurs on most fans, and the proportions of different types of deposits may vary both vertically and in the downslope direction from the fan apex. Each bed of a fan represents a single depositional event that has resulted from one of a wide spectrum of precipitation and erosion events within the source area. The runoff that is supplied to the main stream channel leading to the fan may be the result of rainfall over the entire basin, or of snowmelt runoff from all or part of the basin. Thus, differences in runoff characteristics, source and amount of sediment load, mode of transport, and other factors vary greatly and are reflected in the individual beds preserved in the fan.

The distribution of maximum and mean particle sizes in modern arid-region alluvial fans shows that particle size decreases downslope with increasing distance from the fan apex (Sharp, 1948; Blissenbach, 1954; Beaty, 1963; Drewes, 1963; Bull, 1964a; Ruhe, 1964; Bluck, 1964; and Hooke, 1967). The uniformity of particle size decrease is greatly affected by the amount of temporary channel entrenchment, which causes the loci of deposition to occur in different areas along a given radial line (Buwalda, 1951).

Water-laid Deposits

Two types of water-laid deposits can be found in most alluvial fans, and a third type occurs on fans with certain source-area conditions. Most of the water-laid sediments consist of sheets of sand, silt, and gravel deposited by a network of braided distributary channels. The second type of sediment consists of fillings in stream channels that were temporarily entrenched into the fan. The third type is sieve deposits (Hooke, 1967, p. 453–456) which are formed when surficial fan material is so coarse and permeable that even large discharges infiltrate into the fan completely before reaching the toe. Gravel deposited in the area where the flow infiltrates forms lobate deposits which further decrease the slope and promote additional deposition.

Sheetflood Sediments

Sheets of sediments are deposited by surges of sediment-laden water that spread out from the end of the stream channel on a fan. Deposition is caused by a widening of the flow into shallow bands or sheets and concurrent decrease in depth and velocity of flow, rather than a change in gradient at the end of the stream channel. Depths of water generally are less than a foot. The shallow distributary channels rapidly fill with sediment and then shift a short distance to another location. The resulting deposit commonly is a sheetlike deposit of sand, or gravel, that is traversed by shallow channels that repeatedly divide and rejoin.

The deposits commonly consist of gravel, sand, or silt that contains little visible clay. In general, they are well sorted and may be cross-bedded, laminated, or massive. The characteristics of sediments deposited by braided streams are described in detail by Doeglas (1962).

The low bars and anastomosing channels characteristic of water-laid sheets of sediments on fans are shown in Figure 3A. The photograph is a view of part of an area of water-laid deposition that is 1 mile long. A map of a similar depositional area on the same fan and a summary

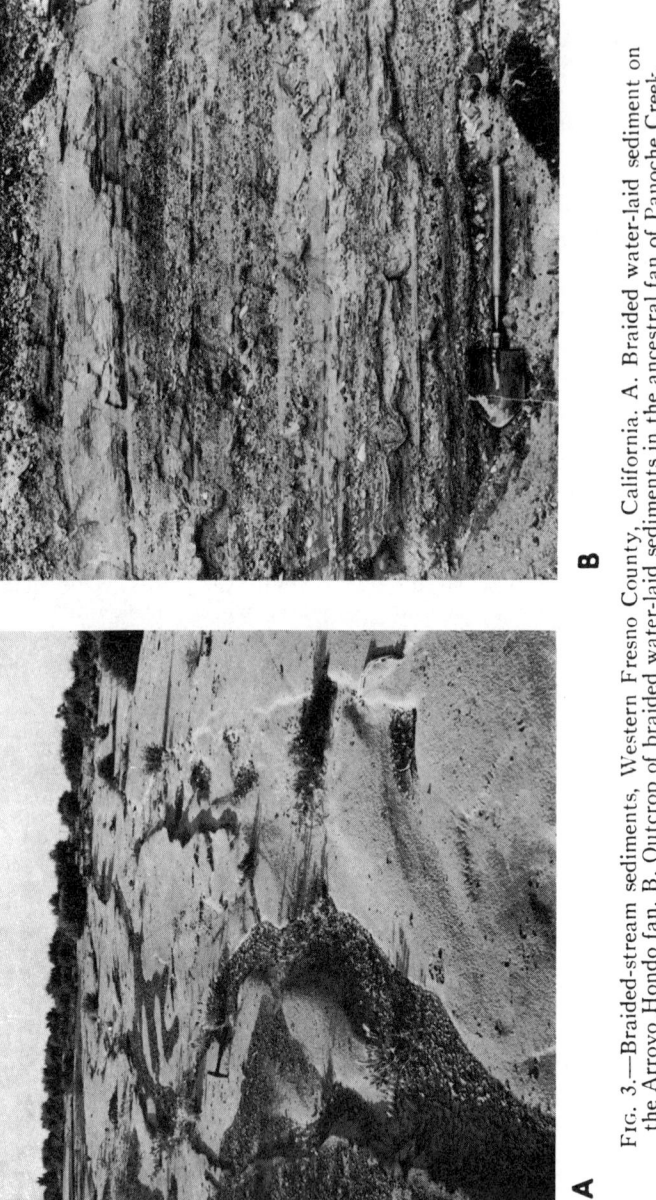

Fig. 3.—Braided-stream sediments, Western Fresno County, California. A. Braided water-laid sediment on the Arroyo Hondo fan. B. Outcrop of braided water-laid sediments in the ancestral fan of Panoche Creek.

FIG. 4.—Cross-section of deposits filling a small stream channel in the Tumey Gulch fan, Western Fresno County, California.

of particle-size distribution of 14 samples are given by Bull (1964a, p. 27).

Lenses caused by deposition as low bars and islands of braided-stream sediments are readily recognized in exposures of older fans that are normal to the direction of flow. The sands, silts, and gravels shown in Figure 3B were deposited near the apex of a Pliocene and Pleistocene alluvial fan that had an area of more than 200 square miles.

Stream-channel Sediments

The deposits that backfill stream channels temporarily entrenched into a fan generally are coarser grained and more poorly sorted than the sheets of water-laid sediments deposited by networks of braided distributary channels. Bedding of the channel-fill sediments may not be as well defined as for the sheetlike deposits. The thickness of individual beds ranges from less than one inch to more than five feet, but bed thicknesses of two inches to two feet are the most common.

A cut-and-fill structure in a canal bank along the fan contour is shown in Figure 4. The adjacent deposits consist mainly of sheets of clayey sand deposited by mudflows, and the channel filling consists of beds of gravel and clayey sand.

Comparison of the sorting of the sediments of braided stream and stream-channel deposits for 36 samples is summarized by Bull (1964a, p. 28). The samples were derived from a source terrain that consists of sandstone and mudstone, and the mean clay content of the samples was 6 percent. The sorting[1] was as follows:

Sieve Deposits

If the source area supplies little sand, silt, and

[1] The three sorting indices used are the Trask sorting coefficient, $S_0 = \sqrt{Q_{75}/Q_{25}}$; the phi quartile deviation,

$$QD_\phi = \frac{\phi 25 - \phi 75}{2}$$

and the phi standard deviation,

$$\sigma_\phi = \frac{\phi 16 - \phi 84}{2}$$

clay to the fan, the deposits may be sufficiently permeable to allow water from a flood discharge to infiltrate entirely before reaching the toe of the fan. Such conditions promote deposition of lobes of gravel. Hooke (1967, p. 453–456) has studied these lobate gravel deposits in detail and has named them "sieve deposits." He stated that because "water passes through rather than over such deposits, they act as strainers or sieves by permitting water to pass while holding back the coarse material in transport."

Thus, unique source-area conditions are responsible for fans composed of sieve deposits. Source areas for sieve deposits are underlain by rocks such as jointed quartzite; and the clasts supplied to the fans characteristically are subangular blocks instead of well-rounded gravel. The excellent sorting of sieve deposits results in a massive bed and poorly defined contacts between beds.

Sieve deposits are much less common than other types of water-laid deposits on alluvial fans but are among the most distinctive.

Sieve deposits on a small, steeply sloping fan are shown in Figure 5. Most of the surface has been darkened by desert varnish coating the particles. The man is standing on a finer-grained, recently deposited sieve lobe.

Debris-flow Deposits

Water flows can selectively deposit part of their sediment load as a result of decrease in velocity or depth of flow. When a flow incorporates sufficient sediment, sediment entrainment becomes irreversible, and the flow behaves more like a plastic mass than a Newtonian fluid. Debris flows have a high density and viscosity compared to stream flows. Because of these traits, debris-flow deposits are poorly sorted, have lobate tongues extending from sheetlike deposits, have well-defined margins, and are capable of transporting boulders weighing many tons. Factors that promote debris flows are abundant water (usually intense rainfall) over short periods of time at irregular intervals, steep slopes having insufficient vegetative cover to prevent rapid erosion, and a source material that provides a matrix of mud. Deposits of high viscosity flows are most common near the fan apexes (Hooke, 1967, p. 452; Crowell, 1954).

The proportions of water-laid and debris-flow deposits vary greatly from fan to fan and may change during the history of accumulation of the deposits of a single fan. Where source-area conditions are not conducive to the production of debris flows, the fan deposits consist entirely of water-laid sediments. Other fans consist mainly of debris flows. Most fans whose source areas produce debris flows also have flood events that result in the deposition of water-laid deposits. Thus, the deposits of many fans consist of inter-

FIG. 5. Sieve deposits on a small fan in Death Valley, California.

FIG. 6. Debris flow on the Sparkplug Canyon fan, west side of the White Mountains, California.

bedded deposits of debris and water flows in varying proportions.

Part of a thick viscous debris flow is shown in Figure 6. Abrupt margins of lobate tongues of debris in the foreground indicate that the flow was highly viscous. The round protuberances further upslope in the flow are 3- to 8-foot, mud-covered boulders. The smooth surface of most of the flow is characteristic of fresh debris flows deposited on alluvial fans. Old debris flows commonly consist of surficial lobes and levees of cobbles and boulders, because rainwash has removed much of their mud matrix. Later water flows may cut through the flow. Debris-flow levees occur adjacent to the channels of some flows (Sharp, 1942). The maximum thickness of the flow shown in Figure 6 is 4 to 6 feet (Beaty, 1968, p. 18).

A mudflow is a type of debris flow that consists mainly of sand-size and finer sediment. Many workers use the term "mud-flow" in a genetic sense for all types of debris flows, because a matrix of mud is the distinguishing feature of this type of flow, as contrasted with water flows.

The mudflow shown in Figure 7 consists of a poorly sorted clayey sand. The viscosity of the flow was not great, as is indicated by the thin margins of the lobate tongue. The flow thickens rapidly away from the margins and is 3 to 4 inches thick at the location of the shovel. Although the flow contained virtually no material larger than 4 millimeters, the cobble train shown at the right side of the photograph was part of a lobate tongue of an earlier debris flow that transported larger particles. Polygonal dessication cracks are characteristic of clayrich mudflows that contain little gravel.

The bedding of debris-flow sequences commonly is not well defined, but upon close examination bedding planes between flows can be discerned in outcrops (Beaty, 1970, fig 4). Where interbedded with water-laid sediments, debris-flow beds are readily apparent.

Several features aid in the recognition of debris flows in the geologic record. The uniform thickness of the central parts of the sheet-like deposits produces beds that are remarkably consistent in thickness when observed in outcrop. An indication of the viscosity of a given flow can be obtained by study of the position and orientation of the larger clasts. A fluid debris flow will have graded bedding and a horizontal or imbricated orientation of the tabular gravel fragments. The more viscous flows have the larger clasts distributed uniformly throughout the thickness of the flow. The most viscous flows not only have uniform distribution of the larger clasts, but the tabular particles commonly have

Fig. 7.—Mudflow on the Santiago Creek fan, north side of the San Emigdio Mountains, California.

a vertical prefered orientation normal to the direction of flow.

Poor sorting is characteristic of debris flows. Many debris-flow deposits are so coarse grained that it is difficult to obtain a representative sample for determining the particle-size distribution of the material. As a result, few particle-size analyses have been made of debris flows (Crawford and Thackwell, 1931; Sharp and Nobles, 1953 p. 556), and even fewer comparisons have been made of the sorting of debris flows and water-laid sediments from the same source areas. Bull (1964a p. 24, 25, 65, 66) made 50 particle-size analyses of mudflows, and the mean sorting characteristics of the suite of samples are summarized in Table 1. The samples are from the same source terrain as noted for the water-laid samples in Table 1. The mean clay content of the mudflow samples was 31 percent.

CM Patterns

Logarithmic plots of the coarsest 1-percentile particle size (C) and the median particle size (M) may make patterns distinctive of the depositional environment according to Passega (1957, 1960). Bull (1962b), using the parameters C and M from the particle-size analyses of 102 surficial and 50 corehole samples of alluvial fan deposits, found distinctive patterns that could be related to the modes of deposition on the alluvial fans.

The mode of deposition was determined in the field for the surface samples, and the CM patterns for these samples are shown in Figure 8. Passega found that a sinuous pattern was definitive of the tractive-current mode of deposition associated with water-laid deposits of perennial rivers, and that a rectilinear pattern roughly parallel to the limit C=M was characteristic of turbidity-current deposits. These modes of deposition apparently have analogs on arid-region alluvial fans.

Three of the four segments of Passega's tractive-current pattern are represented by the wa-

TABLE 1.—SORTING OF ALLUVIAL-FAN DEPOSITS DERIVED FROM THE DIABLO RANGE, CALIFORNIA

	S_o	σ_ϕ	QD_ϕ
Braided-stream deposits			
Range	1.1–2.7	0.48–2.4	0.15–1.4
Mean	1.5	1.0	.56
Stream-channel deposits			
Range	1.3–4.8	0.82–3.4	0.42–23
Mean	2.1	2.0	1.1
Mudflow deposits			
Range	5.0–25.	4.1–6.2	2.3–4.7
Mean	9.7	4.7	3.1

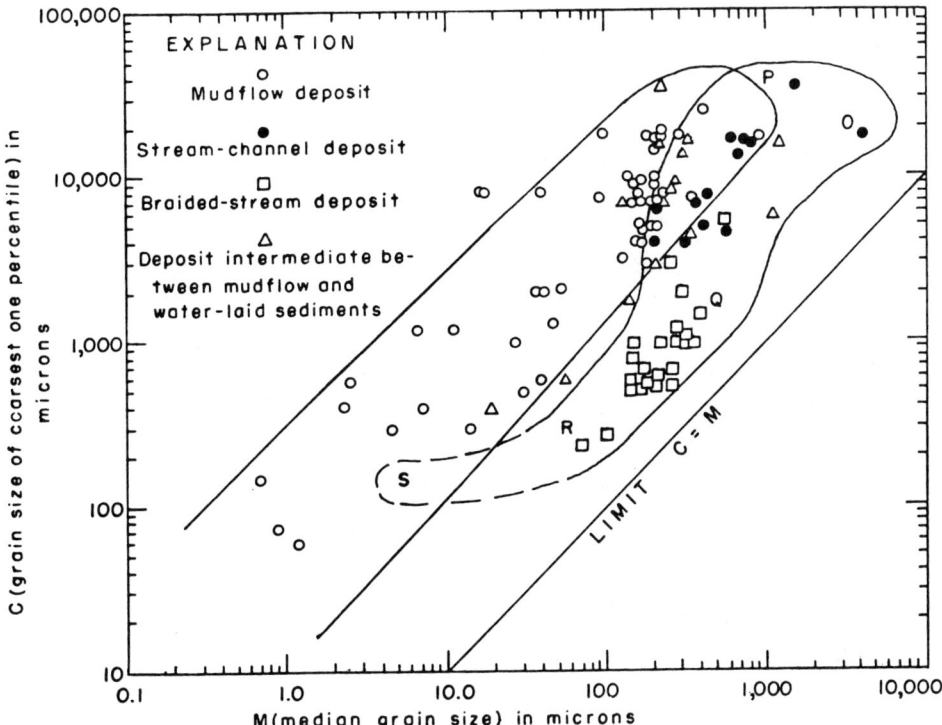

Fig. 8.—CM patterns of surficial alluvial-fan deposits, western Fresno County, California. Letters are discussed in the text.

ter-laid deposits shown in Figure 8. The segment representing the deep protected channel of Passega (RS) is not present, suggesting that this type of deposition is not characteristic of alluvial-fan sediments deposited by shallow ephemeral streams.

The CM pattern for the mudflow deposits is the same type that Passega shows for turbidity currents, which suggests that both turbidity currents and mudflows entrain fine-grained material. The chief difference between the patterns for mudflows and turbidity currents appears to be the sorting of the coarsest half of the deposits. Passega mentions that C is 2.3 to 4.2 times M for points along the axes of his turbidity current patterns. C ranges from about 40 to 80 times M for points along the axis of the mudflow CM pattern. This poorer sorting suggests that the mudflows had a much higher density and viscosity then did the turbidity currents studied by Passega.

CM patterns can be used to ascertain the mode of deposition of alluvial-fan deposits from subsurface samples. Figure 9 shows CM patterns for core-hole samples from two fans adjacent to the Diablo Range in California. About 68 percent of the source area for the Arroyo Ciervo fan is underlain by soft clayey rocks such as mudstone and shale, but only 34 percent of the source area of the Martinez Creek fan is underlain by clayey rocks.

Ten samples from a 70-foot core hole in the Martinez Creek fan are represented by the points with pattern T in Figure 9. Most of the samples are moderately well sorted sand and form a pattern suggestive of deposition by tractive currents.

Forty samples from five core holes in the Arroyo Ciervo fan are represented by the points within pattern M in Figure 9. The wide range of sorting between the 50th and 99th percentile suggests that a mixed depositional environment may be represented. However, all but four of the points can be included in the rectilinear mudflow-type pattern. Points in the vicinity of A probably indicate mudflow deposition, and the four points near C may represent water-laid sediments. C is about 45 times M at points along the axis of the "mudflow" pattern.

STRATIGRAPHY OF ALLUVIAL FANS
Bedding

Bedding of deposits is one of the best methods of identifying the alluvial-fan environment of deposition. Within a single outcrop a variety of strata generally can be observed, each bed representing a particular set of hydraulic conditions that determined the thickness, particle-size and distribution, particle orientation, and type of contact with the underlying bed. Even in fans composed entirely of water-laid sediments, differences in flow result in marked differences in the sedimentological characteristics of the beds (Figure 3B).

Bedding differences are even more striking in the many fans composed of both water-laid and debris-flow deposits. The poorly sorted, massive beds of debris-flow deposits stand out in marked contrast to the beds of water-laid sediments. Thus, one distinctive feature of alluvial-fan deposits is the variety of depositional types that can be observed.

Because the bulk of the fan deposits are deposited as sheets and lobes, uniform thickness for a given bed is common in most outcrops, particularly for debris-flow deposits. The thickness of the bedding of water-laid deposits usually is a function of the amount of relief between the bars and the braided stream channels in the area at the time of deposition, and the degree of erosional modification by post-depositional flows.

An example of bedding variety is shown in Figure 10. The massive thick bed of uniform thickness above the hat consists of clayey gravel and was deposited as a viscous debris-flow. Beneath the debris flow are beds of well-sorted, water-laid sand. A ½-inch bed of water-laid clay immediately above the sands is typical of the waning phase of ephemeral water flooding on a fan, when the competence is sufficient to transport only silt and clay. Beds of poorly sorted, silty gravel occur above the debris-flow bed and beneath the water-laid beds. These beds may be interpreted as low viscosity, debris-flow deposits, or as poorly-sorted, water-laid deposits.

When individual beds can be identified at

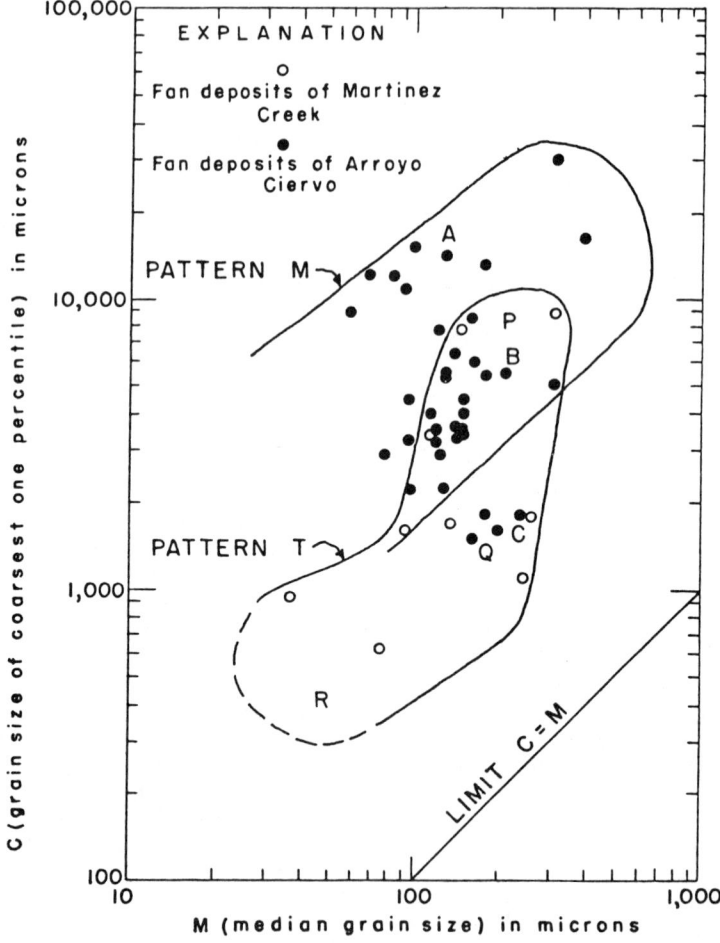

FIG. 9.—CM patterns of alluvial-fan deposits from core holes, western Fresno County, California, Letters are discussed in the text.

Fig. 10.—Bedding of Late Tertiary alluvial-fan deposits, south side of the Santa Catalina Mountains, Arizona.

more than one outcrop, another diagnostic stratigraphic indicator becomes readily apparent—the extensive sheet-like aspect of the individual beds. Sheets of fan deposits commonly are 10 to more than 100 times the width of the channel that transported the material to the fan. Thus, one of the identifying characteristics is the ratio of the trunk channel widths, as shown by cut-and-fill structures, to the width of the former areas of deposition represented by individual beds. This relation is best observed in exposures parallel to the former fan contour.

The sheetlike character of the bedding commonly is not apparent in those fans consisting mainly of gravel. The braided-stream mode of deposition that is characteristic of these gravel deposits results in small-scale, cut-and-fill structures being common in those exposures parallel to the fan contour. Thick, sheetlike beds of water-laid gravel, if present, can be attributed to large floods that reworked large volumes of previously deposited material.

The sheets and tongues of water-laid and debris-flow deposits on modern alluvial fans are generally 5 to 20 times as long as they are wide. The length of the sheets ranges to a few tens of feet to many miles. Exposures usually are not available to indicate the extent of the larger sheets comprising fans. However, by examining the beds that represent flows of small extent, one finds that these commonly are narrow compared to the length of the same bed.

Although local variation of flow direction on a given fan may exceed 30 degrees (Bull, 1971, fig. 52), the lack of meandering channels on most fans results in a high consistency of current flow directions. Statistical studies by Howard (1966, p. 152) and Nilsen (1969, p. 50) indicate that a high consistency of flow directions is characteristic of ancient alluvial-fan environments.

Some of the structures examined by Nilsen have up-current dips that indicate a flow direction that is opposite that of the majority of the orientations. Instead of concluding that the up-current results are in error, Nilsen concludes that the up-current-dipping strata resulted from the preservation of antidune bed forms, such as have been described by Harms and Fahnestock (1965).

Stratigraphic Geometry of a Fan

The overall geometry of an alluvial fan reflects the accumulation of vast numbers of beds of differing extent and thickness and the changes in loci of deposition caused by entrenchment and backfilling of the trunk stream channel. The typical situation is shown diagrammatically in Figure 11. The area portrayed is one where recent uplift along a boundary fault has induced rapid accumulation of alluvial-fan deposits adjacent to the mountain front. The surface of the fan is not entrenched and is traversed by a network of braided distributary streams, one of which is associated with the most recent episode of deposition.

Considerable variety exists in the radial and cross-fan stratigraphic relations. Along the radial sections of a fan, individual beds may be traced for long distances, and channel-fill deposits are rare.

In contrast, the cross-fan sections reveal overlapping beds of limited extent that are interrupted by cut-and-fill structures. Because some channels were entrenched only a short distance downslope from the fan apex and others were entrenched into a midfan area, cut-and-fill structures are most common near the fan apex and rare or absent near the toe of the fan.

Three basic shapes of fans can be preserved in the stratigraphic record. Cross sections of these shapes are shown in Figures 12, 13, and 14. The examples shown in Figures 12 and 13 are actual field examples, where the shape of the body of fan deposits is discussed relative to a time line indicated by the top of a lacustrine clay.

Figure 12 shows a wedge of deposits that is thickest near the mountains. In this area uplift

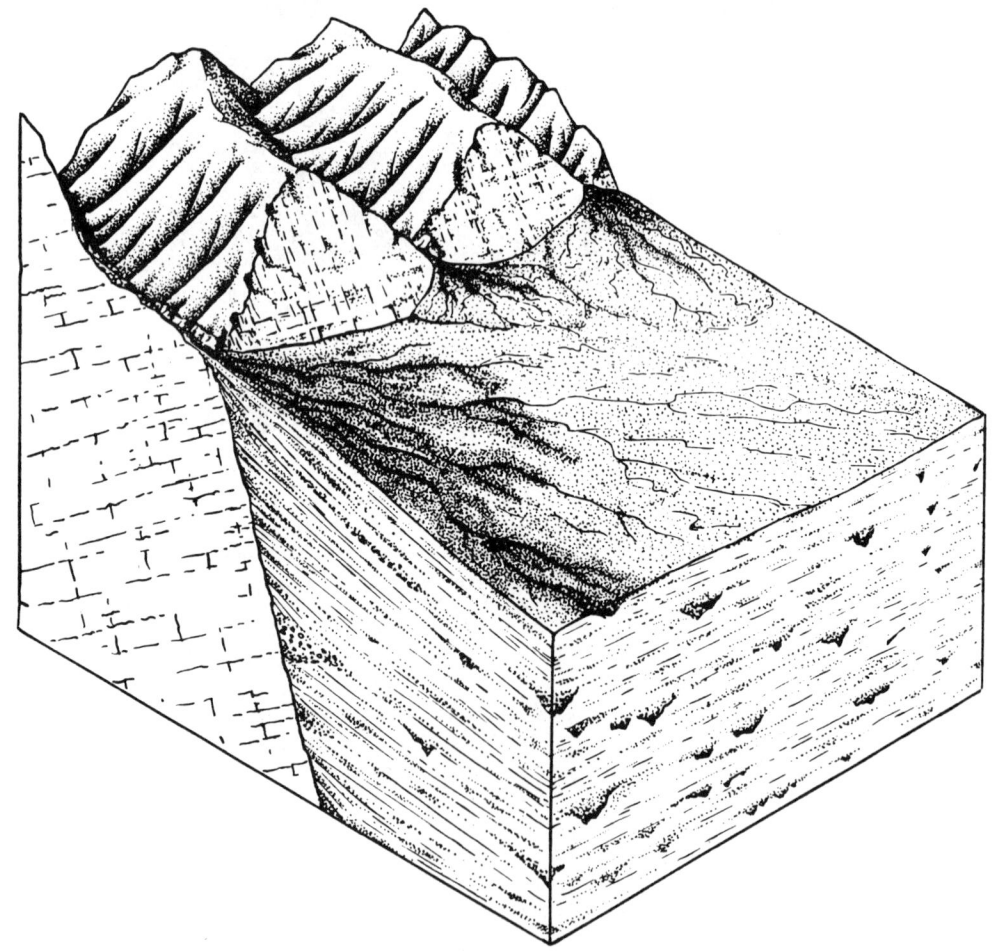

FIG. 11.—Stratigraphic geometry of an alluvial fan.

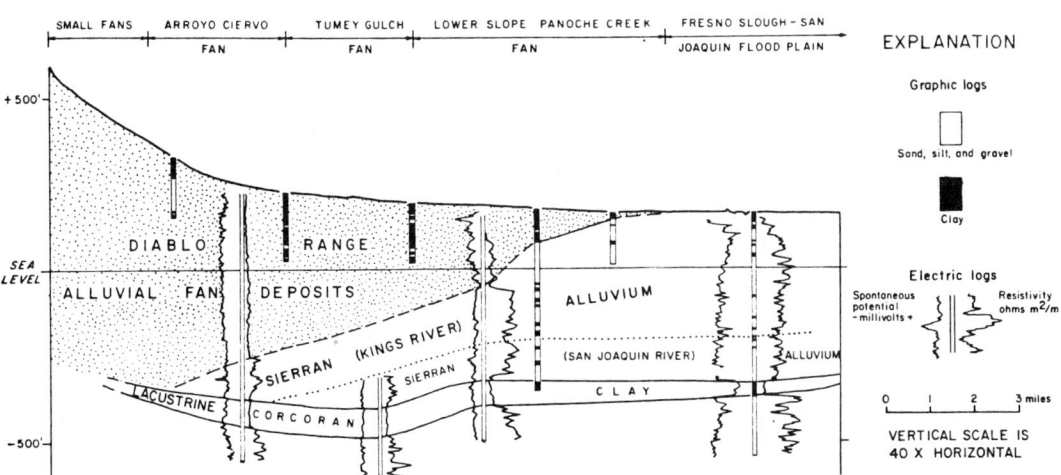

Fig. 12.—Longitudinal section of fan deposits that are thickest adjacent to the mountain front. Modified from Magleby and Klein (1965, Plate 5).

of the mountains occurred mainly before deposition of the Corcoran lake clay, which is only slightly deformed. Uplift of the mountains has increased the sediment yield of the source area and the fans that were only 2½ miles long at the end of the lacustrine period have expanded to 14 miles.

A lense-shaped mass of fan deposits is shown in Figure 13. In this case, the Pleistocene lake clay has been folded along its western extent, as has an undetermined amount of the overlying fan deposits. The largest thicknesses of post-pluvial fan deposition have occurred not adjacent to the mountain front, but in the midfan area. The increase of sediment yield as a result of tectonic uplift has increased the fan areas, thus allowing clayey sands from the sedimentary rocks of the Diablo Range to be deposited on top of the arkosic sands derived from the granitic rocks of the Sierra Nevada.

The relations shown in Figures 12 and 13 reveal the influence of an active orogenic environment—an environment typical of thick accumulations of fan deposits. After tectonic activity

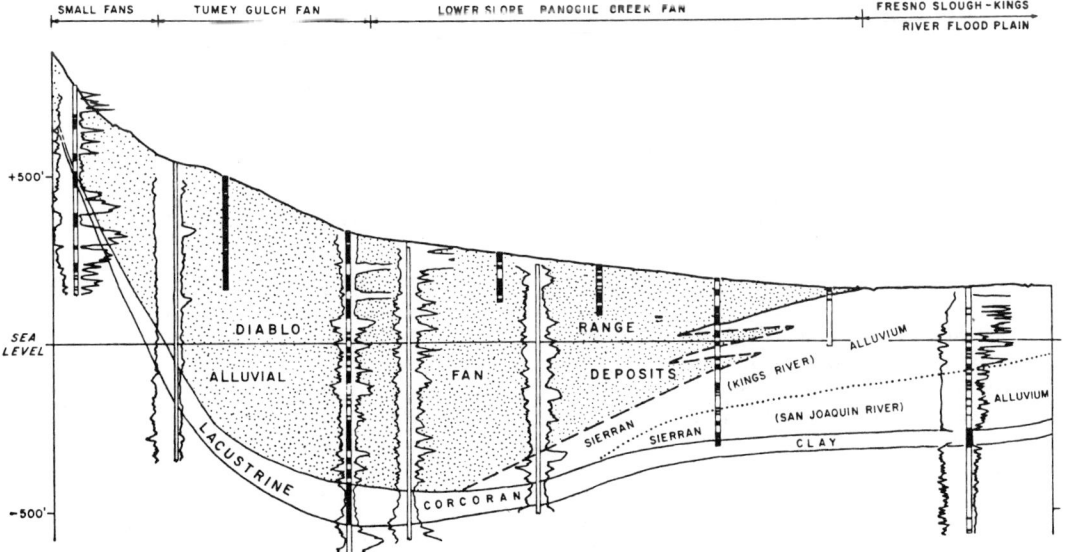

Fig. 13.—Longitudinal section of fan deposits that are lenticular. Modified from Magleby and Klein (1965, Plate 4). See figure 12 for explanation of symbols.

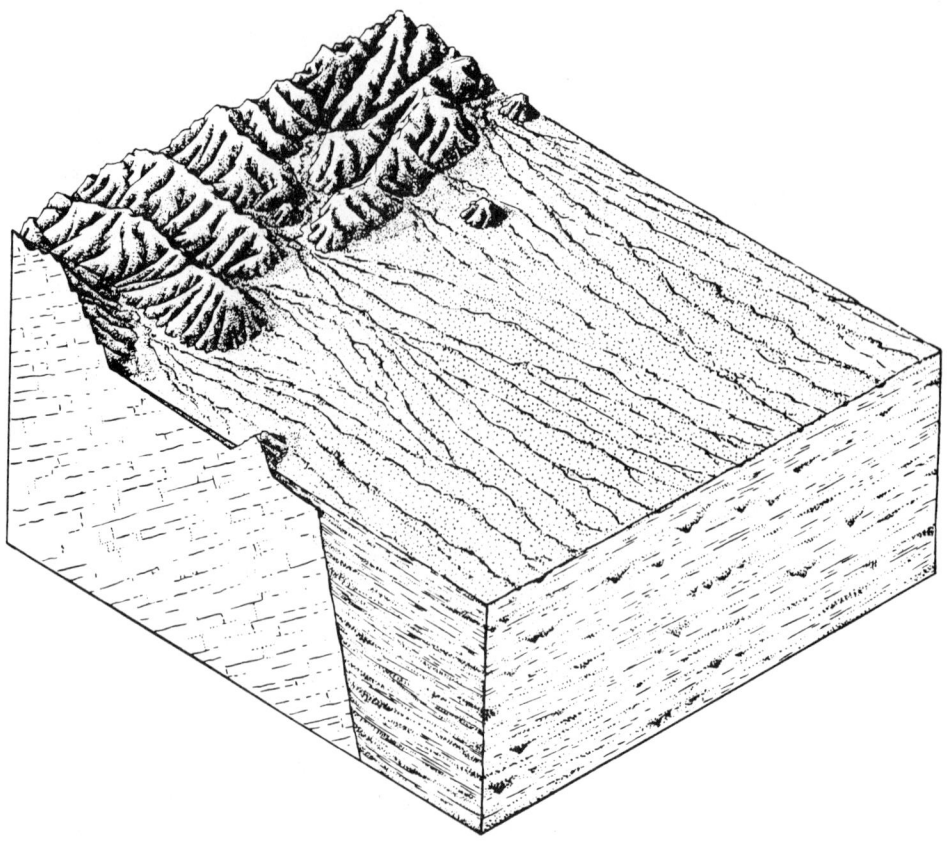

FIG. 14.—Longitudinal section of fan deposits that thicken away from the mountains.

ceases, the streams will continue to downcut within the mountains and eventually will cut below the altitude of the fan apex (Eckis, 1928, p. 237–238). The resulting entrenching of the upslope part of the fan can be regarded as a permanent type of entrenchment that removes the trenched part of the fan as a potential area of deposition. Erosion of the mountain front and the fan deposits above the stream channel continue, and as a result the surface loses much of its aspect of a segment of a cone.

A change from a depositional to an erosional environment may remove tens or even hundreds of feet of surficial fan deposits, and if accompanied by erosional retreat of the mountain front will result in the formation of a pediment upslope from the former mountain front (Figure 14). The stratigraphic relations and depositional characteristics described earlier will still identify the deposits downslope from the former mountain front as being part of a former alluvial fan. However the shape of the remaining deposit will be that of a wedge that thickens away from the former mountain front.

The shape of the overall sequence of deposits that is preserved in the stratigraphic record will in part be a function of the thickness of the deposits that accumulated before erosion of the uppermost deposits occurred, and the thickness of deposits removed by erosion. Part of the body of deposits portrayed in Figure 14 may be the result of redistribution of deposits from the upper to the lower parts of the fan.

Relation of Alluvial-fan Deposits to Adjacent Depositional Environments

Alluvial-fan deposits commonly are in contact with deposits of adjacent alluvial fans or deposits of flood-plain and lacustrine environments.

The deposits of individual alluvial fans interfinger in zones of coalescence. Where core hole and electric-log data are available, the distinctive lithologies between fans derived from differing source-area lithologies are readily discernible. An example is the situation shown in Figure 13, which shows the relations between the "small fans," the Tumey Gulch fan, and the Panoche Creek fan, each of which can be identified by the logs.

Examination of many logs in an area, such as shown in Figures 12 and 13, suggests that the lateral expansion of fans that occurs during the history of accumulation of deposits is minor. Rates of accumulation of deposits on a given fan may increase or decrease with time, but the rates of sediment-yield change in adjacent drainage basins are likely to be similar; hence, little change in the boundaries of adjacent fans occurs with time.

If the downslope edge of the fan is adjacent to a through-flowing stream, the fan deposits are likely to be in contact with floodplain deposits (Figures 12 and 13). If the rate of fan deposition exceeds the rate of floodplain deposition, the fans will expand their area by encroachment over the floodplain depositional area until the rates of deposition in the two areas are approximately equal. In the Western Fresno County area, accelerating rates of sediment yield during the last 600,000 years (the age of the top of the lake clay: Janda, 1965, p. 131) and the tendency toward an equilibrium defined by equal rates of accumulation have resulted in overlap of the fan deposits on the flood-plain deposits.

The tendency toward equal rates of deposition among coalescing alluvial fans and between fans and the playa in a closed basin has been described by Hooke (1968, p. 614–616). Hooke's steady-state model suggests that the areas of accumulation of fan and playa deposits are directly proportional to the volumes of material being supplied to each fan and to the playa, per unit time. Using Hooke's model, the playa-fan contact will be an intertonguing one, as will the contacts between coalescing fans if sediment inputs fluctuate.

In closed basins, pluvial lakes may have formed and inundated parts of the alluvial fans. Lake beds deposited during pluvial intervals form extensive blanket-like deposits that occur as layers in the sequence of fan deposits. Where the areal extent of a lake bed can be defined in the subsurface, the upslope extent of the lake beds will depict not only the shore line, but also the fan contour at that point in the history of accumulation of the fan deposits.

Some workers have described fan deposits that are associated with talus or sedimentary breccias. Blissenbach (1954) notes the interfingering of fan deposits toward the source area with talus deposits, and Drewes (1963, p. 34, 52) describes coarse unbedded sedimentary breccias adjacent to a former mountain front. The lack of bedding, the resistant rock types, and the general lack of fine-grained material noted by these workers suggests that the types of coarse-grained deposits noted by them may be water-laid sieve deposits, such as described by Hooke (1967). Talus can be distinguished in the field from sieve deposits because the particle size increases in the downslope direction in talus, and talus is skewed toward the coarser particle sizes.

Comparison of Alluvial-fan Deposits with Other Coarse-grained Depositional Environments

Some workers regard coarse clastic beds as fan deposits if they appear to be terrestrial deposits. Fans can be readily distinguished from other environments associated with coarse clastics, such as stream-channel gravels, pediment gravels, and marine conglomerates.

Marine conglomerates are extensive in the stratigraphic record, and the transgressive or regressive shorelines associated with their deposition tend to produce sheets of gravel whose interstices are filled with finer-grained material. However, these deposits differ from most fan deposits in the following aspects. (1) The littoral gravels are gray, green, or blue, indicating a reducing environment. (2) The interstices are filled with well-sorted sand instead of a poorly sorted matrix. Gravels filled with sand are found in fan deposits also, but are characteristic of stream-channel deposits and thus generally do not have the lateral extent associated with sheets of littoral gravels. (3) The degree of rounding of both the large and small clasts in marine coarse-grained clastics is much better than found in the alluvial-fan environment.

Most of the differences that can be noted between the marine and alluvial-fan environments also pertain to the littoral facies of lacustrine environments.

Stream-channel and flood-plain deposits, although terrestrial, differ markedly from the deposits of the alluvial-fan environment. Pointbar deposits and channel fills constitute the bulk of the deposits associated with an aggrading river. In contrast alluvial-fan deposits have only a minor amount of stream-channel deposits, the majority of the deposits consisting of lenticular sheets of water-laid sediments and debris flows. Debris flows are not characteristic of most floodplain and stream-channel deposits and floods erode debris-flow deposits deposited within the mountains. The change from an erosional to a depositional environment that occurs at the downslope end of the stream channel terminating on the fan results in conditions that are ideal for the accumulation of hundreds, or thousands, of feet of deposits that consist in part of debris flows.

The fluvial deposits that are found mantling some pediments differ from fan deposits in that they are thin sequences of water-laid sediments that represent a zone of transportation rather than a depositional area. Of course this does not preclude the possibility of the pediment being buried by thick fan deposits (Williams, 1969, p. 191–200). Even some of the temporary accumulations of alluvium on pediments have cross-slope profiles indicative of small alluvial fans (Mabbutt, 1966, fig. 5).

ALLUVIAL FANS IN THE STRATIGRAPHIC RECORD

Thick alluvial fans are orogenic deposits, not only because uplift creates mountainous areas that provide debris and increase stream competence, but also because the loci of deposition on alluvial fans is controlled largely by the rate and magnitude of uplift of the adjacent mountains (Bull, 1964b, 1968). Optimum conditions for accumulation of thick sequences of fan deposits occur where the rate of uplift exceeds the rate of downcutting of the trunk stream channel at the mountain front. The orogenic interpretation of fan deposits applies particularly to the thick sequences of fan deposits found in the stratigraphic record.

Small alluvial fans are found also in areas of nontectonic, base-level change, such as where a river or glacier has eroded its floor at a more rapid rate than have the tributary streams (Suggate, 1963; Carryer, 1966). Changes in the mode and loci of fan deposition ascribed to climatic changes are discussed by Lustig (1965, p. 183–186).

Deposition of thick sequences of fan deposits have occurred during many orogenic periods. Examples are the Late Pre-Cambrian Keweenawan red beds of the Canadian shield, the Devonian fan deposits of Norway that are associated with the post-Caledonian Svalbardian disturbance (Nilsen, 1968a, 1968b, 1969), the Pennsylvanian Fountain Formation of Colorado (Tieje, 1923; Hubert, 1960; Howard, 1966), and the Triassic Newark Group associated with the Appalachian Revolution (Dunbar, 1949, p. 311–316; Krynine, 1950; Reinemund, 1955; and Klein, 1962). It is not the purpose of this article to review the various alluvial-fan sequences that have been described in the stratigraphic record. However, some of the characteristic features of fans can be illustrated by using two suits of fan deposits as examples—the late Pre-Cambrian fans of Scotland and some Pliocene fans of Southern California.

Torridonian Alluvial-fan Deposits of Scotland

Proterozoic fan deposits have been described as overlying Archean gneiss along the northwest coast of Scotland (Maycock, 1962; Williams, 1966, 1969). Maycock (1962, p. 124) describes the Ardheslaig facies as consisting of "thick bedded, very rudely stratified, very coarse, poorly sorted breccia-conglomerate. Interbedded with this material occur varying percentages of thin to thick bedded, medium- to coarse-grained, poorly sorted feldspathic sandstone."

Rare interbedded shale sequences are interpreted as being playa deposits. In Figure 15, playa deposits are preserved as red shales and fine-grained sandstone beds. Overlying the playa deposits are crudely stratified red fanglomerates. Maycock (1962, p. 129) points out that the presumed lack of extensive terrestrial vegetation in Pre-Cambrian times favored the production of debris flows.

Williams (1966, p. 1305) concludes that sheetfloods deposited the tabular conglomerates and that the water-laid sediments consist of (1) braided-stream deposits, and (2) point-bar deposits indicating a channel environment.

By using paleocurrent directions of Maycock and other workers, Williams (1966, fig. 1) concluded that two sets of radiating current directions indicated the existence of two large coalescing Torridonian alluvial fans. Similar paleocurrent studies made by Howard (1966) of the Fountain Formation of Colorado indicated the general positions of fan apexes for a bajada of Pennsylvanian fans.

Sedimentary structures indicative of fluvial deposition have been described in the alluvial fans preserved in several areas. The common sedimentary structures noted in the Precambrian and Paleozoic fan deposits of Norway, Scotland, and Colorado are summarized in Table 2.

TABLE 2—COMMON SEDIMENTARY STRUCTURES DESCRIBED IN ALLUVIAL FANS IN THE STRATIGRAPHIC RECORD

Sedimentary structure	Stream-channel deposit	Sheetflood deposit
Trough cross-stratification	N. S. C.	
Planar cross-stratification	N. S. C.	
Graded bedding	N. S. C.	
Flat-stratification	N.	N. S.
Primary current lineations	N.	N. C.
Flow casts		C.
Ripple-drift bedding		N. S.
Micro-cross lamination		C.

N. Devonian fans of Norway (Nilsen, 1968)
S. Precambrian fans of Scotland (Williams, 1969)
C. Pennsylvanian fans of Colorado (Howard, 1966)

Fig. 15.—Pre-Cambrian Torridonian fanglomerates overlying playa deposits, Scotland. Photo by Ian Maycock, Continental Oil Company.

Ridge Basin Alluvial-fan Deposits, California

Crowell (1954) uses the Ridge Basin area in Southern California to illustrate a closed intermontane basin that was filled with more than 25,000 feet of clastic deposits during Late Miocene and Pliocene time. The deposits of the Ridge Basin Group reveal features that are characteristic of many fan deposits in the stratigraphic record. The deposits exhibit a radial inhomogeneity downslope from the former fan apexes, great thicknesses, and a close association with areas of active tectonic movement.

A diagrammatic section of the Ridge Basin area is shown in Figure 16. The deposits adjacent to the San Gabriel fault zone consist of a poorly bedded cobble and boulder conglomerate with a mudstone matrix that is suggestive of debris-flow deposition. The debris-flow unit grades rapidly away from the fault zone into water-laid conglomerates and sandstone. This is consistent with the laboratory and field studies made by Hooke (1967, p. 452–453) that show that the debris-flow deposition predominates on the upper parts of alluvial fans, and that water-laid deposits predominate on the lower parts of fans because of the ability of water floods to transport debris further downslope.

The conglomerate and sandstone adjacent to the San Gabriel fault scarp has a statigraphic thickness of 27,000 feet (Crowell 1954). Thicknesses of alluvial-fan deposits in the stratigraphic record that exceed 10,000 feet are common. The Newark group in the Appalachians exceeds 20,000 feet, and the Devonian fan deposits in Norway have a maximum thickness of 17,000 feet. Even the remnants of the Pre-Cambrian Torridonian fans in Scotland are 8,000 feet thick.

The great thickness of the Ridge Basin conglomerates, as well as the time span represented by the sequence, provides clear evidence of repeated uplift of the source areas (Crowell, 1954).

The overall sequence of the Ridge Basin deposits from the San Gabriel fault scarp to the center of the depositional basin is from breccia and conglomerate to sandstone to shale, which can be presumed to represent playa deposition in the center of the basin.

SUMMARY OF DIAGNOSTIC CRITERIA OF ALLUVIAL-FAN DEPOSITS

Many alluvial fans have the following diagnostic features that identify them as the products of a distinctive terrestrial depositional environment.

1. Alluvial fans are oxidized deposits that

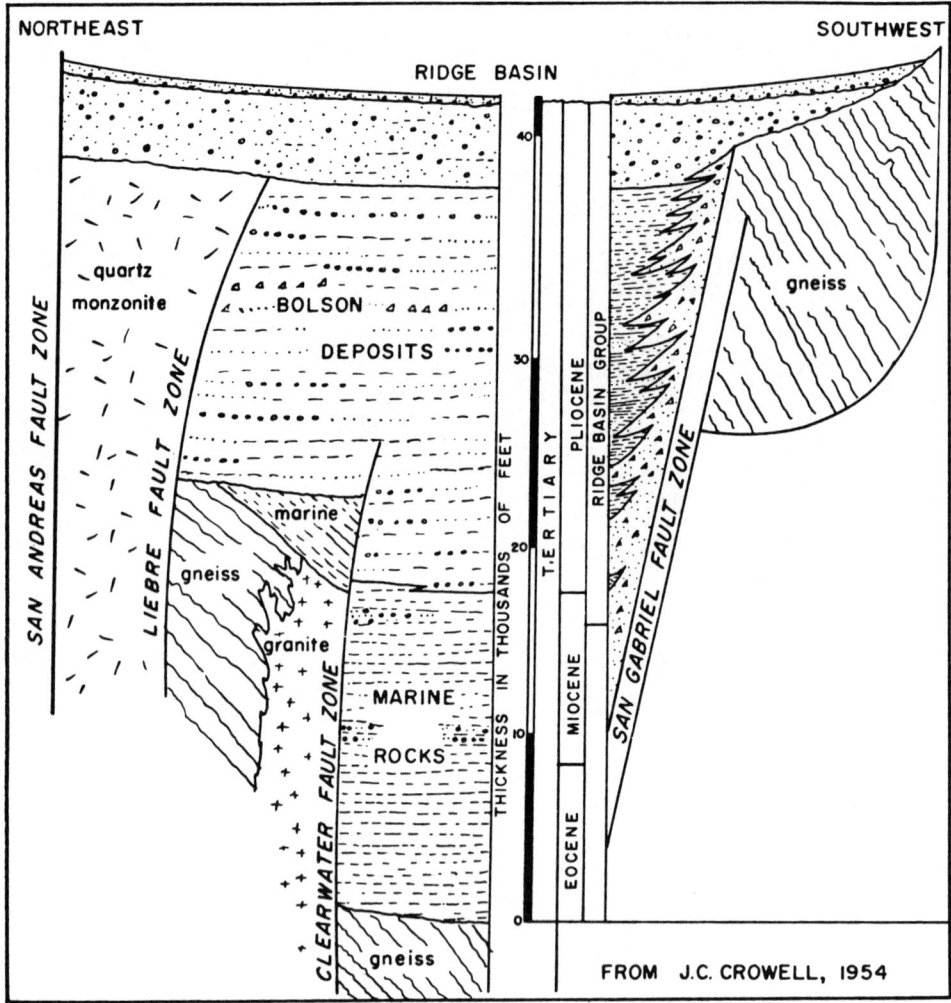

Fig. 16.—Diagrammatic section of the Ridge Basin bolson deposits, California. Modified from Crowell (1954).

rarely contain well-preserved organic material.

2. Alluvial fans commonly consist of thick sequences of water-laid sediments deposited by braided distributary streams and by stream-channel deposition; of mudflow and coarse-grained, debris-flow deposits; or of both water-laid and debris-flow deposits.

3. The bulk of the deposits consist of sheets that have length/width ratios of roughly 5 to 20. Channel-fill deposits comprise a minor proportion of most fans.

4. The proportion of debris-flow deposits decreases downfan from the apex in those fans where both water-laid and debris-flow deposits are present.

5. Particle size decreases downfan from the apex, but the uniformity of particle-size decrease with distance is affected greatly by the amount of temporary channel entrenchment during the history of the fan. Fanhead trenches result in coarse debris being deposited on the middle and lower parts of a fan, instead of being deposited on the upslope part of a fan.

6. Cut-and-fill structures are common near the fan apexes and are rare near the toes of most fans.

7. When compared with other depositional environments, the hydraulics of transport and deposition are greatly different for the individual beds within a sequence of fan deposits. The result is a sequence of beds that vary greatly in particle size, sorting, and thickness.

8. Logarithmic plots of the coarsest 1-percentile particle size (C) and the median particle size (M) of alluvial-fan deposits provide two distinctive patterns. A sinuous pattern representing a tractive-current type of ephemeral-stream deposition is common, and a rectilinear type of pattern that roughly parallels the limit $C = M$ is typical of debris flows. The rectilinear pattern differs from that for turbidity-current deposits in that values of C along the axis of the pattern range from 40 to 80 times M—values that are more than 10 times the values for turbidity-current deposits.
9. Alluvial fans commonly have transgressive or intertonguing relations with the deposits of other depositional environments, such as flood plains or lakes.
10. Depositional structures of fans reflect a radial flow direction from the apex. Individual beds continue for long distances in exposures that are parallel to the radial lines of a fan, but cross-fan exposures reveal overlapping beds of limited extent that are interrupted by cut-and-fill structures. Paleocurrent direction data—such as stream-trough data on horizontal bedding surfaces—produces directions that are suggestive of radial-flow directions on former alluvial-fan surfaces.

REFERENCES

BEATY, C. B., 1963, Origin of alluvial fans, White Mountains, California and Nevada: Ann. Assoc, Amer. Geographers, v. 53, p. 516–535.
———, 1968, Sequential study of desert flooding in the White Mountains of California and Nevada: U.S. Army Natic Laboratories, Earth Science Laboratory, Tech. Rept. 68-31-ES, 96 p.
———, 1970, Age and estimated rate of accumulation of an alluvial fan, White Mountains California, U.S.A.: Amer. Jour. Sci., v. 268, p. 50–77.
BLISSENBACH, ERICH, 1954, Geology of alluvial fans in semiarid regions: Geol. Soc. America Bull., v. 65, p. 175–190.
BLUCK, B. J., 1964, Sedimentation of an alluvial fan in southern Nevada: Jour. Sed. Petrology, v. 34, p. 395–400.
BULL, W. B., 1962a, Relations of alluvial-fan size and slope to drainage-basin size and lithology in western Fresno County, California: U.S. Geol. Survey, Prof. Paper 450-B, p. 51–53.
———, 1962b, Relation of textural (CM) patterns to depositional environment of alluvial-fan deposits: Jour. Sed. Petrology, v. 32, p. 211–216.
———, 1964a, Alluvial fans and near-surface subsidence in western Fresno County, California: U.S. Geol. Survey Prof. Paper 437-A, 70 p.
———, 1964b, Geomorphology of segmented alluvial fans in western Fresno County, California: U.S. Geol. Survey Prof. Paper 352-E, p. 89–129.
———, 1968, Alluvial fans: Jour. Geol. Ed., v. 16, p. 101–106.
———, 1971, Prehistoric near-surface subsidence cracks in western Fresno County, California: U.S. Geol. Survey Prof. Paper 437-C, (in press).
BUWALDA, J. F., 1951, Transportation of coarse material on alluvial fans: Geol. Soc. America Bull., v. 62, p. 1497.
CARRYER, S. J., 1966, A note on the formation of alluvial fans: New Zealand Jour. Geol. and Geophys., v. 9, p. 91–94.
CRAWFORD, A. C., AND THACKWELL, F. E. 1931, Some aspects of the mudflows north of Salt Lake City, Utah: Utah Acad. Sci. Proc., v. 8, p. 97–105.
CROWELL, J. C., 1954, Geology of the Ridge Basin Area: Calif. Div. Mines, Bull. 170, map sheet no. 7.
DENNY, C. S., 1965, Alluvial fans in the Death Valley region, California and Nevada: U.S. Geol. Survey. Prof. Paper, 466, 62 p.
DOEGLAS, D. J., 1962, The structure of sedimentary deposits of braided rivers: Sedimentology, v. 1 p. 167–190.
DREW, FREDERICK, 1873, Alluvial and lacustrine deposits and glacial records of the upper Indus basin: Geol. Soc. London Quart. Jour., v. 29, p. 441–471.
DREWES, HARALD, 1963, Geology of the Funeral Peak Quadrangle, California, on the east flank of Death Valley: U.S. Geological Survey Prof. Paper 413, 78 p.
DUNBAR, C. O., 1949, Historical Geology: New York, John Wiley and Sons; 573 p.
ECKIS, ROLLIN, 1928, Alluvial fans in the Cucamonga district, Southern California: Jour. Geol., v. 36, p. 111–141.
GILE, L. H., PETERSON, F. F. AND GROSSMAN, R. B. 1966, Morphological and genetic sequences of carbonate accumulation in desert soils: Soil Science, v. 101, p. 347–360.
HARMS, J. C. AND FAHNESTOCK, R. K. 1965, Stratification, bed forms, and flow phenomena (with an example from the Rio Grande): In primary sedimentary structures and their hydrodynamic interpretation, Soc. Econ. Paleontologists and Mineralogists Spec. Publ. No. 12, p. 84–115
HAWLEY, J. W., AND WILSON, W. E. 1965, Quaternary geology of the Winnemucca area, Nevada: Nevada Univ. Desert Research Inst. Tech. Rept. No. 5.
HOOKE, R. L. B., 1967, Processes on arid-region alluvial fans: Jour. Geol., vol. 75, p. 438–460.
———, 1968, p. 456, 609–621.
———, 1967, p. 453–456.
———, 1968, Steady-state relationships on arid-region alluvial fans in closed basins: Amer. Jour. Sci., v. 266, p. 609–629.
HOPPE, GUNNAR, AND EKMAN, STIG-RUNE, 1964, A note on the alluvial fans of Ladtjovagge, Swedish Lapland: Geografiska Annaler, v. 46, p. 338–342.

HOWARD, J. D., 1966, Patterns of sediment dispersal in the Fountain Formation of Colorado: Mountain Geologist, v. 3, P. 147–153.
HUBERT, J. F., 1960, Petrology of the Fountain and Lyons Formations, Front Range, Colorado: Colorado School of Mines Quart., v. 55, no. 1, p. 1–242.
JANDA, R. J., 1965, Quaternary alluvium near Friant, California: Internat. Assoc. Quaternary Research, 8th Cong., U.S.A., 1965, Guidebook for field conf. 1, p. 128–133.
KLEIN, G. DeV., 1962, Triassic sedimentation, Maritime Provinces, Canada: Geol. Soc. America Bull., v. 73, p. 1127–1146.
KRYNINE, P. D., 1950, Petrology, stratigraphy, and origin of Triassic sedimentary rocks of Connecticut: Conn. Geol. and Nat. Hist. Survey Bull. 73. 247 p.
LEGGET, R. F., BROWN R. J. E., AND JOHNSTON G. H., 1966, Alluvial-fan formation near Aklavik, Northwest Territories, Canada: Geol. Soc. America Bull., v. 77, p. 15–30.
LUSTIG, L. K., 1965, Clastic sedimentation in Deep Springs Valley, California: U.S. Geol. Survey Prof. Paper 352-F, p. 131–192.
MABBUTT, J. A., 1966, Mantle-controlled planation of pediments: Amer. Jour. Sci., v. 264, p. 78–91.
MAGLEBY, D. C., AND KLEIN, I. E., 1965, Ground-water conditions and potential pumping resources above the Corcoran Clay—an addendum to the ground-water geology and resources definite plan appendix, 1963: U.S. Bur. Reclamation open-file report, 21 Plates.
MAYCOCK, IAN, 1962, The Torridonian Sandstone, Round Loch, Torridon, Wester Ross: Unpub. Ph.D. thesis, Univ. Reading, England, 305 p.
MCKEE, E. D., 1957, Primary structures of some recent sediments: Amer. Assoc. Petroleum Geologists Bull. v. 41, p. 1704–1747.
MEADE, R. H., 1967, Petrology of sediments underlying areas of land subsidence in central California: U.S. Geol. Survey Prof. Paper 497–C, 83 p.
MELTON, M. A., 1965, The geomorphic and paleoclimatic significance of alluvial deposits in southern Arizona: Jour. Geol. v. 73, p. 1–38.
MILLER, R. E., GREEN J. H., AND DAVIS G. H., 1971, Geology of the compacting sediments in the Los Banos-Kettleman City subsidence area, California: U.S. Geol. Survey Prof. Paper 497-E (in press).
MURATA, TEIZO, 1966, A theoretical study of the forms of alluvial fans: Geographical Rept., Tokyo Metropolitan Univ., v. 1, p. 33–43.
NILSEN, T. H., 1968a, Old red sedimentation in the Solund District, western Norway: International Symposium on the Devonian System, Calgary, Canada. Sept. 1967, v. 2, p. 1101–1115.
———, 1968b, The relationship of sedimentation to tectonics in the Solund Devonian district of southwestern Norway. Universitetsforlaget, Olso Norges Geolgiske Undersokelse No. 359, 108 p.
———, 1969, Old Red sedimentation in the Buelandet-Vaerlandet Devonian district, western Norway: Sedimentary Geology, v. 3, p. 35–57.
PASSEGA, RENATO, 1957, Texture as characteristic of clastic deposition: Amer. Assoc. Petroleum Geologists Bull., v. 41, p. 1952–1984.
———, 1960, Sedimentologie et recherche de petrole: Inst. Francais petrole Rev. et Annales combustibles liquides: v. 15, p. 1731–1740.
REINEMUND, J. A., 1955, Geology of the Deep River Coal Field, North Carolina: U. S. Geol. Survey Prof. Paper 246, 159 p.
RUHE, R. V., 1964, Landscape morphology and alluvial deposits in southern New Mexico: Annals. Assoc. Amer. Geographers, v. 54, p. 147–159.
SHARP, R. P., 1942. Mudflow levees: Jour. Geomorphology, v. 5, p. 222–227.
———, 1948, Early Tertiary fanglomerate, Big Horn Mountains, Wyoming: Jour. Geol., v. 56, p. 1–15.
———, AND NOBLES L. H., 1953, Mudflow of 1941 at Wrightwood, Southern California: Geol. Soc. America Bull. v. 64, p. 547–560.
SUGGATE, R. P., 1963, The fan surfaces of the central Canterbury Plain: New Zealand Jour. Geol. and Geophys., v. 6, p. 281–287.
TIEJE, A. J., 1923, The red beds of the Front Range of Colorado: a study in sedimentation: Jour. Geol., v. 31, p. 192–207.
WILLIAMS, G. E., 1966, Paleogeography of the Torridonian Applecross Group: Nature, v. 209, no. 5030, p. 1303–1306.
———, 1969, Characteristics and origin of a Pre-Cambrian pediment: Jour. Geol., v. 77, p. 183–207.
WINDER, C. G., 1965, Alluvial cone construction by alpine mudflow in a humid temperate region: Canadian Jour. Earth Sciences, v. 2, p. 270–277.

AUTHOR CITATION INDEX

Aasheim, S. M., 19, 333
Ackers, P., 238
Alden, W. C., 47
Alexandre, J., 216
Allen, J. R. L., 12, 216, 308
Allen, P., 101
Allwardt, A., 16
Amidon, R. E., 28
Andel, Tj. H. van, 101
Anderson, G. S., 12, 135
Anderson, R. V., 26
Andersson, J. G., 216
Anstey, R. L., 12, 308
Antevs, E., 308
Armstrong, F. C., 12
Ashley, G. M., 13
Atwood, W. W., 47
Averitt, P., 161

Bagnold, R. A., 186
Baker, A. A., 47
Baker, G., 216
Banks, M. R., 218
Barrell, J., 65
Bastin, B., 216
Bauer, C. M., 47
Beaty, C. B., 12, 26, 66, 117, 135, 186, 216, 238, 308, 359
Beaumont, P., 12
Belt, E. S., 12, 308
Bindeman, N. N., 12
Blackwelder, E., 12, 26, 47, 48, 65, 69, 79, 96, 117, 135, 186, 196, 308
Blakeley, R. F., 238
Blatt, H., 278, 339
Blench, T., 238
Blissenbach, E., 65, 69, 96, 101, 102, 117, 136, 186, 216, 308, 339, 359
Blong, R. J., 196
Bloom, A. L., 216
Bluck, B. J., 12, 26, 278, 308, 359
Bohlin, B., 26

Bond, G., 27
Boothroyd, J. C., 12, 13
Bouillet, G., 102
Bout, P., 216
Bradley, W. H., 47
Branson, C. C., 47
Branson, E. B., 47, 48
Brookfield, M. E., 13
Brown, L. F., Jr., 14
Brown, R. J. E., 135, 310, 360
Brush, L. M., 196
Bryan, K., 27, 160
Bryhni, I., 267, 278, 308, 332
Bucher, W. H., 47
Budel, J., 13
Bull, W. B., 13, 27, 117, 136, 186, 217, 238, 278, 308, 309, 332, 339, 359
Burke, K., 309
Buwalda, J. P., 65, 79, 117, 309, 359

Cailleux, A., 20
Caine, N., 217
California Department of Water Resources, 13
Campbell, R. H., 17
Carey, P. J., 310
Carlston, C. W., 238
Carryer, S. J., 13, 309, 359
Carter, W. D., 13
Casagrande, A., 135
Cehrs, D., 13
Chamberlin, R. T., 47
Chappell, J., 218
Charlton, F. G., 238
Chawner, S. J., 309
Chawner, W. D., 13, 27, 65
Chick, N. K., 217
Chitale, S. V., 309
Chorley, R. J., 160
Church, M., 217
Collinson, J. D., 13, 339
Conomos, T. J., 238
Conway, W. M., 27, 96, 136

361

Author Citation Index

Conybeare, C. E. B., 217
Cooke, R. U., 217
Costin, A. B., 217
Cotton, C. A., 27
Crandell, D. R., 96
Crawford, A. C., 359
Croft, M. G., 14
Crook, K. A. W., 217
Crowell, J. C., 14, 186, 278, 309, 359
Curray, J. R., 267
Curry, R. R., 309

Dahlstrom, C. D. A., 117
Dana, J. D., 65
Darton, N. H., 47
Davies, J. L., 217
Davis, G. H., 16, 360
Davis, S. N., 14
Davis, W. M., 14, 27, 65, 81, 96, 136, 160, 309
Dawdy, D. R., 14
Deegan, C. E., 14, 309
Demorest, M., 47
Denny, C. S., 14, 28, 107, 160, 186, 309, 359
Derbyshire, E., 217
Dobbin, C. E., 48
Doeglas, D. J., 359
Douglas, G. R., 18
Dowdall, W. L., 15
Drew, F., 28, 65, 309, 359
Drewes, H., 14, 107, 160, 359
Drysdale, C. W., 47
Duff, P. McL. D., 278
Dunbar, C. O., 359
Dutkiewicz, L., 217
Dutton, C. E., 28, 85

Eardley, A. J., 28, 65
Eckis, R., 14, 65, 69, 102, 136, 160, 186, 217, 309, 359
Eisbacher, G. H., 14
Ekman, S., 15, 217, 310, 359
Eldridge, G. H., 47
Elison, J. H., 19
Embleton, C., 217
Emery, K. O., 218
Enos, P., 309
Ethridge, F. G., 14, 20

Fahnestock, R. K., 196, 359
Fairbridge, R. W., 14, 18
Farengol'ts, Z. D., 12
Fisher, R. V., 217, 309
Fisher, W. L., 14
Flint, R. F., 65, 96, 217
Folk, R. L., 102
Fouch, T. D., 19

Fraser, J. K., 135
Friedman, G. M., 339
Friend, P. F., 12
Fryxell, F. M., 28, 117

Gale, H. S., 47
Galloway, R. W., 217
Ghibaudo, G., 309
Gilbert, C. M., 267
Gilbert, G. K., 28, 65, 68, 186, 238
Gilchrist, J. M., 15
Gile, L. H., 14, 359
Gleason, C. H., 28
Gloppen, T. G., 19, 333
Gole, C. V., 309
Goodlett, J. C., 160
Goreau, T., 309
Grabau, A. W., 65
Graybill, F. A., 238
Green, J. H., 16, 360
Groat, C. G., 16, 309, 310
Gross, W. H., 14
Grossman, R. B., 359
Gualtieri, J. L., 13
Guillien, Y., 216, 217, 218
Gustavson, T. C., 14

Hack, J. T., 160
Hadley, R. F., 15, 196
Hall, F. R., 14
Hallam, A., 278
Hampton, M. A., 310
Hares, C. J., 47
Harms, J. C., 359
Hawley, J. W., 14, 359
Hesjedal, A., 332
Heward, A. P., 15, 309
Hirst, F., 16, 310
Holmes, A., 332
Holmes, D., 332
Holtedahl, O., 267
Holzer, T. L., 15
Hooke, R. LeB., 15, 107, 196, 217, 238, 278, 309, 332, 333, 359
Hoppe, G., 15, 217, 310, 359
Horberg, L., 28, 117
Horne, R. R., 15
Howard, J. D., 15, 28, 244, 310, 360
Howell, J. V., 160
Hubert, J. F., 15, 310, 360
Hueber, F. M., 267
Hulst, H. van, 217
Hunt, C. B., 161, 310
Hussey, K. M., 12, 135

Inglis, C. G., 238
Inman, D. L., 96
Irish, E. J. W., 117
Ives, R. L., 28

Jachens, R. C., 15
Jahn, A., 217
Jahns, R. H., 65, 186, 332
Janda, R. J., 360
Jarvik, E., 267
Jennings, C. W., 161
Jennings, J. N., 217
Jepsen, G. L., 47
Johnson, A. M., 15, 196, 217, 310, 332
Johnson, D. W., 28
Johnston, G. H., 15, 135, 310, 360

Kaizuka, S., 20
Kalander, L. K., 278
Kerr, D. R., 15
Kesseli, J. E., 66
Kiger, J., 267
Kildal, E. S., 278
Kimura, T., 310
King, C. A. M., 217
Klein, G. deV., 15, 310, 360
Klein, I. E., 16, 360
Knechtel, M. M., 161
Knight, J. A., 310
Knight, S. H., 47
Knopf, A., 65, 96
Knowleton, F. H., 47
Kolderup, C. F., 267, 278
Koons, D., 65
Krigström A., 15, 196
Krumbein, W. C., 28, 65, 102, 238, 267
Krynine, P. D., 15, 244, 267, 339, 360
Kuenen, Ph. H., 102

Lahee, F. H., 65
Laming, D. J. C., 16, 310
Langbein, W. B., 161
Larsen, V., 16, 332
Lattman, L. H., 16, 310
Lawson, A. C., 28, 65, 69, 81, 244
Leeder, M. R., 339
Legget, R. F., 135, 310, 360
Leopold, L. B., 186, 196, 238
Link, M. H., 16
Longwell, C. R., 28, 65, 96, 102
Love, J. D., 16, 48
Loveday, J., 218
Lustig, L. K., 16, 161, 186, 218, 278, 310, 360

Maarleveld, G., 101
Mabbutt, J. A., 360

Mabey, D. R., 161, 310
Macar, P., 216
McCann, F. T., 160
McGee, W. J., 28, 65, 81, 96, 136
McGowen, J. H., 16, 310
Mackay, J. R., 135
McKee, E. D., 65, 360
Mackin, J. H., 161, 238
MacPhail, M. K., 218
McPherson, H. J., 16, 310
Maddock, T., Jr., 238
Maehle, S., 19, 278, 312, 333
Magleby, D. C., 16, 360
Malaurie, J., 218
Marchand, D. E., 16
Martini, I. P., 310
Maycock, I., 360
Meade, E. D., 360
Meckel, L. D., 16, 310
Melton, M. A., 360
Merrill, C. L., 135
Miall, A. D., 16, 310
Middleton, G. V., 278, 310, 339
Miller, J. P., 186, 196, 238
Miller, R. E., 16, 360
Miller, R. L., 161
Miller, W. J., 65
Milliman, J. D., 218
Minter, W. E. L., 17
Moore, T. E., 17
Morton, D. M., 17
Murata, T., 360
Murray, R., 278, 339
Mutti, E., 309, 310

Nace, R. L., 48
Nathorst, A. G., 267
Nelson, C. A., 186
Nelson, H., 310
Nemec, W., 333
Nicholson, R. N., 278
Nicolls, K. D., 218
Nikiforoff, C. C., 161
Nilsen, H. R., 19, 278, 312, 333
Nilsen, T. H., 17, 267, 278, 311, 339, 360
Noble, L. F., 161
Nobles, L. H., 19, 28, 73, 79, 96, 117, 136, 186, 311
Normark, W. R., 311
Nummedal, D., 13
Ori, G. G., 17
Osborne, R. H., 16

Pack, F. J., 28, 65, 69, 73, 186
Paige, S., 28
Pappajohn, S., 15

Author Citation Index

Pashley, E. F., Jr., 17
Passega, R., 360
Pearce, A. J., 238
Peterson, F. F., 359
Peterson, G. L., 15
Peterson, J. A., 218
Peterson, J. Q., 79
Pierson, T. C., 17
Pihlainen, J. A., 135
Piper, A. M., 17
Pissart, A., 218
Poland, J. F., 17, 19
Porebski, S., 333
Potter, P. E., 238
Pretorius, D. A., 17, 18
Price, W. E., Jr., 18, 218, 238
Prior, D. B., 18

Rachocki, A. H., 18
Rahn, P. H., 310, 311
Rantz, S. E., 18
Rapp, A., 18, 117
Ray, L. L., 18
Reed, A. A., 15, 310
Reeves, C. C., Jr., 18
Reid, J. C., 18
Reineck, H. E., 339
Reinemund, J. A., 360
Renwick, W. H., 18, 311
Riba, O., 311
Ricci Lucchi, F., 310, 311
Riccio, J. F., 135
Rich, J. L., 28, 48, 161, 244
Rickmers, W. R., 28, 74, 96, 117
Robertson, J. A., 18
Rodine, J. D., 218
Roe, S. L., 19, 312, 333
Roed, M. A., 311
Rohrer, W. L., 107, 238
Roscoe, S. M., 18
Rougerie, G., 102
Rouse, H., 186
Rubey, W. W., 238
Ruhe, R. V., 18, 311, 360
Rupke, N. A., 311
Rust, B. R., 18, 333
Ryder, J. D., 218
Ryder, J. M., 18, 218, 311
Ryder, R. T., 19

Sagri, M., 310, 311
Salisbury, R. D., 48
Sanders, J. E., 339
Sanlaville, P., 19
Schick, A. P., 19, 218
Schluger, P. R., 19, 311

Schumann, H., 19
Schumm, S. A., 19, 161, 196, 238, 311, 312
Scott, H. W., 48
Scott, K. M., 19, 311
Scott, W. B., 65
Selley, R. C., 311
Sestini, G., 311
Sharon, D., 161
Sharp, R. P., 19, 28, 73, 74, 79, 96, 117, 136, 186, 311, 360
Sharpe, C. F. S., 96, 107, 117
Sheh, A., 19
Short, M. N., 65
Singewald, J. T., 196
Singh, I. B., 339
Skjerlie, F. J., 278, 308
Sloss, L. L., 267
Smith, N. D., 278
Sneed, E. D., 102
Soister, P. E., 19
Soons, J. M., 218
Souchez, R., 218
Spearing, D. R., 19
Specht, R. L., 218
Spinnangr, Å., 19, 278, 312, 333
Spry, A. H., 218
Stace, H. C. T., 218
Statham, I., 311
Steel, R. J., 16, 19, 278, 311, 312, 333
Steidtmann, J. R., 19
Stephens, N., 18
Stephenson, C., 20
Strahler, A. N., 161
Suggate, R. P., 360

Taff, J. A., 48
Tanner, W. F., 312
Tator, B. A., 20, 28
Taylor, C. A., 28
Teillard de Chardin, P., 29
Teisseyre, A. K., 20
Thackwell, F. E., 359
Thom, B. G., 218
Thom, W. T., Jr., 47, 48
Thompson, T. B., 14
Thorsteinsson, R., 117
Tieje, A. J., 29, 244, 360
Tolman, C. F., 20, 29, 186, 238
Tolman, C. T., 65
Tourtelot, H. A., 48
Toya, H., 20
Trask, P. D., 65, 96
Tricart, J., 20, 135, 218
Trowbridge, A. C., 48, 69, 79, 117, 186
Troxel, B. W., 312
Troxell, H. C., 29, 65, 79

Tuan, Yi-Fu, 161
Turner, F. J., 267
Twenhofel, W. H., 65
Twidale, C. R., 20

U. S. Geological Survey, 48

Van Houten, F. B., 48
Van de Kamp, P. C., 20
Vanoni, V. A., 186
Vaughan, F. E., 29, 65
Vinogradov, Yu. B., 218
Vliet, A. van, 312
Vogt, Th., 267

Wagner, R. H., 312
Waldron, H. H., 96
Walker, R. G., 312
Walton, E. K., 278
Warren, A., 217
Washburn, A. L., 161, 218
Wasson, R. J., 20, 107, 218, 312

Wasylyk, D. G., 311
Watson, E., 218
Weaver, W., 312
Wegemann, C. H., 47, 48
Wells, S. G., 20
Wescott, W. A., 20
Wessel, J. M., 20
Westgate, L. G., 48
Wiggers, A. J., 101
Williams, G. E., 20, 218, 312, 360
Williams, H., 267
Wilmarth, M. G., 48
Wilson, A. C., 278, 312, 333
Wilson, M. D., 20, 312
Wilson, W. E., 359
Wiman, S., 136
Winder, C. G., 312, 360
Wolman, M. G., 186, 196, 238
Woolley, R. R., 29, 79
Wright, H. E., Jr., 161
Wright, L. A., 161

Yazawa, D., 20

SUBJECT INDEX

Alaska mudflows, 74
Alberta, 108, 340
   central, 108-110
   Hell's Creek, 108-112, 114-116
   Smoky River, 108-111, 113, 115-116
   western, 116
Alluvial cone, 51, 69, 84-85, 108-112, 114-116, 145, 152, 279, 288, 340, 342, 354
Alluvial fans
   apex, 50-51, 53, 56-58, 70-71, 73-74, 82-84, 87, 110-112, 130, 134, 140-141, 145, 203, 210, 219-220, 224, 226, 232-235, 237, 271, 285-286, 295, 299-300, 340, 343, 345-346, 352, 354, 356-359
   axis, 219-220, 224, 227-228, 232-233, 237
   base, 50-51, 54, 56-58, 64, 70, 74, 81, 86, 97, 99, 111, 130-131, 144-145, 162, 178-181, 183, 188, 203, 211-212, 213-216, 220, 271-274, 280, 285-286, 292, 294-296, 299-300, 306, 315, 321, 327-328, 331, 342-343, 346, 352, 358
   coalescing, 86, 137, 139-140, 143, 188, 264-266, 340, 354-355
   compound, 51-54
   fan-bay, 50, 53, 97, 101
   fanhead, 50, 53, 64, 97, 99-101, 140-141, 146, 162, 164, 166-167, 178-180, 182-185, 187, 195, 235, 237, 278-280, 286, 291-296, 299, 302, 304, 340, 342, 358
   fan mesa, 50
   hanging tributary fans, 146
   midfan, 50, 71, 73, 162, 176, 179-181, 183, 210, 232, 321, 327, 352-353
   primary fan, 55, 57, 292, 294-296, 299, 306
   secondary fan, 55, 57, 211, 286, 292, 294-296, 299, 306
   subsurface, 88
   superimposed fans, 54-55
Alps, 51-52
Anticline, 31, 144-115

Antidunes, 351-352
Appalachian Mountains, 151, 245, 266, 356-357
Arctic Circle, 119
Argillite, 285
Arizona, 50-51, 55
   Aubrey Cliffs, Aubrey Valley, 50, 53, 56-57
   Black Hills, 50-51, 53-54, 56-57, 60-61
   Grand Canyon, 50, 54
   Mammoth, 50, 54
   Santa Catalina Mountains, 50-51, 53-63, 351
   Seligman, 50, 58
   Tucson, 50, 53
   Tucson Mountains, 50, 57
Arkose, 30, 34, 38-40, 42, 57, 59, 64, 245, 248-249, 251-252, 353
Artesian water, 59
Australia
   New South Wales, 187
      Lake George, 187-189, 195
      Lake George scarp, 187-189, 196
   Tasmania, 197
      Derwent estuary, 197, 201, 215
      Derwent Valley, Derwent River, 197-199, 201-203, 208-210, 213-216
      Hobart, 201
      Mt. Wellington, 201

Badlands, 138
Bajada, 51, 66, 340, 356
Bar-runnel complex, 190-195
Basalt, 210
Base level, 53
Basin and Range province, 137
Beach deposits, 222
Bear Island, 264
Bedding
   graded, 88-89, 92, 94, 304, 313, 315, 317, 340, 347
   inverse graded, 301-302, 315, 317, 319, 323

lamination, 60, 86, 92, 94–95, 206, 210, 275, 318–319, 321–323, 327, 340, 343
massive, 92, 94, 178, 249, 271–275, 301–302, 305–306, 340, 343, 346, 349, 354
stratification, 39, 41, 45, 59–60, 63–64, 82–83, 92, 94, 97, 130, 153, 162, 164, 179, 193, 205–206, 208, 210–211, 249, 253–256, 259–260, 264–266, 271–276, 279, 281, 286, 301, 305–306, 313, 315, 317, 319, 321, 340, 343, 345, 347, 349–350, 352, 356, 359
Bench gravel, 30, 32, 156
Bioturbation, 303
Block glacis, 204–205
Blue Gate Shale, 156
Bolson deposits, 358
Boulder beds, 34, 40, 42–43, 45
Boulders, 35, 38–41, 43, 45–46, 56, 63–64, 68, 70, 73, 77–84, 98, 108, 110, 112, 115–116, 131–132, 156, 164, 167, 169–170, 173, 177–179, 182, 184, 190, 192–195, 214, 249–250, 256, 323, 340, 346–347, 357
  perched boulder, 36
Breccias, sedimentary, 63, 245, 248–250, 262–263, 265, 355–357
British Columbia, northeastern, 116
Bubble cavities, 86, 91–92, 94–95
Buelandet-Vaerlandet Formation, 245, 248–250, 253–254, 256, 258–259, 261–264, 266
  Melvaer Breccia Member, 245, 248–251, 254, 256–258, 263–265
  Sörlandet Sandstone Member, 245, 248–249, 251–253, 255–256, 261–263, 265
  Vaeroy Conglomerate Member, 245, 248–254, 256–265

Cadomin Conglomerate, 114
Calcite, 212, 249, 251, 253, 342
Caledonian orogen, geosyncline, 245, 247, 263–264, 356
Caliche, 342
California
  Amargosa River, Amargosa Valley, 143–144, 147, 151, 153–154, 156–159
  Bat Mountain fan, 147
  Big Pine, 163
  Bishop, 66
  Black Mountains, 151
  Carson Slough, 145
  Coast Ranges, 86
  Copper Canyon fan, 340
  Copper Fan, 226
  Cucamonga, 69
  Death Valley, Death Valley Junction, 137–140, 143–144, 147, 151–152, 159, 163, 180, 183–184, 198, 226, 233, 298, 300, 332, 341, 346
  Deep Springs Valley, 146, 151, 163, 180, 186
  Diablo Range, 87, 348, 350, 353
  Eureka Valley, 163, 180, 186, 226
  Fresno County, 86–89, 94–95, 298–300, 343–345, 350, 354–355
  Glendora, 288
  Gorak Shep fan, 163–164, 167–168, 171–173, 179–181, 183, 186, 226
  Inyo Mountains, 69
  Los Angeles, 179
  Marble Canyon fan, 226
  Mauve Shadow fan, 226, 236–237
  Mormon fan, 226, 235–236
  Needles, 53
  Owens Valley, 66–67, 69, 77, 79, 82–83, 108, 163
  Panamint Range, 139, 151
  Panamint Valley, 140, 178
  Ridge Basin, 289–290, 357–358
  San Joaquin Valley, 86, 298, 343
  Shadow Rock fan, 163–164, 167, 171, 173, 179–182, 186
  Shadow Mountain fan, 143–147, 153–154, 156, 158
  Sierra Nevada, 46, 68–69
    southern, 50, 55–56, 156, 356–357
  Suprise Canyon fan, 178
  Trail fan, 285
  Trollheim fan, 162–164, 167, 170–171, 179, 182, 186
  Warren fan, 226
  White Mountains, 66–74, 77–85, 198, 279, 296, 347
  Wrightwood, 73
Cambrian, 245, 247–249, 263, 265
Campito Formation, 179, 181
Carbonate, 164, 167, 179, 181, 342
Carboniferous
  Lower, 280
  Stephanian, 279, 300–307
  Upper, 300
  Westphalian, 279, 300–307
Cementation, 39, 41, 57, 59, 97, 251, 253, 260, 342
  calcium carbonate, 59, 97, 251, 253, 342
  limonite, 59
Cenozoic, 158
  late, 40
Channels, 39, 50–51, 54, 60, 62–64, 71, 73–74, 76–84, 88, 92, 94–95, 97, 101, 110, 115–116, 119, 124, 137–140, 144, 146, 148–149, 156–157, 162–164, 167–174, 176–185, 190–193, 211–213, 219–220,

Subject Index

222-225, 227-228, 231-233, 235-238, 276, 279, 281, 284, 286, 292, 294, 299, 303, 319-323, 327, 340-356, 358
Clast shape, 99-101, 207, 258, 261, 284
Clast size
  maximum clast size, 56, 64, 79, 83, 97-98, 101, 190, 249-250, 256, 268, 270-275, 281, 284-286, 289, 291, 294, 299-305, 313, 315-317, 319, 321, 326, 328, 343, 349, 359
  mean clast size, 56, 64, 219, 226-227, 340, 343, 359
Clay, 56, 64, 86, 88-89, 92, 94-96, 110, 112, 116, 119, 127, 130, 152, 169-170, 177, 179, 181, 195, 202, 210, 212-213, 226-227, 233, 236, 249, 251, 254, 264-265, 315, 343, 345-350, 352-353, 355
  kaolinite, 212
  montmorillonite, 89
  palygorskite, 212
Claystone, 153
"Clear Creek gravels," 30
Climate
  arctic, 119-120, 132, 197-198, 203, 207, 215, 288
  arid and semiarid, 50-52, 54, 56, 63-65, 68-69, 86-87, 108, 137, 139-140, 146, 148, 150-154, 156, 158, 160, 288, 308, 340, 342-343, 348
  changes, 299, 331, 340, 356
  humid, 51-52, 54, 56, 64-65, 69, 108-110, 114-116, 137, 139, 149, 153, 159-160, 187-189, 226, 245, 266, 288, 340
CM patterns, 348-350, 359
Coal, 114, 301, 304, 306
Coal fields, 38, 40, 44-45, 279, 301-307
Cobble, 39, 41, 63, 68, 71, 82-83, 109, 144, 167-169, 173, 178-179, 181, 190, 192-195, 206-207, 249-251, 256, 340, 347, 357
Colluvium, 215, 287
Colombia, Bucaramanga fan, Bucaramanga fault, 285, 289
Color, 59, 74, 77, 130, 162, 164, 252, 254, 323, 355-356
Colorado, 356
Colorado Plateau, 84
Colorado River, 53, 288
Compaction, 59
Composition, 34-35, 38-39, 45-46, 56, 57, 59, 69, 212, 246, 250-253, 273, 277-278, 281, 289, 293-294, 299, 306, 343
Conglomerate. 39-41, 63, 87, 97-98, 114, 179, 245-246, 248, 250-251, 254-255, 260, 262, 265, 268-277, 280-281, 289, 292, 295, 301-306, 313-315, 317, 319-322, 326-327, 329, 331, 355-357
  metaconglomerate, 252
Creep, 163, 167, 178, 207
Cretaceous, 40-41, 44, 87, 114, 118, 130
Crossbedding, 39, 60, 92, 94-95, 179, 205, 261-262, 264-266, 274-276, 301, 303, 305-306, 319, 321, 340, 343, 356
Cross-stratification. See Crossbedding
Cut- and-fill structure, 205, 340, 345, 351-352
Cycles, cyclotherms, 268-278
  sequences, megasequences, 279-281, 284, 291-292, 294-297, 299-308, 313, 326-332

Debris flows, 69-85, 162, 164-165, 167, 169-170, 173, 175-179, 182, 184-185, 197-198, 204-205, 208-213, 215, 237-238, 268, 271-273, 279, 286-288, 293, 301-302, 305, 313-320, 322-328, 330, 332, 340, 343, 346-351, 355-359
Debris flow levee, 162, 169-170, 176, 178, 192, 347
Delaware, Brandywine Creek, 224
Deltaic deposits, 284, 300
Desert pavements, 139, 141, 143-149, 154, 158, 164
Desert varnish, 145, 163-164, 167, 181, 266, 346
Devonian, 245-249, 263-264, 266, 268, 273-274, 280, 289, 313-317, 332, 356-357
  Early, 245, 254
  Middle, 245, 254, 280, 313
Diamictite, 313, 315
Dissection, 53-54, 79, 84, 97, 99-101, 140-141, 144, 146, 152, 154, 156-158, 162, 167, 177, 182-185, 187, 190, 195-198, 211, 213-216, 235, 285-286, 289, 291-300, 302, 304, 306, 340, 342-343, 345, 354, 358
Dolomite, 159, 179, 181, 212, 285

Earth flow, 169
Engineering properties, 130, 340
Entrenchment. See Dissection
Eocene, 30-33, 40-42, 44-46, 289, 358
Eolian deposits, 56, 200, 202-203, 208, 212-213, 215, 222
Estuarine deposits, 202-203, 214-215

Fan-delta deposits, 313, 323, 325-326, 330
Fanglomerate, 30, 36, 41-43, 55-56, 62-63, 69, 82, 153, 245, 266, 268, 273-274, 278, 285, 300, 313, 315, 318, 321, 323, 326, 330, 332, 356
Fault scarp, scarp, 139-142, 164, 168-169, 235-236, 265, 274, 279, 281, 290, 294, 297, 303, 313-314
Faults, strike-slip, 277-278, 289, 300, 313, 332

368

Subject Index

Feldspar, 57, 89, 212, 253
Ferntree Group, 205-206
Fires, 214
Fluvial deposits, 56, 58, 63-64, 126, 137-138, 140, 152-153, 155, 159, 192, 200, 202-203, 215, 219, 222-223, 300, 313-314, 330-331, 342, 354-356, 359
Fossils
  bones, 63
  plants, 63, 245, 254, 263-264, 303-304, 306, 342, 358
Fountain Formation, 356
France, 68
Franciscan Formation, 87
Frost wedging, 204, 207

Geometry, 50, 69, 83-84, 219, 268, 273-274, 276, 289-290, 313, 326, 328-332, 340-341, 343, 351-352, 354
Glaciation, glaciers, 30, 32, 45-47, 64-65, 121, 126, 132, 152, 197, 200, 202-209, 213, 215, 266, 288, 356
  moraines, 74
  outwash, 32, 46
Gneiss, 251, 285, 314, 322, 356, 358
Graben, half-graben, 140, 245, 263, 266, 289
Granitic detritus, 146, 353
Granule, 173-174, 180-181, 190, 192-193, 206, 210, 315, 318-321
Gravel, 30, 32, 38-40, 43-46, 55-59, 64, 80, 83, 88-89, 92, 94-95, 108, 118-119, 144-145, 147-148, 152-160, 162-165, 181, 184, 188, 190-194, 200, 206, 207, 209-210, 235-236, 255, 264-266, 315, 330, 340, 342-343, 346-347, 349, 351, 353, 355
  gravel bars, 144, 148, 193-194, 245, 255, 263-265, 321, 342-343, 349
Graywacke, 57, 59, 64
Great Britain, 249
Greenland, 246
Greenschist, 245, 248-250, 258, 265
Grézes litées, 204-208
Ground water, 59, 82, 86, 88, 179, 331, 340, 342
Gypsum, 342

Himalayas, 51-52, 340
Holocene, 197, 200, 214-216
Homocline, 248
Hummocky topography, 164, 322

Iceland, 190
Imbrication, preferred orientation of clasts, fabric, 60, 62-64, 88, 92, 94, 190, 202, 205-206, 209, 245, 249, 251, 259-261, 264-266, 301, 313, 315, 317-319, 321, 340, 347-349
Incision. *See* Dissection
Indus basin, 54, 296
  Kosi River, 288
Intersection point, intersection point deposits, 162, 176, 183, 185, 187-196, 291-295
Italy, 68

Japan, 340
Joints, 207, 249
Jurassic, Upper, 114

Kingsbury conglomerate, 30, 32-33, 36, 41-42, 44-47

Laboratory studies, of alluvial fans, 162, 164-167, 172-177, 219-233, 237
Lacustrine deposits, 64, 188, 190, 275-276, 280, 300-301, 303-305, 313-315, 318-319, 321, 323, 325-327, 330-331, 352-355, 359
Lake, 119, 188, 196
Landslide
  deposits, 44, 75, 88, 108, 116, 188, 191-192, 210-211, 287
  rockfall, 204-205
  slumps, 55, 116, 131, 162, 275, 287
  soft-sediment deformation, 330-331
Laramide orogeny, 30-31, 45, 47
Limestone, 153, 159, 303, 306
Load structures, 303-304
Loam, 204, 206, 212
Log-dam, 191-193
Logs, drifted, 303
Luscar Formation, 110, 114

Mackenzie Delta, 118-120, 122, 126-127, 134
Mackenzie River, 118, 126
Maine, 280
Markov process, 219, 228-233
Massachussetts, 280
Matrix, 34, 58-59, 64, 80, 89, 182, 204-206, 249, 251, 253, 260, 264, 302, 315, 319, 323, 340, 346-347, 355, 357
Megasequences. *See* Cycles
Mesozoic, 31-33, 41-43, 46
Miocene, 44, 357-358
Moncrief gravel, 30, 32-35, 37-47
Monocline, 114
Mud, 55, 58, 64, 70, 73-74, 87-88, 108-109, 114-116, 172, 177-178, 195, 275, 286, 322, 346-347
Mudcracks, 86, 91-92, 94-96, 272, 347
  Arctic polygonal cracks, 126, 131, 203
Mudflow, 52-53, 59-60, 62-64, 69, 73-74, 86,

369

Subject Index

88-90, 92-95, 97-99, 101, 108, 110, 112, 114-116, 184, 210, 237, 245, 254, 263, 265-266, 272, 280, 340, 343, 345, 347-348, 350, 358
    mudflow levees, 74, 108
Mudstone, 55, 87, 127, 131, 276, 280, 304, 306, 345, 350, 357

Nevada
    Arrow Canyon Range, 97, 101
    central, 69
    Fish Lake Valley, 66-67, 83-84
    Las Vegas, 97
    Spring Mountains, 342
    White Mountains, 66-74, 77-85
Newark Series, Newark basins, 63, 245, 266, 356-357
New Brunswick, 280
New Mexico
    central, 156
    Ladron peak, 157-158
    Rio Grande Valley, 137, 153, 156-158
    Rio Puerco, 224
    southern, 50, 56
New Red Sandstone, 280
New Zealand, North Island, 211
Nikinassin Formation, 114
Nivation, 203, 207, 213, 215
Northwest Territories
    Aklavik, 118-121, 123-125, 127-133
    Richardson Mountains, 118-120, 122, 124, 126-127, 130
Norway, 246, 264, 356-357
    Bergen, 246-247
    Buelandet-Vaerlandet district, 245-248, 252, 263-266
    Hädesteinen basin, 247, 266
    Hornelen basin, 246-247, 254, 266, 268-278, 280, 289, 313, 332
    Karkevagge, 110
    Kvamshesten basin, 268, 280
    northern, 108
    Röragen basin, 246-247
    Solund basin, 246-247, 263, 266, 268, 278, 289
    Trondheimsfjord basins, 246-247

Old Red Continent, 245-246
Old Red Sandstone, 249, 280
Oligocene, 44
Ordovician, 40
Overbank deposition, 184-185, 187, 190, 195, 223, 237

Packing, 59, 94, 266, 320
Pakistan
    Hindu Kush, 207
    West Pakistan, 288

Paleocene, 40-41, 45
Paleocurrents, 245, 254-255, 261-262, 268, 272, 274, 276-278, 314, 351, 356, 359
Paleogeography, 246, 263, 266
Paleoslopes, 245, 254, 263-265, 332
paleosol, 205-206, 212
Paleozoic, 31-33, 35, 37-38, 40-41, 43-44, 46, 181, 356
Peat, 130
Pebble, 39, 41, 59, 63, 68, 71, 83, 97-101, 109, 115, 130, 134, 144, 147-148, 169-170, 177-178, 181-182, 188, 190, 193, 206-207, 210, 248, 253-254, 256, 268, 273, 277-278, 302, 305-306, 319, 321-322
Pediment, 53, 55, 137-138, 143-144, 150, 153-158, 160, 340, 354-356
Peneplanation, 63
Pennsylvanian, 356
Permafrost, 119
Permeability, 59, 73, 86, 162, 180, 182, 331, 340
Permian, 206, 280, 289-290
Perry Formation, 280
Piedmont, 30, 32, 40, 46, 51, 66, 137-138, 141-150, 153-154, 156, 158-160, 215, 265-266, 273, 340
Piracy, 141, 143, 145-146, 150-151, 153-154, 158-160, 227, 294, 296
Playa, 137-140, 151, 168, 342, 355-356
Pleistocene, 30, 32, 44-45, 54, 97, 345, 353
Pliocene, 44, 58, 179, 345, 356-358
Pointbar deposits, 355-356
Pollen, 130, 207
Porosity, 59, 86, 94, 163-164, 172, 181-182, 340
Precambrian
    Keweenawan, 63, 356
    Scotland, 280, 356-357
    Wyoming, 30-31, 33-35, 38-39, 41-43, 45-46
Primary current lineation, 255, 265-266, 356
Prince of Wales Island, 280
Pseudotelescope structure, 55

Quartz, 89, 130, 167, 212, 251, 253, 285
    cristobalite, 212
Quartzite, 143, 153, 159, 164, 179, 181, 251-252, 278, 285, 306, 314, 322, 346
Quaternary, 30, 32, 44-45, 63, 203

Radiocarbon dating, 204, 214
Raindrop imprints, 272
Reactivation surface, 319
Recent, 54, 63, 139, 289
Reed Formation, 179
Ridge Basin Group, 179, 357-358

Ripple marks, 60, 303-304, 315, 319, 330, 356
  climbing ripple marks, 275, 356
Rocky Mountains, 45-46, 213
Roundness, 34, 41, 45, 57-58, 63-64, 110, 115, 131, 134, 188, 245, 248-251, 253, 257-259, 301, 318-322, 346, 355
Russia, 74

Sand, 30, 55-56, 58, 64, 68, 71, 74, 80, 83, 89, 92, 94-96, 108, 118-119, 127, 134, 144, 172-174, 179, 181, 190, 192, 202, 204, 209-210, 212-213, 220, 228, 236, 251, 253-254, 264, 278, 302, 321-322, 330-331, 343, 345, 347, 349-350, 353, 355
Sandbars, 94
Sandstone, 34, 38, 40, 42, 55, 87, 114, 118, 124, 127, 130-132, 134, 153, 179, 205, 245-246, 248-250, 253-255, 259-263, 266, 269-276, 280, 285, 301-306, 313-315, 317-321, 326-327, 345, 356-357
Sandur, 193
Schist, 285
Scotland, 290, 356-357
Segmentation, 164, 167, 170-172, 181-184, 226, 235, 237, 274, 288, 294-295, 297-300, 304
Sequences. See Cycles
Shale, 44, 87, 108, 114, 118, 124, 130-132, 134, 154, 156, 159, 179, 301, 306, 350, 356
Sheetfloods, 52-54, 57, 59, 63-64, 80-81, 210, 245, 254, 303-304, 306, 313, 321-324, 327, 340, 343
Sheetflow, 232
Sheetwash, 207
Shoreline deposits, lacustrine, 313, 315, 318-319, 321, 326, 330
Sieve deposits, 162, 164, 166-174, 177, 179-182, 187, 190-191, 193-195, 313, 319, 322-323, 340, 343, 345-346, 355
Silt, 30, 56, 71, 74, 80, 83, 89, 91-92, 94-95, 118-119, 127, 130, 132, 134-135, 144, 147-149, 177, 190, 192, 202, 206, 210, 213-214, 226-227, 233, 236, 251, 265, 275, 315, 323, 330, 343, 345, 349, 353
Siltstone, 34, 38, 87, 108, 114-115, 131-132, 205-206, 207, 253, 280, 285, 304, 306, 323, 326
Silurian, 245, 247-249, 263, 265
Slope, 50, 59, 64, 69, 79, 86-88, 97-98, 101, 131, 139, 143, 153, 156-157, 159-160, 162, 164, 167, 171, 173, 176-179, 181, 183-184, 187, 191-193, 196, 202-203, 207, 209, 211, 219-220, 222, 224-226, 233-234, 286, 294, 300, 330, 332, 343
Soil, 94-95, 118-119, 126-133, 135, 138, 152, 160, 181, 206, 212, 264-265, 286, 294, 342
Sole marks
  flow casts, 356
  grooves, 304
Solifluction mantles, 203-205
Sorting, 57, 59, 63-64, 89, 92-94, 97, 101, 110, 127, 132, 173, 175, 178-179, 187, 195, 202, 207, 209, 253, 284, 292, 301-302, 313, 315, 317, 320-322, 326, 340, 343, 345-350, 355-356
Spain
  Cantabrian Mountains, 300-307
  northern, 279, 300-301
Sphericity, 58, 98-101, 257-259
Spitsbergen, 246, 264
Sponge spicules, 213
Strandflat, 246
Stratigraphy of alluvial fans, 151-152, 178-179, 289, 341, 352, 354, 357
Stream capture. See Piracy
Streamfloods, 52-54, 57, 59, 62-64, 183, 356
Streamflow (waterlaid) deposits, 88-90, 92-95, 97-99, 101, 108, 115, 118, 125, 131, 134, 162, 173, 176-177, 185, 197-198, 209-213, 214-216, 268-276, 278, 280, 293, 305, 313-315, 319, 321, 323-328, 331-332, 340, 342-351, 357-358
Streams
  braided, 52, 92, 94, 134, 144, 166-167, 190, 192-193, 245, 255, 263-266, 273, 275-276, 280, 321, 340, 343-345, 348-349, 351-352, 356, 358
  meandring (gullies), 143, 146-147, 153-154, 223-224, 265, 304
Submarine fans, 279, 281-284, 286, 288-289, 294, 303-304, 315
Subsidence
  of basin floors, 268, 273-274, 276-278, 280-281, 313
  of fans, 96, 318, 323
Svalbardian disturbance, 245, 264-265, 356
Sweden, Lapland, 340

Talus (scree), 50-51, 64, 131, 134, 138, 245, 263, 306, 355
Tectonism, 44-46, 54-55, 268, 272, 274-278, 280-281, 289-290, 292-293, 297, 299-300, 304, 306-307, 313, 318-319, 323, 328, 331-332, 353, 356-357
Telescope structure, 54-55
Terraces
  alluvial, 75-76, 80, 147, 151, 159, 162-164, 169, 172, 202, 209
  colluvial, 75-76, 80
Tertiary, 32, 40, 87, 143, 153
  early, 30-31, 33, 38, 40, 42-43, 45

middle, 46
late, 32, 44-45, 351
Thickness
  of bedding, flows, 60, 73, 75, 77, 88-89, 94, 182, 190, 255, 268, 270-275, 281, 284, 286, 289, 291-292, 299, 302, 304-306, 313, 315, 317-323, 328, 340, 345, 347, 349, 354
  of cycles, 268-278, 280, 301-306, 328, 331
  of fan, 39, 159, 246, 249-250, 268, 270-272, 277-278, 289, 292, 300, 303, 305, 313, 315, 326, 328, 331-332, 352, 354, 357
Thrust fault, 30-31, 33, 35, 44-47, 114
Till, 46
Torridonian Sandstone, 280, 357
Triassic, 245, 266, 274, 280, 289, 356

Unconformity, 41, 276-277, 313-314
  angular, 41-45, 245-246, 249, 276, 300
United Kingdom, 246, 300
Utah, 184
  Fremont River, 154-155
  Henry Mountains, 153-156
  Oak Creek, 156
  southeastern, 78
  Table Mountain, 154-155
  Town Wash, 155-156
  Uinta Mountains, 44
  Wasatch Mountains, 53, 69

Vegetation, on alluvial fans, 64, 68-69, 74-77, 82, 87-89, 108, 110, 114, 116, 118-119, 126, 130, 132, 134, 164, 179, 189, 193-194, 197, 201, 205, 207, 214, 226, 286, 346, 356
Violin Breccia, 179, 289-290
Virginia
  Little River, 151
  Shenandoah Valley, 153, 159
Voids, 86, 89, 92, 94-95, 164, 181, 190, 205

"Wasatch" beds, 30, 39-42, 45
Water-escape structures, 303-304
Weathering
  abrasion, 98, 101, 115, 162-163, 172, 181, 249, 257-259, 263-265
  chemical, 56, 65, 98, 253, 342
  chipping, 98
  frost action, 134
  mechanical, 56, 65, 82, 97-98, 101, 124, 131-132, 134, 137-138, 141-143, 145-149, 152, 159-160, 162-164, 167, 172, 179, 181, 228, 265-266, 286, 294, 331, 342
  splitting of clasts, 97-98, 101
Wind River formation, 39, 45
Wyoming
  Big Horn Mountains, 30-33, 38-41, 43-46
  Buffalo, 31, 33, 39-40
  Buffalo coal field, 40
  Grand Tetons, 108
  Sheridan, 30-31, 40, 44
  Story, 33, 39

Yukon Territory, St. Elias Range, 108

About the Editor

TOR H. NILSEN was born in 1941 in New York City. He received the B.S. degree from the City College of New York in 1962, and the M.S. and Ph.D. degrees from the University of Wisconsin in 1964 and 1967, respectively. He worked briefly for the Shell Oil Company Exploration and Production Research Laboratory in Ventura, California in 1967. During a two-year period of military service from 1967 to 1969, he served as military advisor to the Chief of the Gravity Division, Army Map Service, in Washington, D. C. Since 1969, he has worked for the U.S. Geological Survey in Menlo Park, California. His chief research interest has been the study of coarse clastic sequences in tectonically active regions. He has studied both alluvial fan deposits and deep-sea fan deposits in many areas, including western Norway (see Paper 13 of this volume), Italy, California, Oregon, Idaho, and Alaska. He has also worked extensively on modern and ancient subaerial landslide deposits and modern marine sedimentation in the Norwegian-Greenland Sea, Bering Sea, and Southern California Borderland.